Fundamental Probability

Fundamental Probability

A Computational Approach

Marc S. Paolella

John Wiley & Sons, Ltd

Other Wiley Editorial Offices

John Wiley & Sons Inc., 111 River Street, Hoboken, NJ 07030, USA

Jossey-Bass, 989 Market Street, San Francisco, CA 94103-1741, USA

Wiley-VCH Verlag GmbH, Boschstr. 12, D-69469 Weinheim, Germany

John Wiley & Sons Australia Ltd, 42 McDougall Street, Milton, Queensland 4064, Australia

John Wiley & Sons (Asia) Pte Ltd, 2 Clementi Loop #02-01, Jin Xing Distripark, Singapore 129809

John Wiley & Sons Canada Ltd, 22 Worcester Road, Etobicoke, Ontario, Canada M9W 1L1

Wiley also publishes its books in a variety of electronic formats. Some content that appears
in print may not be available in electronic books.

British Library Cataloguing in Publication Data

A catalogue record for this book is available from the British Library

ISBN-13: 978-0-470-02594-9
ISBN-10: 0-470-02594-8

Typeset in 10/12 Times by Laserwords Private Limited, Chennai, India

Printed and bound by CPI Antony Rowe, Eastbourne

Chapter Listing

Preface   xi

0   Introduction   1

Part I   Basic Probability   7

1   Combinatorics   9
2   Probability spaces and counting   43
3   Symmetric spaces and conditioning   73

Part II   Discrete Random Variables   111

4   Univariate random variables   113
5   Multivariate random variables   165
6   Sums of random variables   197

Part III   Continuous Random Variables   237

7   Continuous univariate random variables   239
8   Joint and conditional random variables   285
9   Multivariate transformations   323

Appendices   343

A   Calculus review   343
B   Notation tables   435
C   Distribution tables   441

References   451
Index   461

Contents

Preface   xi
A note to the student (and instructor)   xvi
A note to the instructor (and student)   xviii

Acknowledgements   xxi

0 Introduction   **1**

Part I   Basic Probability   **7**

1 Combinatorics   **9**

1.1   Basic counting   9
1.2   Generalized binomial coefficients   13
1.3   Combinatoric identities and the use of induction   15
1.4   The binomial and multinomial theorems   18
      1.4.1   The binomial theorem   18
      1.4.2   An extension of the binomial theorem   23
      1.4.3   The multinomial theorem   27
1.5   The gamma and beta functions   28
      1.5.1   The gamma function   28
      1.5.2   The beta function   31
1.6   Problems   36

2 Probability spaces and counting   **43**

2.1   Introducing counting and occupancy problems   43
2.2   Probability spaces   47
      2.2.1   Introduction   47
      2.2.2   Definitions   49
2.3   Properties   58

	2.3.1	Basic properties	58
	2.3.2	Advanced properties	59
	2.3.3	A theoretical property	67
2.4	Problems		68

3 Symmetric spaces and conditioning   **73**

3.1	Applications with symmetric probability spaces		73
3.2	Conditional probability and independence		85
	3.2.1	Total probability and Bayes' rule	87
	3.2.2	Extending the law of total probability	93
	3.2.3	Statistical paradoxes and fallacies	96
3.3	The problem of the points		97
	3.3.1	Three solutions	97
	3.3.2	Further gambling problems	99
	3.3.3	Some historical references	100
3.4	Problems		101

Part II Discrete Random Variables   **111**

4 Univariate random variables   **113**

4.1	Definitions and properties		113
	4.1.1	Basic definitions and properties	113
	4.1.2	Further definitions and properties	117
4.2	Discrete sampling schemes		120
	4.2.1	Bernoulli and binomial	121
	4.2.2	Hypergeometric	123
	4.2.3	Geometric and negative binomial	125
	4.2.4	Inverse hypergeometric	128
	4.2.5	Poisson approximations	130
	4.2.6	Occupancy distributions	133
4.3	Transformations		140
4.4	Moments		141
	4.4.1	Expected value of X	141
	4.4.2	Higher-order moments	143
	4.4.3	Jensen's inequality	151
4.5	Poisson processes		154
4.6	Problems		156

5 Multivariate random variables   **165**

| 5.1 | Multivariate density and distribution | | 165 |
| | 5.1.1 | Joint cumulative distribution functions | 166 |

		5.1.2	Joint probability mass and density functions	168
	5.2	Fundamental properties of multivariate random variables		171
		5.2.1	Marginal distributions	171
		5.2.2	Independence	173
		5.2.3	Exchangeability	174
		5.2.4	Transformations	175
		5.2.5	Moments	176
	5.3	Discrete sampling schemes		182
		5.3.1	Multinomial	182
		5.3.2	Multivariate hypergeometric	188
		5.3.3	Multivariate negative binomial	190
		5.3.4	Multivariate inverse hypergeometric	192
	5.4	Problems		194
6	**Sums of random variables**			**197**
	6.1	Mean and variance		197
	6.2	Use of exchangeable Bernoulli random variables		199
		6.2.1	Examples with birthdays	202
	6.3	Runs distributions		206
	6.4	Random variable decomposition		218
		6.4.1	Binomial, negative binomial and Poisson	218
		6.4.2	Hypergeometric	220
		6.4.3	Inverse hypergeometric	222
	6.5	General linear combination of two random variables		227
	6.6	Problems		232
Part III	**Continuous Random Variables**			**237**
7	**Continuous univariate random variables**			**239**
	7.1	Most prominent distributions		239
	7.2	Other popular distributions		263
	7.3	Univariate transformations		269
		7.3.1	Examples of one-to-one transformations	271
		7.3.2	Many-to-one transformations	273
	7.4	The probability integral transform		275
		7.4.1	Simulation	276
		7.4.2	Kernel density estimation	277
	7.5	Problems		278

8   Joint and conditional random variables   **285**

8.1   Review of basic concepts   285
8.2   Conditional distributions   290
    8.2.1   Discrete case   291
    8.2.2   Continuous case   292
    8.2.3   Conditional moments   304
    8.2.4   Expected shortfall   310
    8.2.5   Independence   311
    8.2.6   Computing probabilities via conditioning   312
8.3   Problems   317

9   Multivariate transformations   **323**

9.1   Basic transformation   323
9.2   The t and F distributions   329
9.3   Further aspects and important transformations   333
9.4   Problems   339

Appendices   **343**

A   Calculus review   **343**

A.0   Recommended reading   343
A.1   Sets, functions and fundamental inequalities   345
A.2   Univariate calculus   350
    A.2.1   Limits and continuity   351
    A.2.2   Differentiation   352
    A.2.3   Integration   364
    A.2.4   Series   382
A.3   Multivariate calculus   413
    A.3.1   Neighborhoods and open sets   413
    A.3.2   Sequences, limits and continuity   414
    A.3.3   Differentiation   416
    A.3.4   Integration   425

B   Notation tables   **435**

C   Distribution tables   **441**

References   **451**

Index   **461**

Preface

Writing a book is an adventure. To begin with, it is a toy and an amusement; then it becomes a mistress, and then it becomes a master, and then a tyrant. The last phase is that just as you are about to be reconciled to your servitude, you kill the monster, and fling him out to the public.

(Sir Winston Churchill)
Reproduced by permission of Little Brown

Like many branches of science, probability and statistics can be effectively taught and appreciated at many mathematical levels. This book is the first of two on probability, and designed to accompany a course aimed at students who possess a basic command of freshman calculus and some linear algebra, but with no previous exposure to probability. It follows the more or less traditional ordering of subjects at this level, though with greater scope and often more depth, and stops with multivariate transformations. The second book, referred to as Volume II, continues with more advanced topics, including (i) a detailed look at sums of random variables via exact and approximate numeric inversion of moment generating and characteristic functions, (ii) a discussion of more advanced distributions including the stable Paretian and generalized hyperbolic, and (iii) a detailed look at noncentral distributions, quadratic forms and ratios of quadratic forms.

A subsequent book will deal with the subject of statistical modeling and extraction of information from data. In principle, such prerequisites render the books appropriate for upper division undergraduates in a variety of disciplines, though the *amount* of material and, in some topics covered, the depth and use of some elements from advanced calculus, make the books especially suited for students with more focus and mathematical maturity, such as undergraduate math majors, or beginning graduate students in statistics, finance or economics, and other fields which use the tools of probability and statistical methodology.

Motivation and approach

The books grew (and grew...) from a set of notes I began writing in 1996 to accompany a two-semester sequence I started teaching in the statistics and econometrics faculty in

Kiel University in Germany.[1] The group of students consisted of upper-division under-graduate and beginning graduate students from quantitative business (50 %), economics (40 %) and mathematics (10 %). These three disparate groups, with their different motivations and goals, complicated the choice of which topics to cover and textbook to use, and I eventually opted to concentrate strictly on my own notes, pursuing a flexible agenda which could satisfy their varying interests and abilities.

The two overriding goals were (i) to emphasize the practical side of matters by presenting a large variety of examples and discussing computational matters, and (ii) to go beyond the limited set of topics and standard, 'safe' examples traditionally taught at this level (more on this below). From my experience, this results in a much more enjoyable and unified treatment of the subject matter (theory, application and computation) and, more importantly, highly successful students with an above average competence, understanding, breadth and appreciation for the usefulness of our discipline.

It must be mentioned, however, that while these two goals certainly complement each other, they gave rise to a substantially larger project than similarly positioned books. It is thus essential to understand that *not everything in the texts is supposed to be (or could be) covered in the classroom.* The teacher has the choice of what to cover in class and what to assign as outside reading, and the student has the choice of how much of the remaining material he or she wishes to read and study. In my opinion, a textbook should be a considerably larger superset of material covered in class, bursting with varied, challenging examples, full derivations, graphics, programming techniques, references to old and new literature, and the like, and not a dry protocol of a lecture series, or a bare-bones outline of the major formulae and their proofs.

With respect to computation, I have chosen Matlab, which is a matrix-based prototyping language requiring very little initial investment to be productive. Its easy structure, powerful graphics interface and enormous computational library mean that we can concentrate on the problem we wish to solve, and avoid getting sidetracked by the intricacies and subtleties of the software used to do it. All these benefits also apply to S-Plus® and R, which certainly could be used in place of Matlab if the instructor and/or student wishes. With that in mind, I have attempted to make the programming code and data structures as simple and universal as possible, so that a line-by-line translation of the numerous programs throughout would often be straightforward.

Numeric methods are used throughout the book and are intentionally woven tightly together with the theory. Overall, students enjoy this approach, as it allows the equations and concepts to 'come to life'. While computational and programming exercises are valuable in themselves, experience suggests that most students are at least as willing to play around on the computer as they are on paper, with the end result that more time is spent actively thinking about the underlying problem and the associated theory, and so increasing the learning effect.

Of course, because of its reliance on modern hardware and software, this approach does not enjoy a long tradition but is quickly gaining in acceptance, as measured by the number of probability and statistics textbooks which either incorporate computational aspects throughout (e.g. Kao, 1996; Childers, 1997; Rose and Smith, 2002; Ruppert,

---

[1] For a bit of trivia, Kiel's mathematics department employed William Feller until 1933, as well as Otto Toeplitz from 1913 to 1928. In 1885, Max Planck became a professor of theoretical physics in Kiel, which is also where he was born.

2004), or are fully dedicated to such issues (e.g. Venables and Ripley, 2002; Dalgaard, 2002; Givens and Hoeting, 2005).

Ready access to complex, numerically intensive algorithms and extraordinarily fast hardware should not be seen as causing a 'break' or 'split' with the classical methodological aspects of our subject, but viewed rather as a driving factor which gradually and continually instigates a progression of ideas and methods of solving problems to which we need to constantly adjust. As a simple example, the admirable and difficult work invested into the Behrens–Fisher problem resulted in several relatively simple closed-form approximations for the distribution of the test statistic which could actually be calculated for a given data sample using the modest computational means available at the time. Now, with more computing power available, bootstrap methodology and MCMC techniques could readily be applied. Similar examples abound in which use of computationally demanding methods can be straightforwardly used for handling problems which are simply not amenable to algebraic (i.e. easily computed) solutions.

This is not to say that certain classical concepts (e.g. central limit theorem-based approximations, sufficient statistics, UMVU estimators, pivotal quantities, UMP tests, etc.) are not worth teaching; on the contrary, they are very important, as they provide the logical foundation upon which many of the more sophisticated methods are based. However, I believe these results need to be augmented *in the same course* with (i) approaches for solving problems which make use of modern theoretical knowledge and, when necessary, modern computing power, and (ii) genuinely useful examples of their use, such as in conjunction with the linear model and basic time series analysis, as opposed to overly simplistic, or even silly, i.i.d. settings because they happen to admit tractable algebraic solutions.

My experience as both a student and teacher suggests that the solid majority of students get interested in our subject from the desire to answer questions of a more applied nature. That does not imply that we should shy away from presenting mathematical results, or that we should not try to entice some of them to get involved with more fundamental research, but we should respect and work with their inherent curiosity and strike a balance between application and theory at this level. Indeed, if the climax of a two-course sequence consists of multivariate transformations, rudimentary uses of moment generating functions, and the 'trinity' of Cramer–Rao, Lehmann–Scheffé and Neyman–Pearson, then the student will leave the course not only with an extremely limited toolbox for answering questions with the slightest hint of realism, but also with a very warped and anachronistic view of what our subject has to offer society at large.

This also makes sense even for people majoring in mathematics, given that most of them will enter private industry (e.g. financial institutions, insurance companies, etc.) and will most likely find themselves making use of simulation techniques and approximate inversion formulae instead of, say, minimal sufficiency theorems associated with exponential families.

Examples

The examples throughout the text serve the purpose of (i) illustrating previously discussed theory with more concrete 'objects' (actual numbers, a distribution, a proof, etc.), and/or (ii) enriching the theory by detailing a special case, and/or (iii) introducing new, but specialized, material or applications. There is quite an abundance of examples throughout the text, and not all of them need to be studied in detail in a first passing, or at all. To help guide the reader, symbols are used to designate the importance of the example, with $\ominus$, $\odot$ and $\circledast$ indicating low, medium and high importance, respectively, where importance refers to *'relevance to the bigger picture'*.

I understand the bigger picture to be: establishing a foundation for successfully pursuing a course in statistical inference (or related disciplines such as biostatistics, actuarial and loss models, etc.), or a course in stochastic processes, or a deeper mathematical treatment of probability theory. In this case, most examples are not highly relevant for the bigger picture. However, an example sporting only the low-importance symbol $\ominus$ could actually be very helpful, or even critical, for understanding the material just presented, or might introduce new, useful material, or detail an interesting application, or could be instructive in terms of Matlab programming. Thus, the low symbol $\ominus$ should certainly not be interpreted as an indication that the example should be skipped! Of course, the high importance symbol $\circledast$ indicates mandatory reading and understanding.

Problems

Some comments on the problem sets provided in the texts are in order. First, I generally avoid overly simplistic exercises of a 'plug and chug' nature, preferring instead to encourage students to test their initial understanding of the material by trying to reproduce some of the more basic examples in the text on their own. Once such a basis is established, the problems can be fruitfully attempted. They range in difficulty, and are marked accordingly (with respect to my subjective beliefs), from rather easy (no stars), to particularly challenging and/or time consuming (three stars).

Second, I try to acknowledge the enormous demands on students' time and the frustration which often arises when attempting to solve homework problems, and thus give them access to the exercise solutions. Detailed answers to all of the problems are provided in a document available from the book's web page www.wiley.com/go/ fundamental. This allows for self study and self correction and, in my experience, actually *increases* students' motivation and time spent working problems because they can, at any time, check their performance.

Third, I advocate encouraging students to solve problems in various ways, in accordance with their professional goals, individual creativity, mathematical expertise, and time budget. For example, a problem which asks for a proof could, on occasion, be 'numerical verified'. Of course, any problem which begins with 'Using one million Cauchy observations...' should be a clear signal that a computer is required! Less trivially, a problem like 'Compare the bias of the three estimators...' can be attacked algebraically or computationally, or both. Use of symbolic computing packages such as Mathematica, Maple or MuPAD straddles the two fields in some senses, and is also strongly encouraged.

Appendices

Appendix A serves to refresh some essential mathematics from analysis used throughout the text. My suggestion is to have the student skim through its entirety to become familiar with its contents, concentrating on those sections where gaps in the student's knowledge become apparent. There was much vacillation regarding its inclusion because of its size, but experience has shown that most students do not readily exhume their previous math books (let alone purchase different, or more advanced ones) and work through them, but they *will* read a shortened, streamlined version which explicitly ties into the rest of the book, and which is physically attached to the main text.

Appendix B consists of various notation tables, including abbreviations, Latin expressions, Greek letters, a list of special functions, math and distributional notation and moments of random variables.

Appendix C collects tables of distributions, including their naming conventions, p.d.f., c.d.f., parameter space, moments and moment generating function (though the latter is first introduced at the beginning of Volume II). Volume II also includes further tables, with generating function notation, inversion formulae and a table illustrating various relationships among the distributions. What you will not find anywhere in these books are tables of standard normal c.d.f. values, chi-square, t and F cut-off values, etc., as they are no longer necessary with modern computing facilities and no statistical analysis or write-up is done without a computer.

Topics covered in Volume I

Instead of writing here a long summary, I opt for a very short one, and invite the reader and instructor to gloss over the table of contents, which – by construction – is the book's outline. Of course, many topics are standard; some are not, and their respective sections could be briefly skimmed. Some highlights include:

- a detailed, self-contained chapter (Appendix A) on the calculus tools used throughout the book, with many proofs (e.g. fundamental theorem of calculus, univariate and multivariate Taylor series expansions, Leibniz' rule, etc.) and examples (convolution of Cauchy r.v.s, resolution of the normal density's integration constant both with and without the use of polar coordinates, etc.);

- numerous tables (Appendices B and C) which collect notational, distributional and other information in a concise, coherent way;

- a strong use of combinatorics and its use in answering probabilistic questions via De Moivre–Jordan's theorem;

- a detailed treatment of the problem of the points and its historical significance, as well as some references to the historical development of probability;

- full coverage of all four 'basic sampling schemes', e.g. derivation of moments for sampling until r successes without replacement (inverse hypergeometric);

- extension of each of the four 'basic sampling schemes' to the multivariate case, with emphasis on random number generation and simulation;

- ample discussion of occupancy distributions, calculation of the p.m.f., c.d.f. and moments using a variety of methods (combinatoric and, in Volume II, characteristic function inversion and saddlepoint approximation);

- a thorough discussion of basic runs distributions, with full derivation of all common results (and a peek at more recent results);

- use of exchangeability for decomposing random variables such as occupancy, hypergeometric and inverse hypergeometric;

- introduction of the common univariate continuous distributions in such a way as to teach the student how to 'approach' a new distribution (recognizing the roles of parameters, major applications, tail behavior, etc.);

- introduction of some basic concepts used in finance, such as utility functions, stochastic dominance, geometric expectation, continuous returns, value at risk and expected shortfall.

A note to the student (and instructor)

Mathematics

The reader will quickly discover that considerable portions of this text are spent deriving mathematically stated results and illustrating applications and examples of such which involve varying degrees of algebraic manipulation. It should thus come as no surprise that a solid command of 'first-year calculus' (univariate differentiation, integration and Taylor series, partial differentiation and iterated integrals) is critical for this course. Also necessary is the ability to work with the gamma and beta functions (detailed from scratch in Section 1.5) and a good working knowledge of basic linear/matrix algebra for Volume II (in which an overview is provided). With these tools, the reader should be able to comfortably follow about 95 % of the material.

The remaining 5 % uses bits of advanced calculus, e.g. basic set theory, limits, Leibniz' rule, exchange of derivative and integral, along with some topics from multivariate calculus, including polar coordinates, multivariate change of variables, gradients, Hessian matrices and multivariate Taylor series. Appendix A provides a streamlined introduction to these topics. Rudiments of complex analysis are required in Volume II, and an overview is provided there.

Thus, with the help of Appendix A to fill in some gaps, the formal prerequisites are very modest (one year of calculus and linear algebra), though a previous exposure to real analysis or advanced calculus is undoubtedly advantageous for two reasons. First, and obviously, the reader will already be familiar with some of the mathematical tools used here. Secondly, and arguably more important, the student is in a much better position to *separate mathematics from the probabilistic concepts and distributional results introduced in the text*. This is very important, as it allows you to avoid getting bogged down in the particulars of the mathematical details and instead you can keep an eye on the concepts associated with the study of probability and the important distributions which are essential for statistical analysis.

The ability to recognize this distinction is quite simple to obtain, once you become conscious of it, but some students ultimately experience difficulty in the course because they do not do so. For example, the probability mass function of a random variable may be easily expressed recursively, i.e. as a difference equation, but whose solution is not readily known. Such a difference equation could have also arisen in, say, oceanography, or any of dozens of scientific fields, and the method of mathematical solution has nothing to do per se with the initial probabilistic question posed. This was another motivation for including the math review in Appendix A: by reading it first, the student explicitly sees precisely those math tools which are used throughout the book – but in a 'pure math' context unburdened by the new concepts from probability. Then, when a messy series of set operations, or a complicated integral, or a Taylor series expression, arises in the main chapters, it can be decoupled from the context in which it arose.

Having now said that, a big caveat is in order: often, a 'mathematical' problem might be more easily solved in the context of the subject matter, so that the two aspects of the problem are actually wedded together. Nevertheless, during a first exposure to the subject, it is helpful to recognize the border between the new concepts in probability and statistics, and the mathematical tools used to assist their study. As an analogy to a similar situation, it seems natural and wise to separate the derivation of a mathematical solution, and the computer language, algorithmic techniques and data structures used to implement its numeric calculation. At a deeper level, however, the form of the solution might be influenced or even dictated by knowledge of the capabilities and limitations of a possible computer implementation.

Motivation and learning habits

It seems safe to say that there are some universally good ways of learning any kind of subject matter, and some bad ways. I strongly advocate an 'active approach', which means grabbing the bull by the horns, jumping right into the cold water (with the bull), and getting your hands dirty. Concretely, this means having pencil, paper and a computer at hand, and picking to pieces any equation, program, derivation or piece of logic which is not fully clear. You should augment my explanations with more detail, and replace my solutions with better ones. Work lots of problems, and for those which you find difficult, read and master the solutions provided, then close the book and do it yourself. Then do it again tomorrow. The same goes for the Matlab programs.

Learn the material with the intention of making it your own, so that you could even teach it with competence and confidence to others. Construct your own questions and write up your own solutions (good opportunity to learn LaTeX), challenge your instructor with good questions (that you have given your own thought to), borrow or buy other textbooks at a lower, similar or higher level, think about applications in fields of your interest, and build learning groups with other motivated students. Now is the time to challenge yourself and learn as much as you can, because once you leave university, the realisms of life (such as family and employment concerns) will confront you and take up most of your time, and learning new things becomes difficult.

Some of the ingredients for success can be put into an acronym: get TIMED! That is, you will need Time, Interest, Mathematical maturity, Energy and Discipline. A deficit in one can often be accommodated by an increase in one or more of the others, though all are required to some extent. If one of these is failing, figure out why, and how to solve it.

A note to the instructor (and student)

Book usage and course content

As mentioned above, the book is ideal for beginning graduate students in the social or mathematical sciences, but is also suited (and designed) for use with upper division undergraduates, provided that the more difficult material (about a quarter of the book) is omitted. Let me first discuss how I use the book in a graduate program, and then in an undergraduate setting.

For teaching graduate students, I begin by quickly covering the math appendix (this has positive externalities because they need the material for other courses anyway), spending about 3 hours (two or three lectures) in total. I just emphasize the major results, and give the students the chance to work on a few basic examples during class (see my comments on teaching style below) which I often take from the easier homework exercises in real analysis textbooks. From the main part of the textbook, I spend disproportionately more time with the first chapter (combinatorics), because (i) students often have problems with this material, (ii) some of it is critical for subsequent chapters, and (iii) it covers mathematics topics which the students can profit from in other courses, e.g. proof by induction, the binomial and multinomial theorems, and the gamma and beta functions. Most of the material in the remaining chapters gets 'covered', but not necessarily in class. In fact, much of it is necessarily read and learned outside of class, and some material can just be skimmed over, e.g. the derivation of the runs distributions and the inverse hypergeometric moments.

For teaching junior or senior undergraduate courses in probability, I also start with the math appendix, but only cover the 'easier' material, i.e. differentiation, the basics of Riemann integration, and the fundamental theorem of calculus, skipping much of the sections on series and multivariate calculus. I also spend a significant amount of time with Chapter 1, for the reasons mentioned above. For the remaining chapters, I skip most of the more 'exotic' material. For example, in Chapter 2, I omit a discussion of occupancy problems, the more advanced set-theoretic material (σ-fields, etc.), and much of the material in the section on advanced properties (e.g. Poincaré's theorem for $n > 3$, the De Moivre–Jordan theorem, and Bonferroni's general inequality), as well as the proof that $\lim_{i \to \infty} \Pr(A_i) = \Pr(\lim_{i \to \infty} A_i)$.

Teaching style

All lecturers have different styles – some like to 'talk and chalk', i.e. write everything on the blackboard, others use overheads or beamer presentations, etc. I personally prefer the use of overheads for showing the main results, which can then be openly discussed (as opposed to having $n + 1$ people quietly acting like human copy machines), and using the blackboard for excursions or 'live' derivations. As many instructors have learned, students can write, or think, but not both at the same time, so my strategy was to prepare a book which contains almost everything I would possibly write down – so the students don't have to. Of course, if a student is not writing, it does not imply that he or she is thinking, and so to help stimulate discussion and liveliness, I spend about a quarter of the lecture time allowing them to work some problems on their own (or in teams) which we subsequently discuss (or I choose a team to present their findings). What I just described accounts for about two thirds of the lecture; the remaining time takes place in the computer room, letting the students actively work with each other and the PC.

I must now repeat what I said earlier: *not everything in the texts is supposed to be (or could be) covered in the classroom.* Parts should be assigned for reading outside of class, while some of the material can of course be skipped. In my opinion, students at this level should be encouraged to start the transition of learning from lectures, to learning from books and research articles. For high school students and beginning undergraduates, the lecture is the main arena for knowledge transfer, and the book is often used only as a secondary resource, if at all. At the other end of the spectrum, doctoral research students, and professionals in and outside academia, attain most of their new knowledge on their own, by reading (and implementing). I 'assist' this transition by taking some exam questions from material which the students should read on their own.

Acknowledgements

I am very fortunate to have had some outstanding teachers throughout my higher education who have made a great impact on my appreciation for this subject and, ultimately, my career choice. These include William Dawes, notorious for teaching probability along the lines of Example 3.17, and Steven Finch, for masterfully conveying the importance of the work of Sir Ronald Fisher. I owe my knowledge of time series analysis, econometrics, and statistical computing to Peter Brockwell and Stefan Mittnik. My notes on probability theory, linear models, design of experiments and survival analysis from graduate courses by Hari Iyer and Yi-Ching Yao are sacred. Ronald Butler exposed me to, and patiently taught me, numerous important topics in statistics, including the Bayesian and likelihood inferential approaches, but most notably saddlepoint methods and their myriad of applications. I have also had many great instructors whom I have never met – the authors of all the numerous books and research articles from which I have learned and drawn. I have made every attempt to cite them whenever their influence crops up, and the reader will quickly see which books have been most influential in this regard.

I am very grateful for the large number of constructive comments and positive feedback provided by several academics who served as anonymous referees for this book. I wish to explicitly thank one reviewer, Daniel Zelterman, for very careful reading and making some terrific and detailed suggestions for improving the presentation in Chapter 8. Also, the editorial team at Wiley has been a pleasure to work with and deserves special mention; in particular, I wish to thank Lucy Bryan, Lyn Hicks, Elizabeth Johnston, Siân Jones, and Simon Lightfoot for their terrific and constant support through the final stages of this project.

Through teaching with the notes that eventually became this book, numerous errors, typos, poor (or missing) explanations, and all kinds of bad things have been pointed out and corrected by scores of students over nearly a decade. I am extremely grateful to all of these individuals, not just for their valuable input, but also for putting up with this project (and my German) while it was still developing and maturing. Special mention goes to Christoph Hartz for very careful proofreading, and also for help with LaTeX, especially with their graphics – he constructed all the non-Matlab ones for me, such as Figures 3.5, 3.6, 3.7 and 3.9, to name just a few. Angelika Reinmüller, Christian Aßmann, Anja Müller, Tina Novotny and Claudine Schnell kindly helped me with all kinds of 'peripheral things', from photocopying articles to ensuring a constant supply of coffee and good moods, but also with some of the typing, in particular, in Chapters 8 and 9.

During my stay in Kiel, two brilliant students emerged who took a great interest in my project and have consistently and unselfishly assisted me with it. These are Markus Haas and Walther Paravicini, both of whom are now themselves budding academics in econometrics and mathematics, respectively. In addition to reading and correcting many parts of the manuscript, both of them have also contributed to it. Markus has added examples and problems related to combinatorics, most notably in Banach's matchbox problem and the Problem of the Points, while Walther has written the entire chapter on generalized inverse Gaussian and hyperbolic distributions (which appears in Volume II), and also added some examples and exercises throughout this text. Their names are in the index so one can locate some of their explicit contributions. I cannot overstate my appreciation for their interest, help and friendship over the years.

In Zurich, my friends and graduate students Simon Broda, Sven-Christian Steude and Yianna Tchopourian went through parts of the text with a fine-tooth comb, exposing errors and making great suggestions for improvements. Their presence will also be explicitly seen in Volume II, where they helped contribute to the chapters on saddlepoint methods and the stable Paretian distribution. In addition, Yianna Tchopourian and our fantastic secretary Gabriela Davenport have spent many hours chasing down the published sources of the quotes in the book – the quotes being a pure luxury, but I hope the reader finds them to be enjoyable and rewarding.

Finally, I am greatly indebted to my family for having put up with all the costs this project has incurred, and for what must have seemed like an eternity. My young children, Kayleigh and Finnley, have done their share of helping as well, by providing me with the most fulfilling and rewarding break from the pressures of this project (and beyond), and serve as a constant reminder of the noble human desire to learn and understand, purely for the sake of learning and understanding.

0

Introduction

> It is remarkable that a science which began with the consideration of games of chance should have become the most important object of human knowledge.
>
> (Pierre Simon Laplace)

As we will be using the words 'probability' and 'statistics' with great frequency, it seems worthwhile to have a brief look at their origins. The word 'probability' is attested from 1551, while 'probable' dates back to 1387, coming from the Latin roots *probare* meaning 'to try, test, or prove worthy' (same origins as 'probe' and 'provable') and *probabilis*, meaning 'credible'. Of course, the rudimentary concept of randomness and its use in various realms of human activity, such as games of chance, religion, etc., dates back to antiquity. Though we will not pursue it here (for space and competence reasons), the development of the subject, from the primitive notions in ancient civilizations, to its modern mathematical edifice and the permeation of its tools and concepts into innumerable branches of science and human endeavor, is quite illuminating and can be read about in several engaging accounts, some of which are listed in Section 3.3.3.

The word 'statistics' derives from the Latin *statisticus*, meaning 'of state affairs', like the Italian 'statista', meaning statesman, or 'one skilled in statecraft'. It was popularized via the German word 'Statistik', used for the first time by the German political scientist (and inadvertent statistician) Gottfried Achenwall (1719–1772), who pioneered the use of statistics for studying the social, political and economic features of a state.

The rest of this small chapter serves to motivate and augment the study of probability spaces by briefly describing some applications, from commonplace facets of everyday life to highly technical endeavors, in which the tools of probability and statistics are fundamental for answering important questions and making rational decisions in the face of complexity and uncertainty.

- *Legal issues and litigation support.* These include a wide variety of situations in which it is necessary to assess the quantitative weight of evidence or the validity of some claim. Examples include analysis of price fixing, employment discrimination, product liability, environmental regulation, appropriateness and fairness of sentences, forensic science, trademark infringement and lost earnings calculations. For further aspects and details on case studies see Gastwirth (2000), DeGroot, Fienberg and Kadane (1994) and Basmann (2003).

- *Effectiveness and safety of new drugs and medical procedures.* These methods of treatment are subjected to a series of tests in order to assess if (or to what extent) the desired claims can be supported. It is rare that a drug can be 'theoretically proven effective' or its side effects predicted with certainty without having tried it, and so evidence in favor (or against) its use is often of a statistical nature. Adding to the already advanced statistical theory required for analyzing an 'ideal' experiment, real life factors complicate matters further (e.g. some rats escape, patients lie or don't always take their medicine at the prescribed time, etc.) and can hinder inference by giving rise to 'biased' results or decreasing the precision of the conclusions.

  This is just one of the many aspects of *biostatistics*. Another example is *survival analysis*, or the modeling of expected lifetimes under particular conditions and treatments. The latter statistical technology can be applied in many settings, an interesting one of which, to various social scientists, is the lengths of unemployment spells.

- *Statistical genetics.* Research in genetics and large-scale DNA sequencing, most notably the human genome project, have significantly and permanently changed the biomedical sciences. The study of heritability is inseparable from the laws of probability, while the modern analysis of genetic variation between individuals and species is very (if not ultimately) statistical in nature.

- *Reliability of a particular device or component in a certain machine or system.* The goal might be to maximize profit of an industrial process: by examining the trade off between the initial, maintenance and repair costs of various machines along with their expected lifetimes and outputs, an 'optimal' strategy (in a sense which would need to be made precise) can be determined. Alternatively, the failure of a component could be life-threatening, for which numerous examples come to mind. The causes of the space shuttle *Challenger* disaster on January 28th, 1986, for example, has been the subject of extensive statistical analysis.[1]

- The previous example of device reliability can be generalized to an entire system such as complex design and manufacturing processes found in numerous industries, city infrastructures (roadways, public transportation, procurement and distribution of water, gas and electricity), airline and airport services, the operations of banking and financial institutions, etc. The body of general principles associated with the study and improvement of such systems is referred to as *total quality management* (TQM). A major part of TQM is *statistical quality control* (SQC), which, as its name suggests, embraces the use of statistical techniques in its implementation. There are many good textbook presentations of TQM and SQC, two of which include Thompson and Koronacki (2001) and Montgomery (2000).

---

[1] See, for example, Poirier (1995, Section 1.3) and Simonoff (2003, pp. 375–380), and the references therein.

Strongly associated with the success and propagation of TQM is the scientist W. Edwards Deming (1900–1994) who, as a consultant to Japan's post World War II industrial sector, played a significant role in their development of products of remarkable quality and affordability and, thus, their enormous economic success. An enjoyable account of Deming's contribution can be found in Salsburg (2001, Chapter 24). To quote from his page 249,

'At that time, the phrase "made in Japan" meant cheap, poor-quality imitations of other countries' products. Deming shocked his listeners by telling them they could change that around within five years. He told them that, with the proper use of statistical methods of quality control, they could produce products of such high quality and low price that they would dominate markets all over the world. Deming noted in later talks that he was wrong in predicting this would take five years. The Japanese beat his prediction by almost two years.'

Reproduced by permission of Henry Holt & Company NY

- *Analysis of insurance policies, risk assessment and actuarial mathematics.* A large insurance company covering several million clients over a large time span will have ample opportunity to observe many 'laws of averages' at work. Indeed, the use of sophisticated probabilistic models and methods of statistical inference is common-place in modern actuarial practice. Moreover, while analysis of 'typical' or average claims is of interest, the occasional catastrophe (which might arise when insuring for earthquakes, fires, floods, nuclear power plant accidents and other natural and man-made disasters) needs to be taken into consideration. Such analysis is one of the primary uses of *extreme-value theory*, or EVT, which is introduced in a later chapter. Recent accounts of these issues are detailed in Embrechts, Kluppelberg and Mikosch (2000) and Klugman, Panjer and Willmot (1998).

- *Financial investment decisions.* The study of portfolio analysis, including aspects of option pricing or, more generally, *financial engineering*, has some similarities to risk analysis in the insurance industry, but focuses more on the optimal allocation of funds in various risky assets. Similar to the occasional 'outlier' in insurance claims, returns on financial assets very often experience large (positive or negative) values which are not at all consistent with the distribution suggested by the bell curve. In addition, instead of occurring more or less haphazardly, these large 'shocks' occa-sionally form clusters of 'mild and wild' periods, giving rise to numerous *stochastic volatility* models which attempt to capitalize on this structure to enhance predictabil-ity. Reliable prediction of the *downside risk* or the left tail of the return distribution associated with currency exchange rates, stock prices and other financial instruments is critical for both investment banks and large companies doing business in global markets. Statistical analysis applied to financial and actuarial issues and data has also given rise to many sub-disciplines, with names such as RiskMetrics, PensionMetrics, CrashMetrics, etc.

- *Marketing.* Marketing studies the purchasing behavior of consumers, and the inter-action between firms and consumers, making use of tools from economics, soci-ology, psychology and statistics. When emphasis is on the latter, one speaks of *quantitative marketing.* In many marketing studies, the type of available data is unique, e.g. scanner data collected at the checkout stand using bar-code readers, clickstream data from a web site, etc., and so requires specialized or even newly developed statistical techniques. In these examples, a huge amount of raw data could be available, so that the techniques and methods of *data mining* become relevant.

Data mining is a rapidly growing field, because of the explosion in data availability and its storage capabilities, the enormous development of hardware and software to process such data, and growing interest in applications in a variety of fields. For example, a new (and potentially Orwellian) offshoot of quantitative marketing is *neuromarketing*, which uses brain-imaging technology to reveal preferences; see *The Economist Technology Quarterly* (June 12th, 2004, pp. 10 and 18), where it is emphasized that 'With over $100 billion spent each year on marketing in America alone, any firm that can more accurately analyse how customers respond to products, brands and advertising could make a fortune.'

For a collection of recent applications in quantitative marketing, see Franses and Montgomery (2002) and the 18 articles of the special issue of the *Journal of Econometrics* (1999, Vol. 89, Issues 1–2), edited by T.J. Wansbeek and M. Wedel, as well as their lengthy and informative introduction.

- *Fraud detection.* Particularly with the advent of online banking and the growth of the internet, the possibilities for credit card fraud are enormous, though 'opportunities' still abound in automobile insurance, telecommunications, money laundering and medical procedures. In many settings, there is a massive amount of data to be processed (major credit card companies, for example, enjoy hundreds of millions of transactions per year), so that case-by-case inspection is clearly impossible. Instead, various statistical methods in conjunction with data mining techniques have been proposed and developed to detect possible fraudulent behavior. See the review article by Bolton and Hand (2002) and the comments and references therein for further information.

- *Analysis of voting and elections.* The United States presidential election in 2000 is by far the most controversial one in recent US history and, soon after the election, prompted a large and ongoing discussion, including numerous statistical analyses of the voting results. In order to make clear, correct arguments of the validity of the election or the impact of the potentially faulty 'butterfly ballot' , the expertise of statisticians was essential, if not paramount. Had such analyses been allowed to play a roll in the outcome of the election, the decision of who became president might well have been different.[2]

  The statistical analysis of voting patterns, elections and electoral systems is far richer than one might initially imagine; excellent survey material, new results and

---

[2] Perhaps it was statistical analysis that the Ohio industrialist and senator Marcus Hanna forgot when he reportedly said in 1895, 'There are two things that are important in politics; the first is money, and I can't remember what the second one is.' The stout, cigar-smoking Hanna had worked for the Republican presidential candidate William McKinley, with obvious success, and is considered responsible for recognizing the relevance of fund raising and campaigning in US politics.

analyses of the 2000 election can be found in the special issue of *Statistical Science*, Vol. 17, 2002. See also Hansen (2003) and the textbook presentation in Simonoff (2003, pp. 157–162).

- *Election Polls.* Before a major political election, a randomly chosen set of several hundred or thousand people can be contacted and asked for whom they intend to vote. This is a special case of the field of *survey sampling*, which is a rich and well-developed field concerned with drawing information in an optimal fashion about particular characteristics of a population without querying all its elements, and how to deal with various associated issues such as bias (people will sometimes exaggerate or downplay their income, for example), nonresponse, asking of sensitive questions, etc. The classic and still highly useful book, familiar to all graduate statistics students, is Cochran (1977), while Thompson (2002) incorporates more recent advances.

  Regarding election polls,[3] various factors need to be taken into account such as the fact that, in many countries, more women and older people are at home. These artifacts can be straightforwardly incorporated by using existing information regarding the sex and age distribution of the target population. However, other factors are considerably more challenging to deal with, such as 'surly people' who do not wish to participate in the survey (and whose proportion in the population, and voting tendency, unlike those for sex or age, is *not* known); or the fact that rich people often have more phone lines than poorer people, or that some people only have cell phones (which, so far, are not used in such polls), or no phone at all (about 5 % of the US population). These and related statistical issues often arise in the research programs of social scientists.

- *Engineering applications* such as signal processing, systems and control theory, guidance of missiles and rockets, image restoration, pattern recognition, and many others, all make heavy use of statistical concepts. The field of cryptography also relies on (or develops its own) tools from probability theory.

There are many more academic disciplines, old and new, which use statistical principles, such as environmetrics, chemometrics, spatial econometrics, psychometrics, labormetrics, cliometrics (the statistical study of economic history) and morphometrics (analysis of the statistical properties of shapes and shape changes of geometric objects like molecules, fossils, bird wings, cars, etc.). This list could be extended almost indefinitely.[4] To illustrate the ideas more concretely, we begin with another real life situation which, however, is far more modest than the above concerns.

Every weekend you cycle to the market and buy, among other things, a dozen eggs. You pack them in your rucksack and cycle home. Your experience suggests that, on average, about two eggs don't survive the trip. Assuming you have no other eggs at

---

[3] This discussion is based on Begley (2004).

[4] The reader might wish to have a look at Chatterjee, Handcock and Simonoff (1994) and Simonoff (2003), which contain a large number of interesting, real data sets and their statistical analyses. Fair (2002) uses statistical and econometric methods to answer questions involving a variety of 'socially interesting' issues, including politics, sex, academic success, sports and drinking.

home, what is the probability that you will be able to follow a recipe which calls for 10 eggs?

There are several further assumptions that need to be made before we can attempt to answer such a question. For instance, if you manage to have a nasty collision with a roller blader on the way home, the chances are higher that more than two eggs will be broken. Thus, we assume that nothing extraordinary has occurred and that past experience is a plausible indicator of how future events might turn out. That rules out, for example, a learning effect, or use of better equipment for transport. Next, let's assume that the status of each egg, broken or intact, does not depend in any way on that of the other eggs. We could then describe the situation as you unpack the eggs as 12 trials, each with a binary outcome, with the probability of being broken equal to the average 'breakage rate' of 2/12. The number of intact eggs is a quantity which depends on the outcome of the 12 trials and which has a finite number of possibilities: $0, 1, 2, \ldots, 12$. Interest centers on computing a probability that the event 'at least 10 eggs are intact' occurs.

Readers familiar with the binomial distribution will immediately recognize this type of problem and know how to answer it. Instead of jumping to the answer, we wish to think it through carefully. Clearly, either no eggs are intact, or exactly one, or exactly two, etc., or exactly 12. Nothing else can occur. These are special events because they *partition* the event of interest and are easier to handle. Consider the even more specific event that precisely the first 10 eggs that you draw out of the bottom of your rucksack are intact. Because we assume that the status of each egg is independent of the others, we might reckon that the probability is given by $(5/6) \times (5/6) \times \cdots (5/6)$, 10 times, multiplied twice by the probability of an egg being damaged, or $1/6$, yielding $p^{10} (1 - p)^2$, where $p = 5/6$. Surely, however, you are not concerned with which particular eggs are intact, but only how many, so that we need to determine in how many ways 10 of the 12 eggs can be intact.

In this example of eggs, we could attach labels to various events: say A_i for the event that exactly i eggs are intact, $i = 0, \ldots, 12$ and R for the event that at least 10 eggs are intact. Event A_7, for instance, would represent the collection of all possible outcomes such that precisely seven eggs are intact. Abbreviating the probability of any particular event, E, as $\Pr(E)$, we can write $\Pr(R) = \Pr(A_{10} \text{ or } A_{11} \text{ or } A_{12})$ or, in set notation, $\Pr(A_{10} \cup A_{11} \cup A_{12})$, or even $\Pr\left(\bigcup_{i=10}^{12} A_i\right)$. Although this is just an abbreviated notation, we might conjecture that a solid understanding of how collections of elements, or *sets*, are mathematically manipulated could be very helpful for answering probability questions. This is indeed the case. A brief overview of what we will need from set theory is provided in Appendix A.1. The second mathematical tool which will be of great use is *combinatorics*, or methods of counting. This also came up in the broken eggs example, which required knowing the total number of ways that 10 objects from 12 had a certain characteristic; the answer is $\binom{12}{10} = 66$. Chapter 1 provides a detailed account of the tools which we will require from combinatoric analysis.

PART I

BASIC PROBABILITY

1

Combinatorics

One cannot escape the feeling that these mathematical formulas have an independent existence and an intelligence of their own, that they are wiser than we are, wiser even than their discoverers, that we get more out of them than was originally put into them. (Heinrich Hertz)

...we have overcome the notion that mathematical truths have an existence independent and apart from our own minds. It is even strange to us that such a notion could ever have existed. (E. Kasner and J. Newman)
Reproduced by permission of Simon and Schuster, NY

A sizable portion of the problems we will be addressing in this book will require some basic tools from *combinatorics*, or the science of counting. These are introduced in this chapter, along with related subjects, such as proof by induction, combinatoric identities, and the gamma and beta functions. Table B.4 provides a list of these and related mathematical functions which we will use throughout the book.

Remember that boxed equations indicate very important material, and most of them should be committed to memory.

1.1 Basic counting

The number of ways that n distinguishable objects can be ordered is given by

$$\boxed{n(n-1)(n-2)\cdots 2\cdot 1 = n!}\,,$$

pronounced 'n factorial'. The number of ways that k objects can be chosen from n, $k \le n$, when order is relevant is

---

$$n\,(n-1)\ldots(n-k+1)=:n_{[k]}=\frac{n!}{(n-k)!}\,,\tag{1.1}$$

which is referred to as the *falling* or *descending factorial*.[1]

If the order of the k objects is irrelevant, then $n_{[k]}$ is adjusted by dividing by $k!$, the number of ways of arranging the k chosen objects. Thus, the total number of ways is

$$\frac{n\,(n-1)\cdots(n-k+1)}{k!}=\frac{n!}{(n-k)!\,k!}=:\binom{n}{k}\,,\tag{1.2}$$

which is pronounced 'n choose k' and referred to as a *binomial coefficient* for reasons which will become clear below. Notice that, both algebraically and intuitively, $\binom{n}{k}=\binom{n}{n-k}$.

Remark: The following is the first numbered example; see the Preface for information about the examples and the meaning of the symbols ($\ominus$, $\odot$, $\circledast$) which accompany them. ∎

$\ominus$ **Example 1.1** Three prizes will be awarded to different individuals from a group of 20 candidates. Assuming the prizes are unique, there are $20\cdot19\cdot18=6840$ possible ways that the awards can be distributed. These can be partitioned into 1140 groups of six, whereby each group contains precisely the same three people, but differs according to the $3\cdot2\cdot1=6$ possible prize allocations. Therefore, if the prizes are all the same, then there are only $6840/3!=1140$ different combinations possible. ∎

$\ominus$ **Example 1.2** From a group of two boys and three girls, how many ways can two children be chosen such that at least one boy is picked? There are $\binom{2}{1}\binom{3}{1}=6$ ways with exactly one boy, and $\binom{2}{2}\binom{3}{0}=1$ way with exactly two boys, or seven altogether. Alternatively, from the $\binom{5}{2}=10$ total possible combinations, we can subtract $\binom{3}{2}=3$ ways such that no boy is picked. One might also attempt to calculate this another way: there are $\binom{2}{1}$ ways to pick one boy; then a person from the remaining one boy and three girls needs to be picked, which ensures that at least one boy is chosen. This gives $\binom{2}{1}\binom{4}{1}=8$ possibilities! Labeling the boys as B_1 and B_2 and, likewise, the girls as G_1,G_2 and

---

[1] Similarly, we denote the *rising* or *ascending factorial*, by

$$n^{[k]}=n\,(n+1)\ldots(n+k-1)\,.$$

There are other notational conventions for expressing the falling factorial; for example, William Feller's influential volume I (first edition, 1950, p. 28) advocates $(n)_k$, while Norman L. Johnson (1975) and the references therein (Johnson being the author and editor of numerous important statistical encyclopedia) give reasons for supporting $n^{(k)}$ for the falling factorial (and $n^{[k]}$ for the rising). One still sees the rising factorial denoted by $(n)_k$, which is referred to as the *Pochhammer symbol*, after Leo August Pochhammer, 1841–1920. We use $n_{[k]}$ and $n^{[k]}$ for the falling and rising factorials in this book as it tends to make it easier to remember what is meant.

G_3, these eight groupings are $\{B_1, G_1\}, \{B_1, G_2\}, \{B_1, G_3\}, \{B_1, B_2\}$ and $\{B_2, G_1\}$, $\{B_2, G_2\}, \{B_2, G_3\}, \{B_2, B_1\}$, which shows that the choice of both boys, $\{B_1, B_2\}$, is counted twice. Thus, there are only $8 - 1 = 7$ possibilities, as before. ∎

⊖ **Example 1.3** For k even, let $A(k) = 2 \cdot 4 \cdot 6 \cdot 8 \cdot \cdots \cdot k$. Then

$$A(k) = (1 \cdot 2)(2 \cdot 2)(3 \cdot 2)(4 \cdot 2) \cdots \left(\frac{k}{2} \cdot 2\right) = 2^{k/2} \left(\frac{k}{2}\right)!.$$

With m odd and $C(m) = 1 \cdot 3 \cdot 5 \cdot 7 \cdot \cdots \cdot m$,

$$C(m) = \frac{(m+1)!}{(m+1)(m-1)(m-3)\cdots 6 \cdot 4 \cdot 2} = \frac{(m+1)!}{A(m+1)} = \frac{(m+1)!}{2^{(m+1)/2}\left(\frac{m+1}{2}\right)!}.$$

Thus,

$$C(2i-1) = 1 \cdot 3 \cdot 5 \cdots (2i-1) = \frac{(2i)!}{2^i i!}, \qquad i \in \mathbb{N}, \tag{1.3}$$

a simple result which we will use below. ∎

A very useful identity is

$$\boxed{\binom{n}{k} = \binom{n-1}{k} + \binom{n-1}{k-1}, \qquad k < n,} \tag{1.4}$$

which follows because

$$\binom{n}{k} = \frac{n!}{(n-k)!\,k!} \cdot 1 = \frac{n!}{(n-k)!\,k!} \cdot \left(\frac{n-k}{n} + \frac{k}{n}\right)$$

$$= \frac{(n-1)!}{(n-k-1)!\,k!} + \frac{(n-1)!}{(n-k)!\,(k-1)!} = \binom{n-1}{k} + \binom{n-1}{k-1}.$$

Some intuition can be added: for any single particular object, either it is among the chosen k or not. First include it in the k choices so that there are $\binom{n-1}{k-1}$ ways of picking the remaining $k-1$. Alternatively, if the object is not one of the k choices, then there are $\binom{n-1}{k}$ ways of picking k objects. As these two situations exclude one another yet cover all possible situations, adding them must give the total number of possible combinations.

By applying (1.4) recursively,

$$\binom{n}{k} = \binom{n-1}{k} + \binom{n-1}{k-1}$$

$$= \binom{n-1}{k} + \binom{n-2}{k-1} + \binom{n-2}{k-2}$$

$$= \binom{n-1}{k} + \binom{n-2}{k-1} + \binom{n-3}{k-2} + \binom{n-3}{k-3}$$

$$\vdots$$

$$= \sum_{i=0}^{k} \binom{n-i-1}{k-i}, \quad k < n. \tag{1.5}$$

In (1.5), replace n with $n + r$, set $k = n$ and rearrange to get

$$\binom{n+r}{n} = \sum_{i=0}^{n} \binom{n+r-i-1}{n-i} = \sum_{i=0}^{n} \binom{i+r-1}{i}, \tag{1.6}$$

which will be made use of on occasion.

If a set of n distinct objects is to be divided into k distinct groups, whereby the size of each group is n_i, $i = 1, \ldots, k$ and $\sum_{i=1}^{k} n_i = n$, then the number of possible divisions is given by

$$\binom{n}{n_1, n_2, \ldots, n_k} := \binom{n}{n_1} \binom{n-n_1}{n_2} \binom{n-n_1-n_2}{n_3} \cdots \binom{n_k}{n_k} = \frac{n!}{n_1! n_2! \cdots n_k!}. \tag{1.7}$$

This reduces to the familiar combinatoric when $k = 2$:

$$\binom{n}{n_1, n_2} = \binom{n}{n_1} \binom{n-n_1}{n_2} = \binom{n}{n_1} \binom{n_2}{n_2} = \binom{n}{n_1} = \frac{n!}{n_1! (n-n_1)!} = \frac{n!}{n_1! n_2!}.$$

⊖ **Example 1.4** A small factory employs 15 people and produces three goods on three separate assembly lines, A, B and C. Lines A and B each require six people, line C needs three. How many ways can the workers be arranged?

$$\binom{15}{6} \binom{15-6}{6} \binom{15-6-6}{3} = \frac{15!}{6!9!} \frac{9!}{6!3!} = \frac{15!}{6!6!3!} = 420\,420. \qquad \blacksquare$$

The last expression in (1.7) also arises in a different context, illustrated in the next example, motivated by Feller (1968, pp. 36–37).

⊖ **Example 1.5** From a set of 12 flags, four blue, four red, two yellow and two green, all hung out in a row, the number of different 'signals' of length 12 you could produce is given by $12! / (4!\,4!\,2!\,2!) = 207\,900$. This is because flags of the same color are (presumably) indistinguishable. $\blacksquare$

⊖ **Example 1.6** From a set of two red and three green marbles, how many different nonempty combinations can be placed into an urn (in other words, the ordering does not count)? Letting R denote red and G denote green, all possibilities are: of size 1, R, G; of size 2, RR, RG, GG; of size 3, RRG, RGG, GGG; of size 4, RRGG, RGGG; and of size 5, RRGGG; for a total of 11. This total can be obtained without enumeration by observing that, in each possible combination, either there are 0,1 or 2 reds, and 0,1,2 or 3 greens, or, multiplying, $(2 + 1) \cdot (3 + 1) = 12$, but minus one, because that would be the zero–zero combination. In general, if there are $n = \sum_{i=1}^{k} n_i$ marbles, where the size of each distinguishable group is n_i, $i = 1, \ldots, k$, then there are $\prod_{i=1}^{k} (n_i + 1) - 1$ combinations ignoring the ordering.   ∎

1.2 Generalized binomial coefficients

It does not matter how they vote. It matters how we count.

(Joseph Stalin)

A generalization of the left-hand sides of (1.1) and (1.2) is obtained by relaxing the positive integer constraint on the upper term in the binomial coefficient, i.e.

$$\binom{n}{k} := \frac{n\,(n-1)\cdots(n-k+1)}{k!}, \qquad n \in \mathbb{R},\ k \in \mathbb{N}. \tag{1.8}$$

The calculations clearly still go through, but the result will, in general, be a real number.

Program Listing 1.1 shows a Matlab program for computing $\binom{n}{k}$. (This is not an ideal program to work with if you are first learning Matlab, and should be returned to after mastering the simpler programs in Chapter 2.) It is designed to work if one or both of the arguments (n and k) are vectors. The generalized binomial coefficients (1.8) of all the values in the vector are computed if any element of vector n is not an integer, or is negative. For the 'usual' binomial coefficient (i.e. n and k integers with $n \geq k \geq 0$), the log of the gamma function is used for calculation to avoid numeric overflow. The gamma function is discussed in Section 1.5.

Example 1.7 For n a positive integer, i.e., $n \in \mathbb{N}$,

$$\binom{-n}{k} = \frac{(-n)\,(-n-1)\cdots(-n-k+1)}{k!} = (-1)^k \frac{(n)\,(n+1)\cdots(n+k-1)}{k!}$$

$$= (-1)^k \binom{n+k-1}{k}. \tag{1.9}$$

For $n = 1$, the result is just $(-1)^k$.   ∎

Example 1.8 For $n \in \mathbb{N}$, the identity

$$\binom{2n}{n} = (-1)^n\, 2^{2n} \binom{-\frac{1}{2}}{n} \tag{1.10}$$

```
function c=c(n,k)

if any(n~=round(n)) | any(n<0),  c=cgeneral(n,k); return,  end

vv=find( (n>=k) & (k>=0) );  if length(vv)==0, c=0; return, end
if length(n)==1, nn=n; else nn=n(vv); end
if length(k)==1, kk=k; else kk=k(vv); end

c=zeros(1,max(length(n),length(k)));
t1 = gammaln(nn+1); t2=gammaln(kk+1); t3=gammaln(nn-kk+1);
c(vv)=round( exp ( t1-t2-t3  )   );

function c=cgeneral(nvec,kvec)
% assumes nvec and kvec have equal length and kvec are positive integers.
c=zeros(length(nvec),1);
for i=1:length(nvec)
  n=nvec(i); k=kvec(i);
  p=1; for j=1:k, p=p*(n-j+1); end
  c(i) = p/gamma(k+1);
end
```

Program Listing 1.1   Computes $\binom{n}{k}$ for possible vector values of n and k

arises in several contexts and can be seen by writing

$$\binom{-\frac{1}{2}}{n} = \frac{\left(-\frac{1}{2}\right)\left(-\frac{3}{2}\right)\cdots\left(-n+\frac{1}{2}\right)}{n!} = \left(-\frac{1}{2}\right)^n \frac{(2n-1)(2n-3)\cdots 3\cdot 1}{n!}$$

$$= \left(-\frac{1}{2}\right)^n \frac{1}{n!} \frac{(2n)!}{(2n)(2n-2)\cdots 4\cdot 2} = (-1)^n \left(\frac{1}{2}\right)^n \frac{1}{n!} \frac{(2n)!}{2^n\, n!}$$

$$= \left(\frac{1}{2}\right)^{2n} (-1)^n \binom{2n}{n}.$$

Noteworthy is also that $(-1)^n \binom{-\frac{1}{2}}{n} = \binom{n-\frac{1}{2}}{n}$, which is simple to verify. ■

⊙ ***Example 1.9***   Let $f(x) = (1-x)^t$, $t \in \mathbb{R}$. With

$$f'(x) = -t(1-x)^{t-1}, \qquad f''(x) = t(t-1)(1-x)^{t-2}$$

and, in general, $f^{(j)}(x) = (-1)^j t_{[j]} (1-x)^{t-j}$, so that the Taylor series expansion (A.132) of $f(x)$ around zero is given by

$$(1-x)^t = f(x) = \sum_{j=0}^{\infty} (-1)^j t_{[j]} \frac{x^j}{j!} = \sum_{j=0}^{\infty} \binom{t}{j} (-x)^j, \quad |x| < 1, \qquad (1.11)$$

i.e. $(1-x)^t = \sum_{j=0}^{\infty} \binom{t}{j}(-x)^j$, $|x| < 1$ or $(1+x)^t = \sum_{j=0}^{\infty} \binom{t}{j}x^j$, $|x| < 1$. For $t = -1$, (1.11) and (1.9) yield the familiar $(1-x)^{-1} = \sum_{j=0}^{\infty} x^j$, while for $t = -n$, $n \in \mathbb{N}$, they imply

$$(1-x)^{-n} = \sum_{j=0}^{\infty} \binom{-n}{j}(-x)^j = \sum_{j=0}^{\infty} \binom{n+j-1}{j} x^j, \quad |x| < 1. \quad (1.12)$$

Taylor's theorem and properties of the gamma function are used to prove the convergence of these expressions (see, for example, Protter and Morrey, 1991, pp. 238–9; Hijab, 1997, p. 91; Stoll, 2001, Theorem 8.8.4). ∎

Example 1.10 Consider proving the identity

$$\sum_{s=0}^{\infty} \frac{m}{m+s} \binom{m+s}{s}(1-\theta)^s = \theta^{-m}, \quad m \in \mathbb{N}, \quad 0 < \theta < 1. \quad (1.13)$$

The result for $m = 1$ is simple:

$$\sum_{s=0}^{\infty} \frac{1}{1+s} \binom{1+s}{s}(1-\theta)^s = \sum_{s=0}^{\infty}(1-\theta)^s = \theta^{-1}.$$

For the general case, observe that

$$\frac{m}{m+s} \frac{(m+s)!}{m!s!} = \frac{(m+s-1)!}{(m-1)!s!} = \binom{m+s-1}{s} = (-1)^s \binom{-m}{s}$$

from (1.9). Using this and (1.12) implies that (1.13) is

$$\sum_{s=0}^{\infty} \binom{-m}{s}(-(1-\theta))^s = (1-(1-\theta))^{-m} = \theta^{-m}. \quad ∎$$

1.3 Combinatoric identities and the use of induction

Equations (1.13) and (1.5) are examples of what are referred to as *combinatoric identities*, of which many exist. Some combinatoric identities can be straightforwardly proven using induction. Recall that this method of proof is applicable for identities which are a function of an integer variable, say n. It first entails verifying the conjecture for $n = 1$ and then demonstrating that it holds for $n + 1$ assuming it holds for n.

Example 1.11 Consider the sum $S_n = \sum_{k=0}^{n} \binom{n}{k}$ for $n \in \mathbb{N}$. Imagine that the objects under consideration are the bits in computer memory; they can each take on the value 0 or 1. Among n bits, observe that there are 2^n possible signals that can be constructed. However, this is what S_n also gives, because, for a given k, $\binom{n}{k}$ is the number of ways of choosing which of the n bits are set to one, and which are set to zero, and we sum this up over all possible k (0 to n) so that it gives all the possible signals that n binary bits can construct. (That $S_n = 2^n$ also follows directly from the binomial theorem, which is discussed below.) To prove the result via induction, assume it holds for $n - 1$, so that, from (1.4),

$$\sum_{k=0}^{n} \binom{n}{k} = \sum_{k=0}^{n} \left[\binom{n-1}{k} + \binom{n-1}{k-1} \right].$$

Using the fact that $\binom{m}{i} = 0$ for $i > m$,

$$\sum_{k=0}^{n} \binom{n}{k} = \sum_{k=0}^{n-1} \binom{n-1}{k} + \sum_{k=1}^{n} \binom{n-1}{k-1} = 2^{n-1} + \sum_{k=1}^{n} \binom{n-1}{k-1},$$

and, with $j = k - 1$, the latter term is

$$\sum_{k=1}^{n} \binom{n-1}{k-1} = \sum_{j=0}^{n-1} \binom{n-1}{j} = 2^{n-1},$$

so that $\sum_{k=0}^{n} \binom{n}{k} = 2^{n-1} + 2^{n-1} = 2\left(2^{n-1}\right) = 2^n$. ∎

⊖ **Example 1.12** To prove the identity

$$\frac{1}{2} = \sum_{i=0}^{n-1} \binom{n+i-1}{i} \left(\frac{1}{2}\right)^{n+i} =: P_n, \tag{1.14}$$

first note that $P_1 = 1/2$ and assume $P_n = 1/2$. Then,

$$2P_{n+1} = \sum_{i=0}^{n} \binom{n+i}{i} \left(\frac{1}{2}\right)^{n+i}$$

$$= \sum_{i=0}^{n} \binom{n+i-1}{i} \left(\frac{1}{2}\right)^{n+i} + \sum_{i=0}^{n} \binom{n+i-1}{i-1} \left(\frac{1}{2}\right)^{n+i}$$

$$= \underbrace{\sum_{i=0}^{n-1} \binom{n+i-1}{i} \left(\frac{1}{2}\right)^{n+i}}_{=P_n=\frac{1}{2}} + \binom{2n-1}{n} \left(\frac{1}{2}\right)^{2n} + \sum_{i=1}^{n} \binom{n+i-1}{i-1} \left(\frac{1}{2}\right)^{n+i}$$

$$\stackrel{j=i-1}{=} \frac{1}{2} + \binom{2n-1}{n} \left(\frac{1}{2}\right)^{2n} + \sum_{j=0}^{n-1} \binom{n+j}{j} \left(\frac{1}{2}\right)^{n+j+1}$$

$$= \frac{1}{2} + \binom{2n-1}{n} \left(\frac{1}{2}\right)^{2n} + P_{n+1} - \binom{2n}{n} \left(\frac{1}{2}\right)^{2n+1}.$$

Now note that

$$\binom{2n-1}{n}\left(\frac{1}{2}\right)^{2n} = \frac{(2n-1)(2n-2)\cdots n}{n!}\left(\frac{1}{2}\right)^{2n}$$

$$= \frac{n}{2n}\frac{2n(2n-1)(2n-2)\cdots(n+1)}{n!}\left(\frac{1}{2}\right)^{2n} = \binom{2n}{n}\left(\frac{1}{2}\right)^{2n+1},$$

or

$$2P_{n+1} = \frac{1}{2} + P_{n+1} \Leftrightarrow P_{n+1} = \frac{1}{2}. \qquad \blacksquare$$

The next example makes use of the so-called indicator function, and it will be used throughout the book. The *indicator function* is defined for every set $M \subseteq \mathbb{R}$ as

$$\mathbb{I}_M(x) = \begin{cases} 1, & \text{if } x \in M, \\ 0, & \text{if } x \notin M. \end{cases} \qquad (1.15)$$

For ease of notation, we will occasionally use the form $\mathbb{I}(x \geq k)$, which is equivalent to $\mathbb{I}_{\{k,k+1,\ldots\}}(x)$ if x takes on only discrete values, and equivalent to $\mathbb{I}_{[k,\infty)}(x)$ if x can assume values from the real line.

Example 1.13 We present three ways of proving

$$\sum_{k=1}^{n} k\binom{n}{k} = n\, 2^{n-1}. \qquad (1.16)$$

1. **By induction.** The relation clearly holds for $n = 1$. Assume the relation holds for $n - 1$. Using (i) the fact that $2^m = \sum_{i=0}^{m}\binom{m}{i}$, which was shown in Example 1.11 above, (ii) the fact that $\binom{m}{i} = 0$ for $i > m$, and setting $j = k - 1$,

$$\sum_{k=1}^{n} k\binom{n}{k} = \sum_{k=1}^{n} k\left[\binom{n-1}{k} + \binom{n-1}{k-1}\right]$$

$$= \sum_{k=1}^{n-1} k\binom{n-1}{k} + \sum_{k=1}^{n} k\binom{n-1}{k-1}$$

$$= (n-1)\, 2^{n-1-1} + \sum_{j=0}^{n-1}(j+1)\binom{n-1}{j}$$

$$= (n-1)\, 2^{n-2} + \sum_{j=0}^{n-1}\binom{n-1}{j} + \sum_{j=0}^{n-1} j\binom{n-1}{j}$$

$$= (n-1)\, 2^{n-2} + 2^{n-1} + (n-1)\, 2^{n-2} = n2^{n-1}.$$

2. **As a committee selection** (Ross, 1988, p. 18). The number of ways of choosing a k-size committee from n people along with a chairperson for the committee is clearly $k\binom{n}{k}$. Summing over k yields all possible sized committees. This is, however, just $n\,2^{n-1}$; there are n possibilities for chairperson and, from the remaining $n-1$ people, either they are in the committee or not.

3. **As an expected value.** Let $f_K(k) = \binom{n}{k}2^{-n}\mathbb{I}_{\{0,1,\,...,n\}}(k)$ be the mass function of random variable K and notice that it is symmetric, so that $\mathbb{E}[K] = n/2$. Equating this and $\mathbb{E}[K] = 2^{-n}\sum_{k=0}^{n} k\binom{n}{k}$ yields (1.16). ∎

⊖ *Example 1.14* The identity

$$\binom{N}{n+i} = \sum_{x=0}^{N-n-i} \binom{n+x-1}{x}\binom{N-n-x}{i} \tag{1.17}$$

appears in Feller (1957, p. 62), who used the identity $\sum_{i=0}^{n}\binom{a}{i}\binom{b}{n-i} = \binom{a+b}{n}$ back and forth to prove it. It was subsequently derived by Dubey (1965) and Wiśniewski (1966) using different methods of proof. Problem 1.14 asks the reader to provide a proof of (1.17) using induction. ∎

Remark: The book by Riordan (1968) provides a plethora of further combinatoric identities, while Petkovšek, Wilf and Zeilberger (1997) detail computer-based methods for deriving and proving combinatorial identities. ∎

1.4 The binomial and multinomial theorems

In comparison to the ancients we are as dwarves sitting on the shoulders of giants.
(Bernard de Chartres, 12th century scholar)

If I have seen further, it is by standing on the shoulders of giants.
(Sir Isaac Newton)[2]

If I have not seen as far as others, it is because giants were standing on my shoulders.
(Hal Abelson)
Reproduced by permission of Hal Abelson

1.4.1 The binomial theorem

The relation

$$(x+y)^n = \sum_{i=0}^{n} \binom{n}{i} x^i y^{n-i} \tag{1.18}$$

___
[2] It is interesting to note that Newton (1643–1727), wrote this in a letter to Robert Hooke (1635–1703), with whom he carried on a lifelong rivalry, and was most likely sarcastic in intent, given that Hooke was practically a midget.

is referred to as (Newton's) *binomial theorem*. It is a simple, yet fundamental result which arises in numerous applications. Examples include

$$(x + (-y))^2 = x^2 - 2xy + y^2, \qquad (x + y)^3 = x^3 + 3x^2y + 3xy^2 + y^3,$$

$$0 = (1 - 1)^n = \sum_{i=0}^{n} \binom{n}{i} (-1)^{n-i}, \qquad 2^n = (1 + 1)^n = \sum_{i=0}^{n} \binom{n}{i}.$$

To prove the binomial theorem via induction, observe first that (1.18) holds for $n = 1$. Then, assuming it holds for $n - 1$,

$$(x + y)^n = (x + y)(x + y)^{n-1} = (x + y) \sum_{i=0}^{(n-1)} \binom{(n-1)}{i} x^i y^{(n-1)-i}$$

$$= \sum_{i=0}^{n-1} \binom{n-1}{i} x^{i+1} y^{n-(i+1)} + \sum_{i=0}^{n-1} \binom{n-1}{i} x^i y^{n-1-i+1}.$$

Then, with $j = i + 1$,

$$(x + y)^n = \sum_{j=1}^{n} \binom{n-1}{j-1} x^j y^{n-j} + \sum_{i=0}^{n-1} \binom{n-1}{i} x^i y^{n-i}$$

$$= x^n + \sum_{j=1}^{n-1} \binom{n-1}{j-1} x^j y^{n-j} + \sum_{i=1}^{n-1} \binom{n-1}{i} x^i y^{n-i} + y^n$$

$$= x^n + \sum_{i=1}^{n-1} \left[\binom{n-1}{i-1} + \binom{n-1}{i} \right] x^i y^{n-i} + y^n$$

$$= x^n + \sum_{i=1}^{n-1} \binom{n}{i} x^i y^{n-i} + y^n = \sum_{i=0}^{n} \binom{n}{i} x^i y^{n-i}$$

proving the theorem.

⊙ *Example 1.15* The binomial theorem can be used for proving

$$\lim_{n \to \infty} \frac{x^n}{n^k} = \infty, \quad x \in \mathbb{R}_{>1}, \ k \in \mathbb{N}. \tag{1.19}$$

As in Lang (1997, p. 55), let $y \in \mathbb{R}_{>0}$ and $x = (1 + y)$. From the binomial theorem,

$$x^n = (1 + y)^n = \sum_{i=0}^{n} \binom{n}{i} y^i. \tag{1.20}$$

For the term with $i = k + 1$,

$$\binom{n}{k+1} = \frac{n(n-1)\cdots(n-k)}{(k+1)!} = \frac{n^{k+1}}{(k+1)!} \left(1 + \frac{c_1}{n} + \cdots \frac{c_{k+1}}{n^{k+1}} \right),$$

where the c_i are constants which do not depend on n. As all the terms in the sum (1.20) are nonnegative,

$$x^n > \binom{n}{k+1} y^{k+1} = \frac{n^{k+1}}{(k+1)!} \left(1 + \frac{c_1}{n} + \cdots \frac{c_{k+1}}{n^{k+1}}\right) y^{k+1}$$

or

$$\frac{x^n}{n^k} > n\left[\frac{y^{k+1}}{(k+1)!}\right]\left(1 + \frac{c_1}{n} + \cdots \frac{c_{k+1}}{n^{k+1}}\right).$$

When $n \to \infty$, the r.h.s. is the product of a term which increases without bound, and one which converges to one, from which (1.19) follows. ∎

◎ **Example 1.16** Let f and g denote functions whose nth derivatives exist. Then, by using the usual product rule for differentiation and an induction argument, it is straightforward to show that

$$[fg]^{(n)} = \sum_{j=0}^{n} \binom{n}{j} f^{(j)} g^{(n-j)}, \tag{1.21}$$

where $f^{(j)}$ denotes the jth derivative of f. This is sometimes (also) referred to as Leibniz' rule. This is not the binomial theorem per se, though it has an obvious association. ∎

⊖ **Example 1.17** Consider computing $I = \int_{-1}^{1} (x^2 - 1)^j \, dx$, for any $j \in \mathbb{N}$. From the binomial theorem and the basic linearity property (A.55) of the Riemann integral,

$$I = \sum_{k=0}^{j} \binom{j}{k} (-1)^{j-k} \int_{-1}^{1} x^{2k} \, dx = \sum_{k=0}^{j} \binom{j}{k} (-1)^{j-k} \frac{2}{2k+1},$$

which is simple to program and compute as a function of j. In fact, as shown in Problem 1.19, I can also be expressed as

$$I = \frac{(-1)^j 2^{2j+1}}{\binom{2j}{j}(2j+1)},$$

thus implying the combinatoric identity

$$\sum_{k=0}^{j} \binom{j}{k} \frac{(-1)^k}{2k+1} = \frac{2^{2j}}{\binom{2j}{j}(2j+1)}, \quad j \in \mathbb{N}, \tag{1.22}$$

as $(-1)^{-k} = (-1)^k$ and cancelling a 2 and $(-1)^j$ from both sides. ∎

The next example illustrates a very useful relation, the derivation of which hinges upon the binomial theorem.

◎ **Example 1.18** Most people are familiar with the childhood story of Carl Friedrich Gauss and the sum of the first 100 integers. This true story, and many others, are enjoyably told in Dunnington's (1955) authoritative biography. Gauss' elementary school teacher (a J. G. Büttner, with whip in hand) had assigned to the class the task of computing the sum of the numbers 1 to 100, to which the young Gauss quickly gave the answer, noting that

$$1 + 2 + \cdots + 99 + 100 = (1 + 100) + (2 + 99) + \cdots + (50 + 51) = 50 \times 101.$$

(Gauss was deservingly spared the whip, though smart, but wrong, attempts at other problems were duly punished.)

A bit of thought reveals the general statement

$$1 + 2 + \cdots + n = \sum_{k=1}^{n} k = \frac{n(n+1)}{2}. \tag{1.23}$$

Upon discovering this, one is tempted to consider extensions to higher integer powers, which also possess closed-form expressions; for example

$$\sum_{k=1}^{n} k^2 = \frac{n(n+1)(2n+1)}{6}, \quad \sum_{k=1}^{n} k^4 = \frac{n(2n+1)(n+1)(3n+3n^2-1)}{30},$$

$$\sum_{k=1}^{n} k^3 = \frac{n^2(n+1)^2}{4}, \quad \sum_{k=1}^{n} k^5 = \frac{n^2(2n+2n^2-1)(n+1)^2}{12}.$$

Once a candidate solution is provided, induction can be used to prove it. Casual inspection of the previous formulae reveals some patterns, but guesswork and trial and error would still be needed to derive further relations. We now detail a simple method for actually *deriving* such equations, as was given by Lentner (1973), but which is at least a century old.[3]

Summing the obvious identity $i^k = i^{k+1} - i^k(i-1)$ gives, with $t = i - 1$,

$$\sum_{i=1}^{n+1} i^k = \sum_{i=1}^{n+1} i^{k+1} - \sum_{i=2}^{n+1} i^k(i-1) = \sum_{i=1}^{n+1} i^{k+1} - \sum_{t=1}^{n}(t+1)^k t.$$

From the binomial theorem, $(t+1)^k = \sum_{j=0}^{k} \binom{k}{j} t^j$ so that

$$\sum_{i=1}^{n+1} i^k = \sum_{i=1}^{n+1} i^{k+1} - \sum_{t=1}^{n} t \sum_{j=0}^{k} \binom{k}{j} t^j$$

$$= \sum_{i=1}^{n+1} i^{k+1} - \sum_{t=1}^{n} \left[\binom{k}{k} t^{k+1} + \sum_{j=0}^{k-1} \binom{k}{j} t^{j+1} \right]$$

$$= \sum_{i=1}^{n+1} i^{k+1} - \sum_{t=1}^{n} t^{k+1} - \sum_{t=1}^{n} \sum_{j=0}^{k-1} \binom{k}{j} t^{j+1} = (n+1)^{k+1} - \sum_{t=1}^{n} \sum_{j=0}^{k-1} \binom{k}{j} t^{j+1}$$

---

[3] As impressively pointed out by Streit (1974), the result appears in a slightly different form in G. Chrystal, *Algebra, an Elementary Textbook*, Vol. I, 2nd edition, 1889, p. 485.

or, as $(n + 1)^k (n + 1) - (n + 1)^k = (n + 1)^k n$,

$$S(n, k) := \sum_{i=1}^{n} i^k = -(n + 1)^k + (n + 1)^{k+1} - \sum_{t=1}^{n} \sum_{j=0}^{k-1} \binom{k}{j} t^{j+1}$$

$$= n(n + 1)^k - \sum_{t=1}^{n} \sum_{j=0}^{k-1} \binom{k}{j} t^{j+1}. \tag{1.24}$$

For $k = 1$, (1.24) reduces to

$$S(n, 1) = \sum_{i=1}^{n} i = n(n + 1) - \sum_{t=1}^{n} t = n(n + 1) - S(n, 1),$$

or $S(n, 1) = n(n + 1)/2$, as in (1.23). To develop a recursive structure, note that

$$\sum_{t=1}^{n} \sum_{j=0}^{k-1} \binom{k}{j} t^{j+1} - \sum_{t=1}^{n} \sum_{j=0}^{k-2} \binom{k}{j} t^{j+1} = \sum_{t=1}^{n} \binom{k}{k-1} t^k = kS(n, k),$$

so that, from (1.24),

$$(1 + k) S(n, k) = S(n, k) + \sum_{t=1}^{n} \sum_{j=0}^{k-1} \binom{k}{j} t^{j+1} - \sum_{t=1}^{n} \sum_{j=0}^{k-2} \binom{k}{j} t^{j+1}$$

$$= n(n + 1)^k - \sum_{t=1}^{n} \sum_{j=0}^{k-2} \binom{k}{j} t^{j+1}$$

$$= n(n + 1)^k - \sum_{j=0}^{k-2} \binom{k}{j} \sum_{t=1}^{n} t^{j+1}$$

$$= n(n + 1)^k - \sum_{j=0}^{k-2} \binom{k}{j} S(n, j + 1). \tag{1.25}$$

With $k = 2$, (1.25) reduces to $3S(n, 2) = n(n + 1)^2 - \binom{2}{0} S(n, 1)$ or

$$S(n, 2) = \frac{n(n + 1)^2}{3} - \frac{n(n + 1)}{6} = \frac{n(n + 1)(2n + 1)}{6},$$

and for $k = 3$, (1.25) gives $4S(n, 3) = n(n + 1)^3 - S(n, 1) - 3S(n, 2)$ or

$$S(n, 3) = \frac{n(n + 1)^3}{4} - \frac{n(n + 1)}{8} - \frac{n(n + 1)(2n + 1)}{8} = \frac{1}{4} n^2 (n + 1)^2.$$

Expressions for other $S(n, k)$ follow similarly. Note that, from the form of (1.25) and $S(n, 1)$, it follows from induction that $S(n, k)$ is a polynomial in n for all $k \in \mathbb{N}$. ∎

Remark: We advocate, and occasionally refer to, use of software with a symbolic computing environment, such as Mathematica, Maple, MuPAD, etc. The previous summation formulae, for example, are known to such packages, but that just scratches the surface of the power and utility of such programs. The following are useful observations on the latter two.

Maple – in the Maple engine accompanying Scientific Workplace version 4.0, attempting to evaluate 0^0 gives an error, but defining $n = 0$ and evaluating 0^n yields 0. Furthermore, with $n = 0$ defined and any value of N, evaluating $\sum_{j=N}^{N} \left(\frac{N-j}{N}\right)^n$ yields an error (but in an older version, yielded 1); similarly, $\sum_{j=N}^{N} \left(\frac{N-j}{N}\right)^0$ yields 1; however, $\sum_{j=N}^{N} \left(\frac{N-j}{N}\right)^k$, for k undefined, yields zero! It is useful to define 0^0 to be one so that the binomial theorem holds for all arguments, i.e. so that

$$0 = (0+0)^n = \sum_{i=0}^{n} \binom{n}{i} 0^i 0^{n-i}$$

is valid and, more generally, that

$$k^n = (0+k)^n = \sum_{i=0}^{n} \binom{n}{i} 0^i k^{n-i}.$$

Via Maple, the latter expression returns zero.

MuPAD – all the calculations mentioned above are correct when using the MuPAD computing engine in Scientific Workplace. With k undefined, the expression $\sum_{j=N}^{N} \left(\frac{N-j}{N}\right)^k$ yields 0^k. MuPAD, however, is not without its problems. In both Maple and MuPAD, evaluating $\sum_{k=0}^{n} \binom{n}{k}$ correctly gives 2^n, but evaluating $\sum_{k=0}^{n-1} \binom{n-1}{k}$ correctly gives 2^{n-1} in Maple, but returns a 0 in MuPAD. ∎

1.4.2 An extension of the binomial theorem

A nonobvious extension of the binomial theorem is

$$(x+y)^{[n]} = \sum_{i=0}^{n} \binom{n}{i} x^{[i]} y^{[n-i]}, \quad x^{[n]} = \prod_{j=0}^{n-1} (x+ja), \tag{1.26}$$

for $n = 0, 1, 2, \ldots$, and x, y, a are real numbers. It holds trivially for $n = 0$, and is easy to see for $n = 1$ and $n = 2$, but otherwise appears difficult to verify, and induction gets messy and doesn't (seem to) lead anywhere. Perhaps somewhat surprisingly, the general proof involves calculus, and is given in Example 1.30. The case with $a = -1$ can be proved directly, as we now do, and is also of great use.

⊛ *Example 1.19* With $a = -1$,

$$k^{[n]} = (k)(k-1)(k-2)\cdots(k-(n-1)) = \frac{k!}{(k-n)!} = \binom{k}{n} n!$$

so that (1.26) reads

$$\binom{x+y}{n} n! = \sum_{i=0}^{n} \binom{n}{i}\binom{x}{i} i! \binom{y}{n-i} (n-i)! = n! \sum_{i=0}^{n} \binom{x}{i}\binom{y}{n-i}$$

or

$$\boxed{\binom{x+y}{n} = \sum_{i=0}^{n} \binom{x}{i}\binom{y}{n-i}}.$$ (1.27)

This is referred to as *Vandermonde's theorem*. Switching to the notation we will use later,

$$\binom{N+M}{k} = \sum_{i=0}^{k} \binom{N}{i}\binom{M}{k-i}$$

$$= \binom{N}{0}\binom{M}{k} + \binom{N}{1}\binom{M}{k-1} + \cdots + \binom{N}{k}\binom{M}{0},$$ (1.28)

for $0 < k \le \min(N, M)$.

Using the standard convention that $\binom{a}{b} \equiv 0$ for $b > a$, then (1.28) also holds for the larger range $0 \le k \le N + M$. For example, with $N = M = 3$,

$$\binom{6}{5} = \binom{3}{0}\binom{3}{5} + \binom{3}{1}\binom{3}{4} + \binom{3}{2}\binom{3}{3} + \binom{3}{3}\binom{3}{2} + \binom{3}{4}\binom{3}{1} + \binom{3}{5}\binom{3}{0}$$

$$= 3 + 3 = 6$$

and

$$\binom{6}{6} = \binom{3}{0}\binom{3}{6} + \binom{3}{1}\binom{3}{5} + \binom{3}{2}\binom{3}{4} + \binom{3}{3}\binom{3}{3} + \binom{3}{4}\binom{3}{2}$$

$$+ \binom{3}{5}\binom{3}{1} + \binom{3}{6}\binom{3}{0}$$

$$= 1.$$

An intuitive proof of (1.28) goes as follows. Assume that $N \ge k$ and $M \ge k$, then either

- all k objects are chosen from the group of size M and none from the group of size N, or

- $k - 1$ objects are chosen from the group of size M and 1 from the group of size N, or

$$\vdots \qquad \vdots \qquad \vdots$$

- no objects are chosen from the group of size M and all k from the group of size N.

Summing these disjoint events yields the desired formula. ∎

⊛ ***Example 1.20*** Vandermonde's theorem (1.28) also follows directly by equating coefficients of x^r from each side of the identity

$$(1+x)^{N+M} = (1+x)^N (1+x)^M$$

and using the binomial theorem (1.18). In particular,

$$(1+x)^{N+M} = \sum_{i=0}^{N+M} \binom{N+M}{i} x^i, \tag{1.29}$$

with the coefficient of x^r, $0 \le r \le N+M$ being $\binom{N+M}{r}$; and

$$(1+x)^N (1+x)^M = \left[\sum_{j=0}^{N} \binom{N}{j} x^j \right] \left[\sum_{k=0}^{M} \binom{M}{k} x^k \right]$$

$$=: \left(a_0 + a_1 x + \cdots a_N x^N \right) \left(b_0 + b_1 x + \cdots b_M x^M \right), \tag{1.30}$$

where $a_j = \binom{N}{j}$ and $b_k = \binom{M}{k}$ are defined for convenience. Observe that the coefficient of x^0 in (1.30) is $a_0 b_0$; the coefficient of x^1 is $a_0 b_1 + a_1 b_0$; and, in general, the coefficient of x^r is $a_0 b_r + a_1 b_{r-1} + \cdots + a_r b_0$, which is valid for $r \in \{0, 1, \ldots, N+M\}$ if we define $a_j := 0$ for $j > N$ and $b_k := 0$ for $k > M$. Thus, the coefficient of x^r is

$$\sum_{i=0}^{r} a_i b_{r-i} = \sum_{i=0}^{r} \binom{N}{i} \binom{M}{r-i},$$

from which it follows that

$$\binom{N+M}{r} = \sum_{i=0}^{r} \binom{N}{i} \binom{M}{r-i}, \quad 0 \le r \le N+M,$$

as was to be shown.

Some further reflection reveals the following derivation, which is related to the Cauchy product discussed in Section A.2.4.3. First observe that

$$S := \left(\sum_{j=0}^{N} a_j x^j \right) \left(\sum_{k=0}^{M} b_k x^k \right) = (a_0 + a_1 x + \cdots)(b_0 + b_1 x + \cdots)$$

$$= a_0 (b_0 + b_1 x + \cdots) + a_1 x (b_0 + b_1 x \cdots) + \cdots$$

$$= \sum_{j=0}^{N} a_j x^j \left(\sum_{k=0}^{M} b_k x^k \right) = \sum_{j=0}^{N} \sum_{k=0}^{M} a_j b_k x^{j+k} = \sum_{j=0}^{L} \sum_{k=0}^{L} a_j b_k x^{j+k},$$

where $L = \max(N, M)$, $a_j = 0$ for $j > N$ and $b_k = 0$ for $k > M$, so that S consists of the sum of the entries of the following table:

		0	1	2	$\cdots$	L
	0	$\circ$	$\bullet$	$\diamond$		
	1	$\bullet$	$\diamond$			
j	2	$\diamond$			$a_j b_k x^{j+k}$	
	$\vdots$			$\swarrow$		
	L					

with typical element $a_j b_k x^{j+k}$. The elements of the table can be summed in any order. In particular, summing the off-diagonals as indicated produces

$$S = \sum_{i=0}^{N+M} \sum_{t=0}^{i} a_t b_{i-t} x^t x^{i-t} = \sum_{i=0}^{N+M} x^i \sum_{t=0}^{i} \binom{N}{t}\binom{M}{i-t}, \qquad (1.31)$$

where i designates the off-diagonal starting from the upper left and t 'travels along' the ith off-diagonal. Comparing the right-hand sides of (1.31) and (1.29) gives the desired result. ∎

See the discussion of the hypergeometric distribution and Problem 4.1 for further probabilistic justification of Vandermonde's theorem.

Remarks:

(a) Relation (1.28) can be expressed in other useful forms. One is obtained simply by replacing N by n, M by $b + r - n - 1$ and k by $b - 1$, giving

$$\binom{b+r-1}{b-1} = \sum_{i=0}^{b-1} \binom{n}{i}\binom{b+r-n-1}{b-1-i}. \qquad (1.32)$$

This is used in Example 3.16.

(b) Another useful form of (1.27) takes $x = k$, $y = m - k$ and $n = k$, to get

$$\binom{m}{k} = \sum_{i=0}^{k} \binom{k}{i}\binom{m-k}{k-i} = \sum_{j=0}^{k} \binom{k}{k-j}\binom{m-k}{j},$$

having used the substitution $j = k - i$ and the (standard) convention that $\binom{a}{b} \equiv 0$ for $b > a$. By omitting the terms in the sum which are identically zero, we can also write

$$\binom{m}{k} = \sum_{j=0}^{\min(k,m-k)} \binom{k}{k-j}\binom{m-k}{j}.$$

Alternatively, it is easy to verify that $\min(k, m - k) \leq \lfloor m/2 \rfloor$, where $\lfloor m/2 \rfloor$ is the floor function, i.e. $\lfloor m/2 \rfloor = m/2$ if m is even and $\lfloor m/2 \rfloor = (m - 1)/2$ if m is odd. Thus,

$$\binom{m}{k} = \sum_{j=0}^{\lfloor m/2 \rfloor} \binom{k}{k-j} \binom{m-k}{j}. \tag{1.33}$$

This will be used in Problem 6.9. ∎

1.4.3 The multinomial theorem

Another important generalization of the binomial theorem, in a different direction than taken in Section 1.4.2, is the *multinomial theorem*, given by

$$\left(\sum_{i=1}^{r} x_i \right)^n = \sum_{\mathbf{n}:\, n_\bullet = n,\, n_i \geq 0} \binom{n}{n_1, \ldots, n_r} \prod_{i=1}^{r} x_i^{n_i}, \quad r, n \in \mathbb{N}, \tag{1.34}$$

where the combinatoric is defined in (1.7), $\mathbf{n}$ denotes the vector $(n_1, \ldots, n_r)$ and $n_\bullet = \sum_{i=1}^{r} n_i$. In words, the sum is taken over all nonnegative integer solutions to $\sum_{i=1}^{r} n_i = n$, the number of which is shown in Chapter 2 to be $\binom{n+r-1}{n}$.

An interesting special case of (1.34) is for $n = 2$; simplifying yields

$$\left(\sum_{i=1}^{r} x_i \right)^2 = \sum_{i=1}^{r} \sum_{j=1}^{r} x_i x_j. \tag{1.35}$$

⊖ **Example 1.21** The expression $(x_1 + x_2 + x_3)^4$ corresponds to $r = 3$ and $n = 4$, so that its expansion will have $\binom{6}{4} = 15$ terms, these being

$$(x_1 + x_2 + x_3)^4 = \binom{4}{4,0,0} x_1^4 x_2^0 x_3^0 + \binom{4}{0,4,0} x_1^0 x_2^4 x_3^0 + \binom{4}{0,0,4} x_1^0 x_2^0 x_3^4$$

$$+ \binom{4}{3,1,0} x_1^3 x_2^1 x_3^0 + \binom{4}{3,0,1} x_1^3 x_2^0 x_3^1 + \cdots.$$

These are

$$(x_1 + x_2 + x_3)^4 = x_1^4 + x_2^4 + x_3^4$$

$$+ 4x_1^3 x_2 + 4x_1^3 x_3 + 4x_1 x_2^3 + 4x_1 x_3^3 + 4x_2^3 x_3 + 4x_2 x_3^3$$

$$+ 6x_1^2 x_2^2 + 6x_1^2 x_3^2 + 6x_2^2 x_3^2$$

$$+ 12x_1^2 x_2 x_3 + 12x_1 x_2^2 x_3 + 12x_1 x_2 x_3^2,$$

obtained fast and reliably from a symbolic computing environment. ∎

The proof of (1.34) is by induction on r. For $r = 1$, the theorem clearly holds. Assuming it holds for $r = k$, observe that, with $S = \sum_{i=1}^{k} x_i$,

$$\left(\sum_{i=1}^{k+1} x_i\right)^n = (S + x_{k+1})^n = \sum_{i=0}^{n} \frac{n!}{i!\,(n-i)!} x_{k+1}^i S^{n-i}$$

$$= \sum_{i=0}^{n} \frac{n!}{i!\,(n-i)!} \sum_{n_1+\cdots+n_k=n-i} \frac{(n-i)!}{n_1!\cdots n_k!} \prod_{i=1}^{k} x_i^{n_i} x_{k+1}^i,$$

from the binomial theorem and (1.34) for $r = k$. By setting $n_{k+1} = i$, this becomes

$$\left(\sum_{i=1}^{k+1} x_i\right)^n = \sum_{n_1+\cdots+n_{k+1}=n} \frac{n!}{n_1!\cdots n_{k+1}!} \prod_{i=1}^{k} x_i^{n_i} x_{k+1}^i,$$

as the sum $\sum_{i=0}^{n} \sum_{n_1+\cdots+n_k=n-i}$ is equivalent to $\sum_{n_1+\cdots+n_{k+1}=n}$ for nonnegative n_i. This is precisely (1.34) for $r = k+1$, proving the theorem.

1.5 The gamma and beta functions

Genius is one per cent inspiration, ninety-nine per cent perspiration.

(Thomas A. Edison)

The gamma function is a smooth function (it is continuous as are all its derivatives) of one parameter, say a, on $\mathbb{R}_{>0}$ which coincides with the factorial function when $a \in \mathbb{N}$. It can also be extended to part of the complex domain, though we will not require this. The gamma function arises notoriously often when working with continuous random variables, so a mastery of its basic properties is essential. The beta function is related to the gamma function, and also appears quite often.

1.5.1 The gamma function

It is adequate for our purposes to take as the definition of the *gamma function* the improper integral expression of one parameter

$$\Gamma(a) := \int_0^\infty x^{a-1} e^{-x} dx, \quad a \in \mathbb{R}_{>0}. \tag{1.36}$$

There exists no closed form expression for $\Gamma(a)$ in general, so that it must be computed using numerical methods. However,

$$\Gamma(a) = (a-1)\,\Gamma(a-1), \quad a \in \mathbb{R}_{>1}, \tag{1.37}$$

and, in particular,

$$\Gamma(n) = (n-1)!, \quad n \in \mathbb{N}. \tag{1.38}$$

For example, $\Gamma(1) = 0! = 1$, $\Gamma(2) = 1! = 1$ and $\Gamma(3) = 2! = 2$. To prove (1.37), apply integration by parts with $u = x^{a-1}$ and $dv = e^{-x}dx$, which gives $du = (a-1)$ $x^{a-2}dx$, $v = -e^{-x}$ and

$$\Gamma(a) = \int_0^\infty x^{a-1}e^{-x}dx = uv|_{x=0}^\infty - \int_0^\infty v\,du$$

$$= -e^{-x}x^{a-1}|_{x=0}^\infty + \int_0^\infty e^{-x}(a-1)x^{a-2}dx$$

$$= 0 + (a-1)\Gamma(a-1).$$

The identity

$$\Gamma(a) = 2\int_0^\infty u^{2a-1}e^{-u^2}du \tag{1.39}$$

is useful (see Example 7.3) and follows directly by using the substitution $u = x^{1/2}$ in (1.36) (recall x is positive), so that $x = u^2$ and $dx = 2udu$. Another useful fact which follows from (1.39) and Example A.79 (and letting, say, $v = \sqrt{2}\,u$) is that

$$\Gamma(1/2) = \sqrt{\pi}. \tag{1.40}$$

⊛ **Example 1.22**   To calculate $I = \int_0^\infty x^n e^{-mx}dx$, $m > 0$, define $u = mx$. Then

$$I = m^{-1}\int_0^\infty (u/m)^n\,e^{-u}du = m^{-(n+1)}\Gamma(n+1), \tag{1.41}$$

which is a simple result, but required for the scaled gamma distribution (see Section 7.1) and the next example.   ∎

⊃ **Example 1.23**   For $r \in \mathbb{N}$ and $x > 1$, (1.41) implies that $\int_0^\infty u^{x-1}e^{-ru}du = \Gamma(x)/r^x$, or

$$\frac{1}{r^x} = \frac{1}{\Gamma(x)}\int_0^\infty u^{x-1}e^{-ru}du.$$

Then, with $\zeta(x) = \sum_{r=1}^\infty r^{-x}$ (the Riemann zeta function), and *assuming* the validity of the exchange of integral and sum,

$$\zeta(x) = \sum_{r=1}^\infty \frac{1}{r^x} = \frac{1}{\Gamma(x)}\sum_{r=1}^\infty \int_0^\infty u^{x-1}e^{-ru}du = \frac{1}{\Gamma(x)}\int_0^\infty u^{x-1}\sum_{r=1}^\infty e^{-ru}du.$$

(It can be justified with the monotone convergence theorem, which is beyond the level of our presentation; see, for example, and Hijab, 1997, p. 168.) Thus, for $x > 1$,

$$\zeta(x)\Gamma(x) = \int_0^\infty u^{x-1}\sum_{r=1}^\infty (e^{-u})^r\,du = \int_0^\infty u^{x-1}\frac{e^{-u}}{1-e^{-u}}du = \int_0^\infty \frac{u^{x-1}}{e^u - 1}du.$$

We will not need this result, but it turns out to be very important in other contexts, for example Havil, 2003, p. 59).   ∎

Remark: In terms of computer software, the gamma function itself is not computed, but rather the natural log of it (see, for example, Macleod, 1989). For instance, in Matlab, the function `gamma` just calls function `gammaln`, where all the computational work is actually done (and, of course, then takes the antilog). This is also advantageous for avoiding numeric overflow when computing binomial coefficient $\binom{n}{k}$ for large n, and was used in Program Listing 1.1. ∎

The gamma function can be approximated by *Stirling's approximation*, given by

$$\Gamma(n) \approx \sqrt{2\pi}\, n^{n-1/2} \exp(-n), \tag{1.42}$$

which, clearly, provides an approximation to $n!$ for integer n. This is a famous result which is important, if not critical, in a variety of contexts in probability and statistics. It is proven in most textbooks on real analysis. We will demonstrate methods of its proof using probabilistic arguments in Volume II, as well as using it in several contexts. The computational method referred to in the previous remark is essentially just the higher order expansion of (the log of) (1.42), also referred to as Stirling's approximation. Its derivation uses the so-called Euler–MacLaurin formula and the Bernoulli numbers; see, for example, Hijab (1997, Section 5.9) for a detailed presentation.

The *incomplete gamma function* is defined as

$$\Gamma_x(a) = \int_0^x t^{a-1} e^{-t} dt, \quad a, x \in \mathbb{R}_{>0} \tag{1.43}$$

and also denoted by $\gamma(x, a)$. The *incomplete gamma ratio* is the standardized version, given by

$$\overline{\Gamma}_z(\alpha) = \Gamma_x(a) / \Gamma(a).$$

In general, both functions $\Gamma(a)$ and $\Gamma_x(a)$ need to be evaluated using numerical methods. However, for integer a, $\Gamma_x(a)$ can be easily computed as follows. Using integration by parts with $u = t^{a-1}$ and $dv = e^{-t} dt$ gives

$$\int_0^x t^{a-1} e^{-t} dt = -t^{a-1} e^{-t} \Big|_0^x + (a-1) \int_0^x e^{-t} t^{a-2} dt$$

$$= -x^{a-1} e^{-x} + (a-1) \Gamma_x(a-1),$$

i.e.

$$\Gamma_x(a) = (a-1) \Gamma_x(a-1) - x^{a-1} e^{-x}, \tag{1.44}$$

so that $\Gamma_x(a)$ can be evaluated recursively, noting that

$$\Gamma_x(1) = \int_0^x e^{-t} dt = 1 - e^{-x}. \tag{1.45}$$

For the general case with $a \in \mathbb{R}_{>0}$, $\Gamma_x(a)$ can be approximated by computing (1.44) and (1.45) for the two integer values $\lfloor a \rfloor$ (floor) and $\lceil a \rceil$ (ceiling) and using linear interpolation for the value at a. More integer values of a and nonlinear curve-fitting techniques could be used to obtain more accurate approximations to $\Gamma_x(a)$. Other ways of approximating $\Gamma_x(a)$ will be considered in Volume II, in the context of saddlepoint approximations and the confluent hypergeometric function. The interested reader should compare the accuracy of these different methods to the procedure in Abramowitz and Stegun (1972, Section 6.5) (or use Matlab's `gammainc` function) which compute the values of $\Gamma_x(a)$ to machine precision.

1.5.2 The beta function

The *beta function* is an integral expression of two parameters, denoted $B(\cdot, \cdot)$ and defined to be

$$B(a, b) := \int_0^1 x^{a-1} (1-x)^{b-1} \, dx, \quad a, b \in \mathbb{R}_{>0}. \tag{1.46}$$

By substituting $x = \sin^2 \theta$ into (1.46) we obtain that

$$B(a, b) = \int_0^{\pi/2} \left(\sin^2 \theta\right)^{a-1} \left(\cos^2 \theta\right)^{b-1} 2 \sin \theta \cos \theta d\theta$$

$$= 2 \int_0^{\pi/2} (\sin \theta)^{2a-1} (\cos \theta)^{2b-1} \, d\theta. \tag{1.47}$$

Closed-form expressions do not exist for general a and b; however, the identity

$$B(a, b) = \frac{\Gamma(a) \Gamma(b)}{\Gamma(a+b)} \tag{1.48}$$

can be used for its evaluation in terms of the gamma function. A probabilistic proof of (1.48) is given in Example 9.11 of Section 9.3, while a direct proof goes as follows. From (1.39) and use of polar coordinates $x = r \cos \theta$, $y = r \sin \theta$, $dx \, dy = r \, dr \, d\theta$ from (A.167),

$$\Gamma(a) \Gamma(b) = 4 \int_0^\infty \int_0^\infty x^{2a-1} y^{2b-1} e^{-\left(x^2 + y^2\right)} dx dy$$

$$= 4 \int_0^{2\pi} \int_0^\infty r^{2(a+b)-2+1} e^{-r^2} (\cos \theta)^{2a-1} (\sin \theta)^{2b-1} \, dr d\theta$$

$$= 4 \left[\int_0^\infty r^{2(a+b)-1} e^{-r^2} dr \right] \left[\int_0^{2\pi} (\cos \theta)^{2a-1} (\sin \theta)^{2b-1} \, d\theta \right]$$

$$= \Gamma(a+b) B(a, b),$$

using (1.39) and (1.47). A direct proof without the use of polar coordinates can be found in Hijab (1997, p. 193).

If $a = b$, then, from symmetry (or use the substitution $y = 1 - x$) and use of (1.48), it follows that

$$\int_0^{1/2} x^{a-1} (1-x)^{a-1} \, dx = \int_{1/2}^1 x^{a-1} (1-x)^{a-1} \, dx = \frac{1}{2} \frac{\Gamma^2(a)}{\Gamma(2a)}, \qquad (1.49)$$

where $\Gamma^2(a)$ is just a shorthand notation for $[\Gamma(a)]^2$. A similar analysis reveals that

$$\Gamma(2a) = \frac{2^{2a-1}}{\sqrt{\pi}} \Gamma(a) \Gamma\left(a + \frac{1}{2}\right), \qquad (1.50)$$

which is referred to as *Legendre's duplication formula for gamma functions*. In particular, from (1.49) with $u = 4x(1-x)$ (and, as $0 \le x \le 1/2$, $x = \left(1 - \sqrt{1-u}\right)/2$ and $dx = 1/(4\sqrt{1-u}) du$),

$$\frac{\Gamma^2(a)}{\Gamma(2a)} = 2 \int_0^{1/2} x^{a-1} (1-x)^{a-1} \, dx = \frac{2}{4^{a-1}} \int_0^{1/2} (4x(1-x))^{a-1} \, dx$$

$$= \frac{2}{4^{a-1}} \int_0^1 u^{a-1} \frac{1}{4} (1-u)^{-1/2} \, du = 2^{1-2a} \frac{\Gamma(a) \Gamma(1/2)}{\Gamma(a+1/2)}.$$

As $\Gamma(1/2) = \sqrt{\pi}$, the result follows. We can use the duplication formula in the following way. From (1.50) with $i \in \mathbb{N}$ and using (1.3),

$$\Gamma\left(i + \frac{1}{2}\right) = \frac{\sqrt{\pi} \Gamma(2i)}{2^{2i-1} \Gamma(i)} = \frac{\sqrt{\pi}}{2^{2i-1}} \frac{(2i-1)!}{(i-1)!} = \frac{\sqrt{\pi}}{2^{2i-1}} \frac{i}{2i} \frac{(2i)!}{i!}$$

$$= \frac{\sqrt{\pi}}{2^{2i}} \frac{(2i)!}{i!} = \frac{\sqrt{\pi}}{2^{2i}} 2^i C(2i-1)$$

$$= \frac{1 \cdot 3 \cdot 5 \cdots (2i-1)}{2^i} \sqrt{\pi}, \qquad (1.51)$$

which is required in Volume II (for noncentral distributions).

We will come across many integrals which can be expressed in terms of the beta function.

⊖ **Example 1.24** To express $\int_0^1 \sqrt{1 - x^4} \, dx$ in terms of the beta function, let $u = x^4$ and $dx = (1/4) u^{1/4-1} \, du$, so that

$$\int_0^1 \sqrt{1 - x^4} \, dx = \frac{1}{4} \int_0^1 u^{-3/4} (1-u)^{1/2} \, du = \frac{1}{4} B\left(\frac{1}{4}, \frac{3}{2}\right). \qquad \blacksquare$$

⊖ **Example 1.25** To compute

$$I = \int_0^s x^a (s-x)^b \, dx, \qquad s \in (0,1), \quad a, b > 0,$$

use $u = 1 - x/s$ (so that $x = (1 - u)\,s$ and $dx = -s\,du$) to get

$$I = \int_0^s x^a \, (s - x)^b \, dx$$

$$= -s \int_1^0 [(1 - u)\,s]^a \, [s - (1 - u)\,s]^b \, du = s^{a+b+1} \int_0^1 (1 - u)^a \, u^b \, du$$

$$= s^{a+b+1} B\,(b + 1, a + 1), \tag{1.52}$$

which we will use in Example 9.12.

⊖ **Example 1.26** To compute

$$I = \int_{-1}^1 \left(1 - x^2\right)^a \, (1 - x)^b \, dx,$$

express $1 - x^2$ as $(1 - x)\,(1 + x)$ and use $u = (1 + x)/2$, (so that $x = 2u - 1$ and $dx = 2du$) to get

$$I = 2^{2a+b+1} \int_0^1 u^a \, (1 - u)^{a+b} \, du = 2^{2a+b+1} B\,(a + 1, a + b + 1).$$

A similar integral arises when working with the distribution of the correlation coefficient. ∎

⊖ **Example 1.27** Recall from (A.46) that $\lim_{n\to\infty} (1 - t/n)^n = e^{-t}$. Following Keeping (1995, p. 392), for $x > 0$ and using $u = t/n$ and (1.48), $\Gamma\,(x)$ is given by

$$\lim_{n\to\infty} \int_0^n \left(1 - \frac{t}{n}\right)^n t^{x-1} dt = \lim_{n\to\infty} n^x \int_0^1 u^{x-1}\,(1 - u)^n \, du = \lim_{n\to\infty} n^x \frac{\Gamma\,(x)\,\Gamma\,(n + 1)}{\Gamma\,(x + n + 1)}.$$

Dividing by $\Gamma\,(x)$ gives

$$1 = \lim_{n\to\infty} \frac{n^x \Gamma\,(n + 1)}{\Gamma\,(x + n + 1)}.$$

But with $x = 1/2$ and using (1.37) and (1.40),

$$\frac{n^x \Gamma\,(n + 1)}{\Gamma\,(x + n + 1)} = \frac{n^{1/2} n!}{\left(n + \frac{1}{2}\right)\left(n - \frac{1}{2}\right)\left(n - \frac{3}{2}\right)\cdots \frac{1}{2}\Gamma\left(\frac{1}{2}\right)} = \frac{n^{1/2} n!}{\left(\frac{2n+1}{2}\right)\left(\frac{2n-1}{2}\right)\left(\frac{2n-3}{2}\right)\cdots \frac{1}{2}\sqrt{\pi}}$$

$$= \frac{n^{1/2} n!}{\frac{\left(\frac{2n+1}{2}\right)\left(\frac{2n}{2}\right)\left(\frac{2n-1}{2}\right)\left(\frac{2n-2}{2}\right)\left(\frac{2n-3}{2}\right)\cdots \frac{1}{2}\sqrt{\pi}}{\left(\frac{2n}{2}\right)\left(\frac{2n-2}{2}\right)\cdots 1}} = \frac{n^{1/2} n!}{\frac{(2n+1)!\sqrt{\pi}}{2^{2n+1} n!}} = \frac{2^{2n+1} n^{1/2}\,(n!)^2}{(2n + 1)!\sqrt{\pi}},$$

or

$$\sqrt{\pi} = \lim_{n\to\infty} \frac{n^{1/2} 2^{2n+1}\,(n!)^2}{(2n + 1)!} = \lim_{n\to\infty} \frac{1}{n^{1/2}} \frac{2n}{(2n + 1)} \frac{2^{2n}\,(n!)^2}{(2n)!}.$$

$$= \lim_{n \to \infty} \frac{1}{n^{1/2}} \frac{2^{2n} (n!)^2}{(2n)!} = \lim_{n \to \infty} \frac{1}{n^{1/2}} \frac{(2n)^2 (2n-2)^2 (2n-4)^2 \cdots}{(2n)(2n-1)(2n-2)(2n-3)\cdots}$$

$$= \lim_{n \to \infty} \frac{1}{n^{1/2}} \frac{(2n)(2n-2)(2n-4)\cdots 2}{(2n-1)(2n-3)\cdots 1},$$

which is Wallis' product (see Example A.42). ■

⊙ **Example 1.28** An interesting relation both theoretically and computationally is given by

$$\sum_{j=k}^{n} \binom{n}{j} p^j (1-p)^{n-j} = \frac{\Gamma(n+1)}{\Gamma(k) \Gamma(n-k+1)} \int_0^p x^{k-1} (1-x)^{n-k} \, dx, \qquad (1.53)$$

for $0 \le p \le 1$ and $k = 1, 2, \ldots$, where $\binom{n}{j}$ is a binomial coefficient, and can be proven by repeated integration by parts. To motivate this, take $k = 1$. From the binomial theorem (1.18) with $x = p = 1 - y$, it follows directly that the l.h.s. of (1.53) is $1 - (1-p)^n$. The r.h.s. is, with $y = 1 - x$,

$$\frac{n!}{(n-1)!} \int_0^p (1-x)^{n-1} \, dx = -n \int_1^{1-p} y^{n-1} dy = y^n \big|_{1-p}^{1} = 1 - (1-p)^n.$$

For $k = 2$, the l.h.s. of (1.53) is easily seen to be $1 - (1-p)^n - np(1-p)^{n-1}$, while the r.h.s. is, using $y = 1 - x$,

$$\frac{n!}{1!(n-2)!} \int_0^p x (1-x)^{n-2} \, dx = -n(n-1) \int_1^{1-p} (1-y) y^{n-2} dy$$

$$= n(n-1) \left(\frac{y^{n-1}}{n-1} \bigg|_{1-p}^{1} - \frac{y^n}{n} \bigg|_{1-p}^{1} \right)$$

$$= 1 - np(1-p)^{n-1} - (1-p)^n,$$

after some rearranging. A probabilistic proof of (1.53) is given in Volume II in the context of order statistics. ■

⊖ **Example 1.29** By substituting $2n - 1$ for n, n for k and taking $p = 1/2$ in (1.53), we get

$$\sum_{i=n}^{2n-1} \binom{2n-1}{i} \left(\frac{1}{2} \right)^{2n-1} = \frac{\Gamma(2n)}{\Gamma^2(n)} \int_0^{1/2} u^{n-1} (1-u)^{n-1} \, du = \frac{1}{2}$$

from (1.49), directly showing (1.14) in Example 1.12. ■

⊙ **Example 1.30** We can now prove the extension of the binomial theorem

$$(x+y)^{[n]} = \sum_{i=0}^{n} \binom{n}{i} x^{[i]} y^{[n-i]}, \qquad x^{[n]} = \prod_{j=0}^{n-1} (x + ja),$$

as was stated in (1.26), which holds for $n = 0, 1, 2, \ldots,$ and x, y, a are real numbers. As

$$\prod_{j=0}^{k} (x + ja) = (x)(x + a) \cdots (x + ka)$$

$$= a^{k+1} \left(\frac{x}{a}\right) \left(\frac{x}{a} + 1\right) \cdots \left(\frac{x}{a} + k\right) = a^{k+1} \frac{\Gamma(k + 1 + x/a)}{\Gamma(x/a)},$$

(1.26) can be written as

$$(x + y)^{[n]} \stackrel{?}{=} \sum_{i=0}^{n} \binom{n}{i} x^{[i]} y^{[n-i]}$$

$$\prod_{j=0}^{n-1} (x + y + ja) \stackrel{?}{=} \sum_{i=0}^{n} \binom{n}{i} \left[\prod_{j=0}^{i-1} (x + ja) \right] \left[\prod_{j=0}^{n-i-1} (y + ja) \right]$$

$$a^n \frac{\Gamma\left[n + (x + y)/a\right]}{\Gamma\left[(x + y)/a\right]} \stackrel{?}{=} \sum_{i=0}^{n} \binom{n}{i} \left[a^i \frac{\Gamma(i + x/a)}{\Gamma(x/a)} \right] \left[a^{n-i} \frac{\Gamma(n - i + y/a)}{\Gamma(y/a)} \right]$$

or

$$\frac{\Gamma\left[n + (x + y)/a\right]}{\Gamma\left[(x + y)/a\right]} \stackrel{?}{=} \sum_{i=0}^{n} \binom{n}{i} \frac{\Gamma(i + x/a)}{\Gamma(x/a)} \frac{\Gamma(n - i + y/a)}{\Gamma(y/a)}$$

or

$$\frac{\Gamma(x/a)\Gamma(y/a)}{\Gamma\left[(x + y)/a\right]} \stackrel{?}{=} \sum_{i=0}^{n} \binom{n}{i} \frac{\Gamma(i + x/a)\Gamma(n - i + y/a)}{\Gamma\left[n + (x + y)/a\right]}$$

$$B\left(\frac{x}{a}, \frac{y}{a}\right) \stackrel{?}{=} \sum_{i=0}^{n} \binom{n}{i} B\left(\frac{x}{a} + i, \frac{y}{a} + n - i\right).$$

But this latter equation is the same as (9.7) on page 327, which is proved there, thus proving (1.26).[4] ∎

Similar to the incomplete gamma function, the *incomplete beta function* is

$$B_x(p, q) = \mathbb{I}_{[0,1]}(x) \int_0^x t^{p-1}(1 - t)^{q-1}\, dt$$

and the normalized function $B_x(p, q)/B(p, q)$ is the *incomplete beta ratio*, which we denote by $\overline{B}_x(p, q)$.

---

[4] The derivation leading up to (9.7) uses probabilistic arguments (the sum of independent gamma random variable with the same scale parameter is also gamma) but the argument directly after (9.7), which involves use of a beta random variable, is actually purely algebraic. It uses properties of integrals and the 'regular' binomial theorem (1.18).

1.6 Problems

Remember that the number of stars provide a rough indication of difficulty and/or amount of work required (see the Preface), with no stars meaning 'pretty easy and quick', one star 'a bit more work and/or time required', etc. up to three stars, which are problems intended to keep you occupied over the weekend. Of course, the solutions are available, but try some of the problems yourself before checking the answer.

> Aim for success, not perfection. (Dr. David M. Burns)
> *Reproduced by permission of E-Newsletter, Pinnacle One*

> Success is getting what you want. Happiness is wanting what you get.
> (Dale Carnegie)
> *Reproduced by permission of Pocket; Reissue edition*

> Try not to be a person of success, but rather a person of virtue.
> (Albert Einstein)
> *Reproduced by permission of Princeton University Press*

> Patience is the greatest of all virtues. (Cato the Elder)

> Patience is the companion of wisdom. (Saint Augustine)

> Wisdom is not a product of schooling but of the life-long attempt to acquire it.
> (Albert Einstein)
> *Reproduced by permission of Princeton University Press*

1.1. I have 10 different books: five are for statistics, three are for econometrics and two are for German.

   (a) How many ways can I arrange them on my shelf?

   (b) Same question, but what if I want to keep books of the same subject together?

1.2. With four different people, say $P_1, \ldots, P_4$, in how many ways

   (a) can they be arranged in a row,

   (b) can they be partitioned into two groups of two people each,
   (1) where the groups are labeled A and B – write out all the possibilities,
   (2) where there are no group labels?

1.3. ★ Assume our statistics institute consists of eight Germans and three foreigners.

   (a) If a name list was made, but only the category 'German / Foreigner' was shown, how many possible combinations could be made?

   (b) If you have to pick five of us to serve on your committee, how many possibilities are there?

   (c) What if you are forced to include one (and only one) foreigner?

(d) What if you are forced to include *at least* one foreigner?

(e) Again, you must pick one (and only one) foreigner, but two of the Germans refuse to serve on the committee together? Explain the following answer:

$$\binom{3}{1}\left[\binom{6}{4} + \binom{2}{1}\binom{6}{3}\right].$$

(f) What if you are forced to include one (and only one) foreigner, and two of the Germans insist on either serving together, or not at all?

1.4. ★ Prove that

$$\sum_{n=0}^{\infty} \binom{n+k}{k} x^n = \frac{1}{(1-x)^{k+1}}, \quad |x| < 1, \quad n \in \mathbb{N}, \tag{1.54}$$

by induction over k.

1.5. ★ Prove

$$\frac{1}{\sqrt{1-4a}} = \sum_{i=0}^{\infty} \binom{2i}{i} a^i.$$

1.6. ★ Prove

$$\sum_{j=i}^{l} (-1)^j \binom{l}{j} = (-1)^i \binom{l-1}{i-1}.$$

1.7. ★ ★ Prove

$$A_{i,n} = \sum_{j=i}^{n} (-1)^{j-i} \binom{j-1}{i-1}\binom{n}{j} p^j = \sum_{j=i}^{n} \binom{n}{j} p^j (1-p)^{n-j} = B_{i,n}$$

for $0 < i \le n$ and $0 < p < 1$ in (at least) the following two ways.

(a) Use (1.18) to express $B_{i,n}$ in terms of powers of p only. Then show that coefficients of like powers of p are the same in both expressions.

(Contributed by Markus Haas)

(b) First verify that $A_{n,n} = B_{n,n}$. Next argue that the result holds if $A_{i,n} - A_{i+1,n} = B_{i,n} - B_{i+1,n}$, $1 \le i < n$; then show it.

(Contributed by Walther Paravicini)

1.8. Show that

$$\sum_{i=\lceil y/2 \rceil}^{y} \binom{2i}{y}\binom{y}{i} (-1)^{i+y} = \sum_{k=0}^{\lfloor y/2 \rfloor} (-1)^k \binom{y-k}{k}\binom{2y-2k}{y-k}.$$

(It is also equal to 2^y; the l.h.s. is (3.31) and the r.h.s. appears in Riordan (1968, p. 37), where it was proved algebraically.)

1.9. ★ Show that

$$\sum_{y=1}^{n} \frac{(n-1)!\, y}{(n-y)!\, n^y} = 1$$

directly, i.e. without induction. In doing so, also prove the identity

$$n^n = \sum_{i=0}^{n-1} \binom{n-1}{i} (n-i)!\, n^i.$$

1.10. Prove the following relations.

(a)
$$\binom{r+m+1}{m} = \sum_{i=0}^{m} \binom{r+i}{i}, \quad m, r \in \mathbb{N},$$

(b)
$$\binom{2n}{n} = \sum_{i=0}^{n} \binom{n}{i}^2,$$

(c) ★
$$\binom{N+M}{k+M} = \sum_{i=0}^{M} \binom{N}{k+i}\binom{M}{i}. \tag{1.55}$$

Prove this by induction on M and also by equating coefficients of x^k in both sides of the simple identity

$$(1+x)^N \left(1+\frac{1}{x}\right)^M = \frac{(1+x)^{N+M}}{x^M}. \tag{1.56}$$

(d) For $b > a$ and $r \in \mathbb{N}$,

$$\sum_{j=0}^{r} \binom{r}{j} a^{r-j} (b-a)^j \frac{1}{j+1} = \frac{b^{r+1} - a^{r+1}}{(b-a)(r+1)}. \tag{1.57}$$

1.11. ★ Prove the following relations.

(a)
$$\binom{r_1 + r_2 + y - 1}{y} = \sum_{i=0}^{y} \binom{r_1 + i - 1}{i}\binom{r_2 + y - i - 1}{y - i}, \tag{1.58}$$

(b)
$$1 = \sum_{k=0}^{N} \binom{2N-k}{N}\left(\frac{1}{2}\right)^{2N-k}, \tag{1.59}$$

(c)
$$\sum_{i=0}^{n-1} \binom{2n}{i} = \sum_{i=0}^{n-1} 2^i \binom{2n-1-i}{n-1-i}.$$

1.12. ★ ★

(a) Evaluate

$$P(n, N) = \sum_{i=1}^{N-1} \binom{N}{i} \left(\frac{N-i}{N}\right)^n (-1)^{i+1}$$

for $n = 0$ and $n = 1$. Verify by induction on n that

$$P(n, N) = 1, \quad 1 \le n \le N - 1, \tag{1.60}$$

but that it does *not* hold for $n \ge N$. How does $P(N, N)$ behave as N grows? A probabilistic explanation of this interesting identity is given in Example 2.13 (with $r = N$) and Section 4.2.6.

(b) Prove that

$$0 = Q(n, N) := \sum_{j=0}^{N} \binom{N}{j} j^n (-1)^{N-j+1}, \quad 1 \le n \le N - 1.$$

Is $Q(N, N)$ close to 0?

(c) Let

$$M(N) := \sum_{j=0}^{N} \binom{N}{j} j^N (-1)^{N-j}.$$

Prove by induction that $M(N) = N!$. As an aside, it appears that, for *any* $r \in \mathbb{R}$,

$$N! = \sum_{j=0}^{N} \binom{N}{j} (j+r)^N (-1)^{N-j}.$$

If you numerically confirm this, watch out for round-off error for large $|r|$.

1.13. Prove the identity

$$\sum_{i=0}^{y} \binom{n+y}{y-i} (-1)^i = \binom{n+y-1}{y} \tag{1.61}$$

for $n \ge 1$ and $y \ge 0$ by induction on y, i.e. first check that it holds for $y = 0$ and $y = 1$, then assume it holds for general $y \in \mathbb{N}$.

1.14. ★ ★ ★ Prove the following relations.

(a) (1.17) (using induction),

(b) $\binom{n+y-1}{y}\binom{N}{n+y} = \sum_{i=0}^{y}\binom{N}{n+i}\binom{N-n-i}{y-i}(-1)^i$

$$= \sum_{i=y}^{N-n}\binom{n+i-1}{i}\binom{i}{y},\qquad\qquad (1.62)$$

(c) $\sum_{i=0}^{m-1}\binom{n+i-1}{i}p^n(1-p)^i = \sum_{j=0}^{N-n}\binom{N}{n+j}p^{n+j}(1-p)^{N-n-j}.$ (1.63)

1.15. ★ We wish to develop a computer program (in Matlab), say `permvec`, which generates a random permutation of the vector $(1, 2, \ldots, N)$, where N is an integer supplied by the user of the program.

Matlab already has such a function built in to accomplish this, called `randperm(n)`. It just executes the statement `[ignore,p]` `=` `sort(rand(1,n));` to return vector p. To understand this clever method, read the help file on the `sort` function. This is certainly not the fastest way to construct a random permutation because of the relatively high cost of sorting. However, it might be the fastest way *in Matlab*, because the sort function, like many functions and matrix operations, are internal, pre-compiled routines which were originally written in a low-level language such as C or assembler. The Matlab programs we write do not get compiled to machine code, but are *interpreted* as they are executed, and so run demonstrably slower in comparison.[5]

Consider the following approach to this problem. It uses the built-in function `unidrnd(N)` to generate a random integer between 1 and N.

```
function y = permvec(N) % first attempt, which does NOT work!
y=zeros(N,1);
for i=1:N
  y(i) = unidrnd(N);
end
```

The problem with the code is that we do not ensure that no duplicates occur in the output vector. One way of fixing things is to remove each value from a 'source vector' as it is chosen, and then pick only values from the remaining elements of this source vector. Listing 1.2 below shows the implementation of this, in which x is the source vector and, after each iteration in the `for` loop, the chosen element of x is removed. Notice that random value p is no longer the actual element of output vector, but rather the *index* into the source array.

(a) A completely different way of accomplishing this is to set up a `for` loop from 1 to N which exchanges each element in turn with a *randomly chosen* element from the vector. Write a program which accomplishes this. If the

---

[5] This discrepancy can now be partially alleviated by using Matlab's built-in compiler, which translates Matlab code to C, and then calls a C compiler, but the speed increase is not impressive. You still can't have your cake and eat it too.

```
function y = permvec(N)
x=1:N; y=zeros(N,1);
for i=1:N
  p = unidrnd(N+1-i); y(i) = x(p);
  x=[x(1:(p-1)) x((p+1):(N-i+1))]; % concatenate the remaining elements
end
```

Program Listing 1.2 Returns a random permutation of vector $(1, 2, \ldots, N)$

algorithm works, then we should get approximately the same number of occurrences of the N elements in each of the N positions. To check this, for a given value of N, run the program many times (say 1000), keeping track of the output. From the output, tabulate how many times each of the numbers 1 through N has occurred in each of the N positions.

(b) Now create a program which delivers a random permutation vector as before, but also ensures that there is no *overlap* with the original vector $(1, 2, \ldots, N)$. In other words, if position i contains the number i, then there is overlap.

1.16. Show that

$$I = \int_0^\infty \frac{dx}{\sqrt{x}\,(1+x)} = \pi.$$

1.17. ★ ★ For $a > 0$ and $n \in \mathbb{N}$, let

$$I_n := \int_0^\infty \frac{1}{\left(x^2 + a\right)^{n+1}}\,dx.$$

Determine a closed-form expression for I_n.

1.18. ★ As in Jones (2001, p. 209(11)), show that

$$I := \int_0^1 \frac{x^{a-1}\,(1-x)^{b-1}}{(x+k)^{a+b}}\,dx = \frac{B\,(a,b)}{(1+k)^a\,k^b}, \qquad a, b, k \in \mathbb{R}_{>0}. \qquad (1.164)$$

(This is used in Example 7.1.) See the footnote for a hint.[6]

1.19. Show that

$$\int_{-1}^1 (x^2 - 1)^j\,dx = \frac{(-1)^j\,2^{2j+1}}{\binom{2j}{j}(2j+1)},$$

thus proving the identity (1.22).

---

[6] Courtesy of Frank Jones, author of the named text – use the substitution $x = (At + B)/(Ct + D)$, and ensure that $x = 0 \Leftrightarrow t = 0$ and $x = 1 \Leftrightarrow t = 1$.

2

Probability spaces and counting

To us, probability is the very guide of life.   (Joseph Butler)[1]

The starting point for understanding the subject of probability theory is a precise description of what is meant by 'the probability of an event'. The details of the associated mathematical structure, called a probability space, and some of its most useful properties, are the main subjects of this chapter. Before addressing probability spaces, however, we begin with methods of counting, which will help us throughout this and subsequent chapters with a variety of probabilistic questions.

2.1 Introducing counting and occupancy problems

Not everything that can be counted counts, and not everything that counts can be counted.

(Albert Einstein)
Reproduced by permission of Princeton University Press

Counting methods are a fundamental tool for the actual calculation of probabilities in numerous problems of both practical and theoretical interest. One of the most standard and useful tools in describing counting methods is the urn, or a large, opaque vessel, from which labeled (or colored) balls which are otherwise identical will be 'randomly' drawn. Randomly removing balls from this proverbial urn are analogous to more real life situations such as randomly choosing a set of people from a larger group, drawing several playing cards from a shuffled deck, picking numbers for a lottery, etc.

To begin, assume our urn is filled with balls labeled $1, 2, \ldots$, up to 100 and we wish to draw five of them. Two issues immediately arise: will the label, or associated number, of each drawn ball be recorded and the ball subsequently replaced into the urn before the next one is drawn (referred to as *drawing with replacement*) or will all five

---

[1] Bishop of Durham, Theologian (1692–1752). In the preface to: The Analogy of Religion, Natural and Revealed, to the Constitution and Course of Nature (often cited simply as 'Butler's Analogy', 1736).

Fundamental Probability: A Computational Approach   M.S. Paolella
© 2006 John Wiley & Sons, Ltd

be removed *without replacement*? This latter case can also be interpreted as removing five balls simultaneously. The second issue is whether the order of the five resulting numbers (or labels) is of importance. For example, if we draw with replacement, a possible scenario is observing, in this order, 97, 29, 4, 60, 29. Do we deem this draw to be equivalent to a different *permutation* of these, say 4, 29, 29, 60, 97? If so, then order is not important and we say that draws are *unordered* – otherwise, draws are *ordered*. Thus, there are four different cases to consider, namely ordered without replacement, ordered with replacement, unordered without replacement and unordered with replacement. We are concerned with the total number of possible arrangements which can occur in each case.

Instead of 100 unique balls, let us assume there are R; and instead of drawing five, we draw n. If we draw without replacement, then we require that $n \leq R$. For the case 'ordered without replacement', the first draw is one of R possibilities; the second is one of $(R - 1)$, ..., the nth is one of $R - n + 1$. This can be expressed using the descending factorial notation as $R_{[n]} = R! / (R - n)!$. For 'ordered with replacement', the first draw is one of R possibilities; the second is also one of R, etc., so that there are R^n possible arrangements. Drawing 'unordered without replacement' is similar to ordered without replacement, but we need to divide $R_{[n]}$ by $n!$ to account for the irrelevance of order, giving 'R choose n', i.e.

$$\frac{R!}{(R - n)!\,n!} = \binom{R}{n},$$

possibilities.

The last case is drawing 'unordered with replacement'. Let the balls be labeled 1, 2, ..., R. Because order does not matter, the result after n draws can be condensed to just reporting the number of times ball 1 was chosen, the number of times ball 2 was chosen, etc. Denoting the number of times ball i was chosen as x_i, $i = 1, \ldots, R$, we are actually seeking the number of nonnegative integer solutions (abbreviated nonis) to $x_1 + x_2 + \cdots + x_R = n$. First let $R = 2$, i.e. we want the number of solutions to $x_1 + x_2 = n$. Imagine placing the n balls along a straight line, like $\bullet\bullet\bullet\bullet\bullet$ for $n = 5$. Now, by placing a partition among them, we essentially 'split' the balls into two groups. We can place the partition in any of the $n + 1$ 'openings', such as $\bullet\bullet\,|\,\bullet\bullet\bullet$, which indicates $x_1 = 2$ and $x_2 = 3$, or as $|\,\bullet\bullet\bullet\bullet\bullet$, which indicates $x_1 = 0$ and $x_2 = 5$, etc. If $R = 3$, we would need two partitions, e.g. $\bullet\bullet\,|\,\bullet\bullet\,|\,\bullet$ or $\bullet\bullet\,|\,\bullet\bullet\bullet\,|$ or $|\,|\,\bullet\bullet\bullet\bullet\bullet$, etc. In general, we need $R - 1$ partitions interspersed among the n balls. This can be pictured as randomly arranging n balls and $R - 1$ partitions along a line, for which there are $(n + R - 1)!$ ways. However, as the balls are indistinguishable, and so are the partitions, there are

$$\frac{(n + R - 1)!}{n!\,(R - 1)!} = \binom{n + R - 1}{n} =: K_{n,R}^{(0)} \tag{2.1}$$

ways to arrange them, or $\binom{n+R-1}{n}$ nonnegative integer solutions to the equation $\sum_{i=1}^{R} x_i = n$ (see also Example 1.5 to see why it is necessary to divide by $n!\,(R - 1)!$). For example, if $n = R = 3$, there are $\binom{5}{3} = 10$ different arrangements; if $n = R = 4$, there are $\binom{7}{4} = 35$ different arrangements, etc. These results lend themselves well to tabular display, as seen in Table 2.1.

Table 2.1 Number of ways of choosing n from R balls

		Replacement	
		Without	With
Ordering	Yes	$\dfrac{R!}{(R-n)!}$	R^n
	No	$\dbinom{R}{n}$	$\dbinom{n+R-1}{n}$

An alternative and equivalent way of viewing the unordered with replacement drawing scheme is to imagine that n *indistinguishable* balls are randomly placed, one at a time, into one of R distinguishable urns, where 'randomly placed' means that each of the R urns is equally likely to be chosen, irrespective of the number of balls it or the other urns already contain. Interest centers on the *occupancy* of each urn. Physicists will recognize the importance of such constructs for analyzing the allocation of atomic particles, i.e. Maxwell–Boltzmann, Bose–Einstein and Fermi–Dirac statistics. More generally, such a mental construction will be useful for visualizing and solving a wide variety of problems.[2] Because we will use this calculation several times, we denote the number of possibilities in (2.1) as $K_{n,R}^{(0)}$.

Continuing this development, assume $n \geq R$ and consider the number of possible arrangements when each of the R urns is restricted to containing at least one ball, denoted $K_{n,R}^{(1)}$. This is the same situation as above, but with R less balls, i.e.

$$K_{n,R}^{(1)} = K_{n-R,R}^{(0)} = \binom{n-1}{R-1}. \tag{2.2}$$

This can also be seen by drawing n balls in a row and choosing $R-1$ of the $n-1$ gaps between the objects into which a partition is placed. This forms R 'blocks'. Thus, the number of positive integer solutions (nopis) of $\sum_{i=1}^{R} x_i = n$ is $K_{n,R}^{(1)}$.

Note that relation (2.2) can also be obtained by writing $\sum_{i=1}^{R} x_i = n$ as $\sum_{i=1}^{R} (x_i + 1) = n + R$, with $y_i = x_i + 1 \geq 1$. Then, the number of positive integer solutions to $\sum_{i=1}^{R} y_i = n + R$ is given by

$$K_{n+R,R}^{(1)} = \binom{n+R-1}{R-1} = \binom{n+R-1}{n} = K_{n,R}^{(0)},$$

as before.

---

[2] To ease the understandable suspense, one such application which is discussed later is the following classic example. As part of a popular sales campaign, each breakfast cereal box of a certain brand contains one of 10 prizes, the entire set of which the consumer is inclined to want to collect. Assuming the prizes are equally likely to appear in a cereal box (i.e. no prize is 'rarer' than any other), how many boxes of cereal can the consumer expect to purchase in order to get one (or n) of each prize? How many if they wish to be '99 % certain' of getting at least seven different prizes? These questions will be addressed in Section 4.2.6.

Now consider the nonis to the equation $\sum_{i=1}^{r} x_i \leq n$. This is given directly by $\sum_{j=0}^{n} K_{j,r}^{(0)}$ or as the nonis to $\sum_{i=1}^{r+1} x_i = n$, yielding the relation

$$\binom{n+r}{n} = \sum_{j=0}^{n} \binom{j+r-1}{j},$$

which is, in fact, useful. It is also given in (1.6) where it was verified algebraically.

Next, note that the number of ways in which exactly j out of r urns are empty is given by $\binom{r}{j} K_{n,r-j}^{(1)}$. A bit of thought then reveals that

$$K_{n,r}^{(0)} = \sum_{j=0}^{r-1} \binom{r}{j} K_{n,r-j}^{(1)},$$

i.e.

$$\binom{n+r-1}{r-1} = \sum_{j=0}^{r-1} \binom{r}{j} \binom{n-1}{r-j-1}. \tag{2.3}$$

Returning to our point of departure, recall that we had set out to count the number of possible arrangements for unordered draws with replacement. This led to consideration of the occupancy of R cells (or urns), which fostered an easy solution and was of interest in itself for more general situations. One might now ask: can the occupancy model be extended such that the n balls are distinguishable? For example, with $R = n = 2$, the four possible situations are $\{12 \mid \cdot\}, \{\cdot \mid 12\}, \{1 \mid 2\}, \{2 \mid 1\}$, where $\cdot$ denotes an empty cell. (If the balls were indistinguishable, then only the $K_{2,2}^{(0)} = \binom{n+R-1}{n} = 3$ cases $\{\bullet\bullet \mid \cdot\}, \{\cdot \mid \bullet\bullet\}, \{\bullet \mid \bullet\}$ would arise.) This is the same logic as the 'ordered with replacement' drawing strategy, for there are R choices for the first ball, R for the second, etc. yielding R^n possibilities.

From the nature of how the balls are placed into the cells, each of the R^n configurations is equally likely to occur. Likewise, with indistinguishable balls, the $K_{n,R}^{(0)}$ cases are clearly *not* equally likely to occur, e.g. for the $R = n = 2$ case, configurations $\{\bullet\bullet \mid \cdot\}$ and $\{\cdot \mid \bullet\bullet\}$ each occur with probability 1/4, while $\{\bullet \mid \bullet\}$ occurs with probability 1/2. The number of unique cases could be further reduced by taking the *cells* to be indistinguishable, so that, for the $R = n = 2$ case, only $\{\bullet\bullet \mid \cdot\}$ or $\{\bullet \mid \bullet\}$ will be observed. These two happen to be equally likely, but such a result will certainly not be true in general.

For $R = 3$ and $n = 2$, if the n balls are distinguishable, then there are nine (equally likely) possibilities, given in Table 2.2.

If balls are indistinguishable, there are only $K_{n,R}^{(0)} = 6$ cases, which are separated by double lines in the above table, and are not equally likely to occur. The entries could

Table 2.2

			Cases								
	1	1	2	1	2	·	·	12	·	·	
Cell	2	2	1	·	·	1	2	·	12	·	
	3	·	·	2	1	2	1	·	·	12	

be further reduced by taking the cells to be indistinguishable, resulting in the two cases $\{\bullet \mid \bullet \mid \cdot\}$ and $\{\bullet\bullet \mid \cdot \mid \cdot\}$, with respective probabilities 2/3 and 1/3.

Example 2.1 Consider the case with $R = n = 3$. There are $3^3 = 27$ possibilities, outlined as follows. With two cells empty, there are $R = 3$ possibilities. With no cells empty, there are $n! = 6$ possibilities (or arrangements of the three balls 'in a row').

The case with one cell empty requires a bit more thought. Assume for the moment that the balls are indistinguishable. There are $\binom{R}{1} = 3$ ways to pick the empty cell. For the nonempty cells, either one has one ball or two balls, in general, there are

$$K_{n,R-1}^{(1)} = \binom{n-1}{R-2}$$

possible arrangements. Now factoring in the uniqueness ('distinguishability') of the balls, there are $n! = 6$ ways of arranging the balls 'in a row', but we adjust this (i.e. divide) by 2! because the order of the balls in the cell with two balls is irrelevant. (Equivalently, there are $\binom{3}{1}$ ways of choosing 'the lone ball'.) Multiplying gives 18 unique arrangements with one empty cell. Summing gives the 27 possibilities.

This decomposition also shows that, with three indistinguishable balls and three indistinguishable cells, either $\{\bullet\bullet\bullet \mid \cdot \mid \cdot\}$ or $\{\bullet \mid \bullet \mid \bullet\}$ or $\{\bullet\bullet \mid \bullet \mid \cdot\}$ occurs, with respective probabilities 3/27, 6/27 and 18/27. ∎

In general, when the n balls are distinguishable, the number of possible constellations leaving exactly m of the r cells empty is

$$S_{m,r} = \binom{r}{m} \sum_{i=0}^{r-m} (-1)^i \binom{r-m}{i} (r-m-i)^n, \tag{2.4}$$

and $r^n = \sum_{m=0}^{r-1} S_{m,r}$. Example 2.13 below shows that (2.4) is quite straightforward to prove, once we have at our disposal some useful tools for working with several events, as will be developed in Section 2.3.2. A more direct derivation is provided by Feller (1957, p. 58). At this point, the reader is encouraged to verify that for values $r = n = 3$, as were used in Example 2.1 above, (2.4) yields $S_{0,3} = 6$, $S_{1,3} = 18$ and $S_{2,3} = 3$. Also, based on (2.4), $S_{3,3} = 0$, as it should.

2.2 Probability spaces

2.2.1 Introduction

Does anyone believe that the difference between the Lebesgue and Riemann integrals can have physical significance, and that whether say, an airplane would or would not fly could depend on this difference? If such were claimed, I should not care to fly in that plane.

(Richard W. Hamming)

The formal description of a probability space is inherently 'mathematical' and, to a large extent, abstract. Its roots, however, lie in a 'common-sense' frequentistic description. To illustrate, we use the example of drawing a card from a (well-shuffled) standard deck of 52 playing cards. Each of the cards is equally likely to be drawn, so that the probability of getting, for example, the ace of spades, is just 1/52, which we denote by $\Pr(\clubsuit A) = 1/52$. This assignment of probability appears quite reasonable. It can also be thought of as the 'long run' fraction of times the ace of spades is drawn when the experiment is repeated indefinitely under the same conditions (i.e. the deck has all 52 cards, is well-shuffled, and a single card is 'randomly' drawn in such a way that all 52 cards are equally likely to be picked).

Each card represents one of the 52 possible outcomes and is said to be one of the *sample points* making up the *sample space*. We could arbitrarily assign labels to each of the sample points, e.g. $\omega_1 = \clubsuit A$, $\omega_2 = \clubsuit 2$, ... , $\omega_{13} = \clubsuit K$, $\omega_{14} = \spadesuit A$, ... , $\omega_{52} = \heartsuit K$ and denote the sample space as $\Omega = \{\omega_i, i = 1, \ldots, 52\}$. To determine the probability that the card drawn is, say, an ace, we sum the number of relevant sample points $(\omega_1, \omega_{14}, \omega_{27}, \omega_{40})$ and divide by 52, i.e. $4/52 = 1/13$. Any such case of interest is termed an *event*. The event that a picture card is drawn has probability $12/52 = 3/13$, etc. Some thought reveals that there are $2^{52} \approx 4.5 \times 10^{15}$ possible events for this sample space and experiment; this totality of possible events is simply called the *collection of events*.

Another example of a sample space is that of drawing two cards without replacement (and order relevant), with sample points consisting of all $52 \cdot 51 = 2652$ possibilities, all of which are equally likely. A possible event is that both cards have the same suit. Two ways of computing its probability are as follows.

- There are 52 ways of drawing the first card; its suit determines the 12 cards which can be drawn next so that both have the same suit. This gives a probability of

$$\Pr(\text{both cards have same suit}) = \frac{52}{52}\frac{12}{51} = \frac{12}{51}.$$

- There are $\binom{4}{1}$ possible suits to 'choose from' times the number of ways of getting two cards from this suit, or $\binom{13}{2}$. This gets doubled because order matters. Dividing by the $52 \cdot 51$ total possible ways,

$$\Pr(\text{both cards have same suit}) = \frac{2 \cdot \binom{4}{1} \cdot \binom{13}{2}}{52 \cdot 51} = \frac{12}{51}.$$

The number of events for this experiment is now astronomical, i.e. $2^{2652} \approx 2.14 \times 10^{798}$, but is still finite. This intuitive method of assigning probabilities to events is valid for any finite sample space in which each sample point is equally likely to occur and also works for non-equally likely sample points. If we denote the sample space as the set Ω with elements, or possible outcomes, $\omega_i, i = 1, \ldots, N$, then Ω and a function Pr with domain Ω and range $[0, 1]$ such that $\sum_{i=1}^{N} \Pr(\omega_i) = 1$ is referred to as a *finite probability space*. If $\Pr(\omega_i)$ is the same for all i (in which case $\Pr(\omega_i) = 1/N$), then the finite probability space $(\Omega, \Pr)$ is also called a *symmetric probability space*.

For some experiments of interest, it might be the case that the sample space Ω is *denumerable* or *countably infinite*, i.e. each $\omega_i \in \Omega$ can be put in a $1-1$ correspondence with the elements of $\mathbb{N}$. For example, consider tossing a fair coin until a tail appears. While highly unlikely, it is theoretically possible that the first 100, 1000 or even

10 000 trials will result in heads. Indeed, a tail may never occur! By letting ω_i be the total number of required tosses, $i = 1, 2, \ldots$, we see that Ω is countable. If there is a function Pr with domain Ω and range [0, 1] associated with Ω such that $\sum_{i=1}^{\infty} \Pr(\omega_i) = 1$, then $(\Omega, \Pr)$ is referred to as a *probability space*. For example, taking $\Pr(\omega_i) = (1/2)^i$ is valid, as the reader should verify (see also Example 2.4 below).

Matters increase in complexity when the sample space is *uncountable*. Examples include the sample space of 'random' numbers between zero and one, or the time until an electrical component fails, or the measurement of a patient's blood pressure using an analog device. In these cases, one could argue that a finite number of significant digits ultimately has to be used to measure or record the result, so that the sample space, however large, is actually finite. This is valid but becomes cumbersome. By allowing for a continuum of sample points, the powerful techniques of calculus may be employed, which significantly eases many computations of interest. However, it is no longer conceptually clear what probability is assigned to a 'random' number between 0 and 1.

To see this, let R be a random number drawn from the interval whose decimal representation is then truncated (not rounded) at, say, the third place. There are then 1000 possible outcomes, $0.000, 0.001, \ldots, 0.999$, each equally likely, so that we have a symmetric sample space. The probability that $R \leq 0.400$ is then 401/1000. As the number of digits in the decimal expansion is increased, the probability of any particular outcome goes to zero and, in the limit, it is zero. The calculation of $\Pr(R \leq 0.4)$, however, approaches 0.4. Similarly, for any interval $\mathcal{I}$ of [0, 1], $\Pr(R \in \mathcal{I})$ will approach the length of $\mathcal{I}$ as the number of digits retained in the decimal expansion of R is increased.

2.2.2 Definitions

It is customary to begin courses in mathematical engineering by explaining that the lecturer would never trust his life to an aeroplane whose behaviour depended on properties of the Lebesgue integral. It might, perhaps, be just as foolhardy to fly in an aeroplane designed by an engineer who believed that cookbook application of the Laplace transform revealed all that was to be known about its stability.   (T. W. Körner, 1988, p. 406)
Reproduced by permission of Cambridge University Press

There is no need to labour the point that the [Lebesgue theory of measure and integration] is useful, for this is self-evident to any practising analyst. What is perhaps less obvious and less widely accepted is the conviction implicit in what follows that the Lebesgue theory is not only more complete and satisfying in itself than the classical Riemann approach, but is also easier to use, and that its practical usefulness more than repays the initial effort of learning it.
                                                    (H. R. Pitt, 1985, p. v)
Reproduced by permission of Oxford University Press

Continuing the above discussion, the approach taken with an uncountable sample space is to assume that probabilities have been (appropriately) preassigned to each possible subset of Ω. For the random number between 0 and 1, each point in [0, 1] is assigned probability zero while each interval of $[0, 1]$ is assigned probability corresponding to its length.

Recall that a *set* can be described as a well-defined collection of objects; but this is a poor definition, as the word 'collection' is just a synonym for 'set', and our intuition is required for at least an initial, workable understanding of what a set represents. A similar situation holds when attempting to precisely define 'number' (try to without using the word number or any of its synonyms). In a similar vain, we refrain from providing elaborate, potentially tautological, and often useless, definitions of certain concepts required in probability, and instead describe them as usefully as possible in the context in which they are required. In particular, a *realization* is the result of some well-defined *trial* or *experiment* performed under a given *set of conditions*, whose *outcome* is not known in advance, but belongs to a *set of possibilities* or *set of outcomes* which are known in advance.

The set of possible outcomes could be *countable* (either *finite* or *denumerable*, i.e. *countably infinite*) or *uncountable*. Denote the *sample space* as Ω, the set of all possible outcomes, with individual outcomes or *sample points* $\omega_1, \omega_2, \ldots$. A *subset* of the sample space is known as an *event*, usually denoted by a capital letter, possibly with subscripts, i.e. A, B_1, etc., the totality of which under Ω will be denoted $\mathcal{A}$, and forms the *collection* of events – also called the *collection of measurable events*.

Note that an outcome $\omega \in \Omega$ may belong to many events, always belongs to the *certain event* Ω, and never to $\emptyset$, the *empty set* or *impossible event*. If outcome ω belongs to event A, then it does not belong to the *complement* of A, which is also an event, denoted A^c. In symbols, this is written $\omega \in A \Rightarrow \omega \notin A^c$, and it should be clear that $\omega \in A \Leftrightarrow \omega \notin A^c$. The usual operations in set theory can be applied to two events, i.e. intersection, union, difference, symmetric difference, inclusion, etc. As in general set theory, two events are mutually exclusive if $A \cap B = \emptyset$. If a particular set of events A_i, $i \in J$, are such that $\bigcup_{i \in J} A_i \supseteq \Omega$, then they are said to cover, or exhaust, the same space Ω. If events A_i, $i \in J$, are such that (a) $A_i \subseteq \Omega$, (b) they exhaust Ω with $\bigcup_{i \in J} A_i = \Omega$ and (c) are mutually exclusive, then they *partition* Ω, i.e. one and only one of the A_i will occur on a given trial.

A *probability function* or *(real) probability measure* is a *set function* which assigns a real number $\Pr(A)$ to each event $A \in \mathcal{A}$ such that

- $\Pr(A) \geq 0$,

- $\Pr(\Omega) = 1$, and

- for a countable infinite sequence of mutually exclusive events A_i,

$$\Pr\left(\bigcup_{i=1}^{\infty} A_i \right) = \sum_{i=1}^{\infty} \Pr(A_i).$$

$$(2.5)$$

Requirement (2.5) is known as *(countable) additivity*. If $A_i \cap A_j = \emptyset$, $i \neq j$ and $A_{n+1} = A_{n+2} = \cdots = \emptyset$, then $\Pr\left(\bigcup_{i=1}^{n} A_i\right) = \sum_{i=1}^{n} \Pr(A_i)$, which is referred to as *finite additivity*. The triplet $\{\Omega, \mathcal{A}, \Pr(\cdot)\}$ refers to the *probability space* or *probability system* with sample space Ω, collection of measurable events $\mathcal{A}$ and probability measure $\Pr(\cdot)$.

Remark: Not all subsets of Ω can be assigned probabilities. The class of subsets for which assignment is possible is termed a σ-*field* and containing essentially all subsets relevant for practical problems. A precise description of such classes is better conducted in a measure-theoretic framework and is not pursued in any detail here. We just briefly mention here a few important definitions which are intrinsic to a rigorous elaboration of probability theory.

A *field* $\mathcal{A}$ is a collection of subsets of (nonempty) Ω such that (i) $\Omega \in \mathcal{A}$, (ii) $A \in \mathcal{A} \Rightarrow A^c \in \mathcal{A}$, (iii) $A, B \in \mathcal{A} \Rightarrow A \cup B \in \mathcal{A}$. Induction shows (iii'): if $A_1, \ldots, A_n \in \mathcal{A}$, then $\bigcup_{i=1}^{n} A_i \in \mathcal{A}$. Also, if $A_1, \ldots, A_n \in \mathcal{A}$, then, from (ii), $A_i^c \in \mathcal{A}$, $i = 1, \ldots, n$, so that, from (iii'), $\bigcup_{i=1}^{n} A_i^c \in \mathcal{A}$. However, from (ii), $\left(\bigcup_{i=1}^{n} A_i^c\right)^c \in \mathcal{A}$ and, as $\left(\bigcup_{i=1}^{n} A_i^c\right)^c = \bigcap_{i=1}^{n} A_i$ from De Morgan's law, it follows that, if $A_1, \ldots, A_n \in \mathcal{A}$, then $\bigcap_{i=1}^{n} A_i \in \mathcal{A}$.

If, in addition, $\mathcal{A}$ is such that (iii') holds for countable collection $A_1, A_2, \ldots$, then $\mathcal{A}$ is a σ-*field*, i.e. a σ-*field* is a field with the property that $A_1, A_2, \ldots \in \mathcal{A} \Rightarrow \bigcup_{i=1}^{\infty} A_i \in \mathcal{A}$. Let $\mathcal{A}_0$ be a particular collection of subsets of Ω and let $\sigma(\mathcal{A}_0)$ be the intersection of all σ-fields which contain $\mathcal{A}_0$. Then $\sigma(\mathcal{A}_0)$ is *the smallest σ-field containing $\mathcal{A}_0$* and referred to as the σ-field *generated by* $\mathcal{A}_0$. The most prominent and important example is with $\Omega = \mathbb{R}$ and $\mathcal{A}_0$ taken to be the collection of intervals $(a, b]$, $a, b \in \mathbb{R}$. In this case, $\sigma(\mathcal{A}_0)$ is denoted the *Borel σ-field*, after Émile Borel (1871–1956),[3] and is of great importance when working with random variables.

To see that $\sigma(\mathcal{A}_0)$ is itself a σ-field which contains $\mathcal{A}_0$, let $\mathcal{C}$ be the collection of all σ-fields which contain $\mathcal{A}_0$, so that $\sigma(\mathcal{A}_0) = \bigcap_{A \in \mathcal{C}} A$. As each σ-field in $\mathcal{C}$ contains $\mathcal{A}_0$ by definition, $\mathcal{A}_0 \in \sigma(\mathcal{A}_0)$. For (i), as each element of $\mathcal{C}$ is a σ-field, Ω is contained in each element of $\mathcal{C}$, so that $\Omega \in \bigcap_{A \in \mathcal{C}} A$, i.e. $\Omega \in \sigma(\mathcal{A}_0)$. For (ii), if $A \in \sigma(\mathcal{A}_0)$, then A must be in every element of $\mathcal{C}$, so that, as each element of $\mathcal{C}$ is a σ-field, A^c is in every element of $\mathcal{C}$, so that $A^c \in \sigma(\mathcal{A}_0)$. Lastly, to show (iii) with a countable collection $A_1, A_2, \ldots$, if $A_i \in \sigma(\mathcal{A}_0)$ for $i = 1, 2, \ldots$, then A_i is in each element of $\mathcal{C}$, which implies that $\bigcup_{i=1}^{\infty} A_i$ is in each element of $\mathcal{C}$ and, hence, $\bigcup_{i=1}^{\infty} A_i \in \sigma(\mathcal{A}_0)$. ∎

Example 2.2 A six-sided dice is rolled once and the number of dots inscribed on the upper face is observed. The countable (finite) set of outcomes is $\Omega = \{1, 2, 3, 4, 5, 6\}$ with a natural ordering of the sample points given by $\omega_i = i$, $i = 1, \ldots, 6$. If the die is 'fair', then each outcome is equally likely, or $\Pr(\omega_i) = 1/6$. Possible events include $E = \{2, 4, 6\}$, $O = \{1, 3, 5\}$ and $A_i = \{1, 2, \ldots, i\}$. Note that E and O partition the sample space, i.e. $E^c = O$ and $\Omega = E \cup O$, while $A_i \subseteq A_j$, $1 \le i \le j \le 6$. Events A_i exhaust Ω because $\bigcup_{i=1}^{6} A_i = \Omega$, but do not partition the sample space. Defining events

---

[3] We refrain, grudgingly, from adding further biographical information to the scientists mentioned, only because of the great amount of information easily available on the internet. See, for example, the internet site from the University of St Andrews, http://www-history.mcs.st-andrews.ac.uk/history/BiogIndex.html. Admittedly, learning about the historical development of mathematics and the personalities behind the discoveries will not help in the understanding of the material in question, but it often makes for very interesting reading, and helps to put the theory into a broader historical context.

$B_1 = A_1 = \omega_1$ and $B_i = A_i \setminus A_{i-1} = \omega_i$, it follows that B_i, $i = 1, \ldots, 6$, partition Ω because precisely one and only one B_i can occur on a given trial and $\bigcup_{i=1}^{6} B_i = \Omega$.

Furthermore, the collection of all events or measurable sets, $\mathcal{A}$, has 64 elements. This is seen as follows. Any event in $\mathcal{A}$ specifies which of the six ω_i are included. By associating a binary variable with each of ω_i, say 1 if present, 0 otherwise, we see that the number of possible elements in $\mathcal{A}$ is the same as the number of possible codes available from a six-length string of binary numbers, or 2^6. For example, 000000 would denote the empty set $\emptyset$, 111111 denotes the certain event Ω, 010101 denotes event E, etc. ∎

⊖ *Example 2.3* Imagine an automobile factory in which all new cars coming off a particular assembly line are subjected to a battery of tests. The proportion of cars that pass all the tests is a random quantity with countable sample space $\Omega^0 = \{m/n; m \in \{0 \cup \mathbb{N}\}, n \in \mathbb{N}, m \leq n\}$, i.e. all rational numbers between and including 0 and 1. If m and n are typically quite large, it is usually mathematically more convenient to use the uncountable sample space $\Omega = \{x : x \in [0, 1]\}$. Then $\mathcal{A}$, the collection of measurable events, can be described as all interval subsets of $[0, 1]$ along with their complements and their finite unions and intersections. If A is any such event, we can take $\Pr(A) = \int_A f$ for a suitable function f. For example, with $f(x) = 20x^3(1-x)$ and $A = \{\text{proportion greater than } 0.8\}$, $\Pr(A) = \int_{0.8}^{1} f(x) \, dx \approx 0.263$. Observe also that $\Pr(A) \geq 0 \; \forall \; A \in \mathcal{A}$ and $\Pr(\Omega) = \int_0^1 f(x) \, dx = 1$. ∎

The next example is typical of countably infinite sample spaces.

⊙ *Example 2.4* Your newly married colleague has decided to have as many children as it takes until she has a son. Exclude for simplicity any factors which would prevent her from having an unlimited number of children and denote a boy as B and a girl as G. The set of possible outcomes can be listed as $\omega_1 = B$, $\omega_2 = GB$, etc., i.e. the collection of all sequences which end in a boy but have no previous occurrence of such. The sample space Ω is clearly countable. Events of interest might include $A_i = \{\text{at most } i \text{ children}\} = \{\omega_1, \omega_2, \ldots, \omega_i\}$ and $O = \{\text{odd number of children}\} = \{\omega_1, \omega_3, \ldots\}$. Events O and O^c partition the sample space, but events A_i do not (they do exhaust it though). If we define $B_1 = A_1$ and $B_i = A_i \setminus A_{i-1}$, $i \geq 2$, we see that $B_i = \omega_i$, $i = 1, 2, \ldots$, so that the B_i partition the sample space. Let $\Pr(\omega_i) = 2^{-i}$; then $\Pr(A_i) = \sum_{j=1}^{i} \Pr(\omega_j) = 1 - 2^{-i}$ and $\Pr(\Omega) = \sum_{j=1}^{\infty} \Pr(\omega_j) = 1$. ∎

In the previous example, we might also be interested in the total number of children born. Like $\Pr(\cdot)$, this is also a set function, but with domain Ω and range $\{1, 2, \ldots, \}$. It is referred to as a *random variable*. With such a function, questions such as 'if many couples carried out such a family planning strategy, how many children in total might we expect to be born?' can be answered. Random variables and the calculation of such quantities are the subject of Chapter 4, although we are already implicitly using them to a small extent.

⊙ *Example 2.5* Not to be outdone by your colleague, you and your spouse decide to have as many children as it takes until you have three daughters in a row. Exclude for simplicity any constraints that might interfere with matters. The set of possible outcomes is the collection of all sequences which end in three girls but have no previous occurrence of such. Events of interest might include the total number of sons, the number of 'near misses', i.e. two girls in a row followed by a boy, etc.

As in the previous example, the total number of children 'required' is of particular interest. Let f_n be the probability that $n \geq 3$ children will be born under this family planning strategy.

Assume that p is the probability of getting a girl and $q := 1 - p$ is that of a boy. Clearly, $f_1 = f_2 = 0$, while $f_3 = p^3$, i.e. all three children are girls or, in obvious notation, GGG is obtained. For $n = 4$, the first child must be a boy and the next three are girls, i.e. BGGG, with $f_4 = qp^3 = qf_3$. For $n = 5$, the situation must end as BGGG but the first child can be either B or G, i.e. either BBGGG or GBGGG occurs, so that $f_5 = q^2 f_3 + qpf_3 = qf_4 + qpf_3$. For $n = 6$, either BBBGGG or BGBGGG or GBBGGG or GGBGGG can occur. The first two of these start with a boy and are accounted for in the expression qf_5. Series GBBGGG corresponds to qpf_4 and series GGBGGG to $qp^2 f_3$, i.e. $f_6 = qf_5 + qpf_4 + qp^2 f_3$.

This may seem to be getting complicated, but we are practically done. Notice so far, that, for $4 \leq n \leq 6$,

$$f_n = qf_{n-1} + qpf_{n-2} + qp^2 f_{n-3}. \tag{2.6}$$

However, this holds for all $n \geq 4$ because either a boy on the first trial occurs and is followed by one of the possibilities associated with f_{n-1}; or the first two trials are GB followed by one of the possibilities associated with f_{n-2}; or GGB followed by f_{n-3}. That's it!

This recursive scheme is easy to implement in a computer program. To compute and plot (2.6) in Matlab, the following code can be used:

```
p=0.5; q=1-p; f=zeros(60,1);
f(3)=p^3;
for i=4:60
  f(i)=q*f(i-1)+q*p*f(i-2)+q*p^2*f(i-3);
end
bar(f)
```

The output from the code is shown in Figure 2.1(a), while Figure 2.1(b) shows $f_n(3)$ based on $p = 0.3$. In both cases (and for any $0 < p < 1$), $\sum_{n=3}^{\infty} = 1$.

Some further comments are:

- Solution (2.6), along with the *initial conditions* $f_1 = f_2 = 0$, $f_3 = p^3$, is our first example of a *difference equation*. We will encounter several situations in which a solution is comfortably arrived at in terms of a difference equation. In general, difference equations arise in many modeling contexts and so are a very important technique to be familiar with. See the next chapter for more examples, in particular, Example 3.6 (which gives some references), Section 3.3 and Problems 3.21 and 3.19.

- Another quantity of interest is the 'average' number of children a family can expect if they implement such a strategy. Assuming $p = 1/2$, i.e. that girls and boys are equally likely, Example 8.13 shows the solution to this latter question to be 14.

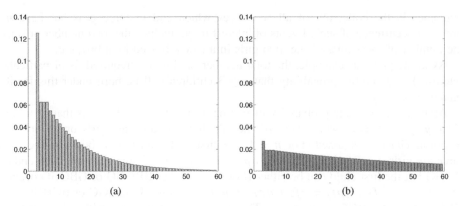

Figure 2.1 Recursion (2.6) for (a) $p = 0.5$ and (b) $p = 0.3$

- If instead of three girls in a row, m are desired, $m \geq 1$, then

$$f_n(m) = \begin{cases} 0, & \text{if } n < m, \\ p^m, & \text{if } n = m, \\ q \sum_{i=0}^{m-1} p^i f_{n-i-1}, & \text{if } n > m. \end{cases} \qquad (2.7)$$

Using recursion (2.7), $f_n(m)$ can be computed for any $n, m \in \mathbb{N}$.

- Solution (2.7) is appealing because it is concise, exact and computable. Note, however, that it is not a closed-form expression, which is usually more desirable, though recursive expressions are often more intuitive. ∎

It is not difficult to extend the previous example in such a way that the solution is either quite complicated, or even intractable. In such cases, *simulation* could be used either to corroborate a tentative answer, or just to provide an approximate, numeric solution if an analytic solution is not available. According to Webster's online dictionary, simulating means 'To assume the mere appearance of, without the reality; to assume the signs or indications of, falsely; to counterfeit; to feign.' This rather negative connotation is misleading in this context; indeed, other dictionaries also mention that it can mean 'to create a representation or model of a physical system or particular situation', which is what we have in mind.

In the case of Example 2.5 with $m = 3$, simulation involves getting the computer to imitate the family planning strategy numerous times (independently of one another), and then tabulate the results. More concretely, a stream of independent random numbers 'uniformly distributed' between zero and one are desired, say $u_1, u_2, \ldots$, from which it is easy to form a binary value indicating boy (if $u_i \geq p$) or girl. In Matlab, this is done with the unpretentiously named `rand` command. Before discussing how this is done, the next example shows how it can be used to simulate the problem at hand.

⊚ **Example 2.6** (Example 2.5 cont.) A program to simulate $f_n(m)$ is given in Program Listing 2.1. It is significantly more complicated compared to the few lines of code given on page 53, but is very instructive, as it illustrates very typical programing methods. Just as with the solution to an algebraic problem, we highly recommend first

```
function len = consecutive (p,m,sim)
len=zeros(sim,1);
for i=1:sim
  if mod(i,100)==0, i, end
  cnt=0; mcount=0;
  while mcount<m
    r=rand; cnt=cnt+1;
    if r>p, mcount=0; else mcount=mcount+1; end
  end
  len(i)=cnt;
end
hist(len,max(len)-m+1)
```

Program Listing 2.1 Simulates the required number of independent trials with probability p of success until m successes in a row first occurs; this is conducted `sim` times, from which a histogram can be produced

understanding how the program works and then, without looking at it, try to reproduce it (interactively with Matlab, not on paper!)[4]

Figure 2.2 shows the output from the program based on $m = 3$ 'girls' and using 10 000 'family replications'. It indeed agrees well with the exact values of $f_n(3)$ in Figure 2.1(a). ∎

The code in Program Listing 2.2 can be used to change the appearance of the default histogram produced in Matlab. It is important to realize that these commands are specific to Matlab, and have nothing to do with general concepts and techniques of programming.

```
[histcount, histgrd] =hist(data,num_bins); % collect the graphics data
h1=bar(histgrd,histcount);      % make histogram and get the 'handle'
set(h1,'facecolor',[0.9 0.9 0.9], ...
    'edgecolor',[1 0 0],'linewidth',1.2)
axis([0 60 0 1300])
set(gca,'fontsize',14)
```

Program Listing 2.2 Commands specific to Matlab for changing graphical appearance

Returning now to the issue regarding computer generation of uniform random-variables (r.v.s), one might have the tendency to consider this to be an issue of only secondary interest, and perhaps even somewhat misplaced in a discussion which has, up to this point, concentrated on 'theoretical' issues. Nothing could be further from the truth. The ability to generate a stream of uniform r.v.s quickly and devoid of dependence of any sort among them is essential for all kinds of simulation exercises. In particular, a uniform r.v. is (at least in theory) sufficient to generate an r.v. from *any*

---

[4] It is worth emphasizing the obvious: if you are new to Matlab or, more critically, to programming, then you need to take off your kid gloves and play around with the code to see how it works. The Matlab commands `help` and `why` should also be of great immediate use, while more substantial sources of assistance can be found in the form of several books on Matlab, as well as numerous tutorials available on the internet.

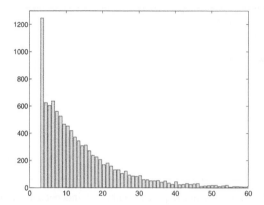

Figure 2.2 Simulated number of children (assuming equal probability of getting a boy or girl) until one obtains three girls in a row

type of distribution! This will be formally shown in Section 7.4, and also demonstrated at various places in this and subsequent chapters.

It is not hard to see that a digital computer, inherently deterministic, will not be able to generate truly random numbers by software means. Instead, a sizeable amount of research has gone into developing methods which generate *pseudo-random* sequences of numbers which appear random, but are actually completely deterministic. We opt not to go into details about how this is done, but instead point the reader to the excellent discussions provided by Ripley (1987, Chapter 2), Bratley, Fox and Schrage (1987, Chapter 6) and Press, Teukolsky, Vetterling and Flannery (1989, Section 7.1),[5] where references to the original literature (most notably Donald Knuth's landmark publication *The Art of Computer Programming*) and computer code can also be found.

Most software platforms have built-in routines to generate a stream of (pseudo) uniform random numbers; as already mentioned Matlab, for example, has the function `rand`, which seems adequate for the instruction purpose of this book.

However, it must be stressed that:

> Many users of simulation are content to remain ignorant of how such numbers are produced, merely calling standard functions to produce them. Such attitudes are dangerous, for random numbers are the foundations of our simulation edifice, and problems at higher levels are frequently traced back to faulty foundations. Do not yield to the temptation to skip [understanding how your computer generates such numbers], for many of the random number generators (at the time of writing) have serious defects. (Brian D. Ripley, 1987, p. 14)
> *Reproduced by permission of John Wiley & Sons, Ltd*

⊖ ***Example 2.7*** (Example 2.5 cont.) Your neighbors are somewhat more family planning conscious and decide to have exactly seven children. There are numerous sample spaces

---

[5] A downloadable p.d.f. version of the latter book (the C version) is available online at `http://www.library.cornell.edu/nr/cbookcpdf.html`.

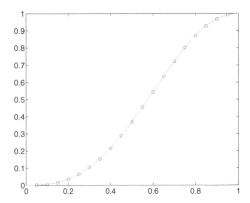

Figure 2.3 Probability of having at least three girls in a row when having seven children, versus p, where p is the probability of having a girl at each 'trial'. Solid line is the exact probability $5p^3 - 4p^4 - p^6 + p^7$; the dots are the empirically determined probabilities obtained from simulated values using the program in Program Listing 2.3 based on 250 000 replications for each value of p shown

which might be of interest, e.g. $\Omega_1 = \{x : x \in \{0, 1, \dots, 7\}\}$, which could be the total number of girls, or the total number of children measuring over 30 cm at birth; or

$$\Omega_2 = \left\{ \mathbf{w} = (w_1, w_2, \dots, w_7) : \mathbf{w} \in \mathbb{N}^7 \right\},$$

the vector of the childrens' respective weights rounded off to the nearest gram, etc.

You (as the watchful neighbor), however, are far more interested in the sequence of genders of the children, i.e. Ω consists of the 2^7 possible binary sequences of B and G. In particular, you are focused on knowing the chances that they have (at least) three girls in a row. Example 3.11 will show this to be about 0.367, assuming that boys and girls are equally likely. More generally, we will see that, if the probability of having a girl is p, $0 < p < 1$, then at least three girls occur with probability $P = 5p^3 - 4p^4 - p^6 + p^7$. This is plotted as the solid line in Figure 2.3. For example, some trial and error reveals that $P = 0.5$ for $p \approx 0.576$.

This problem is also quite amenable to simulation. The program in Program Listing 2.3 shows one possible way of doing this, and also introduces some further programing concepts which the reader is encouraged to understand and use; in particular, Boolean variables (after George Boole, 1815–1864), and the logical AND and OR operators.

The simulation output is shown in Figure 2.3 as circles, which agree well with the theoretical values. The following code was used to generate the plot.

```
pvec=0.05:0.05:0.95; ff=zeros(length(pvec),1);
for i=1:length(pvec), p=pvec(i); ff(i)=threegirls(p,250000); end
plot(pvec,ff,'ro'), p=0:0.02:1; f=5*p.^3 - 4*p.^4 - p.^6 + p.^7;
hold on, plot(p,f,'g-'), hold off, set(gca,'fontsize',16)
```

See also Problem 2.14 for a related question. ∎

```
function f=threegirls(p,sim,kids,inarow)

if nargin<3  % nargin is a pre-defined variable in Matlab
   kids=7;   %    which indicates the Number of ARGuments INput
end          %    to the function.
if nargin<4  % It is used to help assign default values
   inarow=3; %    to variables if they are not passed to the function
end

sum = 0;
for i=1:sim
   k = binornd(1,p,kids,1);
   bool=0;   % boolean variable, i.e., either true or false
   for j=1:kids-inarow+1;
      bool = ( bool | all(k(j:j+inarow-1) == ones(inarow,1)) );
      if bool
         sum=sum+1;
         break % BREAK terminates the inner FOR loop, which
      end       % is useful because once 3 girls in a row are
   end           % found, there is no further need to search.
end
f = sum/sim;
```

Program Listing 2.3   Assume you have `kids` children with probability p of having a girl. This simulates the probability of having at least `inarow` girls in a row. The Matlab function `binornd` is used to generate a `kids` × 1 vector of independent *binomial random variables*, which are discussed in Section 4.2.1. When its first argument is one, this reduces to a sequence of *Bernoulli random variables*, i.e. an independent sequence of zeros and ones, with probability *p* of getting a one

2.3   Properties

2.3.1   Basic properties

From the definition of a probability measure, many properties of $(\Omega, \mathcal{A}, \Pr)$ are intuitive and easy to see, notably so with the help of a *Venn diagram* (after John Venn, 1834–1923), a typical one being given in Figure 2.4.

The following six facts are of fundamental importance.

(i) $\Pr(\emptyset) = 0$;

(ii) if $A \subset B$, then $\Pr(A) \leq \Pr(B)$;

(iii) $\Pr(A) \leq 1$;

(iv) $\Pr(A^c) = 1 - \Pr(A)$;

(v) $\Pr\left(\bigcup_{i=1}^{\infty} A_i\right) \leq \sum_{i=1}^{\infty} \Pr(A_i)$;

(vi) $\Pr(A_1 \cup A_2) = \Pr(A_1) + \Pr(A_2) - \Pr(A_1 A_2)$.

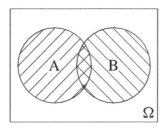

Figure 2.4 Venn diagram of two overlapping events

From (i) and (ii), it follows that $\Pr(A) \geq 0$. Result (v) is equivalent to saying that $\Pr(\cdot)$ is *subadditive* and is also referred to as Boole's inequality (also after George Boole); if the A_i are disjoint, then equality holds. It can also be written as $\Pr\left(\bigcap_{i=1}^{n} A_i^c\right) \geq 1 - \sum_{i=1}^{n} \Pr(A_i)$. These results are proven in Problem 2.3.

Another useful result which is particularly clear from the Venn diagram is that event A can be partitioned into two disjoint sets AB and AB^c, i.e. $\Pr(A) = \Pr(AB) + \Pr(AB^c)$ or $\Pr(A \setminus B) = \Pr(AB^c) = \Pr(A) - \Pr(AB)$. If $B \subset A$, then $\Pr(AB) = \Pr(B)$, so that

$$\Pr(A \setminus B) = \Pr(A) - \Pr(B), \quad B \subset A. \tag{2.8}$$

2.3.2 Advanced properties

We now state four less obvious results which build on (i) to (vi) above. These are followed by several examples either proving or illustrating their use. The following notation will be useful. For events $A_1, \ldots A_n$, let

$$S_j = \sum_{i_1 < \cdots < i_j} \Pr\left(A_{i_1} \cdots A_{i_j}\right), \quad S_0 = 1.$$

For example, $S_1 = \sum_{i=1}^{n} \Pr(A_i)$ and $S_2 = \sum_{i=1}^{n} \sum_{j=i+1}^{n} \Pr\left(A_i A_j\right) = \sum_{i<j}^{n} \Pr\left(A_i A_j\right)$.

- *Bonferroni's inequality*, after Carlo Emilio Bonferroni (1892–1960). For events $A_1, \ldots, A_n$,

$$\Pr(A_1 A_2) \geq \Pr(A_1) + \Pr(A_2) - 1 \tag{2.9}$$

and, more generally,

$$\Pr\left(\bigcap_{i=1}^{n} A_i\right) \geq \sum_{i=1}^{n} \Pr(A_i) - (n-1) = 1 - \sum_{i=1}^{n} \Pr\left(A_i^c\right). \tag{2.10}$$

Bonferroni's inequality is of great importance in the context of *joint confidence intervals* and *multiple testing*.

- *Poincaré's theorem*, after Jules Henri Poincaré (1854–1912), and also called the *inclusion–exclusion principle*. Generalizing property (vi) above,

$$\Pr\left(\bigcup_{i=1}^{n} A_i\right) = \sum_{i=1}^{n} (-1)^{i+1} S_i \;,$$

i.e.

$$
\begin{aligned}
\Pr\left(\bigcup_{i=1}^{n} A_i\right) = & \sum_{i=1}^{n} \Pr(A_i) - \sum_{i<j} \Pr\left(A_i\, A_j\right) \\
& + \sum_{i<j<k} \Pr\left(A_i A_j A_k\right) - \cdots \\
& + (-1)^{r+1} \sum_{i_1<i_2<\cdots<i_r} \Pr\left(A_{i_1} A_{i_2} \cdots A_{i_r}\right) \\
& + \cdots + (-1)^{n+1} \Pr(A_1 A_2 \cdots A_n).
\end{aligned}
\tag{2.11}
$$

- *De Moivre–Jordan*[6] *theorem.* For events $B_1, \ldots, B_n$, the probability that *exactly m* of the B_i occur, $m = 0, 1, \ldots, n$, is given by

$$p_{m,n} = p_{m,n}(\{B_i\}) = \sum_{i=m}^{n} (-1)^{i-m} \binom{i}{m} S_i \;.\tag{2.12}$$

The probability that *at least m* of the B_i occur is given by

$$P_{m,n} = P_{m,n}(\{B_i\}) = \sum_{i=m}^{n} (-1)^{i-m} \binom{i-1}{i-m} S_i \;.\tag{2.13}$$

The two are trivially related by

$$P_{m,n} = \sum_{i=m}^{n} p_{i,n}, \qquad \text{and} \qquad p_{m,n} = P_{m,n} - P_{m+1,n}.$$

---

[6] These results were obtained by Abraham De Moivre (1667–1754), by generalizing results from Montmort, but only for the special (but important) case that the events $B_1, \ldots, B_n$ are exchangeable. In 1867, Marie Ennemond Camille Jordan (1838–1922), proved the validity of the formulae in the nonexchangeable case. See Johnson, Kotz and Kemp (1993, Section 10.2) for further details and references dealing with their historical development.

- *Bonferroni's general inequality.* Bonferroni showed[7] in 1936 that $p_{m,n}$ from (2.12) is bounded by

$$\sum_{i=0}^{2k-1} (-1)^i \binom{m+i}{i} S_{m+i} \leq p_{m,n} \leq \sum_{i=0}^{2k} (-1)^i \binom{m+i}{i} S_{m+i}, \qquad (2.14)$$

for any $k = 1, 2, \ldots, \lfloor (n-m)/2 \rfloor$. For example, with $k = 1$, this gives

$$S_m - (m+1) S_{m+1} \leq p_{m,n} \leq S_m - (m+1) S_{m+1} + \frac{(m+2)(m+1)}{2} S_{m+2}.$$

⊛ *Example 2.8 (**Proof of Bonferroni's inequality**)*   For $n = 2$, (2.9) follows directly from properties (iii) and (vi):

$$\Pr(A_1 A_2) = \Pr(A_1) + \Pr(A_2) - \Pr(A_1 \cup A_2) \geq \Pr(A_1) + \Pr(A_2) - 1. \qquad (2.15)$$

Now assume (2.10) holds for n and define $C = \bigcap_{i=1}^{n} A_i$. Using (2.15),

$$\Pr\left(\bigcap_{i=1}^{n+1} A_i\right) = \Pr(CA_{n+1}) \geq \Pr(C) + \Pr(A_{n+1}) - 1$$

$$\geq \sum_{i=1}^{n} \Pr(A_i) - (n-1) + \Pr(A_{n+1}) - 1 = \sum_{i=1}^{n+1} \Pr(A_i) - n,$$

so that the inequality follows from induction.   ■

⊚ *Example 2.9 (**Proof of Poincaré's theorem**)*   By assuming that Ω is denumerable, the proof is very easy. Let a particular ω be in j of the A_i, $1 \leq j \leq n$, say $A_1, A_2, \ldots, A_j$ and observe that ω 'gets counted' once on the l.h.s. and

$$T = j - \binom{j}{2} + \binom{j}{3} - \cdots + (-1)^{j+1} \binom{j}{j}$$

times on the r.h.s. of (2.11). That $T = 1$ follows directly from the binomial theorem:

$$0 = (1-1)^j = \sum_{i=0}^{j} \binom{j}{i} (-1)^i = 1 - T.$$

---

[7] See Alt (1982) for the two Bonferroni references (in Italian). Schwager (1984) provides a discussion on the sharpness of the bounds. An entire monograph on Bonferroni-type inequalities has recently been written by Galambos and Simonelli (1996), who provide three different methods of proof as well as a brief historical summary of the development of the results and references to further historical accounts.

The theorem without the denumerable assumption can be shown by induction (see, e.g. Port, 1994, p. 27–28 or Roussas, 1997, p. 17–19), while Problem 5.4 provides another method of proof. Problem 2.4 shows the derivation of

$$\Pr\left(\bigcap_{i=1}^{n} A_i\right) = \sum_{i=1}^{n} (-1)^{i+1} R_i, \qquad R_j = \sum_{i_1 < \cdots < i_j} \Pr\left(A_{i_1} \cup A_{i_2} \cdots \cup A_{i_j}\right), \quad (2.16)$$

which can be thought of as the 'dual' of Poincaré's theorem. ■

Of the four advanced properties, we will make the most use of Poincaré's theorem (the inclusion–exclusion principle) and the De Moivre–Jordan theorem. The following examples, and those in Section 3.1, should demonstrate the usefulness of these important results.

⊖ **Example 2.10**   To warm up, we first illustrate the De Moivre–Jordan theorem for the case $n = 4$, $m = 2$. From (2.12), the probability that exactly two of the four B_i occur is

$$
\begin{aligned}
p_{2,4} &= \binom{2}{0} S_2 - \binom{3}{1} S_3 + \binom{4}{2} S_4 \\
&= \sum_{i<j} \Pr\left(B_i B_j\right) - 3 \sum_{i<j<k} \Pr\left(B_i B_j B_k\right) + 6 \Pr\left(B_1 B_2 B_3 B_4\right), \quad (2.17)
\end{aligned}
$$

where $\sum_{i<j} \Pr\left(B_i B_j\right)$ is given by

$$\Pr\left(B_1 B_2\right) + \Pr\left(B_1 B_3\right) + \Pr\left(B_1 B_4\right) + \Pr\left(B_2 B_3\right) + \Pr\left(B_2 B_4\right) + \Pr\left(B_3 B_4\right)$$

and

$$\sum_{i<j<k} \Pr\left(B_i B_j B_k\right) = \Pr\left(B_1 B_2 B_3\right) + \Pr\left(B_1 B_2 B_4\right) + \Pr\left(B_1 B_3 B_4\right) + \Pr\left(B_2 B_3 B_4\right).$$

Abbreviating event B_i to i, and B_i^c to $\bar{i}$, $i = 1, 2, 3, 4$, $p_{2,4}$ can also be written as

$$\Pr\left(12\bar{3}\bar{4}\right) + \Pr\left(1\bar{3}2\bar{4}\right) + \Pr\left(1\bar{4}2\bar{3}\right) + \Pr\left(2\bar{3}1\bar{4}\right) + \Pr\left(2\bar{4}1\bar{3}\right) + \Pr\left(3\bar{4}1\bar{2}\right),$$

but its equality with (2.17) is not immediate.

Similarly, from (2.13), the probability that at least two of the four B_i occur is

$$P_{2,4} = \binom{1}{0} S_2 - \binom{2}{1} S_3 + \binom{3}{2} S_4 = S_2 - 2S_3 + 3S_4.$$

This latter expression could also be arrived at directly using Poincaré's theorem. Again abbreviating event B_i as i,

$$P_{2,4} = \Pr\left(\text{at least two of four events occur}\right) = \Pr\left(12 \cup 13 \cup 14 \cup 23 \cup 24 \cup 34\right).$$

Then, with $A_1 = B_1 B_2 = 12$, $A_2 = B_1 B_3 = 13$, etc., the first sum in (2.11) is just $\Pr(12) + \Pr(13) + \cdots + \Pr(34)$. The second sum involves taking pairs and gives the $\binom{6}{2} = 15$ terms

$$123, 124, 1234, 134, 234,$$

each three times. The third sum involves three at a time for $\binom{6}{3} = 20$ terms; namely 1234, 16 times, and 123, 124, 134, 234, each one time. Next, we have $\binom{6}{4} = 15$ terms, each of which is 1234 (note there is no way to avoid B_1 when taking four terms of 12, 13, 14, 23, 24, 34, and similarly for B_2, B_3 and B_4). The next $\binom{6}{5} = 6$ terms are all 1234; the last is also 1234. Putting this together, we get

$$\sum_{i<j} \Pr\left(B_i B_j\right) - 2 \sum_{i<j<k} \Pr\left(B_i B_j B_k\right) + 3 \Pr\left(B_1 B_2 B_3 B_4\right),$$

which agrees with result from the De Moivre–Jordan theorem. Also, as before,

$$P_{2,4} = 1 - \Pr\left(\overline{1}\,\overline{2}\,\overline{3}\,\overline{4}\right) - \Pr\left(12\,\overline{3}\,\overline{4}\right) - \Pr\left(\overline{1}\,23\,\overline{4}\right) - \Pr\left(\overline{1}\,2\,3\overline{4}\right) - \Pr\left(\overline{1}\,\overline{2}\,34\right),$$

using complements. ∎

Remark: Because of the importance of the De Moivre–Jordan theorem, one should be aware of how to prove it. Problem 3.5(e) details a proof of the De Moivre–Jordan theorem (2.12) when the events are *exchangeable*, i.e. letting $\{i_1, i_2, \dots, i_j\}$ be any of the $\binom{n}{j}$ subsets of $\{1, 2, \dots, n\}$,

$$\Pr\left(A_{i_1}, \dots A_{i_j}\right) = \Pr\left(A_1, \dots A_j\right),$$

$j = 1, \dots, n$. An essentially equivalent, but much more concise proof of (2.12) is given in Grimmett and Stirzaker (2001a, p. 143(13)), where it is referred to as *Waring's theorem* (see also Johnson and Kotz, 1977, p. 31). Feller (1968, p. 106–7) proves (2.12) using basic combinatoric arguments. Port (1994, p. 289) and Stirzaker (1994, Section 1.4) provide elegant proofs using generating functions and properties of the expectation operator. Finally, the result also appears in Wilf (1994, Section 4.2), again using generating functions, but in a vastly larger context. ∎

Example 2.11 As might be expected, Bonferroni's inequality (2.10) follows from Bonferroni's general inequality (2.14). For $m = 0$, (2.14) reduces to

$$\sum_{i=0}^{2k-1} (-1)^i S_i \leq p_{0,n} \leq \sum_{i=0}^{2k} (-1)^i S_i,$$

for any $k = 1, 2, \dots, \lfloor n/2 \rfloor$. With $k = 1$, this gives $1 - S_1 \leq p_{0,n} \leq 1 - S_1 + S_2$, recalling that $S_0 = 1$. This is useful, for example, when higher-order S_i are not available. Note that, from the De Moivre–Jordan theorem (2.12), Poincaré's theorem and De Morgan's law,

$$p_{0,n} = \sum_{i=0}^{n} (-1)^i S_i = 1 - \Pr\left(\bigcup_{i=1}^{n} B_i\right) = \Pr\left(\bigcap_{i=1}^{n} B_i^c\right),$$

so that $1 - S_1 \leq p_{0,n}$ can be expressed as

$$\Pr\left(\bigcap_{i=1}^{n} B_i^c\right) \geq 1 - \sum_{i=1}^{n} \Pr(B_i),$$

which is, with $A_i = B_i^c$, (2.10). ∎

⊙ ***Example 2.12 (De Montmort's problem of coincidences)***   If n objects, labeled 1 to n are randomly arranged in a row, the probability that the position of exactly m of them coincide with their label, $0 \leq m \leq n$, is

$$c_{m,n} = \frac{1}{m!}\left[\frac{1}{2!} - \frac{1}{3!} + \frac{1}{4!} - \cdots + \frac{(-1)^{n-m}}{(n-m)!}\right]. \qquad (2.18)$$

The solution is credited to Pierre Rémond De Montmort (1678–1719). For large $n - m$, note that (2.18) is approximately $e^{-1} / m!$. Johnson, Kotz and Kemp (1993, p. 409) provide further references and some generalizations, while Takacs (1980) details the history of the problem and numerous generalizations.

The problem is actually a straightforward application of Poincaré's theorem and the De Moivre–Jordan theorem. Let A_i denote the event that the ith object is arranged correctly and observe that $\Pr(A_1) = (n-1)!/n!$, $\Pr(A_1) = \cdots = \Pr(A_n)$; $\Pr(A_1 A_2) = (n-2)!/n!$, etc. and, in general,

$$\Pr\left(A_{i_1} A_{i_2} \cdots A_{i_k}\right) = \frac{(n-k)!}{n!}.$$

For $m = 0$, the probability of no coincidences is, from Poincaré's theorem,

$$c_{0,n} = 1 - \Pr(\text{at least one}) = 1 - \Pr\left(\cup_{i=1}^{n} A_i\right)$$

$$= 1 - \left[\binom{n}{1}\frac{(n-1)!}{n!} - \binom{n}{2}\frac{(n-2)!}{n!} + \cdots + (-1)^{n+1}\frac{0!}{n!}\right]$$

$$= 1 - 1 + \frac{1}{2!} - \frac{1}{3!} + \cdots + \frac{(-1)^n}{n!}.$$

For the general case, use (2.12) to get

$$c_{m,n} = p_{m,n}(\{A_i\}) = \binom{n}{m}\frac{(n-m)!}{n!} - \binom{m+1}{m}\binom{n}{m+1}\frac{[n-(m+1)]!}{n!}$$

$$+ \binom{m+2}{m}\binom{n}{m+2}\frac{[n-(m+2)]!}{n!} - \cdots \pm \binom{n}{m}\frac{1}{n!},$$

which easily simplifies to (2.18). ∎

⊖ ***Example 2.13*** Assume that each of n indistinguishable balls is randomly placed in one of r distinguishable cells, with $n \geq r$. Letting U_i be the event that cell i is empty, $i = 1, \ldots, r$, it is easy to see that

$$p_1 := \Pr(U_i) = \left(\frac{r-1}{r}\right)^n, \quad i = 1, \ldots, r,$$

$$p_2 := \Pr(U_i U_j) = \left(\frac{r-2}{r}\right)^n, \quad 1 \leq i < j \leq r,$$

etc., so that Poincaré's theorem gives the probability of at least one empty cell as

$$P_{1,r} = r p_1 - \binom{r}{2} p_2 + \cdots + (-1)^r \binom{r}{r-1} p_{r-1} = \sum_{i=1}^{r-1} (-1)^{i+1} \binom{r}{i} p_i$$

$$= \sum_{i=1}^{r-1} (-1)^{i+1} \binom{r}{i} \left(\frac{r-i}{r}\right)^n = r^{-n} \sum_{i=1}^{r-1} (-1)^{i+1} \binom{r}{i} (r-i)^n, \quad (2.19)$$

noting that $\Pr(U_1 U_2 \cdots U_r) = 0$. Expression (2.19) can be used to confirm the results for $R = n = 2$ (for which $P_{1,r} = 1/2$) and for $R = n = 3$ (for which $P_{1,r} = 21/27$), as detailed at the end of Section 2.1.

It follows immediately from (2.19) that

$$\Pr(\text{no empty cells}) = 1 - P_{1,r} = r^{-n} \sum_{i=0}^{r-1} (-1)^i \binom{r}{i} (r-i)^n. \quad (2.20)$$

As there are r^n total possible arrangements, (2.20) implies that $\sum_{i=0}^{r-1} (-1)^i \binom{r}{i} (r-i)^n$ arrangements are such that no cell is empty. This proves (2.4) for $m = 0$. (It is important to keep in mind that $\Pr(\text{no empty cells}) \neq K_{n,r}^{(1)} / K_{n,r}^{(0)}$ because the possible configurations are not equally likely.)

In order to calculate $p_{m,r}$, the probability that exactly m of the r cells are empty, observe that there are $\binom{r}{m}$ ways of choosing the m empty cells and $(r-m)^n$ ways of arranging the n balls into the $r - m$ cells. However, only $(r-m)^n \left(1 - P_{1,r-m}\right)$ of these are such that no cell is empty, so that the number of arrangements with exactly m of the r cells empty is

$$S_{m,r} = \binom{r}{m} (r-m)^n \left(1 - P_{1,r-m}\right) = \binom{r}{m} \sum_{i=0}^{r-m} (-1)^i \binom{r-m}{i} (r-m-i)^n,$$

which proves (2.4) for all m. It then follows that the probability of getting exactly m of the r cells empty is

$$p_{m,r} = \frac{S_{m,r}}{r^n} = r^{-n} \binom{r}{m} \sum_{i=0}^{r-m} (-1)^i \binom{r-m}{i} (r-m-i)^n$$

$$= r^{-n} \sum_{j=m}^{r} (-1)^{j-m} \binom{j}{m} \binom{r}{j} (r-j)^n,$$

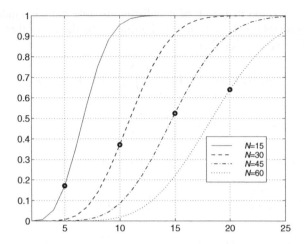

Figure 2.5 Probability of at least one empty cell, $P_{1,r}$, versus r for different values of N

where $j = i + m$ and $\binom{r}{m}\binom{r-m}{j-m} = \binom{j}{m}\binom{r}{j}$. The latter expression for $p_{m,r}$ is, however, precisely that as given by (2.12) using the events $U_1, \ldots, U_r$, i.e.

$$p_{m,r}(\{U_i\}) = \Pr(\text{exactly } m \text{ of the } r \text{ cells are empty})$$

$$= r^{-n} \sum_{i=m}^{r-1} (-1)^{i-m} \binom{i}{m}\binom{r}{i} (r-i)^n, \quad m = 0, 1, \ldots, r, \quad (2.21)$$

again noting that $\Pr(U_1 U_2 \cdots U_r) = 0$. Clearly, $p_{r,r}$ is zero. Similarly, from (2.13), the probability that at least m of the r cells are empty is

$$P_{m,r}(\{U_i\}) = r^{-n} \sum_{i=m}^{r-1} (-1)^{i-m} \binom{i-1}{i-m}\binom{r}{i} (r-i)^n, \quad m = 0, 1, \ldots, r. \quad (2.22)$$

Note that, for $m = 1$, (2.22) reduces to (2.19).

As an illustration, Figure 2.5 plots $P_{1,r}$ for $N = 15$, 30, 45 and 60 over a grid of r values. The large dots indicate $r = N/3$ for each N; from these, we can see how $P_{1,r}$ increases for a fixed proportion of r to N as N increases. ∎

⊖ ***Example 2.14*** (Example 2.13 cont.) Looking again at the proof of Poincaré's theorem for denumerable Ω, it becomes clear that the result holds with the cardinality set function #, i.e. the number of members ω in a set, replacing the set function $\Pr(\cdot)$. Continuing Example 2.13 with U_i the event that cell i is empty, $i = 1, \ldots, r$, Poincaré's theorem can be applied to get $S = \#(\cup_{i=1}^r U_i)$, i.e. S is the number of ways that at least one cell winds up empty. From symmetry,

$$s_1 := \#(U_1) = \#(U_2) = \cdots = \#(U_r) = K_{n,r-1}^{(0)} = \binom{n + (r-1) - 1}{(r-1) - 1}.$$

Similarly, for $s_2 := \#\left(U_i \cap U_j\right)$, $i \neq j$, for which there are $\binom{r}{2}$ different pairs,

$$s_2 = \#\left(U_1 \cap U_2\right) = K_{n,r-2}^{(0)} = \binom{n + (r-2) - 1}{(r-2) - 1}.$$

Continuing (and observing that event $U_1 \cap \cdots \cap U_r$ cannot occur), Poincaré's theorem gives

$$S = rs_1 - \binom{r}{2}s_2 + \cdots + (-1)^r \binom{r}{r-1}s_{r-1}$$

$$= \sum_{i=1}^{r-1}(-1)^{i+1}\binom{r}{i}s_i = \sum_{i=1}^{r-1}(-1)^{i+1}\binom{r}{i}\binom{n+r-i-1}{n};$$

but, more simply, $S = K_{n,r}^{(0)} - K_{n,r}^{(1)}$, which implies the combinatoric identity

$$\binom{n+r-1}{r-1} - \binom{n-1}{r-1} = \sum_{i=1}^{r-1}(-1)^{i+1}\binom{r}{i}\binom{n+r-i-1}{n}, \quad n \geq r. \quad \blacksquare$$

2.3.3 A theoretical property

If you have built castles in the air, your work need not be lost. That is where they should be. Now put the foundation under them.

(Henry David Thoreau)

We end this section by examining a continuity property of $\Pr(\cdot)$ which is not only of theoretical interest, but will be required for the development of distribution functions in Chapter 4.

If $A_1, A_2, \ldots$ is a monotone sequence of (measurable) events, then

$$\lim_{i \to \infty} \Pr(A_i) = \Pr\left(\lim_{i \to \infty} A_i\right), \tag{2.23}$$

i.e. the limiting probability of A_i is the same as the probability of the limiting A_i.

The proof hinges on countable additivity. Two cases need to be considered, monotone increasing and monotone decreasing. For the former, i.e. $A_1 \subset A_2 \subset \cdots$, express A_i as the sum of i disjoint 'telescoping' events

$$A_i = A_1 + (A_2 - A_1) + (A_3 - A_2) + \cdots + (A_i - A_{i-1})$$
$$= A_1 \cup (A_2 \setminus A_1) \cup (A_3 \setminus A_2) \cup \cdots \cup (A_i \setminus A_{i-1}),$$

with $\lim_{i \to \infty} A_i = A_1 \cup (A_2 \setminus A_1) \cup \cdots$. From the countable additivity property (2.5),

$$\Pr\left(\lim_{i \to \infty} A_i\right) = \Pr(A_1) + \Pr(A_2 \setminus A_1) + \cdots$$

$$= \lim_{i \to \infty} \left[\Pr(A_1) + \Pr(A_2 \setminus A_1) + \cdots + \Pr(A_i - A_{i-1})\right]$$

$$= \lim_{i \to \infty} \Pr\left[A_1 + (A_2 - A_1) + (A_3 - A_2) + \cdots + (A_i - A_{i-1})\right]$$

$$= \lim_{i \to \infty} \Pr(A_i).$$

For monotone decreasing A_i, i.e. $A_1 \supset A_2 \supset \cdots$, note that $A_1^c \subset A_2^c \subset \cdots$, so that

$$\lim_{i \to \infty} \Pr\left(A_i^c\right) = \Pr\left(\lim_{i \to \infty} A_i^c\right)$$

from the previous result. However,

$$\lim_{i \to \infty} \Pr\left(A_i^c\right) = \lim_{i \to \infty} (1 - \Pr(A_i)) = 1 - \lim_{i \to \infty} \Pr(A_i)$$

and, from (A.1), De Morgan's law (A.3) and (A.2),

$$\Pr\left(\lim_{i \to \infty} A_i^c\right) = \Pr\left(\bigcup_{i=1}^{\infty} A_i^c\right) = 1 - \Pr\left(\bigcap_{i=1}^{\infty} A_i\right) = 1 - \Pr\left(\lim_{i \to \infty} A_i\right),$$

so that

$$1 - \lim_{i \to \infty} \Pr(A_i) = 1 - \Pr\left(\lim_{i \to \infty} A_i\right).$$

2.4 Problems

Nothing in the world can take the place of persistence. Talent will not; nothing is more common than unsuccessful men with talent. Genius will not; unrewarded genius is almost a proverb. Education will not. The world is full of educated derelicts. Persistence and determination alone are omnipotent. The slogan 'Press on,' has solved and always will solve the problems of the human race.

(Calvin Coolidge)

If at first you don't succeed, try, try, and try again. Then give up. There's no use being a damned fool about it.

(William Claude Dunkenfield)

2.1. Prove the following.

(a) If $\Pr(A) \leq \Pr(B)$, then $\Pr(B^c) \leq \Pr(A^c)$.

(b) If $A \subset B$, then $B^c \subset A^c$.

(c) If $A \cap B$ imply C, then $\Pr(C^c) \leq \Pr(A^c) + \Pr(B^c)$.

2.2. Given $\Pr(A) = 3/4$ and $\Pr(B) = 3/8$, show that

(a) $\Pr(A \cup B) \geq 3/4$,

(b) $1/8 \leq \Pr(AB) \leq 3/8$.

2.3. ★ Using only the definition of the probability function, prove the basic properties (i) to (vi) of $(\Omega, \mathcal{A}, \Pr)$.

2.4. ★ Prove the 'dual' of Poincaré's theorem given in (2.16), i.e.

$$\Pr\left(\bigcap_{i=1}^{n} A_i\right) = \sum_{i=1}^{n} (-1)^{i+1} R_i, \qquad R_j = \sum_{i_1 < \cdots < i_j} \Pr\left(A_{i_1} \cup A_{i_2} \cdots \cup A_{i_j}\right).$$

2.5. An equivalent proof that $\lim_{i \to \infty} \Pr(A_i) = \Pr(\lim_{i \to \infty} A_i)$ for monotone increasing A_i starts by defining $B_1 = A_1$ and $B_i = A_i \setminus (A_1 \cup A_2 \cup \cdots \cup A_{i-1})$. Then,

$$\Pr\left(\bigcup_{i=1}^{\infty} A_i\right) = \Pr\left(\bigcup_{i=1}^{\infty} B_i\right) = \sum_{i=1}^{\infty} \Pr(B_i)$$

$$= \lim_{n \to \infty} \sum_{i=1}^{n} \Pr(B_i) = \lim_{n \to \infty} \Pr\left(\bigcup_{i=1}^{n} B_i\right)$$

$$= \lim_{n \to \infty} \Pr\left(\bigcup_{i=1}^{n} A_i\right) = \lim_{n \to \infty} \Pr(A_n). \qquad (2.24)$$

Justify each step in (2.24).

2.6. A man has n keys only one of which will open his door. If he tries the keys at random ('for reasons which can only be surmised'), what is the probability that he will open the door on the kth try, assuming that[8]

(Feller, 1968, p. 48, 55(11))

(a) he discards those keys which do not work, or

(b) he does not discard those keys which do not work?

2.7. Six people enter a train with three compartments. Each sits in a random compartment, independent of where anyone else sits.

(a) How many different ways, N, can the six people be seated into the three compartments?

(b) What is the probability that exactly two people sit in the first compartment?

---

[8] Parts (a) and (b) are special cases of the random variables discussed in Section 4.2.4 and Section 4.2.3, respectively.

2.8. Three men, M_i, $i = 1, \ldots, 3$, and three women, W_i, $i = 1, \ldots, 3$, sit down randomly in a row of six chairs. Calculate the following.

(a) Pr(three men sit together and three women sit together).

(b) Pr(all three men sit together).

(c) ★ Pr(every woman sits next to at least one man).

2.9. A cake recipe calls for b coffee beans. After mixing all the ingredients well and baking the cake, it is cut into b pieces. Calculate the probability that each piece has one bean. Why isn't

$$\frac{K_{b,b}^{(1)}}{K_{b,b}^{(0)}} = \frac{\binom{b-1}{b-1}}{\binom{b+b-1}{b}} = \frac{b! \, (b-1)!}{(2b-1)!}$$

correct?

2.10. How many children, n, should you plan to have if you want to be $100\alpha\%$ sure, $0 < \alpha < 1$, of having at least one boy *and* at least one girl?

(Cacoullos, 1989, p. 21(88))

2.11. A bakery makes 80 loaves of bread daily. Ten of them are underweight. An inspector weighs at most five random loaves, stopping if an underweight loaf is discovered. What is the probability that he discovers one?

(Cacoullos, 1989, p. 10(36))

2.12. ★ A lottery consists of 100 tickets, labeled $1, 2, \ldots, 100$, three of which are 'winning numbers'. You buy four tickets. Calculate the probability, p, that you have at least one winning ticket. Try deriving two solutions via Poincaré's theorem (2.11) and four solutions using basic combinatoric and probabilistic arguments.

2.13. If n balls are individually dropped into r urns such that the probability of landing in urn i is p_i, $i = 1, \ldots, r$, with $\sum_{i=1}^{r} p_i = 1$, what is the probability p that no urn is empty?

2.14. A family plans to have n children. Let $p = \Pr(\text{girl})$.

(a) If $n = 2$, what is the probability that they have *at least* one girl?

(b) ★ ★ For general $n \geq k$, what is the probability $P(p, n, k)$ that they have *at least* k girls, and *all* these girls are born in a row? Verify that $P(p, 2, 1) = p(2 - p)$ drops out as a special case. Simplify the result for $P(p, 7, 3)$ and compare with the result in Example 2.7.

2.15. After having written n job application letters and addressing the n corresponding envelopes, your child does you the kind favor of inserting one letter into each of the envelopes and mailing them. However, she thought all the letters were the same, so that they were placed into the envelopes in a completely random fashion.

(a) What is the probability that all companies get the correct letter?

(b) What is the probability that exactly k are correct?

(c) What is the probability that at least one is correct?

2.16. ★ Recall Problem 1.15, which dealt with random permutations. Write a program which simulates many random permutations of vector $(1, 2, \ldots, N)$ and, for each, computes the number of positions of the vector such that position i contains the number i. Keep track of these numbers and then plot their empirical (i.e. simulated) frequencies, along with the exact probability of their occurrence.

2.17. ★ In the beginning of a game of bridge, each of four players randomly receives 13 cards from a standard 52 card playing deck. Compute the probability that one of these hands contains at least one ace and king from the *same* suit.

(Ross, 1988, p. 57(32); see also Feller, 1968, p. 57, 100)

2.18. Your new neighbors have two children, and the older is a boy. What is the probability that the other is also a boy?

2.19. Your new neighbors have two children, and at least one is a boy. What is the probability that the other is also a boy? (Gardner, 1961)

2.20. An urn contains three red and seven black balls, and your friend wants to play a game with you. One at a time, you each draw one ball out, without replacement, and the first person to get a red ball wins. Are your chances at winning better if you draw first, or your friend?

2.21. A game is played as follows. Of three same-sized cards, one has a black dot on both sides, one has a red dot on both sides, and one a red dot on one side and a black dot on the other. I pick a card at random, and show you one side of it. What's the probability that the other side of the card has a different colored dot? What do you think of the 'solution': If I see a red dot, then I *know* that the card is either the red–red one or the red–black one, so my chances are 50–50?

(Ross, 1988, pp. 70–1; Pitman, 1997, pp. 34–5)[9]

2.22. Two persons, say A and B, alternate between playing a video game against the computer. Person A has constant probability p_1, $0 < p_1 < 1$ of winning, while person B has p_2, $0 < p_2 < 1$. First A plays, then B, then A, etc., until someone wins against the computer. Assume that each game is independent from the previous one, i.e. there is, for example, no 'learning effect'. Let A and B be the events that persons A and B win, respectively.

(a) Write closed form expressions for $\Pr(A)$ and $\Pr(B)$. Define $q_i = 1 - p_i$, $i = 1, 2$.

---

[9] This is just a variant of *Bertrand's box problem* (from Joseph Bertrand's *Calcul des probabilites*, 1889) in which a box has three draws, one with two gold coins, one with two silver, and one with a silver and gold. A draw is randomly chosen and a coin from the chosen draw is randomly picked.

(b) A game between two people is said to be fair if $\Pr(A) = \Pr(B)$. Give an algebraic expression for p_2 in terms of p_1 such that the game is fair and examine how p_2 behaves for various p_1.

2.23. Prove that, for $m \geq 0$, $b > 0$ and any $k \geq 0$,

$$\sum_{j=0}^{b-1} \binom{b}{j} \binom{m-1}{b-j-1} = \sum_{j=0}^{b-1} \binom{k}{j} \binom{m-1+b-k}{b-j-1},$$

assuming $\binom{a}{b} = 0$ for $a < b$. (Hint: instead of proving this directly, look for similar formulae in the text.)

2.24. Verify the algebraic equivalence of the expressions for $p_{2,4}$ and $P_{2,4}$ arrived at by use of the De Moivre–Jordan theorem and by use of complements for the important special case in which the events A_i are independent with equal probability of occurring, say p.

2.25. ★ Decompose the $R^n = 4^3 = 64$ possible ways of randomly placing each of $n = 3$ distinguishable balls into one of $R = 4$ distinguishable cells and then list the possible occurrences and their respective probabilities if both balls and cells are indistinguishable.

3

Symmetric spaces and conditioning

I've done the calculation and your chances of winning the lottery are identical whether you play or not.   (Fran Lebowitz)

This chapter begins by using the tools developed in Chapter 2 to solve some more advanced applications. Then the immensely important concept of conditional probability is introduced in Section 3.2, which will allow us to straightforwardly tackle significantly harder problems. Section 3.3 brings these concepts together to solve the problem which is considered to be the historical starting point of modern probability.

3.1 Applications with symmetric probability spaces

Example isn't another way to teach, it is the only way to teach.
                                                                  (Albert Einstein)

Chapter 2 introduced methods for counting the ways in which balls can be distributed into cells. Recall how the various configurations of distinguishable balls allocated randomly among distinguishable cells were taken to be equally likely; it is an example of a symmetric probability space. If the sample space consists of N equally likely outcomes, i.e. $\Pr(\omega_1) = \Pr(\omega_2) = \cdots = \Pr(\omega_N)$, then, either from the definition of a symmetric probability space or the more general definition of a probability measure, $\Pr(\omega_i) = N^{-1}$, $i = 1, \ldots, N$. For event A, $\Pr(A) = N(A)/N$, where $N(A)$ denotes the number of sample points contained in A. The following examples further serve to illustrate the use of such sample spaces and some of the types of problems that can be addressed with them. The end-of-chapter problems contain extensions and further applications.

Example 3.1 A fair, six-sided die is rolled t times. The probability that a one or two never appears is easy: on each roll, only one of four choices occur. As rolls are

Fundamental Probability: A Computational Approach   M.S. Paolella
© 2006 John Wiley & Sons, Ltd

independent, $p_t = (4/6)^t$. If the question is slightly altered such that *any* two sides never appear, then matters are more complicated. In fact, the question is ambiguous; either all of the remaining four sides must appear or not. In the former case, this is precisely the statement of the De Moivre–Jordan theorem for an exact number of events occurring. That is, with $m = 2$, $n = 6$, events $B_i = \{$side i does not appear$\}$ and

$$S_j = \sum_{i_1 < i_2 < \cdots < i_j} \Pr\left(B_{i_1} \cdots B_{i_j}\right) = \binom{6}{j} \Pr\left(B_1 \cdots B_j\right) = \binom{6}{j}(1 - j/6)^t,$$

$p_t = \Pr\,(\text{exactly two sides do not appear})$

$$= \sum_{i=2}^{6} (-1)^{i-2} \binom{i}{2}\binom{6}{i}(1 - i/6)^t. \tag{3.1}$$

Note also that, in this case, p_t must be zero for $t \le 3$, which is indeed true for the above expression.

If all of the remaining four sides do not necessarily have to appear, then we have the De Moivre–Jordan theorem for at least a certain number of events occurring, i.e.

$$P_t = \Pr\,(\text{at least two sides do not appear})$$

$$= \sum_{i=2}^{6} (-1)^{i-2} \binom{i-1}{i-2}\binom{6}{i}(1 - i/6)^t. \tag{3.2}$$

Directly with $t = 5$, the complement is that 0 or 1 sides do not appear, i.e. that five different sides occur, so that

$$P_5 = 1 - \frac{5}{6}\frac{4}{6}\frac{3}{6}\frac{2}{6} = \frac{49}{54},$$

which agrees with (3.2). Also, (3.2) is one for $t = 0, 1, 2, 3, 4$, as it should be.

One might think that, as there are $\binom{6}{2} = 15$ ways of picking the two sides to remain empty, the solution is also given by $P'_t = 15\,(4/6)^t$ for $t > 4$. The fact that $P'_5 > 1$ and $P'_6 > 1$ already hints that something is wrong. It is too large for all $t > 4$ because it overcounts the possibilities. For example, with $t = 5$ and outcome $\{2, 1, 1, 2, 3\}$, P'_5 would count this event three times: once with sides four and five 'chosen' not to occur, once with four and six and once with five and six. However, (3.2) can be expressed (easily with Maple) as

$$p_t = 15 \left(\frac{2}{3}\right)^t - \frac{40}{2^t} + \frac{45}{3^t} - \frac{24}{6^t},$$

from which it is clear that, as t increases, $p_t \to 15\,(4/6)^t$. In fact, the same argument holds for (3.1), i.e. it can be expressed as $15\,(2/3)^t - 60/2^t + 90/3^t - 60/6^t$.

As a last question, consider the probability that all six sides occur at least once. Clearly, for $t < 6$, $p_t = 0$. Otherwise, p_t is given by (2.20), i.e. with $r = 6$, $n = t$ and using Maple,

$$a_t = 6^{-t} \sum_{i=0}^{5} (-1)^i \binom{6}{i} (6-i)^t$$

$$= 1 - 6\left(\frac{5}{6}\right)^t + 15\left(\frac{2}{3}\right)^t - \frac{20}{2^t} + \frac{15}{3^t} - \frac{6}{6^t}, \quad t \geq 6.$$

(In fact, this expression is zero for $t = 1, 2, 3, 4, 5$.) Also, note that, using basic principles for $t = 6$,

$$a_6 = \frac{6}{6}\frac{5}{6} \cdots \frac{1}{6} = \frac{6!}{6^6} = \frac{5}{324}.$$

See the next example and Problem 3.1 for further related calculations.   ∎

⊖ **Example 3.2** A fair, six-sided die is rolled t times. We are interested in p_t, the probability that all six sides occur at least twice. Clearly, for $t < 12$, $p_t = 0$. For $t = 12$, from basic principles,

$$p_{12} = \frac{\binom{12}{2,2,2,2,2,2}}{6^{12}} = \frac{12!}{2^6\,6^{12}} = \frac{1925}{559872} \approx 3.438 \times 10^{-3}.$$

For general t, with $B_i = \{\text{side } i \text{ occurs zero or one time}\}$,

$$p_t = \Pr(\text{all six sides occur at least twice}) = 1 - \Pr\left(\bigcup_{i=1}^{6} B_i\right),$$

and use of Poincaré's theorem will solve the problem if an expression for $\Pr(B_1 \cdots B_j)$ can be obtained. Letting $N_i = $ the number of times side i occurs, event B_i can be decomposed into the two disjoint events $\{N_i = 0\}$ and $\{N_i = 1\}$, i.e. $\{\text{side } i \text{ does not occur}\}$ and $\{\text{side } i \text{ occurs once}\}$, so that

$$\Pr(B_i) = \left(\frac{5}{6}\right)^t + \binom{t}{1}\left(\frac{1}{6}\right)\left(\frac{5}{6}\right)^{t-1}.$$

Next, partition event $B_1 B_2$ into four disjoint components

$$\{N_1 = 0,\ N_2 = 0\},\ \{N_1 = 0,\ N_2 = 1\},\ \{N_1 = 1,\ N_2 = 0\} \quad \text{and} \quad \{N_1 = 1,\ N_2 = 1\},$$

so that $\Pr(B_1 B_2)$ is given by

$$\left(\frac{4}{6}\right)^t + \binom{t}{1}\left(\frac{1}{6}\right)\left(\frac{4}{6}\right)^{t-1} + \binom{t}{1}\left(\frac{1}{6}\right)\left(\frac{4}{6}\right)^{t-1} + \binom{t}{1,1,t-2}\left(\frac{1}{6}\right)^2\left(\frac{4}{6}\right)^{t-2}.$$

Notice that the binomial coefficient of the latter term is not $\binom{t}{2}$, because rolling a one and rolling a two are not indistinguishable. As a check, with $t = 2$, the only way in which $B_1 B_2$ does not occur is if both rolls are ones, or both rolls are twos. These disjoint events each occur with probability 1/36, so that $\Pr(B_1 B_2) = 1 - 2/36 = 17/18$, which agrees with the computation from the above expression.

Similarly, in abbreviated notation for the N_i, event $B_1 B_2 B_3$ can be decomposed as {000}, {001}, {010}, {100}, {011}, {101}, {110} and {111}, so that $\Pr(B_1 B_2 B_3)$ is given by

$$\left(\frac{3}{6}\right)^t + 3\binom{t}{1}\left(\frac{1}{6}\right)\left(\frac{3}{6}\right)^{t-1} + 3\binom{t}{1, 1, t-2}\left(\frac{1}{6}\right)^2\left(\frac{3}{6}\right)^{t-2} + t_{[3]}\left(\frac{1}{6}\right)^3\left(\frac{3}{6}\right)^{t-3},$$

where

$$t_{[3]} = t\,(t-1)\,(t-2) = \binom{t}{1, 1, 1, t-3}.$$

Checking with $t = 2$ again, the only way in which $B_1 B_2 B_3$ cannot occur is if both rolls are ones, or both rolls are twos, or both rolls are threes, i.e. $1 - 3/36 = 11/12$, which agrees with the computation from the above expression.

It is easy to see that, in general, for $j = 1, 2, \ldots, 5$,

$$\Pr(B_1 \cdots B_j) = \sum_{i=0}^{j}\left(\frac{1}{6}\right)^i\binom{j}{i}t_{[i]}\left(\frac{6-j}{6}\right)^{t-i},$$

which also holds for $j = 6$, as event $B_1 B_2 B_3 B_4 B_5 B_6$ cannot occur (for $t \le 5$). From symmetry, for any permutation $\{i_1, \ldots, i_j\}$ of $\{1, \ldots, j\}$, $\Pr(B_{i_1} \cdots B_{i_j}) = \Pr(B_1 \cdots B_j)$, so that

$$\sum_{i_1 < i_2 < \cdots < i_j} \Pr(B_{i_1} \cdots B_{i_j}) = \binom{6}{j}\Pr(B_1 \cdots B_j).$$

Thus, using $t_{[i]} = \binom{t}{i}i!$, Poincaré's theorem gives

$$p_t = \sum_{j=1}^{5}(-1)^{j+1}\binom{6}{j}\sum_{i=0}^{j}\left(\frac{1}{6}\right)^i\binom{j}{i}\binom{t}{i}i!\left(\frac{6-j}{6}\right)^{t-i}.$$

Also, for $t = 12$, this yields

$$p_{12} = \frac{1925}{559872} \approx 3.438 \times 10^{-3},$$

which agrees with the answer using basic principles. Problem 3.2 examines the similar, but more complicated case in which all six sides are to occur at least three times. ∎

⊙ **Example 3.3** Consider ordering n people (or any distinguishable objects) around a table which can be arbitrarily 'rotated'. Instead of $n!$ possible permutations of the people (as would be the case if they were arranged in a row), there are only $(n - 1)!$. To see this, observe that, for one person, there is one way; for two people, one way; for three people, there are two ways: of the $3! = 6$ ways of arranging three objects in a row,

$$123 \ 231 \ 312 \ 132 \ 213 \ 321,$$

when these are formed into a circle, the first three become the same (in the sense that, for each 'shift and roll' permutation, each person always has the same left and right neighbors) and similarly for the second three. With four people, imagine three people already at the table; there are three 'spaces' into which the fourth person can sit, i.e. there are three times as many possibilities, or $3 \times 2 = 3! = (4 - 1)!$. In general, with n people, the nth person has $n - 1$ spaces to choose from, implying that there are $(n - 1)!$ ways of seating n individuals around a table. ∎

Example 3.4 Let $2n$ individuals, paired as n couples, be seated completely at random around a table and define K to be the number of couples which happen to be sitting together. Assume the couples are numbered from 1 to n and let C_k denote the event that couple k sits together, $k = 1, \ldots, n$. From the De Moivre–Jordan theorem (2.12), it follows that

$$\Pr(K = k) = \sum_{i=k}^{n} (-1)^{i-k} \binom{i}{k} S_i, \quad \text{where} \quad S_i = \binom{n}{i} \Pr(C_1 \cdots C_i).$$

Problem 3.4(b) shows that

$$\Pr(C_1 \cdots C_i) = 2^i \frac{(2n - i - 1)!}{(2n - 1)!}, \tag{3.3}$$

yielding

$$\Pr(K = k) = \sum_{i=k}^{n} (-1)^{i-k} \binom{i}{k} \binom{n}{i} 2^i \frac{(2n - i - 1)!}{(2n - 1)!}. \tag{3.4}$$

Using the function for the binomial coefficient given in Listing 1.1 and Matlab's elementwise vector operations (. * and . /), (3.4) is easily computed for each k, as shown in Program Listing 3.1.

It would be nice if the output vector f were such that its first element, f(0), corresponded to Pr($K = 0$), and f(1) corresponded to Pr($K = 1$), etc. but Matlab does not allow an array index less than or equal to zero. Thus, f is such that its first element, f(1), corresponds to Pr($K = 0$), up to its last element, f(n+1), which corresponds to Pr($K = n$).

To simulate the values of $\Pr(K = k)$, we construct a vector with the numbers 1 to n, which represent the n men, and append to it another such vector, which represents the women, say v=[1:n 1:n]. Then a random permutation of this $2n$-length vector is constructed using Matlab's built-in randperm function (but see also Problem 1.15); this represents the table. Then, we 'go around the table' and count how many adjacent

```
function f=coupleprob(n)
f=zeros(n+1,1);
for k=0:n
  i=k:n; temp = (-1).^(i-k) .* c(i,k).*c(n,i) .* (2.^i);
  f(k+1)=sum(temp .* gamma(2*n-i-1+1) ./ gamma(2*n-1+1) );
end
```

Program Listing 3.1   Computes (3.4)

'seats' have the same numeric value, which means also checking the first and last values in the array to account for the circular table structure. This is repeated a large number of times, and the results are tabulated using Matlab's `tabulate` function, which is perfect for this purpose, but has the minor drawback that it only can tabulate positive integer values. This is easily remedied by adding one to all the values. (This idiosyncracy has been surmounted in Matlab version 7.) The program is given in Program Listing 3.2

```
function f=coupleprobsim(n)
sim=500000; match=zeros(sim,1); v=[1:n 1:n];
for s=1:sim
  tab=v(randperm(2*n)); % this is the "table"
  mcnt = (tab(1)==tab(end)); % account for table wrapping
  for i=1:(2*n-1)
     if tab(i)==tab(i+1), mcnt=mcnt+1; end
  end
  match(s)=mcnt;
end
g=tabulate(match+1); f=g(:,end)/100;
```

Program Listing 3.2   Simulates the values of $\Pr(K = k)$ in (3.4).

For $n = 3$ couples, the true probabilities are $\Pr(K = 0) = 4/15$, $\Pr(K = 1) = 2/5$, $\Pr(K = 2) = 1/5$ and $\Pr(K = 3) = 2/15$, which each agree to three significant digits with the values obtained by simulating half a million replications. Figure 3.1 plots $\Pr(K = k)$ for $n = 10$ as well as the function $e^{-1}/k!$, which are the *limiting* or *asymptotic* values of $\Pr(K = k)$ as $n \to \infty$. These values are the mass function of the Poi(1) distribution, which is discussed in Section 4.2.5.

Problem 3.5 details other methods of deriving $\Pr(K = k)$ without use of the De Moivre–Jordan theorem, and Problem 3.6 looks at some related questions. Later, Example 6.2 will illustrate how the mean and variance of K can be computed (the mean and variance are examples of *moments*, which are introduced in Section 4.4). ∎

⊖ **Example 3.5**   Your sock draw contains n different pairs of socks all mixed up. Fumbling in the dark while you hastily pack your suitcase, you randomly draw r of the $2n$ socks with the fleeting hope that some will actually match. Assume that r is even and define $p_m = \Pr(\text{exactly } m \text{ matches})$, $m \le r/2$.

For p_0, the probability of no matches, clearly $p_0 = 1$ for $r = 0$ and $p_0 = 0$ for $r > n$. For general r, define events A_i to be that the ith pair of socks gets chosen, $i = 1, \ldots, n$.

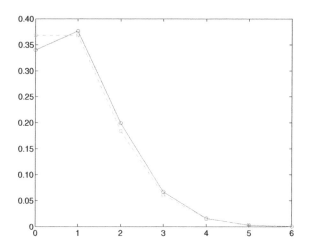

Figure 3.1 Exact values of $\Pr(K = k)$ for $n = 10$ (solid) and their asymptotic values as $n \to \infty$ (dashed)

If $r \geq 2$, then

$$\Pr(A_1) = \binom{2}{2}\binom{2n-2}{r-2} \bigg/ \binom{2n}{r} = \Pr(A_i), \quad i = 1, \ldots, n;$$

for $r \geq 4$,

$$\Pr(A_1 A_2) = \binom{4}{4}\binom{2n-4}{r-4} \bigg/ \binom{2n}{r} = \Pr(A_i A_j), \quad 1 \leq i < j \leq n;$$

and, in general, for $1 \leq j \leq r/2$,

$$\Pr\left(A_{i_1} \cdots A_{i_j}\right) = \binom{2j}{2j}\binom{2n-2j}{r-2j} \bigg/ \binom{2n}{r}. \tag{3.5}$$

Poincaré's theorem then gives

$$p_0 = 1 - \Pr\left(\bigcup_{i=1}^{n} A_i\right) = 1 - \binom{2n}{r}^{-1} \sum_{i=1}^{r/2} (-1)^{i+1} \binom{n}{i}\binom{2n-2i}{r-2i},$$

where the summand is zero for $r/2 < i \leq n$. Figure 3.2 plots p_0 versus r for various n.

Now for p_m, it helps to picture the socks organized in a $2 \times n$ array with the jth pair in the jth column, with the top (bottom) row for the left (right) socks. Starting with $m = 1$,

$$p_1 = \Pr(\text{exactly one match}) = \binom{n}{1} \Pr\left(A_1 \bigcap_{i=2}^{n} A_i^c\right),$$

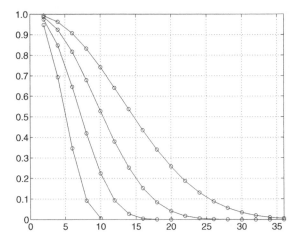

Figure 3.2 p_0 versus r for, from left to right, $n = 10, 20, 40$ and 80

with the number of possibilities for event $A_1 \cap_{i=2}^{n} A_i$ given as follows. The first column (A_1) is chosen in $\binom{2}{2} = 1$ way; $r - 2$ pairs are selected in $\binom{n-1}{r-2}$ ways; from each, either the left or right sock is picked, in 2^{r-2} different ways. Thus,

$$p_1 = n \binom{n-1}{r-2} 2^{r-2} \bigg/ \binom{2n}{r}.$$

In general, there are $\binom{n}{m}$ choices for which complete pairs are obtained. Of the remaining $n - m$ pairs and $r - 2m$ draws, there are $\binom{n-m}{r-2m}$ pairs which can be selected; from these, either the left or right sock is drawn, so that

$$p_m = \binom{n}{m} \binom{n-m}{r-2m} 2^{r-2m} \bigg/ \binom{2n}{r}$$

(see Figure 3.3 for illustration).

Finally, from the De Moivre–Jordan theorem (2.12) and recalling (3.5), it follows that

$$p_m = p_{m,n}(\{A_i\}) = \binom{2n}{r}^{-1} \sum_{i=m}^{n} (-1)^{i-m} \binom{i}{m} \binom{n}{i} \binom{2i}{2i} \binom{2n-2i}{r-2i}.$$

Thus, we arrive at the identity

$$\binom{n}{m} \binom{n-m}{r-2m} 2^{r-2m} = \sum_{i=m}^{n} (-1)^{i-m} \binom{i}{m} \binom{n}{i} \binom{2n-2i}{r-2i}, \tag{3.6}$$

which may or may not be useful, but is certainly not obvious. ∎

⊙ **_Example 3.6_** Let f_n denote the number of ways of tossing a fair coin n times such that successive heads never appear. Also let $f_0 = 1$ and observe that $f_1 = 2$. For higher

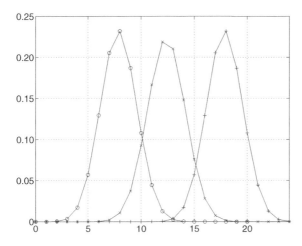

Figure 3.3 p_m versus m for $n = 50$ and, from left to right, $r = 40$, 50 and 60

n, consider the two possibilities for the first toss. Denoting a head by H and a tail by T, if the first toss is tails, we have the situation

$$T \underbrace{\_ \_ \_ \cdots \_}_{f_{n-1}} \qquad \text{and, if heads,} \qquad H\,T \underbrace{\_ \_ \cdots \_}_{f_{n-2}},$$

i.e. the second toss *must* be tails to avoid a match, and then one essentially 'starts over from scratch'. Assuming heads and tails are equally likely, it follows that $f_n = f_{n-1} + f_{n-2}$, which is the Fibonacci sequence with starting values $f_0 = 1$ and $f_1 = 2$. This difference equation has solution[1] $f_n = c_1 \lambda_1^n + c_2 \lambda_2^n$ with λ_i, $i = 1, 2$, the solution to the characteristic equation $\lambda^2 - \lambda - 1 = 0$, or $\lambda_{1,2} = \left(1 \pm \sqrt{5}\right)/2$. The two c_i are determined from the two initial conditions $c_1 + c_2 = 1$ and $c_1\lambda_1 + c_2\lambda_2 = 2$. Solving,

$$f_n = \frac{3 + \sqrt{5}}{2\sqrt{5}} \left(\frac{1 + \sqrt{5}}{2} \right)^n - \frac{3 - \sqrt{5}}{2\sqrt{5}} \left(\frac{1 - \sqrt{5}}{2} \right)^n.$$

The derivation shows that f_n is always an integer, which is not otherwise obvious. For n large, $\left((1 - \sqrt{5})/2\right)^n$ approaches zero, so that the second term in the above expression becomes negligible.

An alternative solution is obtained as follows. Assume that n tosses yields H heads, $h_1, \ldots, h_H$, which partition the $n - H$ tails, shown as t:

$$\underbrace{tt \cdots t}_{x_1} h_1 \underbrace{tt \cdots t}_{x_2} h_2 \cdots h_H \underbrace{tt \cdots t}_{x_{H+1}}$$

with the constraints that $x_1, x_{H+1} \geq 0$, $x_2, \ldots, x_H > 0$ and $\sum_{i=1}^{H+1} x_i = n - H$. Thus,

---

[1] See, for example, Elaydi (1999, p. 67) or Kelly and Peterson (2001, p. 63).

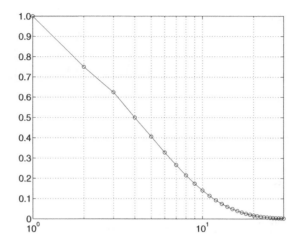

Figure 3.4 Probability p_n of no successive heads versus n

from the discussion on occupancy in Section 2.1, there are

$$K^{(0,1,\ldots,1,0)}_{n-H,H+1} = \binom{n-H+1}{H}$$

possibilities. Now observe that, for n even, $H > n/2$ and, for n odd, $H > (n+1)/2$ heads will ensure that two heads appear in a row. Thus,

$$f_n = \sum_{h=0}^{U} \binom{n-h+1}{h}, \quad \text{where} \quad U = \begin{cases} n/2, & \text{if } n \text{ even,} \\ \frac{n+1}{2}, & \text{if } n \text{ odd.} \end{cases}$$

To compute the probability, p_n, that successive heads never appear in n tosses, divide f_n by 2^n. A plot of p_n on a log scale is shown in Figure 3.4. ∎

◎ ***Example 3.7***   From a standard, shuffled deck of 52 playing cards, imagine that the cards are dealt out one at a time, face down, so that their identity is not revealed. What is the probability that the nth card is an ace, $1 \le n \le 52$? As the nth card is equally likely to be any one of the 52 cards and there are four aces in total, the answer is just 1/13.

What is the probability that no aces occur in the first 48 draws? To compute this, observe that, of the 52! ways of ordering all the cards, there are 48! ways of ordering the 48 cards omitting the aces and 4! ways of arranging those four aces. Hence, the probability is

$$\frac{48! \cdot 4!}{52!} = \frac{1}{\binom{52}{4}} = \frac{1}{270\,725} \approx 3.7 \times 10^{-6}.$$

The expression $1/\binom{52}{4}$ could have been arrived at directly by noting that, of all $\binom{52}{48} = \binom{52}{4}$ ways of choosing the first 48 cards, only one is such that no aces are among them.

What is the probability that the first ace you encounter occurs on the nth draw, $1 \le n \le 49$? As there are four aces, there are 48 ways of drawing the first card, 47 for the second, etc. down to $(48 - n + 2)$ ways of drawing the $(n-1)$th card. Recalling that $x_{[n]}$ denotes $x \cdot (x-1) \cdots \cdots (x - n + 1)$, the number of ways of drawing the first $n-1$ cards is $48_{[n-1]}$. There are four possibilities for the nth position and the remaining $(52 - n)$ cards can be arranged in any of $(52 - n)!$ ways. Thus, dividing by the total number of ways of arranging 52 cards, we have

$$\frac{4 \cdot 48_{[n-1]} \cdot (52 - n)!}{52!} = \frac{4 \cdot 48_{[n-1]}}{52_{[n]}}, \quad 1 \le n \le 49.$$

Alternatively, the probability can be expressed as the number of ways that the last three aces could be found among the last $52 - n$ cards, divided by the total number of ways that the four aces could be found among all 52 cards, or $\binom{52-n}{3} / \binom{52}{4}$. This might also be seen as follows. Imagine partitioning the deck into three pieces: the first $n-1$ cards (which contain no aces), the nth card (which is an ace), and the remaining $52 - n$ cards (which must contain three aces). Then the probability is

$$\frac{\binom{n-1}{0}\binom{1}{1}\binom{52-n}{3}}{\binom{52}{4}}.$$

One should algebraically verify that this agrees with the previous solution. Such a formulation is useful for more general partitions.

What is the probability that the first ace occurs sometime after the nth draw? For example, if $n = 5$, we want to know the probability that the first ace occurs on the sixth draw, or on the seventh, ..., or on the 49th. One solution is to use the previous question: the answer must be $\binom{52}{4}^{-1} \sum_{i=n+1}^{49} \binom{52-i}{3}$. One could then (try to) algebraically simplify this.

Another way is as follows. Using your 'combinatoric thinking', the first n drawn may be any of 48 non-ace cards, for which there are $\binom{48}{n}$ ways. Now, you might continue with 'and there are $52 - n$ possible cards for the $(n-1)$th position, which might or might not be an ace, ...'. This is correct, but tough and unnecessary. Instead of thinking about all the remaining combinations and then dividing by 52!, we can restrict both the numerator and denominator to the first n cards. That is, we want the number of ways that the first n cards come from a set of 48, divided by the number of ways that the first n cards come from the whole set of 52. We don't have to think about the remaining cards. Thus, the probability is $\binom{48}{n}/\binom{52}{n}$.

It is easy to verify algebraically that $\binom{48}{n}/\binom{52}{n} = \binom{52-n}{4}/\binom{52}{4}$. This latter expression leads, in fact, to another method of arriving at the answer: the number of ways four aces can be distributed among the last $52 - n$ positions divided by the number of ways of distributing four aces among all 52 positions.

Putting these together and generalizing to the situation with a aces among a total of N cards, we have derived the combinatoric equality

$$\binom{N}{a}^{-1} \sum_{i=n+1}^{N-a+1} \binom{N-i}{a-1} = \frac{\binom{N-a}{n}}{\binom{N}{n}} = \frac{\binom{N-n}{a}}{\binom{N}{a}}, \quad a < N, \quad n \le N - a. \qquad \blacksquare$$

⊖ ***Example 3.8*** Six married couples decide to take a vacation together in the sunny northern German city of Kiel, and will be staying in the same hotel, one couple per room, six rooms. Each person receives one room key; assume that the two keys for any particular room are distinguishable. One evening, while enjoying the famous night life and having perhaps too much to drink, they accidentally mix up their 12 room keys. Each person then takes a random key, staggers back separately to the hotel, and stays in the room for which the key works.

1. What is the probability that k of the people, $1 \leq k \leq 12$, get back *precisely their own, original* room key? This (otherwise difficult) question is just (2.18) with $N = 12$.

2. How many ways can the 12 people find themselves paired together? There are clearly 12! ways of distributing the keys among themselves, but two factors need to be considered. The first is that the keys come in pairs, so that the number of ways is reduced to $12!/2^6$. Another way of seeing this is by observing that the 12 keys are distributed as six groups of two, which is given by

$$\binom{12}{2, 2, 2, 2, 2, 2} = \frac{12!}{(2!)^6}.$$

The second factor is that the actual room which the two people are in does not matter. There are 6! ways of ordering the six rooms, so the final answer is given by

$$\frac{12!}{(2!)^6 \, 6!} = 10\ 395.$$

3. What is the probability that every married couple finds themselves together? Only one of the 10 395 ways can give rise to this, because we already took into account the key duplication and the irrelevancy of the room order.

4. What is the probability that each room has a man/woman pair, possibly, but not necessarily, married? This can be solved as follows. The first man can be placed with one of six women, the second man can be placed with one of five, etc. so that there are 6! ways this can occur. Thus, the probability is $6! \, / \, 10\ 395 \approx 6.9264 \times 10^{-2}$.

5. What is the probability that all the men are paired with men, and all the women are paired with women? There are $\binom{6}{2,2,2} / 3! = 15$ unique ways of placing six men into three unordered groups of two. Likewise, there are 15 ways of placing six women into three unordered groups of two and, hence, the probability is $15^2 \, / \, 10\ 395 = 2.1645 \times 10^{-2}$.
   One might wonder about the room ordering. Once the men and women are paired off into groups, there are 6! ways of arranging them in the rooms, which would suggest the answer

$$\frac{\binom{6}{2,2,2}^2 / 6!}{10\ 395} = \frac{\binom{6}{2,2,2}^2}{10\ 395 \cdot 6!}$$

which does not agree with the previous one because $3!3! \neq 6!$. We neglected to account for the fact that there are $\binom{6}{3}$ ways of assigning three rooms to the men. This gives

$$\frac{\binom{6}{2,2,2}^2 \binom{6}{3}}{10\ 395 \cdot 6!} = \frac{\binom{6}{2,2,2}^2}{10\ 395 \cdot 3!3!}.$$

Notice that, in all (correct) solutions, both the numerator and denominator are consistent with one another in that they take extraneous room ordering into account. ∎

3.2   Conditional probability and independence

In most applications, there will exist information which, when taken into account, alters the assignment of probability to events of interest. As a simple example, the number of customer transactions requested per hour from an online bank might be associated with an *unconditional* probability which was ascertained by taking the average of a large collection of hourly data. However, the *conditional* probability of receiving a certain number of transactions might well depend on the time of day, the arrival of relevant economic or business news, etc. If these events are taken into account, then more accurate probability statements can be made. Other examples include the number of years a manufacturing product will continue to work, conditional on various factors associated with its operation, and the batting average of a baseball player conditional on the opposing pitcher, etc.

We begin with the most basic notion of conditioning, which involves just two events. If $\Pr(B) > 0$, then the *conditional probability of event A given the occurrence of event B*, or just the *probability of A given B*, is denoted by $\Pr(A \mid B)$ and defined to be

$$\Pr(A \mid B) = \frac{\Pr(AB)}{\Pr(B)}. \qquad (3.7)$$

This definition is motivated by observing that the occurrence of event B essentially reduces the relevant sample space, as indicated in the Venn diagram in Figure 2.4. The probability of A given B is the intersection of A and B, scaled by $\Pr(B)$. If $B = \Omega$, then the scaling factor is just $\Pr(\Omega) = 1$, which coincides with the unconditional case.

Example 3.9   (Example 3.8 cont.) Regarding our six married couples, what is the probability that each room has a man/woman pair, but no married couples? Denote as A the event that no pair of people is married, and by B the event that each room has one man and one woman, possibly, but not necessarily, married. Then, from (3.7), the answer is given by $\Pr(AB) = \Pr(A \mid B) \Pr(B)$, where

$$\Pr(A \mid B) = \sum_{i=2}^{6} \frac{(-1)^i}{i!} = 0.36806$$

corresponds to (2.18) with $n = 6$ and $m = 0$, and $\Pr(B)$ is precisely question 4 in Example 3.8.[2] ■

If the occurrence or 'nonoccurrence' of event B does not influence that of A, and vice versa, then the two events are said to be *independent*, i.e. $\Pr(A \mid B) = \Pr(A)$ and $\Pr(B \mid A) = \Pr(B)$. It follows from (3.7) that, if events A and B are independent, then $\Pr(AB) = \Pr(A)\Pr(B)$. This is also referred to as *pairwise* independence. In general, events A_i, $i = 1, \ldots, n$, are *mutually* or *completely* independent if, and only if, for every collection $A_{i_1}, A_{i_2}, \ldots, A_{i_m}, 1 \le m \le n$,

$$
\Pr\left(A_{i_1} A_{i_2} \cdots A_{i_m}\right) = \prod_{j=1}^{m} \Pr\left(A_{i_j}\right).
\tag{3.8}
$$

For $n = 3$, this means that $\Pr(A_1 A_2) = \Pr(A_1)\Pr(A_2)$, $\Pr(A_1 A_3) = \Pr(A_1)\Pr(A_3)$, $\Pr(A_2 A_3) = \Pr(A_2)\Pr(A_3)$ and $\Pr(A_1 A_2 A_3) = \Pr(A_1)\Pr(A_2)\Pr(A_3)$.

That pairwise independence does not imply mutual independence is referred to as *Bernstein's paradox*. Simple examples of the paradox are provided by Rohatgi (1976, p. 48) and Feller (1968, p. 143(26)). Further, letting B_i be either A_i or A_i^c, $i = 1, \ldots, n$, the events B_i are mutually independent. While this might appear obvious, the reader is invited to prove it formally (or see Roussas, 1997, p. 30–32, for a detailed proof).

⊖ **Example 3.10** Independent events $\{A, B, C\}$ occur with $\Pr(A) = a$, $\Pr(B) = b$, $\Pr(C) = c$. Let E be the event that at least one of $\{A, B, C\}$ occur. Using the complement, we can write

$$
\Pr(E) = 1 - \Pr\{\text{neither A nor B nor C occur}\}
$$
$$
= 1 - \Pr\left(A^c B^c C^c\right) = 1 - (1-a)(1-b)(1-c)
$$

from the independence of A, B and C. Alternatively, from Poincaré's theorem (2.11),

$$
\Pr(E) = \Pr(A \cup B \cup C) = a + b + c - ab - ac - bc + abc.
$$

A third method results by observing that $A \cup B \cup C$ is the same as $A \cup A^c B \cup A^c B^c C$, as best seen using the Venn diagram in Figure 3.5. As the three events A, $A^c B$ and $A^c B^c C$ are nonoverlapping,

$$
\Pr(E) = \Pr(A) + \Pr\left(A^c B\right) + \Pr\left(A^c B^c C\right) = a + (1-a)b + (1-a)(1-b)c.
$$

Straightforward algebra shows the equivalence of all the solutions. ■

---

[2] One might be tempted to reason as follows. Similar to question 4, there are five (not six) ways that the first woman can 'choose' her room partner, and four ways for the second woman, etc. so that the solution is $5!/10395$ or only $1/6$ of $\Pr(B)$ instead of 0.36806. This reasoning is wrong, however. Label the women 1 to 6, say $W_1, \ldots, W_6$, and their respective husbands get the same labeling, $M_1, \ldots, M_6$. The first woman, W_1, indeed has five choices; $M_2, \ldots, M_6$. If she chooses, say, M_3, then W_2 has four choices; M_1, M_4, M_5, M_6. If, instead, W_1 chooses M_2, then W_2 has five choices, $M_1, M_3, \ldots, M_6$. While it is possible to enumerate all the possible permutations, the solution based on $\Pr(A \mid B)$ is far easier.

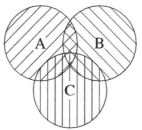

Figure 3.5 Venn diagram of $\Pr(A \cup B \cup C)$

3.2.1 Total probability and Bayes' rule

When you come to a fork in the road .... Take it.   (Yogi Berra)
Reproduced by permission of Hyperion

From a Venn diagram with (overlapping) events A and B, event A may be partitioned into mutually exclusive events AB and AB^c, so that, from (3.7),

$$\Pr(A) = \Pr(AB) + \Pr\left(AB^c\right) \tag{3.9}$$
$$= \Pr(A \mid B)\Pr(B) + \Pr\left(A \mid B^c\right)(1 - \Pr(B)).$$

This is best understood as expressing $\Pr(A)$ as a weighted sum of conditional probabilities in which the weights reflect the occurrence probability of the conditional events. In general, if events B_i, $i = 1, \ldots, n$, are exclusive and exhaustive, then the *law of total probability* states that

$$\Pr(A) = \sum_{i=1}^{n} \Pr(A \mid B_i)\Pr(B_i). \tag{3.10}$$

⊖ ***Example 3.11***   (Example 2.7 cont.) Interest centers on the probability of having at least three girls in a row among seven children. Assume that each child's gender is independent of all the others and let $p = \Pr(\text{girl on any trial})$. Denote the event that three girls in a row occur as R and the total number of girls as T. Then, from the law of total probability,

$$\Pr(R) = \sum_{t=0}^{7} \Pr(R \mid T = t)\Pr(T = t).$$

Clearly, $\Pr(R \mid T = t) = 0$ for $t = 0, 1, 2$ and $\Pr(R \mid T = 6) = \Pr(R \mid T = 7) = 1$. For $T = 3$, there are only five possible configurations, these being

$$gggbbbb, \ bgggbbb, \ \ldots, \ bbbbggg,$$

so that

$$\Pr(R \mid T = 3) = \frac{5p^3(1-p)^4}{\binom{7}{3}p^3(1-p)^4} = \frac{5}{35}.$$

For $T = 4$, we outline the possible configurations:

- either the first three children are girls, denoted as [123], which can happen in four ways, i.e. there are four unique permutations of the underlined group in $\overline{gg}\underline{gbbbg}$, while the overlined group stays fixed;

- or [234] occurs, i.e. children 2,3 and 4 are girls, in which case we have sequence $b\underline{ggg}\overline{bbg}$, for which there are three permutations (which are distinct from the [123] cases);

- or [345], i.e. the three permutations of $\underline{bb}\overline{ggg}bg$;

- or [456], i.e. the three permutations of $\overline{gbb}\underline{ggg}\overline{b}$;

- or [567], i.e. $gbbb\underline{ggg}$ which has three permutations.

Thus, we have a total of 16 possibilities and $\Pr(R \mid T = 4) = 16/\binom{7}{4} = 16/35$.

For $T = 5$:

- with [123], $\overline{ggg}\underline{bbgg}$ has $\binom{4}{2} = 6$ permutations;

- or [234], i.e. the three permutations of $\overline{bgggbgg}$;

- or [345], i.e. the three permutations of $g\overline{bgggbg}$;

- or [456], i.e. the three permutations of $gg\underline{bggg}\underline{b}$;

- or [567], i.e. the three permutations of $gg\underline{bb}\overline{ggg}$.

Thus, $\Pr(R \mid T = 5) = 18/\binom{7}{5} = 18/21$.

Putting this all together,

$$\Pr(R) = 0 + 0 + 0$$

$$+ \frac{5}{35}\binom{7}{3}p^3(1-p)^4 + \frac{16}{35}\binom{7}{4}p^4(1-p)^3 + \frac{18}{21}\binom{7}{5}p^5(1-p)^2$$

$$+ \binom{7}{6}p^6(1-p) + p^7$$

$$= 5p^3 - 4p^4 - p^6 + p^7,$$

which was plotted in Figure 2.3. For $p = 1/2$, $\Pr(R) \approx 0.367$. ∎

From (3.7) and (3.9), *Bayes' rule* is given by

$$\Pr(B \mid A) = \frac{\Pr(AB)}{\Pr(A)} = \frac{\Pr(A \mid B)\Pr(B)}{\Pr(A \mid B)\Pr(B) + \Pr(A \mid B^c)\Pr(B^c)} \tag{3.11}$$

while, for mutually exclusive and exhaustive events B_i, $i = 1, \ldots, n$, the *general Bayes' rule* is given by

$$Pr(B \mid A) = \frac{Pr(A \mid B) Pr(B)}{\sum_{i=1}^{n} Pr(A \mid B_i) Pr(B_i)} \qquad (3.12)$$

from (3.10).

A classic, but very important, example of Bayes' rule is the following.

⊛ *Example 3.12*   A test for a disease possesses the following accuracy. If a person has the disease (event D), the test detects it 95 % of the time; if a person does *not* have the disease, the test will falsely detect it 2 % of the time. Let d_0 denote the *prior probability* of having the disease before the test is conducted. (This could be taken as an estimate of the proportion of the relevant population believed to have the disease.) Assume that, using this test, a person is detected as having the disease (event +). To find the probability that the person actually has the disease, given the positive test result, we use Bayes' rule (3.12):

$$Pr(D \mid +) = \frac{0.95 d_0}{0.95 d_0 + 0.02 (1 - d_0)}. \qquad (3.13)$$

For a rare disease such that $d_0 = 0.001$, $Pr(D \mid +)$ is only 0.045. The answer may also be graphically depicted via a tree diagram, as shown in Figure 3.6. The probability of an end-node is the product of the 'branch' probabilities starting from the left. Then $Pr(D \mid +)$ is obtained as the ratio of the end-node $\{D+\}$ and all branches (including $\{D+\}$) leading to +. ∎

Note that, in the previous example, the question has little meaning without knowledge of d_0. Furthermore, instead of using the entire population proportion for d_0, referred to as an *unconditional* prior probability, a more accurate assessment of $Pr(D \mid +)$ could be obtained by using a *conditional* value. For example, with regard to a test for lung cancer, d_0 might be a function of *exogenous* variables (those which do not influence the performance of the test itself) such as age, gender, how much the person smokes, when he or she started smoking, etc.

⊃ *Example 3.13*   In January 1966, at 30 000 feet near the coast of Palomares, Spain, a B-52 bomber collided with an air tanker during a refueling operation, and lost its payload of four hydrogen bombs. Three were subsequently recovered, but the other was missing at sea. This started a race between the American and Soviet navy to find it, the latter viewing this as an immensely important cold war treasure. As brilliantly told by Sontag and Drew (1998) in their book about American submarine espionage, the ocean engineer John P. Craven, under enormous pressure from President Lyndon Johnson, led the search team and located the missing H-bomb by using Bayes' rule. As explained by Sontag and Drew (1998, p. 55):

Craven applied [Bayes' rule] to the search. The bomb had been hitched to two parachutes. He took bets on whether both had opened, or one, or none. He went

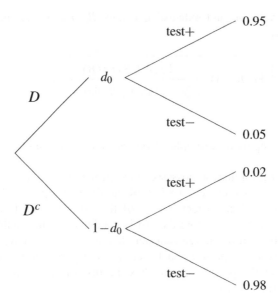

Figure 3.6   Tree diagram for Bayes' rule

through the same exercise over each possible detail of the crash. His team of mathematicians wrote out possible endings to the crash story and took bets on which ending they believed most. After the betting rounds were over, they used the odds they created to assign probability quotients to several possible locations. Then they mapped those probabilities and came up with the most probable site and several other possible ones.

Imagine, for arguments' sake, that there are three possible areas where the bomb could be located, and it is initially believed that the bomb is in area i with probability 0.2, 0.3 and 0.5, for $i = 1, 2, 3$. Obviously, area 3 will be searched first, but, even if the bomb is there, a single deep-water search will find it only with probability (say) of 0.7. So, if the first search fails, which area should be subsequently searched, and if that fails, where next? The details of this important application are left for the reader in Problem 3.11. ∎

The next example illustrates another true story, this one being highly publicized (also making it into the Finance and Economics section of *The Economist*, Feb. 20, 1999) and which stumped many people, including some academics, but which is easily dealt with via an application of Bayes' rule.

◎ ***Example 3.14***   For her popular question-and-answer column 'Ask Marylin' of Parade magazine, Marylin vos Savant received the following question which pertained to the game show with Monty Hall.

Suppose you're on a game show, and you're given the choice of three doors. Behind one door is a car, behind the others, goats. You pick a door, say number 1, and the host, who knows what's behind the doors, opens another door, say number 3, which has a goat. He says to you, 'Do you want to pick door number 2?' Is it to your advantage to switch your choice of doors?

Craig F. Whitaker, Columbia, MD

Reproduced by permission of St. Martin's Press

The wrong answer is that, as you are faced with two doors, one of which hides the car, there is a 50–50 chance, so that switching does not increase or decrease your odds of winning. Marylin's (correct) response, that you should switch, provoked a huge response from her readers, among them mathematically trained people, claiming her answer to be wrong. She stuck to her answer and several iterations of correspondence ensued, until the truth of what is now sometimes referred to as the *Monty Hall dilemma* or the *Monty Hall puzzle* prevailed. More details can be found in vos Savant (1996).

Before proceeding with the solution, it is important to note one of the valid mis-understandings which can arise: knowing whether Monty Hall *always* presents a door with a goat, or only occasionally. That is, upon seeing that you initially pick the correct door, Monty will have an incentive to show you a goat-door in the hope of enticing you to switch, whereas if you pick a goat-door, the odds of him giving you the possi-bility of switching would be smaller. This underscores the general necessity of having a well-defined problem. The answer presented here assumes that Monty Hall *always* presents you with a goat-door and the offer to switch. We present three solutions.

1. *Basic logic* Your first guess is correct $1/3$ of the time, and wrong $2/3$ of the time. If your first guess is correct, then you lose if you switch and, if it is wrong, then you win if you switch. So, by switching, you double your chances of winning from $1/3$ to $2/3$.

2. *Bayes' rule* Label the doors A, B and C. The unconditional probability that the car is behind any of the three doors is clearly $1/3$. Let A be the event that the car is behind door A, and let $\overset{\circ}{B}$ be the event that Monty Hall opens door B. Assume you pick door A. Then $\Pr(\overset{\circ}{B} \mid A) = 1/2$, $\Pr(\overset{\circ}{B} \mid B) = 0$ and $\Pr(\overset{\circ}{B} \mid C) = 1$, so that the law of total probability gives

$$\Pr(\overset{\circ}{B}) = \Pr(\overset{\circ}{B} \mid A)\Pr(A) + \Pr(\overset{\circ}{B} \mid B)\Pr(B) + \Pr(\overset{\circ}{B} \mid C)\Pr(C) = \frac{1}{6} + 0 + \frac{1}{3} = \frac{1}{2}.$$

Now, Bayes' rule gives

$$\Pr(A \mid \overset{\circ}{B}) = \frac{\Pr(\overset{\circ}{B} \mid A)\Pr(A)}{\Pr(\overset{\circ}{B})} = \frac{1/6}{1/2} = \frac{1}{3},$$

while

$$\Pr(C \mid \overset{\circ}{B}) = \frac{\Pr(\overset{\circ}{B} \mid C)\Pr(C)}{\Pr(\overset{\circ}{B})} = \frac{1/3}{1/2} = \frac{2}{3}.$$

Thus, if you initially pick door A, you should switch. Clearly, from the symmetry of the problem, you should switch whatever door you initially choose.

3. *As a maximization problem* Let s be the probability that you switch, and we wish to find that value of s such that the probability of winning, W, is maximized. We have

$$\Pr(W) = \Pr(\text{first guess wrong}) \cdot s + \Pr(\text{first guess right}) \cdot (1 - s)$$

$$= \frac{2}{3}s + \frac{1}{3}(1 - s) = \frac{1}{3}s + \frac{1}{3},$$

which attains a maximum at $s = 1$.

Further details and historical references can be found in Chun (1999), who details various approaches to solving a generalized version of, as he puts it, 'this wonderfully confusing little problem'. ∎

The next example explicitly illustrates the different notions of conditional and unconditional probabilities.

⊛ **Example 3.15** An urn contains b black and r red balls from which balls are drawn without replacement. The first ball is drawn (and discarded), but its color is not observed. A second ball is drawn. What is $\Pr(B_2)$, the probability that it is black? In general, let B_i and R_i be the events that the ith ball drawn is black and red, respectively, and define $T = b + r$.
Clearly, $\Pr(B_1) = b/T$ and

$$\Pr(B_2) = \Pr(B_2 \mid B_1)\Pr(B_1) + \Pr(B_2 \mid R_1)\Pr(R_1)$$

$$= \frac{b-1}{T-1}\frac{b}{T} + \frac{b}{T-1}\frac{r}{T} = \frac{b}{b+r},$$

so that $\Pr(B_1) = \Pr(B_2)$.
This result can be formulated by saying that, while the conditional probability of B_2 depends on B_1, the *unconditional* probability does not. ∎

The result in the previous example is actually quite plausible: Although we know a ball was drawn, not having observed its outcome implies that we can say nothing more about the second draw than we knew initially. One might surmise that this intuitive result holds in general, i.e. that $\Pr(B_i) = \Pr(B_1)$, $i = 1, 2, \ldots, T$. This is indeed true and is proven in the next example. However, to do so, it is convenient to use the – not yet introduced – notion of random variables (see Chapter 4). The reader can return to the example after reading Chapter 4 if he or she wishes.

⊚ **Example 3.16** (Example 3.15 cont.) Let the random variable Y_n be the total number of black balls drawn up to and including the nth trial, $0 \leq Y_n \leq \min(n, b)$. Then, for $j \leq b$,

$$\Pr(B_{n+1} \mid Y_n = j) = \frac{b - j}{T - n}$$

and, from the hypergeometric distribution (4.14),

$$\Pr(Y_n = j) = \frac{\binom{b}{j}\binom{r}{n-j}}{\binom{T}{n}},$$

so that

$$\Pr(B_{n+1}) = \sum_{j=0}^{b} \Pr(B_{n+1} \mid Y_n = j)\Pr(Y_n = j) = \sum_{j=0}^{b} \frac{b-j}{T-n} \frac{\binom{b}{j}\binom{r}{n-j}}{\binom{T}{n}}$$

from (3.10). By expanding and rearranging the factorial expressions, we get that

$$\frac{b-j}{T-n} \frac{\binom{b}{j}\binom{r}{n-j}}{\binom{T}{n}} = \frac{\binom{n}{j}\binom{T-n-1}{b-j-1}}{\binom{T}{b}}$$

or

$$\Pr(B_{n+1}) = \frac{1}{\binom{b+r}{b}} \sum_{j=0}^{b} \binom{n}{j}\binom{b+r-n-1}{b-j-1} = \frac{\binom{b+r-1}{b-1}}{\binom{b+r}{b}} = \frac{b}{b+r}$$

from (1.32) (see also Problem 3.9). ∎

3.2.2 Extending the law of total probability

The law of total probability $\Pr(A) = \sum_{i=1}^{n} \Pr(A \mid B_i)\Pr(B_i)$ can be extended by making all events conditional on a further event, say C, or

$$\Pr(A \mid C) = \sum_{i=1}^{n} \Pr(A \mid B_i C)\Pr(B_i \mid C),$$

which simplifies to

$$\Pr(A \mid C) = \Pr(A \mid BC)\Pr(B \mid C) + \Pr\left(A \mid B^c C\right)\Pr\left(B^c \mid C\right) \qquad (3.14)$$

for $n = 2$, i.e. when B and B^c partition the sample space. The latter follows from writing $\Pr(A \mid C)$ as

$$\Pr(A \mid C) = \frac{\Pr(AC)}{\Pr(C)} = \frac{\Pr(ABC)}{\Pr(C)} + \frac{\Pr(AB^c C)}{\Pr(C)}$$

along with

$$\Pr(A \mid BC)\Pr(B \mid C) = \frac{\Pr(ABC)}{\Pr(BC)}\frac{\Pr(BC)}{\Pr(C)} = \frac{\Pr(ABC)}{\Pr(C)}$$

and, similarly, $\Pr(A \mid B^c C)\Pr(B^c \mid C) = \Pr(AB^c C)/\Pr(C)$. Multiplying (3.14) by $\Pr(C)$ yields

$$\Pr(AC) = \Pr(A \mid BC)\Pr(B \mid C)\Pr(C) + \Pr(A \mid B^c C)\Pr(B^c \mid C)\Pr(C)$$
$$= \Pr(A \mid BC)\Pr(C \mid B)\Pr(B) + \Pr(A \mid B^c C)\Pr(C \mid B^c)\Pr(B^c).$$

An 'application' which makes use of these results is given next.

⊖ **Example 3.17**   You're in jail in the Middle Ages and your only chance to get out is by picking the key from a special box created by the jailer. The box has 12 different compartments, the contents of which you cannot see, and one of them contains the key. Six others each contain a tarantula spider, three others each contain a black widow spider and two are empty. Assume if you pick a spider, they always bite; and the probability that you die from a tarantula bite is 0.1, and that of the black widow is 0.5. Assume further that 'death probabilities' add so that, for example, if you pick two tarantulas, the probability of death is 0.2, etc.

Denote by D the event that you die, T that you get a tarantula, B that you get a black widow, E that the box is empty, and K that you get the key.

Assume first that one item is chosen randomly from the box. Then Pr (you get the key) is just $\Pr(K) = \frac{1}{12}$; Pr (you die) is, from the law of total probability (3.10) with $n = 2$,

$$\Pr(D) = \Pr(D \mid B)\Pr(B) + \Pr(D \mid T)\Pr(T) = \frac{1}{2}\frac{3}{12} + \frac{1}{10}\frac{6}{12} = \frac{7}{40} = 0.175;$$

and Pr (you had picked a black widow, given that you died) is

$$\Pr(B \mid D) = \frac{\Pr(D \mid B)\Pr(B)}{\Pr(D)} = \frac{\frac{1}{2}\frac{3}{12}}{\frac{1}{2}\frac{3}{12} + \frac{1}{10}\frac{6}{12}} = \frac{5}{7}.$$

Now assume (independent of the previous question) that two items are picked. Use the notation $\{T, B\}$ to denote that, from the two draws, one was T and one was B, but order is irrelevant. For example,

$$\Pr(\{T, B\}) = \binom{6}{1}\binom{3}{1}\binom{3}{0} \Big/ \binom{12}{2} = \frac{3}{11}$$

$$= \Pr(T)\Pr(B \mid T) + \Pr(B)\Pr(T \mid B) = \frac{6}{12}\frac{3}{11} + \frac{3}{12}\frac{6}{11}$$

and

$$\Pr(\{T, T\}) = \frac{6}{12}\frac{5}{11} = \binom{6}{2}\binom{6}{0} \Big/ \binom{12}{2}.$$

1. *Pr(you get the key)*. From the law of total probability (3.10) with $n = 2$ and defining K_i to be the event that the ith draw is the key, $i = 1, 2$,

$$\Pr(K) = \Pr(K_1 \cup K_2)$$
$$= \Pr(K_1) + \Pr(K_2) - \Pr(K_1 K_2)$$

$$= \Pr(K_1) + \Pr(K_2 \mid K_1)\Pr(K_1) + \Pr\left(K_2 \mid K_1^c\right)\Pr\left(K_1^c\right) - \Pr(K_1 K_2)$$

$$= \frac{1}{12} + 0 + \frac{1}{11}\frac{11}{12} - 0 = \frac{1}{6}.$$

Alternatively, the answer is obtained directly from the combinatoric

$$\Pr(K) = \frac{\binom{11}{1}\binom{1}{1}}{\binom{12}{2}} = \frac{1}{6}.$$

2. *Pr(you die)*. Again from the law of total probability, either you get two tarantulas, two black widows, one tarantula (and one of the three 'safe' boxes, empty or key), or one black widow (and one of the three safe boxes). Let the event $[E, K]$ mean $E \cup K$, the (disjoint) union of E and K. Then

$$\Pr(D) = \Pr(D \mid \{T, T\})\Pr(\{T, T\})$$
$$+ \Pr(D \mid \{B, B\})\Pr(\{B, B\})$$
$$+ \Pr(D \mid \{T, B\})\Pr(\{T, B\})$$
$$+ \Pr(D \mid \{T, [E, K]\})\Pr(\{T, [E, K]\})$$
$$+ \Pr(D \mid \{B, [E, K]\})\Pr(\{B, [E, K]\})$$
$$= \binom{12}{2}^{-1} \left\{ \begin{array}{c} (2 \cdot 0.1)\binom{6}{2} + (2 \cdot 0.5)\binom{3}{2} + (0.1 + 0.5)\binom{6}{1}\binom{3}{1} \\ +0.1\binom{6}{1}\binom{3}{1} + 0.5\binom{3}{1}\binom{3}{1} \end{array} \right\}$$
$$= 0.35.$$

3. *Pr(you get the key, but die as well)*. This is $\Pr(DK)$, which can be written as

$$\Pr(DK) = \Pr(D \mid T)\Pr(\{T, K\}) + \Pr(D \mid B)\Pr(\{B, K\})$$
$$= 0.1\frac{\binom{6}{1}\binom{1}{1}}{\binom{12}{2}} + 0.5\frac{\binom{3}{1}\binom{1}{1}}{\binom{12}{2}} = 0.03\overline{18},$$

or as $\Pr(K)\Pr(D \mid K)$, with

$$\Pr(D \mid K) = \Pr(D \mid TK)\Pr(T \mid K) + \Pr(D \mid BK)\Pr(B \mid K)$$

from (3.14). Note here that the sample space is partitioned into three pieces: tarantula (T), black widow (B) and neither T nor B. However, the occurrence of D implies that either T or B must have occurred, so that events T and B partition the space. Event $T \mid K$ means 'pick a tarantula (on the second draw), given that (on the first draw) the key was picked', so that $\Pr(T \mid K) = \binom{6}{1}/\binom{11}{1} = \frac{6}{11}$. Likewise, $\Pr(B \mid K) = \frac{3}{11}$, so that $\Pr(D \mid K) = 0.1 \cdot \frac{6}{11} + 0.5 \cdot \frac{3}{11} = 0.19\overline{09}$ and $\Pr(DK) = (1/6) \cdot 0.19\overline{09} = 0.03\overline{18}$.

4. *Pr(you got the key, given that you died).*

$$Pr(K \mid D) = \frac{Pr(DK)}{Pr(D)} = 0.\overline{09}.$$ ∎

3.2.3 Statistical paradoxes and fallacies

Only two things are infinite, the universe and human stupidity, and I'm not sure about the former. (Albert Einstein)

Some scientists claim that hydrogen, because it is so plentiful, is the basic building block of the universe. I dispute that. I say that there is more stupidity than hydrogen, and that is the basic building block of the universe. (Frank Zappa)

Recall Bernstein's paradox regarding pairwise and mutual independence. It is referred to as a paradox not because there is anything wrong with the definitions or mathematics behind it, but rather because it defies (initially at least) common sense. This brief section serves to mention another well-known statistical paradox and bring to the readers' attention the existence of many such cases.

For events A, B and C, it would seem intuitive that, if

$$Pr(A \mid BC) \geq Pr\left(A \mid B^c C\right) \quad \text{and} \quad Pr\left(A \mid BC^c\right) \geq Pr\left(A \mid B^c C^c\right),$$

then

$$Pr(A \mid B) \geq Pr\left(A \mid B^c\right).$$

That this is *not* the case is referred to as *Simpson's paradox* (Simpson, 1951) though, as noted in Grimmett and Stirzaker (2001b, p. 19), the result was remarked by Yule in 1903. In addition to those references, see Dudewicz and Mishra (1988, pp. 55–57), Aitkin (1998), Lloyd (1999, Section 3.6), Simonoff (2003, Section 8.1.1), Rinott and Tam (2003), and the references therein for further details, as well as Beck-Bornholdt and Dubben (1997, pp. 181–5) for a very detailed illustration (in German). A nice discussion of the paradox in the context of the implicit function theorem of multivariate calculus is provided by Simon and Blume (1994, Section 15.6).

While it is easy to construct artificial data sets for which the paradox holds, it actually does not arise all that often with real data. Morrell (1999), however, demonstrates the paradox using a data set based on growth rates of children in South Africa.

Simpson's paradox is also discussed in Székely (1986, p. 58) and Romano and Siegel (1986, p. 11), which are compilations of many interesting (and accessible) paradoxes in probability and statistics. A collection of mathematically more advanced counterexamples in probability theory is given in Stoyanov (1987).

3.3 The problem of the points

Biographical history, as taught in our public schools, is still largely a history of boneheads: ridiculous kings and queens, paranoid political leaders, compulsive voyagers, ignorant generals – the flotsam and jetsam of historical currents. The men who radically altered history, the great scientists and mathematicians, are seldom mentioned, if at all. (Martin Gardner)

Reproduced by permission of Mc Graw Hill Inc.

Let two people, say A and B, repeatedly play a game, where each game is independent and $p = \text{Pr}(A \text{ wins}) = 1 - \text{Pr}(B \text{ wins})$. The probability that A wins n times before B wins m times, say $\text{Pr}(A_{n,m})$, is desired. If the contest is prematurely ended without a winner, knowledge of $\text{Pr}(A_{n,m})$ can be used to appropriately 'split the winnings'.

The *problem of the points* was originally posed in the late 15th century by the mathematician and tutor to Leonardo da Vinci, Luca Paccioli, but is often associated with the French nobleman and gambler, Antoine Gombauld, the Chevalier de Méré, who challenged his friend and Parisian mathematician Blaise Pascal in 1654 to solve it. Pascal's ensuing correspondence with Pierre de Fermat and eventual solution of the problem is often considered to mark the beginning of probability theory. As Hacking writes regarding their correspondence:

Poisson is perfectly right to attribute the foundation of the probability calculus to these letters, not because the problems were new, nor because other problems had not been solved by earlier generations, but because here we have a completely new standard of excellence for probability calculations.

(Ian Hacking, 1975, p. 60)

Reproduced by permission of Cambridge University Press

3.3.1 Three solutions

We now demonstrate the three solutions which were developed by Pascal and Fermat and derive some related results along the way.

1. Observe that there can be at most $n + m - 1$ rounds and that the occurrence of event $A_{n,m}$ implies that A wins the last round played. If A wins within n rounds, then A must win n in a row, with probability p^n. If A wins in $n + 1$ rounds, then A wins the last one and, in the first n rounds, A wins $n - 1$ and B wins one time. This occurs with probability $np^n(1 - p)$, because there are $\binom{n}{1}$ ways that B can win exactly one of the first n rounds. Similarly, A wins with probability $\binom{n+1}{2}p^n(1-p)^2$ if there are $n + 2$ rounds, etc., and A wins with probability

$$\binom{n + m - 2}{m - 1} p^n (1 - p)^{m-1} \tag{3.15}$$

if all $n + m - 1$ rounds are played. Summing the probabilities of these disjoint events,

$$\Pr\left(A_{n,m}\right) = \sum_{i=0}^{m-1} \binom{n+i-1}{i} p^n \left(1 - p\right)^i . \tag{3.16}$$

Using substitution $j = n + i$ and recalling that $\binom{n+i-1}{n-1} = \binom{n+i-1}{i}$, (3.16) can also be written as

$$\Pr\left(A_{n,m}\right) = \sum_{j=n}^{n+m-1} \binom{j-1}{n-1} p^n \left(1 - p\right)^{j-n} . \tag{3.17}$$

The case in which $n = m$ is of natural interest, and is developed further in the next example and Problem 3.15.

⊖ ***Example 3.18***   Intuitively, if $n = m$ and $p = 0.5$, then $\Pr\left(A_{n,n}\right)$ must be 0.5. Using $p = 1/2$ in (3.16), this implies the combinatoric identity

$$\frac{1}{2} = \sum_{i=0}^{n-1} \binom{n+i-1}{i} \left(\frac{1}{2}\right)^{n+i} ,$$

which was also given in (1.14), where it was proven via induction.   ■

2. This method of solution hinges on the fact that the event of interest is the same as when A wins *at least* n times in the first $n + m - 1$ rounds, even if all $n + m - 1$ are not played out. (To see this, note that B cannot win if this is the case.) Because the probability of A winning k times in $n + m - 1$ trials is $\binom{n+m-1}{k} p^k \left(1 - p\right)^{n+m-1-k}$ (i.e. A's number of wins in all $n + m - 1$ trials is binomially distributed, see Chapter 4),

$$\Pr\left(A_{n,m}\right) = \sum_{k=n}^{n+m-1} \binom{n+m-1}{k} p^k \left(1 - p\right)^{n+m-1-k}$$

$$= \sum_{j=0}^{m-1} \binom{n+m-1}{n+j} p^{n+j} \left(1 - p\right)^{m-1-j} , \tag{3.18}$$

the latter obtained by substituting $j = k - n$.

Note that this gives rise to the combinatoric identity (14c), i.e.

$$\sum_{i=0}^{m-1} \binom{n+i-1}{i} p^n \left(1 - p\right)^i = \sum_{j=0}^{N-n} \binom{N}{n+j} p^{n+j} \left(1 - p\right)^{N-n-j} , \tag{3.19}$$

which was far more difficult to prove algebraically.

3. This solution is in terms of the difference equation

$$\Pr\left(A_{n,m}\right) = p \Pr\left(A_{n-1,m}\right) + (1-p) \Pr\left(A_{n,m-1}\right). \tag{3.20}$$

This follows because, if A wins the first trial (with probability p), then A only needs $n-1$ more wins to reach n wins before B reaches m wins, and similarly for B. Clearly, $\Pr\left(A_{n,0}\right) = 0$ and $\Pr\left(A_{0,m}\right) = 1$, so that one can recursively compute $\Pr\left(A_{n,m}\right)$ for any $n, m \in \mathbb{N}$.

⊖ **Example 3.19** In light of the last solution, it must be the case that both closed–form solutions satisfy (3.20). To verify this for (3.16), with $q = 1 - p$,

$$\Pr\left(A_{n,m}\right) = \sum_{i=0}^{m-1} \binom{n+i-1}{i} p^n q^i$$

$$= \sum_{i=0}^{m-1} \binom{n+i-2}{i} p^n q^i + \sum_{i=1}^{m-1} \binom{n+i-2}{i-1} p^n q^i$$

$$= \sum_{i=0}^{m-1} \binom{n+i-2}{i} p^n q^i + \sum_{i=0}^{m-2} \binom{n+i-1}{i} p^n q^{i+1}$$

$$= p \sum_{i=0}^{m-1} \binom{n+i-2}{i} p^{n-1} q^i + q \sum_{i=0}^{m-2} \binom{n+i-1}{i} p^n q^i$$

$$= p \Pr\left(A_{n-1,m}\right) + q \Pr\left(A_{n,m-1}\right).$$

For (3.18),

$$\Pr\left(A_{n,m}\right) = \sum_{i=0}^{m-1} \binom{n+m-1}{n+i} p^{n+i} q^{m-1-i}$$

$$= p \sum_{i=0}^{m-1} \binom{n+m-2}{n+i-1} p^{n+i-1} q^{m-1-i} + q \sum_{i=0}^{m-2} \binom{n+m-2}{n+i} p^{n+i} q^{m-2-i}$$

$$= p \Pr\left(A_{n-1,m}\right) + q \Pr\left(A_{n,m-1}\right).$$

In fact, any algebraic expression $W(n,m)$ which satisfies the recursion and for which $W(n,0) = 0$ and $W(0,m) = 1$ will be equivalent to (3.16) and (3.18). This follows because, for all n and m, any such formula when 'unraveled' produces one for $n = 0$ and zero for $m = 0$ and must give the same numerical value for $\Pr\left(A_{n,m}\right)$ (see Problems 3.12, 3.13 and 3.14). ∎

3.3.2 Further gambling problems

[Upon proving that the best betting strategy for 'gambler's ruin' was to bet all on the first trial.] It is true that a man who does this is a fool. I have only proved that a man who does anything else is an even bigger fool.

(Julian Lowell Coolidge)

There are numerous games of chance (and gambling schemes) which can be analyzed with the tools so far developed. Epstein (1977) is an excellent and entertaining source of scores of such games and includes both historical commentary and probabilistic analysis, while volume I of Feller (1957, 1968) is still considered to be the definitive textbook source of applied probability. We mention just two more, with their solutions. Both are similar to the problem of the points, in that we let two people, say A and B, repeatedly play a game, where each game is independent and $p = \Pr(A \text{ wins}) = 1 - \Pr(B \text{ wins})$.

The *gambler's ruin problem* goes as follows. If on any given round, A wins, she collects one dollar from B; if B wins, he gets one dollar from A. There is a total of T dollars at stake, and person A starts with i dollars and person B starts with $T - i$ dollars; play continues until someone loses all their money. The probability that A winds up with all the money (and B goes bankrupt, or is *ruined*) is, with $r = (1 - p)/p$,

$$\frac{1 - r^i}{1 - r^T}, \quad p \neq 1/2, \quad \text{and} \quad \frac{i}{T}, \quad p = 1/2, \tag{3.21}$$

and, similarly, the probability that A goes bankrupt is

$$\frac{r^i - r^T}{1 - r^T}, \quad p \neq 1/2, \quad \text{and} \quad \frac{T - i}{T}, \quad p = 1/2. \tag{3.22}$$

Notice that the probability of neither A nor B winning is, irrespective of T, zero. This is a classic example in the theory of stochastic processes (in particular, random walks) and is presented (in varying degrees of complexity) in virtually all introductory accounts of probability and stochastic processes. Problem 3.19 walks the reader through the derivation of (3.22). Johnson, Kotz and Kemp (1993, pp. 445–446) discuss the mass function and moments (see Chapter 4 below) of the number of games lost by the ruined gambler and its association to other random variables of interest in stochastic processes. See Edwards (1983) for a discussion of the history of the gambler's ruin problem (it too goes back to an exchange between Pascal and Fermat).

Another interesting scheme is where A wins the contest if she wins n *consecutive* rounds before B wins m *consecutive* rounds. The probability that A wins is, with $q = 1 - p$,

$$\frac{p^{n-1}(1 - q^m)}{p^{n-1} + q^{m-1} - p^{n-1}q^{m-1}}. \tag{3.23}$$

See, for example, Feller (1968, pp. 197, 210(2,3)), Ross (1988, pp. 87–89) or Grimmett and Stirzaker (2001a, p. 146(21)) for derivation. Figure 3.7 illustrates (3.23) for $n = m$.

3.3.3 Some historical references

It is the height of irony that the initiator of measure theory should have failed to see the natural application of his own creation to probability theory.

(Nicholas H. Bingham, 2000)

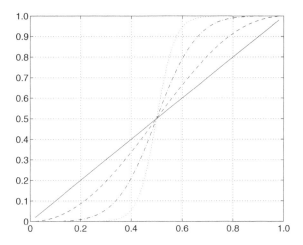

Figure 3.7 Probability (3.23) versus p for $n = m = 1$ (solid), $n = m = 2$ (dashed), $n = m = 4$ (dash-dot) and $n = m = 8$ (dotted)

In addition to the brief overview provided by Johnson, Kotz and Kemp (1993, p. 203), full-length accounts of the problem of the points and the historical development leading up to it can be found in David (1962), Hacking (1975), Bernstein (1996) and Franklin (2002). A short and light history of probability and randomness starting from antiquity is given by Bennett (1998).

A particularly detailed account of the development of probability theory in the 19th century can be found in Stigler (1986), while Bingham (2000) discusses the period from 1902 to 1933, in which measure theory became the foundation of probability theory. Finally, the collection of papers in Hendricks, Pedersen and Jørgensen (2001) are concerned with the philosophical and mathematical significance of the measure theoretic definition of probability and the role of probability theory in other scientific disciplines.

3.4 Problems

Obstacles are those frightening things that become visible when we take our eyes off our goals.   (Henry Ford)

All successful people understand that discomfort is a sign of progress, and are therefore not averse to seeking out discomfort or experiencing growing pains. Because the majority of people prefer to remain firmly within their comfort zone, they are not successful. It is this tolerance of discomfort, more than anything else, which separates the successful from the unsuccessful.

(Iain Abernethy, 2005, p. 50)

3.1. Recall Example 3.1, in which a fair, six-sided die is rolled t times and

$$p_t = \Pr\left(\text{exactly two sides do not appear}\right).$$

(a) ★ Calculate p_t using basic principles for $t = 5$ and $t = 6$.

(b) ★ ★ Derive the expression for p_t in (3.1) without using the De Moivre–Jordan theorem (use Poincaré's theorem instead).

3.2. A fair, six-sided die is rolled t times.

(a) ★ Let $t = 6n$ and compute $p_t(n)$, the probability that all six sides occur exactly n times. Approximate the result using Stirling's approximation $n! \approx (2\pi)^{1/2} n^{n+1/2} e^{-n}$ from (1.42) and compute the associated relative percentage error for $n = 1, 2, \ldots, 20$.

(b) ★ ★ ★ For general t, compute the probability p_t that all six sides occur at least *three* times. Construct a program which computes p_t and also simulates the result. Plot the exact p_t versus $t = 18, 20, \ldots, 60$ along with the simulated results for several values of t.

3.3. Consider our 12 travelers in Example 3.8. At breakfast the next morning, they seat themselves randomly around a round table with exactly 12 chairs.

(a) How many seating arrangements are possible?

(b) What is the probability that each married couple sits next to one another?

3.4. Consider ordering n people (distinguishable objects) around a table which can be arbitrarily 'rotated'.

(a) If $n > 4$ people are randomly seated around a table, what is the probability that persons A and B sit next to one another?

(b) If $n > 4$ people are randomly seated around a table, what is the probability that persons A and B sit next to one another, and persons C and D sit next to one another?

(c) If two boys (B) and two girls (G) randomly sit at a table (with four chairs), what is the probability that they alternate, i.e. as in Figure 3.8? What do you think of the 'solution': because the only possibilities are the two shown in Figure 3.8, the probability is 1/2?

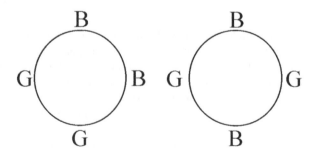

Figure 3.8 Possible arrangements of two boys (B) and two girls (G) around a table

(d) If n boys and n girls randomly sit at a table (with $2n$ chairs), what is the probability that they alternate?

3.5. ★ ★ ★ Recall Example 3.4 in which $2n$ people as n couples are randomly seated around a table. There are other interesting ways of determining $\Pr(K = k)$ without the explicit use of the De Moivre–Jordan theorem. As before, let the couples be numbered 1 to n and define events $C_k = \{\text{couple } k \text{ sits together}\}$, $k = 1, \dots, n$. (Contributed by Walther Paravicini)

(a) Derive $\Pr(K = n)$ and $\Pr(K = n - 1)$ directly from basic principles.

(b) Derive $\Pr(K = 0)$ and $\Pr(K = 1)$ directly using Poincaré's theorem.

(c) Let $C_{k,n}$ denote the number of ways in which k couples can be seated together when n couples are sitting around the table, so that

$$\Pr(K = k) = \frac{C_{k,n}}{(2n - 1)!}.$$

Instead of obtaining a closed-form expression for $C_{k,n}$, express $C_{k,n}$ recursively as

$$C_{k,n} = \alpha_1 C_{k-1,n-1} + \alpha_2 C_{k,n-1} + \alpha_3 C_{k+1,n-1} + \alpha_4 C_{k+2,n-1},$$

where $\alpha_1, \dots, \alpha_4$ are functions of k and n. (If we envisage a fixed-size round table with $2n$ chairs as opposed to one which 'expands as needed', then we define a couple sitting together to be when nobody else sits between them, even if they are separated by many empty chairs.) For instance, with $n - 1$ couples already seated such that $k + 2$ pairs are sitting together, in order for couple n to seat themselves such that only k pairs result is to have the 'husband' sit between one of the $k + 2$ couples and the 'wife' to sit between a different pair, i.e. $\alpha_4 = (k + 2)(k + 1)$. In addition, the 'starting values' $C_{0,1}$, $C_{1,1}$, $C_{0,2}$, $C_{1,2}$ and $C_{2,2}$ will be required, e.g. $C_{0,1} = 0$ and $C_{1,1} = 1$.
Finally, program the algorithm for computing $\Pr(K = k)$ via the $C_{k,n}$.

(d) Let E_k be the event that couples 1 to k, *and only* couples 1 to k, are seated together and let L_k be the event that *at least* couples 1 to k are seated together. Note that $\Pr(L_k)$ is given by (3.3). Explain the formula

$$\Pr(L_k) = \sum_{i=k}^{n} \binom{n - k}{i - k} \Pr(E_i), \tag{3.24}$$

express it as

$$\Pr(E_k) = \Pr(L_k) - \sum_{i=k+1}^{n} \binom{n - k}{i - k} \Pr(E_i), \tag{3.25}$$

and show how it may be used recursively to obtain $\Pr(E_k)$. With $\Pr(E_k)$ computable, find the relationship between $\Pr(K = k)$ and $\Pr(E_k)$.

(e) Argue that event L_k can be written as the disjoint union of E_k and

$$L_k \cap (C_{k+1} \cup \ldots \cup C_n) = (L_k \cap C_{k+1}) \cup \ldots \cup (L_k \cap C_n).$$

By applying Poincaré's theorem to the latter event, show

$$\Pr(E_k) = \sum_{i=k}^{n} (-1)^{i-k} \binom{n-k}{n-i} 2^i \frac{(2n-i-1)!}{(2n-1)!} \qquad (3.26)$$

and then derive (3.4).

(f) Use the recursive formula (3.25) to verify (3.26).

3.6. ★ ★ Recall Example 3.4, in which $2n$ individuals, who are paired as n couples, are randomly seated around a table.

(a) Use simulation to determine the probability that s of the $2n$ people have left and right table neighbors of the same gender. Plot $\Pr(S = s)$ for several values of n. What appears to be the limiting distribution?

(b) *Problème des ménages* This is the more popular (and more difficult) problem associated with sitting next to one's spouse at a round table. There are n married couples at a dinner party, and the women sit themselves in alternating seats around a table with $2n$ seats. Then, each man is randomly assigned to an open seat. The problème des ménages is concerned with the probability that no one sits next to his/her spouse. More generally, let W_n be the number of husbands who are seated next to their wives. Then

$$\Pr(W_n = w) = \sum_{r=w}^{n} (-1)^{r-w} \binom{r}{w} \frac{1}{r!} \frac{2n}{2n-r} \frac{\binom{2n-r}{r}}{\binom{n}{r}}, \qquad w = 0, 1, \ldots, n,$$

$$(3.27)$$

and zero otherwise. See Holst (1991), and the references therein for derivation of (3.27).[3]

You should: (i) program the exact probabilities (3.27), (ii) construct a program to simulate the result, (iii) plot both for several n and confirm that they are arbitrarily close for a large enough number of replications, and (iv) compare (3.4) and (3.27). As the probabilities for both must sum to one, how might you expect the distribution of the probabilities to change?

3.7. Regarding the 12 travelers from Example 3.8, assume that, on the plane ride home, they are the only 12 passengers, and the choice for dinner is either beef or fish. The kitchen has nine beef dishes, and nine fish dishes, and the probability that any passenger eats beef is 0.5.

(a) Explain why the probability that 10 people order beef is $\binom{12}{10} (0.5)^{10} (0.5)^2$.

---

[3] Holst (1991) also derives a recursive expression for $\Pr(W_n = w)$, moments of W_n and the asymptotic distribution as $n \to \infty$ (this being Poisson with $\lambda = 2$.) Holst also shows that $\mathrm{Bin}(2n, n^{-1})$ is a better approximation for the p.m.f. of W_n than $\mathrm{Poi}(2)$.

(b) What is the probability that at least one person will have to eat something different from what they ordered?

(c) Repeat part (b) if the probability that someone orders beef is only 0.1 (and, hence, the probability of ordering fish is 0.9).

3.8. ★ Your new neighbors have two children, and you happen to see one of them playing in their garden, and note that it is a boy. Assuming that girls and boys are equally likely to play in the garden, what is the probability that the other child is also a boy?

3.9. An urn initially contains s black balls and r red balls. Define $n = s + r$. A ball is randomly drawn and replaced, along with c balls of the same color. Define S_k to be the event that one obtains a black ball on the kth draw, and R_k to be the event that one obtains a red ball on the kth draw. This is referred to as Pólya's urn scheme, after the prolific George Pólya (1887–1985). A detailed discussion of this urn scheme is provided by Chung and AitSahlia (2003, Section 5.4).

(a) Compute $\Pr(S_1 \cap S_2)$.

(b) Compare $\Pr(S_2 \mid R_1)$ and $\Pr(S_1 \mid R_2)$.

(c) ★ Compare $\Pr(S_3 \mid R_1)$ and $\Pr(S_1 \mid R_3)$.

3.10. n cards are randomly chosen from a deck of 52 cards without replacement. Let P_n be the conditional probability that the first card is a heart, given that cards two to n are hearts. Compute P_n.

3.11. An object is hidden in one of three boxes. If the object is in box i, $i = 1, 2, 3$, and if box i is searched once, then the object is found with probability 0.7. Suppose that it is initially believed that the object is in box i with probability 0.2, 0.3 and 0.5, for $i = 1, 2, 3$. One always searches the box with the highest current probability; thus one would begin with box 3. If the first search fails, which box should be searched next? What do you think of the 'solution': multiplying the prior probability of box 3, i.e. 0.5, by the probability of not finding it if it were there, 0.3, gives 0.15. Comparing this value with the original probabilities of 0.2 and 0.3 (for boxes 1 and 2) implies that box 2 should be searched?

3.12. Consider the baseball World Series between teams A and B. The first team to win four games wins the series, and there are no ties. Assume independence of games and that the probability of team A winning against team B in any particular game is 0.6. Calculate the probability that A wins the series.

(Rohatgi, 1984, p. 41(6), 65(14), 363)

3.13. ★ In the problem of the points, solve Fermat's recursion for $n = 1$ and compare the answer to the closed-form solution (3.16).

3.14. ★ ★ Consider the problem of the points.

(a) Program Fermat's recursive solution (3.20) to the problem of the points.

(b) Program $\Pr(A \mid n, m, p)$ using one of the closed-form solutions.

(c) Now set $n = m = g$.

    i) Write a program which finds g^*, the smallest number of games g such that $\Pr(A \mid g, g, p) \geq \alpha$, for given $p > 0.5$ and $\alpha > 0.5$. What is the interpretation of g^*?

    ii) Write a program to plot vector of g^* values corresponding to a given vector of values p and cut-off α. If α is not specified, use default value of 0.90, and if a vector of α values is provided, construct a three-dimensions plot, for example, `plotgmin(0.65:0.01:0.9,0.8:0.05:0.9)`.

3.15. ★   For the problem of the points, let $P_n := \Pr(A_{n,n})$ and $r := p(1-p)$. Use (3.16) and (1.4) to show that

$$P_n = p + \left(p - \frac{1}{2}\right) \sum_{i=1}^{n-1} \binom{2i}{i} r^i. \tag{3.28}$$

This formula is intuitively appealing in that it directly shows that $P_n = 1/2$ for all n if $p = 1/2$ while it increases (decreases) with growing n if p is larger (smaller) than $1/2$. An alternative representation of (3.28) is easily seen to be

$$\frac{1}{2} + \frac{1}{2}(p - q) \sum_{i=0}^{n-1} \binom{2i}{i} (pq)^i,$$

where $q = 1 - p$, which is also an attractive form.

(Contributed by Markus Haas)

3.16. ★   In the problem of the points, let T be the total number of rounds played between A and B until the contest is decided. Show that $\Pr(T = t)$ is given by

$$\binom{t-1}{n-1} p^n (1-p)^{t-n} \mathbb{I}_{\{n,n+1,\ldots\}}(t) + \binom{t-1}{m-1} (1-p)^m p^{t-m} \mathbb{I}_{\{m,m+1,\ldots\}}(t),$$

for $\min(n, m) \leq t \leq n + m - 1$, and zero otherwise. Give an expression for $\Pr(T \mid A_{n,m})$, i.e. the conditional probability that $T = t$ rounds were played *given* that A has won, and the conditional probability that A wins *given* that $T = t$ rounds were played. Finally, construct a plot of $\Pr(T = t \mid A_{10,10})$ versus t for $p = 0.3, 0.5$ and 0.7, and a plot of $\Pr(A_{n,18} \mid T = t)$ versus t for $p = 0.4$ and $n = 13, 15$ and 17.

3.17. ★ ★   In the problem of the points, let $P_n := \Pr(A_{n,n})$, and consider the case when $p < 1/2$. Based on the nature of the game, $\lim_{n \to \infty} P_n = 0$. This can be used to compute the series $S = S(p) := \lim_{n \to \infty} S_n(p)$, where

$$S_n(p) := \sum_{i=1}^{n} \binom{2i}{i} \left[p(1-p)\right]^i. \tag{3.29}$$

Show that the sum in (3.29) is absolutely convergent for $p < 1/2$. (The quantity $S_n(p)$ is not convergent for $p = 1/2$, but a closed-form solution does exist for finite n. In the context of another well-known example in combinatoric probability, Problem 4.13 will show that

$$S_n(1/2) = \binom{n+\frac{1}{2}}{n} - 1 \approx -1 + \frac{2}{\sqrt{\pi}}\sqrt{n},$$

with the approximation holding for large n.)

Use (3.28) to show that

$$S = \sum_{i=1}^{\infty} 2^i p^i,$$

from which

$$\sum_{i=1}^{\infty} \binom{2i}{i} [p(1-p)]^i = \sum_{i=1}^{\infty} 2^i p^i \qquad (3.30)$$

follows. Use this to show the identity

$$2^y = \sum_{i=1}^{y} \binom{2i}{i}\binom{i}{y-i}(-1)^{i+y} = \sum_{i=\lceil y/2 \rceil}^{y} \binom{2i}{y}\binom{y}{i}(-1)^{i+y}, \qquad (3.31)$$

where $\lceil x \rceil = \text{ceil}(x)$. (Contributed by Markus Haas)

3.18. ★ Referring to (3.29), this exercise provides a direct proof that

$$S(p) = \sum_{n=1}^{\infty} \binom{2n}{n} [p(1-p)]^n = \frac{2p}{1-2p}, \qquad p < \frac{1}{2},$$

which is straightforward once the solution is known. Define

$$g(p) := S(p) + 1 = \sum_{n=0}^{\infty} \binom{2n}{n} [p(1-p)]^n$$

so that

$$g(p) = \frac{2p}{1-2p} + 1 = \frac{1}{1-2p}.$$

Next, let

$$h(p) := g(p)(1-2p) = 1,$$

so that the relation for $S(p)$ is equivalent to showing that $h(p) = 1$. Clearly

$$S(0) = \sum_{n=1}^{\infty} \binom{2n}{n} 0^n$$

is zero, so that

$$h(0) = g(p)(1 - 2p) = (S(0) + 1)(1) = 1.$$

Thus, if $h'(p) = g'(p)(1 - 2p) - 2g(p)$ can be shown to be zero, it follows that $h(p) = 1$ for all $p \in [0, 1/2)$ so that, in turn, the proof of the closed-form solution to $S(p)$ is established for all $p \in [0, 1/2)$. Show that $g'(p)(1 - 2p) = 2g(p)$.

(Contributed by Markus Haas)

3.19. ★ We wish to derive (3.22), the probability that B wins (i.e. that A goes bankrupt) in the gambler's ruin problem.

(a) First assume $p \neq 1/2$. Let A be the event that person A is ruined, let W be the event that A wins the first round played, let $q = 1 - p$, and define

$$s_i := \Pr_i(A) := \Pr(A \mid A \text{ starts with } i \text{ dollars and B starts}$$
$$\text{with } T - i \text{ dollars}).$$

Use the law of total probability to derive the difference equation

$$s_i = p s_{i+1} + q s_{i-1}, \qquad 1 \leq i \leq T.$$

With $r = q/p$ and $d_i = s_{i+1} - s_i$, show that $d_i = r^i d_0$. Then determine the boundary conditions s_0 and s_T and use that $s_0 - s_T = \sum_{i=0}^{T-1} d_i$ to derive an expression for d_0. Finally, write a similar telescoping expression for $s_j - s_T$, from which the answer follows.

(b) Derive the result for $p = 1/2$, either by similar steps as above, or with l'Hôpital's rule.

3.20. Consider Example 3.12 with d_0, the prior probability of having the disease, a function of ones age a, and quantity of cigarettes smoked daily, c. Assume that d_0 has been estimated to be given approximately by

$$d_0(a, c) = (c + 1) \frac{\tanh\left(\frac{a}{20} - 3\right) + 1}{100}.$$

Construct a graph with two plots: the top plot showing $\Pr(D \mid +)$ for $0 \leq a \leq 80$ and $c = 10$, while the bottom plot is $\Pr(D \mid +)$ for $a = 45$ and $0 \leq c \leq 40$.

3.21. ★ Consider f_n (2) given in (2.7) of Example 2.5, i.e.

$$f_n = q f_{n-1} + q p f_{n-2}, \quad f_1 = 0, \ f_2 = p^2.$$

This is a second-order homogeneous difference equation for which a closed-form solution can be easily found. Do so, and also check that $\sum_{i=1}^{\infty} f_i = 1$ and $\sum_{i=1}^{\infty} i f_i = p^{-1} + p^{-2}$ (use of a symbolic software package is recommended!). The latter is the expected value, discussed in Chapter 4.

PART I

DISCRETE RANDOM VARIABLES

4

Univariate random variables

> While writing my book [Stochastic Processes] I had an argument with Feller. He asserted that everyone said 'random variable' and I asserted that everyone said 'chance variable'. We obviously had to use the same name in our books, so we decided the issue by a stochastic procedure. That is, we tossed for it and he won.
>
> (J. Doob, quoted in Snell, 1997)
> *Reproduced by permission of the Institute of Mathematical Statistics*

The previous two chapters covered much ground in terms of introducing all the basic probabilistic concepts, methods for manipulating events and illustrating numerous examples. This was done without the notion of random variables (r.v.s), which we will see in this chapter is fundamental for expanding the realm of possible applications in probability and statistics. The reason is simple: a random variable X is a function from the sample space Ω to the real line. As such, the well-developed mathematical machinery of real analysis can be implemented for studying the properties of random variables.

This chapter also introduces the major discrete sampling schemes and their associated r.v.s and distributions. The primary reference on univariate discrete r.v.s is the encyclopedic work of Johnson, Kotz and Kemp (1993), while Zelterman (2004) gives a very readable and modern textbook presentation of univariate and multivariate discrete r.v.s and their use in the health sciences.

4.1 Definitions and properties

4.1.1 Basic definitions and properties

From the general probability space $\{\Omega, \mathcal{A}, \Pr(\cdot)\}$, the function X is defined to be a *univariate random variable* relative to the collection of measurable events $\mathcal{A}$ if and only if it is a function with domain Ω and range the real number line and such that, for every $x \in \mathbb{R}$, $\{\omega \in \Omega \mid X(\omega) \leq x\} \in \mathcal{A}$. That is, the random variable X, or r.v. X for short, is a mapping from the sample space Ω to the real line such that, for any

Fundamental Probability: A Computational Approach   M.S. Paolella
© 2006 John Wiley & Sons, Ltd

real number x, the set of points in Ω which get mapped to (or are associated with) real numbers less than or equal to x belong to $\mathcal{A}$, or are measurable. The important mapping $X(\omega) = \omega$ is referred to as the *identity function*.[1]

For a given Ω and r.v. X, just as not all subsets of Ω belong to the domain of $\Pr(\cdot)$, not all subsets of the real line can be assigned a probability. The class of subsets of $\mathbb{R}$ for which this can be done is, like $\mathcal{A}$, a σ-field but given the special name a *Borel σ-field*, denoted $\mathcal{B}$, the elements (sets) of which are referred to as *Borel sets*. As mentioned in Chapter 2, $\mathcal{B}$ is generated by the collection of intervals $(a, b], a, b \in \mathbb{R}$.

Because random variable X maps Ω to the real line, X is said to *induce* the probability space $\{\mathbb{R}, \mathcal{B}, \Pr(\cdot)\}$. Although $\Pr(\cdot)$ is used in both the original and induced probability spaces, it will be clear in each context which is meant because the r.v. will be explicitly mentioned as $\Pr(X \in A)$ for some Borel set A. As the set $A \in \mathcal{B} \in \mathbb{R}$ is typically a point or interval, the notation $\Pr(a_1 \leq X \leq a_2)$ or $\Pr(X = a)$, etc. will often be seen.

The most important function associated with an r.v. X and its induced probability space $\{\mathbb{R}, \mathcal{B}, \Pr(\cdot)\}$ is the *cumulative distribution function*, or *c.d.f.* usually denoted $F_X(\cdot)$ or, when clear from the context, just $F(\cdot)$, and defined to be $\Pr(X \leq x)$ for some point $x \in \mathbb{R}$. A c.d.f. F has the following properties:

(i) $0 \leq F(x) \leq 1$ for all $x \in \mathbb{R}$;

(ii) F is nondecreasing, i.e. if $x_1 < x_2$, then $F(x_1) \leq F(x_2)$;

(iii) F is *right continuous*, i.e. $\lim_{x \to x_0^+} F(x) = \lim_{x \downarrow x_0} F(x) = F(x_0)$ for all $x_0 \in \mathbb{R}$;

(iv) $\lim_{x \to -\infty} F(x) = 0$ and $\lim_{x \to \infty} F(x) = 1$.

To see these, first note that (i) follows from (ii) and (iv). Define event $E_a = \{X \leq a\}$ for finite a, so that $\Pr(E_a) = F_X(a)$. To prove (ii), take a and b such that $-\infty < a < b < \infty$ so that $E_a \subset E_b$ and, thus, $\Pr(E_a) < \Pr(E_b)$. Then, as $F(b) - F(a) = \Pr(E_b) - \Pr(E_a) > 0$, it follows that $F(x)$ is a nondecreasing function. To prove (iv), observe that $E_{-1} \supset E_{-2} \supset \cdots$ is a monotone decreasing sequence with $\lim_{a \to \infty} E_a = \emptyset$. Then, from (2.23),

$$F(-\infty) = \lim_{a \to \infty} F(-a) = \lim_{a \to \infty} \Pr(E_{-a}) = \Pr\left(\lim_{a \to \infty} E_{-a}\right) = \Pr(\emptyset) = 0.$$

Likewise, for the increasing sequence $E_1 \subset E_2 \subset \cdots$, $\lim_{a \to \infty} E_a = \mathbb{R}$ so that (2.23) implies

$$F(\infty) = \lim_{a \to \infty} F(a) = \lim_{a \to \infty} \Pr(E_a) = \Pr\left(\lim_{a \to \infty} E_a\right) = \Pr(\mathbb{R}) = 1.$$

Finally, property (iii) is just a result of how $F(\cdot)$ is defined, namely $\Pr(X \leq a)$ instead of $\Pr(X < a)$. More formally, let $x_1, x_2, \ldots$ be a *decreasing* sequence with

---

[1] It is commonly noted that X is neither random nor a variable, but a (deterministic) function of the sample space, the outcomes of which are, however, random. The misleading notation is, of course, so well established, but one should nevertheless keep the functional notion of a random variable in mind.

$\lim_{i\to\infty} x_i = x$, so that E_{x_i}, $i = 1, 2, \ldots$, is a monotone decreasing sequence with limit E_x. From (2.23),

$$\lim_{i\to\infty} F(x_i) = \lim_{i\to\infty} \Pr\left(E_{x_i}\right) = \Pr\left(\lim_{i\to\infty} E_{x_i}\right) = \Pr\left(E_x\right) = F(x),$$

i.e. F is right continuous.

⊚ **Example 4.1**   A fair, six-sided die is thrown once with the value of the top face being of interest, so that $\Omega = \{1, 2, \ldots, 6\}$ with each element being equally likely to occur. If the random variable X is this value, then $X(\omega) = \omega$ and F_X is given by the right continuous step function depicted below.

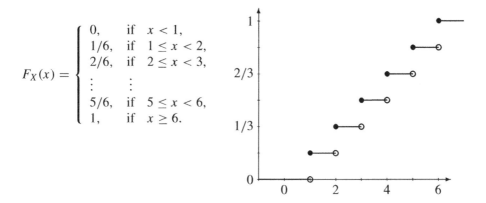

$$F_X(x) = \begin{cases} 0, & \text{if } x < 1, \\ 1/6, & \text{if } 1 \le x < 2, \\ 2/6, & \text{if } 2 \le x < 3, \\ \vdots & \vdots \\ 5/6, & \text{if } 5 \le x < 6, \\ 1, & \text{if } x \ge 6. \end{cases}$$

Note that lower case x refers to a particular point on the real line, while X is the random variable of interest; this is the usual notation used. Let x_i be a bounded decreasing sequence with $\lim_{i\to\infty} x_i = x_0 = 2$. Then $\lim_{i\to\infty} F(x_i) = 2/6 = F(2)$. It is easy to see that $\lim_{x\downarrow x_0} F(x) = F(x_0)$ for all $x_0 \in \mathbb{R}$, i.e. that F_X is right continuous. It is not, however, left continuous: if x_i is an increasing sequence with $\lim_{i\to\infty} x_i = 2$, then $\lim_{i\to\infty} F(x_i) = 1/6 \ne 2/6 = F(2)$.

As an aside, some thought reveals that F_X could also be expressed as

$$F_X(x) = \min\left[\max\left(0, \lfloor x \rfloor /6\right), 1\right],$$

where $\lfloor \cdot \rfloor$ is the *floor* function, which rounds its argument to the nearest integer towards minus infinity, e.g. $\lfloor 2.7 \rfloor = 2$ and $\lfloor -2.3 \rfloor = -3$. ∎

A random variable is said to be *discrete*, and has a *discrete distribution* if it takes on either a finite or countably infinite number of values. Thus, r.v. X in the previous example is discrete. The *probability mass function*, or *p.m.f.*, of discrete r.v. X is given by $f_X(x) = f(x) = \Pr(X = x)$. In the previous example, $f_X(x) = 1/6$, $x \in \Omega$, and zero otherwise. It is commonplace to speak of the distribution of a random variable, and be referring to the p.m.f. (or the p.d.f. in the continuous case, see below). For example, we might say 'consider the set of discrete distributions', or 'for those distributions which are symmetric about zero', or 'compute the distribution of the number of heads in 12 tosses of a fair coin', etc.

The *support* of a discrete r.v. X (or the support of the distribution of X) is the subset $x \in \mathbb{R}$ such that $f_X(x) > 0$, and is denoted as $\mathcal{S}_X$. Note that $f_X(x) = 0$ for any $x \notin \mathcal{S}$ and that the p.m.f. sums to unity:

$$\sum_{x \in \mathcal{S}} f_X(x) = 1.$$

If there exists a number $B > 0$ such that $|x| < B$ for all $x \in \mathcal{S}_X$, then X is said to have *bounded support*, otherwise it has unbounded support. If a discrete distribution (i.e. the p.m.f. of a discrete r.v.) has support only on a set of equidistant points, then it is more precise to refer to it as a *lattice distribution*. In fact, most of the discrete r.v.s one encounters (and all of them in this book) will have lattice distributions. Also, the support is virtually always a subset of $\mathbb{Z}$.

⊖ *Example 4.2*   An urn contains n balls, one of which is red, the others are white. A ball is drawn; if it is white, a red ball is replaced back into the urn; if it is red, sampling stops. If Y is the number of draws, then

$$f_Y(y) = \Pr(Y = y) = \frac{(n-1)(n-2)\cdots(n-y+1)\, y}{n^y} = \frac{(n-1)!\, y}{(n-y)!\, n^y},$$

for $y = 1, 2, \ldots, n$, and zero otherwise. Thus, $\mathcal{S} = \{1, 2, \ldots, n\}$, which depends on the constant n. That $\sum_{y=1}^{n} f_Y(y) = 1$ is not immediately obvious and is shown in Problem 1.9. Random variable Y is said to follow *Naor's distribution* – see Johnson, Kotz and Kemp (1993, pp. 447–448) for references, extensions and applications.   ∎

If the c.d.f. of r.v. X is absolutely continuous and differentiable, i.e. there exists a function f such that, for all $x \in \mathbb{R}$,

$$F(x) = \int_{-\infty}^{x} f(t)\, \mathrm{d}t \qquad \text{and} \qquad f(x) = \frac{\mathrm{d}}{\mathrm{d}x} F(x) = \frac{\mathrm{d}}{\mathrm{d}x} \int_{-\infty}^{x} f(t)\, \mathrm{d}t, \qquad (4.1)$$

then X is a *continuous random variable* and function f is denoted a *probability density function*, or *p.d.f.*, of X. Function f is not unique, although, for our purposes, there will always exist a p.d.f. which is 'most simple' and will be referred to as 'the' p.d.f. of X. For continuous r.v.s, the support can also be taken to be the subset $x \in \mathbb{R}$ such that $f_X(x) > 0$, though it is more precise to define $\mathcal{S}$ to be the smallest closed subset of $\mathbb{R}$ such that $\Pr(\mathcal{S}) = 1$. As in the discrete case, if there exists a number $B > 0$ such that $|x| < B$ for all $x \in \mathcal{S}_X$, then X has bounded support, otherwise the support is unbounded.

Remark: There are some contexts for which a random variable is neither discrete nor continuous, but rather a mix of the two. They are such that their c.d.f.s have a countable number of discrete 'jumps', but which are otherwise continuous, and are referred to as *mixed* r.v.s. See, for example, Lukacs (1970, Section 1.1) for a rigorous account, and Bean (2001, Section 4.1.6) for a practical introduction.   ∎

4.1.2 Further definitions and properties

President Dwight Eisenhower expressed astonishment and alarm on discovering that fully half of all Americans had below average intelligence.

The *median* of a random variable X, denoted med(X), is any value $m \in \mathbb{R}$ such that the two conditions

$$\Pr(X \leq m) \geq \frac{1}{2} \quad \text{and} \quad \Pr(X \geq m) \geq \frac{1}{2} \tag{4.2}$$

are satisfied. If X is a continuous r.v., then the median is unique and given by $F_X(m) = 0.5$.

Example 4.3 (Example 4.1 cont.) For r.v. X, every element $m \in [3, 4]$ is a median. To see this, note that $\Pr(X \leq m) = \Pr(X \in \{1, 2, 3\}) = 1/2$ and $\Pr(X \geq m) = \Pr(X \in \{4, 5, 6\}) = 1/2$. From the definition, it follows that m is a median of X.

Now consider an experiment with outcomes $\{1, 2, 3, 4, 5\}$ with each number having probability $1/5$. Let Y denote the random variable connected with this experiment. Now, simple calculations yield that 3 is the unique median of Y. ∎

Recall the definition of the indicator function from (1.15) or from Table B.6.

Example 4.4 The random variable X used in Examples 4.1 and 4.3 is a special case of what is referred to as the *(standard) discrete uniform*, written $X \sim \text{DUnif}(\theta)$, with mass function

$$f_X(x) = \theta^{-1} \mathbb{I}_{\{1,2,\ldots,\theta\}}(x) \tag{4.3}$$

and c.d.f.

$$F_X(t) = \min\left[\max\left(0, \frac{\lfloor t \rfloor}{\theta}\right), 1\right] = \begin{cases} 0, & \text{if } t < 1, \\ \lfloor t \rfloor / \theta, & \text{if } 1 \leq t \leq \theta, \\ 1, & \text{if } t > \theta. \end{cases} \tag{4.4}$$

If the desired support is instead given by $\{\theta_1, \theta_1 + 1, \ldots, \theta_2\}$ for $\theta_1, \theta_2 \in \mathbb{N}$, $\theta_1 < \theta_2$, then we say X is a *(shifted) discrete uniform* r.v., denoted $\text{DUnif}(\theta_1, \theta_2)$, with $\theta_1, \theta_2 \in \mathbb{Z}$ such that $\theta_1 < \theta_2$. Generalizing the previous results,

$$f_X(x) = (\theta_2 - \theta_1 + 1)^{-1} \mathbb{I}_{\{\theta_1, \theta_1 + 1, \ldots, \theta_2\}}(x)$$

and

$$F_X(t) = \min\left[\max\left(0, \frac{\lfloor t \rfloor - \theta_1 + 1}{\theta_2 - \theta_1 + 1}\right), 1\right] = \begin{cases} 0, & \text{if } t < \theta_1, \\ \frac{\lfloor t \rfloor - \theta_1 + 1}{\theta_2 - \theta_1 + 1}, & \text{if } \theta_1 \leq t \leq \theta_2, \\ 1, & \text{if } t > \theta_2. \end{cases}$$

Let $m = (\theta_2 + \theta_1)/2$. As in Example 4.3, the median is m if $\theta_2 - \theta_1 + 1$ is odd, and the closed, unit-length interval centered at m if $\theta_2 - \theta_1 + 1 > 0$ is even. ■

The *quantile* ξ_p of continuous r.v. X is defined to be the value such that $F_X(\xi_p) = p$ for given $0 < p < 1$. The median is thus a special case of a quantile. For discrete r.v.s, a definition similar to (4.2) can be made.

⊛ **Example 4.5** The p.d.f. of the *(standard) uniform distribution* is

$$f_{\text{Unif}}(x) = \mathbb{I}_{(0,1)}(x),\tag{4.5}$$

from which it follows that the c.d.f. is

$$F_{\text{Unif}}(x) = \int_0^x \mathbb{I}_{(0,1)}(t)dt = x\mathbb{I}_{(0,1)}(x) + \mathbb{I}_{[1,\infty]}(x).$$

It should be clear that the median is 0.5, while, more generally, quantile ξ_p is just p, for $0 < p < 1$.

A possible interpretation of a uniform random variable, say X, is that it can take on any real value between zero and one, and each value in $(0, 1)$ is equally likely. As already touched upon at the end of Section 2.2.1, this is a delicate issue, because there are an infinite number of possible points, implying that the probability of any single one of them (or even a countably infinite subset of them!), is zero.

This interpretation can be 'operationalized' by arguing that, in the real world (as opposed to the artificial world of mathematical constructs), any phenomenon (that being an observable fact or event) can only be measured with finite precision so that, if X is a standard uniform random variable, then it can, realistically speaking, take on only a finite number of different values, equally spaced throughout $(0, 1)$, and these values are all equally likely. Thus, ultimately, X is a discrete uniform random variable as in Example 4.4 but, *for practical purposes*, when the number of values in the genuine support of X is large, we envisage f_X to be the idealized, mathematical limiting case (4.5). This also accords with the computer generation of uniform random variables, as was discussed just before and after Example 2.6, in the sense that the numbers generated have a finite precision and, thus, can take on only a finite number of values.

A better interpretation, from a mathematical viewpoint, is that, for any $k \in [0, 1)$ and $\epsilon \in (0, 1)$ such that $0 \leq k < k + \epsilon \leq 1$, the probability that $X \in (k, k + \epsilon)$ is ϵ.■

⊛ **Example 4.6** The p.d.f. of the *exponential distribution* is given by

$$f_{\text{Exp}}(x; \lambda) = \lambda \exp(-\lambda x)\,\mathbb{I}_{[0,\infty)}(x),\tag{4.6}$$

for a given value of $\lambda \in (0, \infty)$. The c.d.f. is

$$F_{\text{Exp}}(x; \lambda) = \int_0^x \lambda \exp(-\lambda t)\,dt = 1 - \exp(-\lambda x),$$

so that $p = F_X(\xi_p) = 1 - e^{-\lambda \xi_p}$ gives the closed-form solution for the quantile as $\xi_p = -\lambda^{-1} \ln(1 - p)$. The median is given by $x = \lambda^{-1} \ln 2$.

The exponential distribution arises often in the study of lifetimes of certain mechanical or electronic components. The difference between a uniform and exponential r.v. is the distribution of probability: values near zero are more likely than values further away from zero (how much depends on λ). Similar comments to those made about the uniform random variable apply: an exponentially distributed r.v. is an idealized fiction, as X will ultimately only take on one of a finite number of values. Another fictitious aspect is that the support is the entire real half line. This is a mathematical construct. With respect to the modeling of lifetimes, we all know that, in the real world, 'nothing lasts forever'. This fact need not deter us from using the exponential distribution to model the lifetime of a set of new pocket PCs, because models are *supposed to be* just approximations to reality. ∎

If there exists a p.d.f. f_X such that, for all a, $f_X(m+a) = f_X(m-a)$, then density $f_X(x)$ and random variable X are said to be *symmetric* about m. This is equivalent to the condition that $\Pr(X \le m-a) = \Pr(X \ge m+a)$ for all a. For example, a uniform random variable with p.d.f. (4.5) is symmetric about $1/2$, while an exponential r.v. with p.d.f. (4.6) is not symmetric (about any point) for any value of $\lambda \in (0, \infty)$.

A family of distributions indexed by parameter vector $\boldsymbol{\theta} = (\theta_1, \dots, \theta_k)$ belongs to the *exponential family* if it can be algebraically expressed as

$$f(x; \boldsymbol{\theta}) = a(\boldsymbol{\theta}) b(x) \exp\left[\sum_{i=1}^{k} c_i(\boldsymbol{\theta}) d_i(x)\right], \tag{4.7}$$

where $a(\boldsymbol{\theta}) \ge 0$ and $c_i(\boldsymbol{\theta})$ are real-valued functions of $\boldsymbol{\theta}$ but not x; and $b(x) \ge 0$ and $d_i(x)$ are real-valued functions of x but not $\boldsymbol{\theta}$. The exponential distribution in Example 4.6 is easily seen to be a member of the exponential family. Members of the exponential family enjoy certain properties which are valuable for statistical inference.

⊛ **Example 4.7** The p.d.f. of the *standard normal* distribution is

$$f_N(x; 0, 1) = (2\pi)^{-1/2} \exp\left(-\frac{x^2}{2}\right), \tag{4.8}$$

while the more general form, sometimes referred to as the *Gaussian* distribution, is given by

$$f_N(x; \mu, \sigma) = \frac{1}{\sqrt{2\pi}\sigma} \exp\left[-\frac{1}{2}\left(\frac{x-\mu}{\sigma}\right)^2\right] \tag{4.9}$$

$$= \underbrace{\frac{1}{\sqrt{2\pi}\sigma} \exp\left(-\frac{1}{2}\frac{\mu^2}{\sigma^2}\right)}_{a(\mu,\sigma)} \exp\left(\underbrace{-\frac{1}{2\sigma^2}}_{c_1(\mu,\sigma)}\underbrace{x^2}_{d_1(x)} + \underbrace{\frac{\mu}{\sigma^2}}_{c_2(\mu,\sigma)}\underbrace{x}_{d_2(x)}\right).$$

With $b(x) = 1$ we see that the normal family is a member of the exponential family. Also, $f_N(x; \mu, \sigma)$ is symmetric about μ, which is also its median. ∎

Occasionally, certain values or ranges of values of random variable X cannot be observed, and this gives rise to a *truncated distribution*. The density is normalized to sum to unity, i.e. if X is truncated on the left at a and on the right at b, $a < b$, then the density is given by

$$\frac{f_X(x)}{F_X(b) - F_X(a)} \mathbb{I}_{(a,b]}(x). \tag{4.10}$$

The reason for omitting the point a from the indicator function arises only in the discrete case. To illustrate this, let X be the r.v. in Example 4.1 with c.d.f. F_X and let Y be X truncated at $a = 2$ and $b = 3$. This appears to leave two equally likely possibilities in Ω, namely 2 and 3 but, according to (4.10) with $F_X(a) = 2/6$ and $F_X(b) = 3/6$,

$$f_Y(y) = \frac{1/6\, \mathbb{I}_{\{1,\dots,6\}}(y)}{3/6 - 2/6} \mathbb{I}_{(2,3]}(y) = \mathbb{I}_{\{3\}}(y),$$

which may not have been desired. If outcomes 2 and 3 should belong to the support of Y, then a should be chosen as, say, 2^-, meaning any value arbitrarily close to but smaller than 2. In this case, any $1 \le a < 2$ will suffice, so that $F_X(a) = 1/6$ and

$$f_Y(y) = \frac{1/6\, \mathbb{I}_{\{1,\dots,6\}}(y)}{3/6 - 1/6} \mathbb{I}_{(a,3]}(y) = \frac{1}{2} \mathbb{I}_{\{2,3\}}(y).$$

The first half of Table B.5 summarizes the notation we use for events and r.v.s, while Table B.6 contains the most important notation for quantities associated with r.v.s.

4.2   Discrete sampling schemes

Consider two sampling schemes whereby objects are consecutively and randomly drawn from a known population, either *with* or *without replacement*. In general, stating that sampling is conducted at random and with replacement means that probabilities associated with all possible outcomes remain the same across draws (or trials). Each object or element in the population can be classified as belonging to one of a finite number of distinguishable classes or groups and is otherwise indistinguishable. For example, an urn containing four red, five white and six blue marbles has 15 elements and three classes, within which the marbles are completely identical.

In all cases considered in this chapter, we assume that each element is equally likely to be drawn. More specifically, on the ith trial, each element is drawn with probability $1/T_i$, where T_i is the total number of elements remaining in the population. Sampling with replacement implies that $T_i = T_1$, i.e. T_i remains constant for all i, while sampling without replacement means $T_i = T_{i-1} - 1 = T_1 - i + 1$, $i = 2, 3, \dots$. That the elements are equally likely to be drawn is not a restriction of great interest: if, for example, a sampling scheme were desired in which $\Omega = \{\text{red, white, blue}\}$ and such that red is 1.4 times as likely as either white or blue, then an urn with 14 red, 10 white and 10 blue marbles could be used.[2]

---

[2] Notice that an infinite number of marbles might be necessary in order to accommodate the desired probability structure, e.g. if red should occur $\sqrt{2}$ times as often as white or blue, etc.

For both sampling schemes, either a fixed number of draws, n, is set in advance, or trials continue until given numbers of objects from each class are obtained. This gives rise to the four major sampling schemes. Of course, all kinds of sampling schemes can be devised and analyzed. For example, the method which gave rise to Naor's distribution in Example 4.2 is not one of the four major schemes.

In addition to developing the p.m.f.s of the four sampling schemes, we will also discuss how to *simulate* the scheme, i.e. how to (let the computer) generate realizations of random variables from the respective distributions. Tables C.2 and C.3 summarize the major properties of the discrete distributions we use in this and other chapters.

4.2.1 Bernoulli and binomial

A *Bernoulli*[3] random variable, say X, has support $\{0, 1\}$ and takes on the value one ('success') with probability p or zero ('failure') with probability $1 - p$:

$$\Pr(X = x) = \begin{cases} p, & \text{if } x = 1, \\ 1 - p, & \text{if } x = 0, \end{cases}$$

and zero otherwise. By using the indicator function, the p.m.f. can be written as

$$\Pr(X = x) = f_X(x) = p^x (1 - p)^{1-x} \, \mathbb{I}_{\{0,1\}}(x). \tag{4.11}$$

Because a Bernoulli random variable represents only a single draw, there is no need to specify whether trials are conducted with or without replacement. For notational convenience, the abbreviation $X \sim \text{Ber}(p)$ means 'the random variable X is Bernoulli distributed with parameter p'.

If an urn contains N white and M black marbles, and n of them are randomly withdrawn *with* replacement, then $X = $ *the number of white marbles drawn* is a random variable with a *binomial* distribution. Letting $p = N/(N + M)$ be the probability of drawing a white one (on any and all trials), we write $X \sim \text{Bin}(n, p)$ with p.m.f.

$$f_X(x) = f_{\text{Bin}}(x; n, p) = \binom{n}{x} p^x (1 - p)^{n-x} \, \mathbb{I}_{\{0,1,\dots,n\}}(x). \tag{4.12}$$

This follows from the independence of trials and noting that there are $\binom{n}{x}$ possible orderings with x white and $n - x$ black marbles.

Taking $a(p) = (1 - p)^n$, $b(x) = \binom{n}{x}\mathbb{I}_{\{0,1,\dots,n\}}(x)$, $c(p) = \ln\left(\frac{p}{1-p}\right)$ and $d(x) = x$ in (4.7) shows that $f_X(x)$ in (4.12) is a member of the exponential family when n is considered a known constant and $\theta = p$. If $n = 1$, then $X \sim \text{Ber}(p)$. The p.m.f. of

---

[3] After the Swiss mathematician Jacob Bernoulli (1654–1705), also known as Jacques or James. He is remembered for his masterpiece *Ars Conjectandi*, published posthumously in 1713 which, among other things, dealt with (what were subsequently called) Bernoulli trials and Bernoulli numbers, the latter first referred to as such by De Moivre and Euler.

$X \sim \text{Bin}(n, p)$ can be calculated either directly with (4.12), or by using the easily-verified recursion

$$\frac{f_X(x)}{f_X(x-1)} = \frac{p}{1-p} \frac{n-x+1}{x},$$

which is useful, for example, when computing the c.d.f. of X.

⊖ **Example 4.8** The situation described in the beginning of Chapter 0 regarding the dozen purchased eggs fulfills the requirements of Bernoulli trials, so that the probability of getting at least 10 intact eggs is, with $p = 5/6 = 1 - q$,

$$\binom{12}{10} p^{10} q^2 + \binom{12}{11} p^{11} q^1 + \binom{12}{12} p^{12} \approx 0.68. \qquad \blacksquare$$

⊙ **Example 4.9** Consider a tournament between A and B in which rounds, or games, are repeatedly played against one another. Assume the probability that, in a particular round, A wins is 0.3, that B wins is 0.2, and that a tie results is 0.5. If 10 rounds are played, the probability that there are exactly five ties can be computed by noting that the number of ties is binomial with $p = 0.5$ and $n = 10$, yielding

$$\binom{10}{5} 0.5^5 0.5^5 = \frac{252}{1024} \approx 0.246. \qquad \blacksquare$$

The following is a useful result when working with sampling schemes conducted with a fixed number of trials with replacement, but with possibly more than two possible outcomes. Let A_i be the event that outcome type i occurs on any particular trial, $i = 1, 2$. Then Problem 4.7 shows that, in a sequence of trials, the probability that A_1 occurs *before* A_2 occurs is given by

$$\boxed{\Pr(A_1 \text{ before } A_2) = \frac{\Pr(A_1)}{\Pr(A_1) + \Pr(A_2)}} \qquad (4.13)$$

If there are only two possible outcomes for each trial, then this reduces to just $\Pr(A_1)$.

⊙ **Example 4.10 (Example 4.9 cont.)** Persons A and B conduct their tournament in a 'sudden death' manner, i.e. they repeatedly play games against one another until one person wins a round, with the probability that A wins a particular round 0.3, that B wins is 0.2 and that a tie results is 0.5. The probability that A wins the tournament is then given by (4.13) as $0.3/(0.3 + 0.2) = 0.6$. What if, instead of sudden death, they play 20 rounds. What is the probability that A won seven rounds, given that 10 of 20 ended in a tie? Similarly, the answer follows from the binomial quantity

$$\binom{10}{7} (0.6)^7 (0.4)^3 \approx 0.215.$$

Alternatively, the conditional probability is given by

$$\frac{\Pr(\text{A wins seven} \cap \text{B wins three} \cap 10 \text{ ties})}{\Pr(10 \text{ ties in } 20 \text{ rounds})}.$$

The numerator is similar to binomial probabilities, but there are three possible outcomes on each trial instead of just two. Generalizing in an obvious way,

$$\frac{\text{Pr}\,(A \text{ wins seven} \cap B \text{ wins three} \cap 10 \text{ ties})}{\text{Pr}\,(10 \text{ ties in } 20 \text{ rounds})} = \frac{\binom{20}{7,3,10}(0.3)^7\,(0.2)^3\,(0.5)^{10}}{\binom{20}{10}(0.5)^{20}}$$

$$= \binom{10}{7}(0.6)^7\,(0.4)^3\,.$$

This is actually a special case of the *multinomial distribution* which will be detailed in Chapter 5. ∎

4.2.2 Hypergeometric

If an urn contains N white and M black balls, and n balls are randomly withdrawn *without* replacement, then $X =$ *the number of white balls drawn* is a random variable with a *hypergeometric* distribution; we write $X \sim \text{HGeo}\,(N, M, n)$. Notice that, on any trial i, $i = 1, \dots, n$, the probability of drawing a white ball depends on the outcomes of trials $1, 2, \dots, i - 1$.

We have already encountered several situations using hypergeometric random variables, without having called them that (see, in particular, Problems 2.11 and 2.12, and Example 3.5). Those derivations imply that

$$f_X\,(x) = f_{\text{HGeo}}\,(x; N, M, n) = \frac{\binom{N}{x}\binom{M}{n-x}}{\binom{N+M}{n}}\mathbb{I}_{\{\max(0,n-M),1,\dots,\min(n,N)\}}\,(x)\,. \quad (4.14)$$

Mass function (4.14) cannot be expressed as (4.7) and therefore does not belong to the exponential family. That $\sum_x f_X\,(x) = 1$ follows directly from (1.28). The range of x follows from the constraints in the two numerator combinatorics, i.e. $0 \le x \le N$ together with $0 \le n - x \le M \Leftrightarrow n - M \le x \le n$. These are very intuitive; if, say, $N = M = 5$ and $n = 6$, then x has to be at least one and can be at most five.

Using our vectorized function for the binomial coefficients from Program Listing 1.1, the entire mass function is easily computed with the code in Program Listing 4.1.

```
function [xvec,f] = hypergeoden(N,M,n)
xlo=max(0,n-M); xhi=min(n,N); xvec=xlo:xhi;
num1=c(N,xvec); num2=c(M,n-xvec); denom=c(N+M,n);
f=num1.*num2/denom;
```

Program Listing 4.1 Computes (4.14) over the whole support

Another way of computing the mass function is to use a recursion. With $x + 1$, the numerator of (4.14) is

$$\binom{N}{x+1}\binom{M}{n-x-1} = \frac{N!}{(x+1)!\,(N-x-1)!} \frac{M!}{(n-x-1)!\,(M-n+x+1)!}$$

$$= \frac{N!\,(N-x)}{(x+1)\,(x)!\,(N-x)!}$$

$$\times \frac{(n-x)\,M!}{(n-x)!\,(M-n+x+1)\,(M-n+x)!}$$

$$= \frac{(N-x)}{(x+1)} \frac{(n-x)}{(M-n+x+1)} \binom{N}{x}\binom{M}{n-x},$$

so that

$$f_X(x+1) = \frac{(N-x)}{(x+1)} \frac{(n-x)}{(M-n+x+1)} f_X(x). \qquad (4.15)$$

When the entire p.m.f. is to be computed, (4.15) involves less computation, and should thus be faster. However, for the reasons discussed in Problem 1.15, the vectorized method in Listing 4.1 will actually be faster in Matlab. Both methods, however, share the problem that, for large enough N, M and n, the finite precision calculations will break down. For example, some trial and error reveals that, with $N = 800$, $M = 700$ and $n = 274$, both methods return correct p.m.f. values (with the use of (4.15) taking about three times as long), while for $n = 275$, both methods fail. This can be remedied by doing all nine factorial calculations of (4.14) in terms of logs (via Matlab's gammaln function). The program in Program Listing 4.2 does this for the first p.m.f. value, and then calculates the rest via the recursion (4.15).

```
function [xvec,f] = hypergeoden(N,M,n)
xlo=max(0,n-M); xhi=min(n,N); xvec=xlo:xhi;
len=length(xvec); f=zeros(1,len);
%%% f(1)= c(N,xlo) * c(M,n-xlo) / c(N+M,n);
f(1)=exp( gammaln(N+1) - gammaln(xlo+1) - gammaln(N-xlo+1)  ...
     + gammaln(M+1) - gammaln(n-xlo+1) - gammaln(M-n+xlo+1) ...
      - gammaln(N+M+1) + gammaln(n+1) + gammaln(N+M-n+1) );
for x=xlo:(xhi-1) % now compute the pmf for x+1
  f(x-xlo+1+1) = (N-x)/(x+1) * (n-x)/(M-n+x+1) * f(x-xlo+1);
end
```

Program Listing 4.2 Computes (4.15) over the whole support

A different method of calculating the p.m.f. (using Chebyshev polynomials) is developed in Alvo and Cabilio (2000), where further references to the literature on the computation of the hypergeometric p.m.f. can be found.

4.2.3   Geometric and negative binomial

The random variable X follows a *geometric* distribution with parameter $p \in (0, 1)$ and
p.m.f.

$$f_X (x) = f_{Geo}(x; p) = p (1 - p)^x \, \mathbb{I}_{\{0,1,\dots\}} (x) \tag{4.16}$$

if it represents the number of 'failed' Bernoulli trials required until (and not including)
a 'success' is observed. This is denoted $X \sim \text{Geo} (p)$. A straightforward calculation
reveals that, with $q = 1 - p$,

$$F_X (x) = \sum_{k=0}^{x} f_X (k) = p \sum_{k=0}^{x} q^k = 1 - q^{x+1}. \tag{4.17}$$

Alternatively, X can be defined as the number of trials which are observed until (and
including) the first 'success' occurs, i.e.

$$f_X (x) = p (1 - p)^{x-1} \, \mathbb{I}_{\{1,2,\dots\}} (x) . \tag{4.18}$$

Recall Example 2.4 in which your colleague will keep having children until the first
son. The total number of children can be modeled with mass function (4.18) with, say,
$p = 0.5$, yielding, for example, a one in 16 chance that she will have more than four
children.

A generalization of the geometric distribution (4.16) is the *negative binomial*, which
represents the number of 'failures' observed until r 'successes' are observed. For
$X \sim \text{NBin} (r, p)$, the p.m.f. of X is given by

$$f_{NBin} (x; r, p) = \Pr (X = x) = \binom{r + x - 1}{x} p^r (1 - p)^x \, \mathbb{I}_{\{0,1,\dots\}} (x) \tag{4.19}$$

$$= \underbrace{p^r}_{a(p)} \underbrace{\binom{r + x - 1}{x} \mathbb{I}_{\{0,1,\dots\}} (x)}_{b(x)} \exp\{\underbrace{\log (1 - p)}_{c(p)} \underbrace{x}_{d(x)}\}$$

which is also a member of the exponential family when r is treated as a given constant
and $\theta = p$. Function $f_{NBin} (x; r, p)$ takes its form from the independence of trials and
noting that the $(r + x)$th trial must be a success and that $r - 1$ successes must occur
within the first $r + x - 1$ trials. That

$$\sum_{x=0}^{\infty} f_{NBin} (x; r, p) = 1$$

follows from (1.12). The name 'negative binomial' is used because of the similarity
of its p.m.f. to the binomial via relation (1.9).

Similarly, if we wish to model the total number of trials until r successes are
observed, then

$$f_X (x; r, p) = \binom{x - 1}{r - 1} p^r (1 - p)^{x-r} \, \mathbb{I}_{\{r,r+1,\dots\}} (x) , \tag{4.20}$$

which generalizes (4.18). The c.d.f. of a negative binomial r.v. is related to that of a binomial. Letting $X \sim \mathrm{NBin}\,(r, p)$ with p.d.f. (4.20) and $B \sim \mathrm{Bin}\,(n, p)$, note that

$$\Pr\,(B \geq r) = \Pr\,(X \leq n),$$

because $\{B \geq r\}$ is the event that r or more successes occur in n trials, while $\{X \leq n\}$ is the event that at most n trials will be needed to obtain the first r successes; these two are equivalent, provided that the success probability p is the same for both r.v.s.

⊖ **Example 4.11**   Recall the problem of the points from Section 3.3, in which event $A_{n,m}$ is 'A wins n rounds before B wins m rounds' and rounds are independent with p the probability that A wins any particular one. From the negative binomial distribution, the probability that exactly t trials are required until a total of n successes are observed is given by (4.20), i.e.

$$\binom{t-1}{n-1} p^n \,(1-p)^{t-n}\,.$$

Note that event $A_{n,m}$ is equivalent to the event 'nth success occurs *before* trial $m+n$' so that

$$\Pr\left(A_{n,m}\right) = \sum_{t=n}^{n+m-1} \binom{t-1}{n-1} p^n \,(1-p)^{t-n}\,,$$

as was given in (3.17).   ∎

⊖ **Example 4.12 (Example 4.10 cont.)**   Now assume that, instead of a fixed number of games, A and B play until one of them wins g times. What is $\Pr\,(A)$, the probability that A wins the tournament? We can ignore ties and take the probability that A wins a round to be $0.3/(0.3+0.2) = 0.6$. Then, from the problem of the points solution (3.17) and (3.18) with $p = 0.6$ and $n = m = g$,

$$\Pr\,(A) = \sum_{j=g}^{2g-1} \binom{j-1}{g-1} 0.6^g 0.4^{j-g} = \sum_{k=g}^{2g-1} \binom{2g-1}{k} 0.6^k 0.4^{2g-1-k}. \qquad (4.21)$$

Clearly, for $g = 1$, $\Pr\,(A) = 0.6$. For $g = 4$, $\Pr\,(A) = 0.710208$. Intuitively, we expect $\Pr\,(A)$ to be an increasing function of g; it is shown in Figure 4.1.   ∎

⊙ **Example 4.13 (Banach's matchbox problem)**   The mathematician Stefan Banach (1892–1945) kept two matchboxes, one in each pocket, each originally containing N matches. Whenever he wanted a match, he randomly chose between the boxes (with equal probability) and took one out. Upon discovering an empty box, what is the probability that the other box contains $K = k$ matches, $k = 0, 1, \ldots, N$ (see Feller, 1968, p. 166)?

Assume that he discovers the right-hand pocket to be empty (r.h.p.e.). Because trials were random, X, the number of matches that were drawn from the left, can be

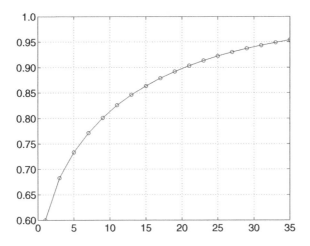

Figure 4.1 $\Pr(A)$ as given in (4.21) versus g

thought of as 'failures' from a negative binomial-type experiment with $p = 1/2$ (and support only $\{0, 1, \ldots, N\}$ instead of $\{0 \cup \mathbb{N}\}$), where sampling continues until $r = N + 1$ 'successes' (draws from the right pocket) occur. Thus $\Pr(K = k \cap \text{r.h.p.e.}) = \Pr(X = x \cap \text{r.h.p.e.})$, where $X = N - K$, $x = N - k$ and

$$\Pr(X = x \cap \text{r.h.p.e.}) = \binom{r + x - 1}{x} p^r (1 - p)^x = \binom{2N - k}{N} \left(\frac{1}{2}\right)^{2N+1-k}.$$

With $\Pr(X = x) = \Pr(X = x \cap \text{r.h.p.e.}) + \Pr(X = x \cap \text{l.h.p.e.})$ and from the symmetry of the problem,

$$f(k; N) = \Pr(K = k \mid N) = 2 \Pr(X = x \cap \text{r.h.p.e.}) = \binom{2N - k}{N} \left(\frac{1}{2}\right)^{2N-k}.$$

That the mass function sums to one was shown directly in Problem 1.11b. It can also be expressed as

$$\Pr(K = k) = \binom{2N - k}{N} \left(\frac{1}{2}\right)^{2N-k} \tag{4.22}$$

$$= 2 \frac{N - (k - 1)}{2N - (k - 1)} \binom{2N - (k - 1)}{N} \left(\frac{1}{2}\right)^{2N-(k-1)}$$

$$= \frac{N - (k - 1)}{N - \frac{k-1}{2}} \Pr(K = k - 1),$$

from which it is directly seen that $\Pr(K = 0) = \Pr(K = 1)$ and that $\Pr(K = k)$ decreases in k, $k \geq 1$. This also provides a numerically accurate way of calculating $\Pr(K = k)$ for large N (see Problem 4.12 for some extensions). ∎

An interesting generalization of the geometric r.v. which is somewhat more sophisticated than the negative binomial is presented in Problem 4.6.

4.2.4 Inverse hypergeometric

The only sampling scheme left to consider is that for which trials continue until k 'successes' are obtained, but sampling is without replacement. This is referred to as the *inverse hypergeometric* distribution and arises in many useful applications (Guenther, 1975, Johnson, Kotz and Kemp 1993, p. 239, and Terrell, 1999). If an urn contains w white and b black balls, the probability that a total of x balls needs to be drawn to get k white balls, $1 \leq k \leq w$, is given by

$$f_X(x; k, w, b) = \Pr(X = x) = \binom{x-1}{k-1} \frac{\binom{w+b-x}{w-k}}{\binom{w+b}{w}}$$

$$\times \mathbb{I}_{\{k, k+1, \dots, b+k\}}(x), \quad 1 \leq k \leq w, \tag{4.23}$$

denoted $X \sim \text{IHGeo}(k, w, b)$ and is not a member of the exponential family.

To see this for general k and x, where $1 \leq k \leq w$ and $k \leq x \leq b + k$, the xth draw must be the kth white ball, while the previous $x - 1$ trials must have produced $k - 1$ white balls and $x - k$ black balls (in any order). This latter requirement is hypergeometric; thus,

$$\Pr(X = x) = \frac{w - (k-1)}{w + b - (x-1)} \frac{\binom{w}{k-1}\binom{b}{x-k}}{\binom{w+b}{x-1}}$$

which, upon rewriting, is precisely (4.23). To verify (4.23) directly for $k = 1$, note that

$$\Pr(X = 1 \mid k = 1) = \frac{w}{w+b},$$

$$\Pr(X = 2 \mid k = 1) = \frac{b}{w+b} \frac{w}{w+b-1},$$

$$\Pr(X = 3 \mid k = 1) = \frac{b}{w+b} \frac{b-1}{w+b-1} \frac{w}{w+b-2}$$

and, in general, for $1 \leq x \leq b + 1$,

$$\Pr(X = x \mid k = 1) = \frac{b}{w+b} \frac{b-1}{w+b-1} \cdots \frac{b-(x-2)}{w+b-(x-2)} \frac{w}{w+b-(x-1)}$$

$$= w \frac{b_{[x-1]}}{(w+b)_{[x]}}.$$

This agrees with (4.23) because

$$w \frac{b_{[x-1]}}{(w+b)_{[x]}} = \frac{w!}{(w-1)!} \frac{b!}{[b-(x-1)]!} \frac{(w+b-x)!}{(w+b)!} = \frac{\binom{w+b-x}{w-1}}{\binom{w+b}{w}}.$$

The probability mass function (4.23) is easily programmed; Program Listing 4.3 shows this and some accompanying plot commands (most of which we have seen already in Program Listing 2.2). Graphical output is shown in Figure 4.2, which plots the p.m.f. for various parameter values.

```
function [x,pmf] = invhyp(k,w,b,doplot)
if nargin<4, doplot=0; end
x=k:(b+k); pmf = c(x-1,k-1) .* c(w+b-x,w-k) / c(w+b,w);
if doplot>0
  figure, h=bar(x,pmf);
  set(h,'facecolor',[0.9 0.7 0.5],'edgecolor',[1 0 0],'linewidth',1.2)
  set(gca,'fontsize',14), axis([k-1 b+k+1 0 1.05*max(pmf)])
  text(k,0.95*max(pmf),['k=',int2str(k)],'fontsize',16)
  text(k,0.87*max(pmf),['w=',int2str(w)],'fontsize',16)
  text(k,0.79*max(pmf),['b=',int2str(b)],'fontsize',16)
end
```

Program Listing 4.3 Computes the inverse hypergeometric p.m.f. and, if `doplot` is passed as a nonzero value, plots it

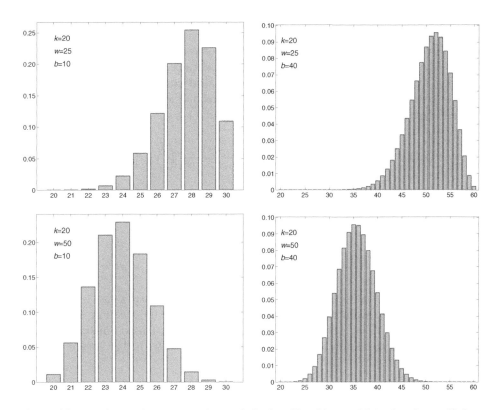

Figure 4.2 The inverse hypergeometric p.m.f. for $k = 20$; with $w = 25$ (top) and $w = 50$ (bottom), and $b = 10$ (left) and $b = 40$ (right)

4.2.5 Poisson approximations

Life is good for only two things, discovering mathematics and teaching mathematics.

(Siméon Poisson)

The last of the 'common' discrete distributions is the Poisson, after Siméon Poisson (1781–1840). For $\lambda > 0$, the mass function of $X \sim \text{Poi}(\lambda)$ is

$$f_X(x) = f_{\text{Poi}}(x; \lambda) = \Pr(X = x \mid \lambda) = \frac{e^{-\lambda}\lambda^x}{x!} \mathbb{I}_{\{0,1,\dots\}}(x) \qquad (4.24)$$

$$= \underbrace{e^{-\lambda}}_{a(\lambda)} \underbrace{\frac{1}{x!} \mathbb{I}_{\{0,1,\dots\}}(x)}_{b(x)} \exp[\underbrace{\log(\lambda)}_{c(\lambda)} \underbrace{x}_{d(x)}]$$

and is also a member of the exponential family. The recursion

$$\frac{f_X(x)}{f_X(x-1)} = \frac{\lambda}{x}$$

can also be used to compute the p.m.f.

There is no obvious sampling scheme which gives rise to this distribution. Instead it turns out that, under certain conditions, all four previous schemes asymptotically behave like a Poisson distribution. This has great value when the parameters of a sampling experiment are 'close enough' to the ideal conditions, in which case the Poisson distribution becomes an accurate approximation involving less computation. First consider the binomial. If $X \sim \text{Bin}(n, p)$ and $np = \lambda$, then

$$\Pr(X = x) = \binom{n}{x} p^x (1-p)^{n-x} = \frac{n!}{(n-x)!x!} \left(\frac{\lambda}{n}\right)^x \left(1 - \frac{\lambda}{n}\right)^{n-x}$$

$$= \underbrace{\frac{n(n-1)\cdots(n-x+1)}{n^x}}_{\to 1} \frac{\lambda^x}{x!} \underbrace{\left(1 - \frac{\lambda}{n}\right)^n}_{\to e^{-\lambda}} \underbrace{\left(1 - \frac{\lambda}{n}\right)^{-x}}_{\to 1},$$

as $n \to \infty$ and $p \to 0$, so that, for large n and small p, $\Pr(X = x) \approx e^{-\lambda}\lambda^x/x!$.

⊖ **Example 4.14** A new border patrol is set up to inspect 1000 cars a day. If it is known that the probability of finding a violator of some sort is 0.001, we can approximate the binomial r.v. X, the number of violators caught, as a Poisson distribution with $\lambda = np = 1$. The probability of finding at least three violators is then approximately

$$1 - \left(\frac{e^{-1}1^0}{0!} + \frac{e^{-1}1^1}{1!} + \frac{e^{-1}1^2}{2!}\right) = 1 - \frac{5}{2}e^{-1} \approx 0.0803014.$$

The exact answer is 0.0802093. ∎

⊙ *Example 4.15* An online bank has a team of workers to receive phone calls for conducting customer transactions. If the bank has a total of 100 000 customers and the probability of receiving an order in a particular one-minute interval is (taken to be) 0.005, how many workers should the bank have working so that a customer has no more than a 5 % chance of getting a busy signal?

The number of incoming calls, say X, is binomially distributed and might well be approximated by the Poisson distribution with $\lambda = 500$. We need to find the smallest value x such that $\Pr(X \geq x) \leq 0.05$ or $F_X(x) = \Pr(X \leq x) \geq 0.95$. This turns out to be $x = 537$ as $\Pr(X \leq 536) = 0.9479$ and $\Pr(X \leq 537) = 0.9524$.[4] Observe that the use of the Poisson approximation is not much of a numeric help in this case. However, $x \approx \lambda + 2\sqrt{\lambda} = 545$. This is not a coincidence and will be discussed later in conjunction with the central limit theorem. ∎

⊙ *Example 4.16* Imagine an area of a city consisting of a grid of 15×15 streets, thus forming 225 blocks. Within each block are $20 \times 20 = 400$ apartment households. Each household has a constant probability $p = 1/500$ of having a person with a particular rare disease D, and households are mutually independent. Thus, for any given block, the total number of households with D is a Bin(400, p) r.v., with mean 4/5. Because 400 is large and p is small, the Poisson distribution with $\lambda = 4/5$ should approximate this well.

The program in Program Listing 4.4 simulates each of the $(15 \times 20)^2$ households as a Bernoulli r.v., plots a point on a grid if the disease is present for a particular household, and then calculates and overlays onto the graph the number of households with the disease in each of the 225 blocks. This results in Figure 4.3. As the numbers should approximately be realizations of a Poisson r.v. with mean $\lambda = 4/5$, we can compare the empirically obtained percentages with those according to the Poisson distribution. They are given in the following table.

#	Poisson	Empirical
0	44.93	47.11
1	35.95	34.67
2	14.38	12.00
3	3.83	4.89
4	0.77	1.33

The agreement is clearly quite good. Over 80 % of the blocks have one or less cases of the disease, while 11 blocks have three cases, and three blocks have four cases. If one of these latter blocks happens to be downwind of a garbage recycling plant or a nuclear facility, some of the neighbors might feel inclined to believe that the relatively high presence of the disease is not by chance. ∎

Next, let $X \sim \text{NBin}(r, p)$ with mass function (4.19) and define $\lambda = r(1 - p)$. Problem 4.15 shows that $X \stackrel{\text{asy}}{\sim} \text{Poi}(\lambda)$ when $r \to \infty$ and $p \to 1$ such that λ stays constant. It should also be intuitively clear that both sampling schemes without replacement asymptotically behave like their with-replacement counterparts when the number of

---

[4] In Matlab, for example, this can be calculated as `binoinv(0.95,100000,0.005)`.

```
function leuk
n=300;      % n X n total households
p=0.002;    % probability of getting the disease
D=zeros(n,n); % boolean matrix of disease
for i=1:n,
  for j=1:n
    s= binornd(1,p,1,1); % s is Bernoulli(p)
    if s>0, plot(i,j,'r.'); hold on, D(i,j)=s; end
  end
  if mod(i,20)==0, i, end % watch progress and also activate
                      %  the keyboard so we can ctrl c if needed
  axis([0 n 0 n]), pause(0.01), refresh % Watch the graph develop!
end
r=20; % the 'resolution' of the grid. There are n/r blocks

% put axis grid exactly at the block points, and label with block #
set(gca,'XTick',0:r:n), set(gca,'YTick',0:r:n)
set(gca,'XTicklabel',(0:r:n)/r), set(gca,'YTicklabel',(0:r:n)/r)

% Now count the total number of hits per block
T=zeros(n/r,n/r);
for i=1:r:(n-r+1), ui=i+r-1;
  for j=1:r:(n-r+1), uj=j+r-1;
    s=0;
    for ii=i:ui, for jj=j:uj, s=s+D(ii,jj); end, end
    p1=(i-1)/r + 1; p2=(j-1)/r + 1; T(p1,p2)=s;

    % write the number onto the little square
    pos1=i+floor(r/2)-5; pos2=j+floor(r/2)-1;
    text(pos1,pos2,int2str(s),'fontsize',18,'Color',[0 1 0])
  end
end
hold off

% calculate Poisson and empirical probabilities
tt=tabulate(T(:)+1); emp = tt(:,3)
v=0:max(T(:)); v=v'; lam=r^2 * p;
pp = exp(-lam) * lam.^v ./ gamma(v+1); pois = 100 * pp;
[v pois emp] % print them out
```

Program Listing 4.4   Simulates the presence of a rare disease in many blocks of households

elements in the population tends to infinity such that the proportion of successes stays constant. A nice discussion of the Poisson approximation to the inverse hypergeometric is provided by Terrell (1999, Sections 6.8 and 11.10); see also Port (1994, p. 354(31.13)). There are also other situations for which the Poisson is the limiting distribution; Example 3.4 is one case, although it was only empirically observed and not algebraically proven.

There are a vast number of applications associated with the Poisson distribution; see Johnson, Kotz and Kemp (1993, Chapter 4) for historical overview and some extensions of the Poisson distribution. In Section 4.5, we introduce the Poisson process, which is related to the Poisson distribution.

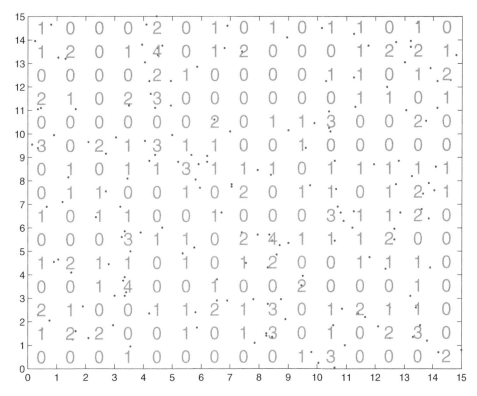

Figure 4.3 Dots indicate the presence of the disease in one of the 300^2 households; overlaid numbers are the total number of diseases within a 20×20 block

4.2.6 Occupancy distributions

> It is impossible to trap modern physics into predicting anything with perfect determinism because it deals with probabilities from the outset.
>
> (Sir Arthur Eddington)
> *Reproduced by permission of Simon and Schuster, NY*

Recall the occupancy model introduced in Section 2.1, whereby n indistinguishable balls are randomly distributed into r urns. This refers to the unordered with replacement sampling scheme. Two primary questions of interest are the number of urns with at least one ball and the number of balls required until each urn has at least one ball. The study of these problems dates back at least to De Moivre (*De Mensura Sortis*, 1712) and Laplace (*Theorie Analytique des Probabilités*, 1812). These are often now referred to as 'coupon-collecting' questions because of the application to collecting each element of a particular set. For instance, imagine that boxes of a certain breakfast cereal each contain one of r different prizes, and that (i) equal numbers of the r prizes were distributed out, (ii) randomly so, and (iii) the number of cereal boxes manufactured is so large that purchases (sampling) can be adequately assumed to be conducted with replacement. The two questions then become:

(1) If you buy n boxes of cereal, how many unique prizes can you expect?

(2) How many boxes should you buy in order to be 95 % sure of getting at least one of each of the r prizes?

These are now answered in turn. Table C.1 summarizes the results.

(1) Let the random variable X denote the number of unique prizes (or the number of nonempty urns) and $Z = r - X$ the number of missing prizes (or empty urns). It is clear that $f_X(x) = \Pr(X = x) = \Pr(Z = r - x)$; but, from (2.21) with $m = r - x$, this is just

$$f_X(x) = r^{-n} \sum_{i=r-x}^{r-1} (-1)^{i-(r-x)} \binom{i}{r-x} \binom{r}{i} (r-i)^n$$

$$= r^{-n} \binom{r}{x} \sum_{j=0}^{x-1} (-1)^j \binom{x}{j} (x-j)^n$$

using the easily-verified fact that

$$\binom{j}{r-x} \binom{r}{j} = \binom{r}{x} \binom{x}{r-j}$$

and setting $j = i - (r - x)$. Similarly, $\Pr(X \le x) = \Pr(Z \ge r - x)$ so that, from (2.22) with $m = r - x$, using $j = i - (r - x)$ and that

$$\binom{j+r-x-1}{j} \binom{r}{j+r-x} = \binom{r}{x} \binom{x}{j} \frac{r-x}{j+r-x},$$

the c.d.f. of X can be expressed as

$$F_X(x) = \Pr(X \le x) = r^{-n} \binom{r}{x} \sum_{j=0}^{x-1} (-1)^j \binom{x}{j} \frac{r-x}{j+r-x} (x-j)^n .$$

(2) Let the random variable Y denote the number of cereal boxes it is necessary to purchase in order to get at least one of each prize, i.e. such that there are no empty urns. Clearly, $\Pr(Y = n) = 0$ for $n \le r - 1$. For $n \ge r$, note that $\Pr(Y > n)$ is the probability that at least one urn is empty (after having placed n balls), i.e. from (2.19),

$$\Pr(Y > n) = r^{-n} \sum_{i=1}^{r-1} (-1)^{i+1} \binom{r}{i} (r-i)^n \qquad (4.25)$$

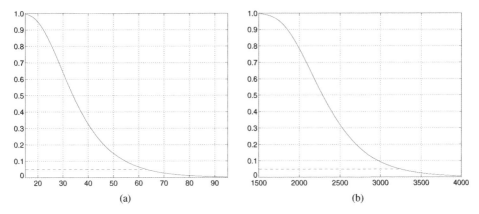

Figure 4.4 $\Pr(Y > n)$ versus n for (a) $r = 12$ and (b) $r = 365$; dashed line indicates a probability of 0.05, which corresponds to (a) $n = 63$ and (b) $n = 3234$, the minimum number of balls required such that all urns have at least one ball with at least 95 % probability

and

$$F_Y(n) = \Pr(Y \le n) = 1 - \Pr(Y > n) = \sum_{i=0}^{r-1}(-1)^i \binom{r}{i}\left(\frac{r-i}{r}\right)^n. \quad (4.26)$$

Figure 4.4 plots $\Pr(Y > n)$ for two values of r. From these, we can ascertain the minimum number of balls required such that all urns have at least one ball with at least, say, 95 % probability, as is indicated by the dashed lines. The c.d.f. values were computed with the program in Program Listing 4.5 below, by calling it with

```
nvec=1500:20:4000; cdf = occupancycdf(nvec,365,365);
plot(nvec,1-cdf), grid
```

```
function cdf = occupancycdf(nvec,r,k)
nl=length(nvec); cdf=zeros(nl,1); m=r-k+1;
for nloop=1:nl
  n=nvec(nloop); s=0; i=m:(r-1); sgn=(-1).^(i-m);
  c1=c(i-1,i-m); c2=c(r,i); frc=( (r-i)/r ).^n;
  s=sum(sgn.*c1.*c2.*frc); cdf(nloop)=1-s;
end
```

Program Listing 4.5 Computes the c.d.f. of Y_k; for $r > 30$, it begins to fail for small values of n, i.e. in the left tail of the distribution

The way we have it programmed, both (4.25) and (4.26) are prone to fail for small values of n. In particular, for $r = 30$, the program returns a negative c.d.f. value at $n = 30$ (but is positive for $r = n \le 29$). One might think that expressing the terms in (4.25) as

$$\binom{r}{i} \left(\frac{r-i}{r}\right)^n = \exp\left[g\,(r+1) - g\,(i+1) - g\,(r-i+1)\right.$$
$$\left. + n \ln (r-i) - n \ln (r)\right],$$

where $g\,(\cdot)$ is the log gamma function, would improve matters, but that is not where the problem lies. Using it does return a positive value for $r = n = 30$, but it too fails at $r = n \geq 31$. The problem is the magnitude of the

$$(-1)^i \binom{r}{i} \left(\frac{r-i}{r}\right)^n$$

terms, which can be very large when n is relatively small. Because Matlab uses finite precision arithmetic with about 15 significant digits, adding them results in round-off error, which becomes catastrophic as r grows and n is relatively close to r. Figure 4.5 plots these terms for $r = n = 100$. Adding a number with magnitude of at least 10^{11} to one with magnitude at most 10^{-7} will fail to maintain all the significant digits.[5]

As we see from Figure 4.4, this problem is fortunately not relevant for values of n large enough to be of any interest. If, for some reason, the c.d.f. (4.26) were desired for a problematic value of n, then one could (i) attempt to add the terms of the sum in a judicious way (note in Figure 4.5 that adjacent terms have a similar magnitude), or (ii) use a software package which supports variable precision arithmetic. Two more ways which will be detailed in Volume II include: (iii) numerically invert the characteristic function, and (iv) use the saddlepoint approximation.

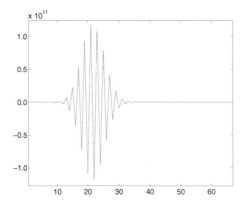

Figure 4.5 Terms $(-1)^i \binom{r}{i} \left(\frac{r-i}{r}\right)^n$ for $r = n = 100$ as a function of i

[5] As a concrete example, consider the sample variance, the formula for which you undoubtedly recall from an earlier statistics class. In Matlab, it is computed with the `var` function. Compare the calculation of the following three statements, which, in theory, should result in the same value, namely one: `var([1 2 3])`, `var([1 2 3]+12e15)` and `var([1 2 3]+12e16)`.

The mass function of Y is just

$$f_Y(n; r) = \Pr(Y = n) = \Pr(Y > n - 1) - \Pr(Y > n). \tag{4.27}$$

It can be computed by calling

```
nvec=1000:20:4000; r=365; cdfa = occupancycdf(nvec,r,r);
cdfb = occupancycdf(nvec-1,r,r); pmf = max(0,cdfa-cdfb);
```

Figure 4.6 plots (4.27) for $r = 365$ and $r = 1000$. The dashed line shows the normal p.d.f. with the same mean and variance as Y (see Section 4.4); we see it is not particularly good, even for large values of r, because of the large amount of asymmetry (right skewness, see Section 4.4.2).

Random variable Y can be generalized as follows. Let Y_k be the number of cereal boxes it is necessary to purchase in order to get at least one of k different prizes, $k = 2, \ldots, r$. (For $k = 1$, Y_1 is degenerate, with $Y_1 = 1$.) Then $\Pr(Y = n) = 0$ for $n \leq k - 1$ and, for $n \geq k$, $\Pr(Y_k > n)$ is the same as the probability that at least $r - k + 1$ of the r prizes have not yet been purchased (i.e. those 'cells' are still empty). Thus, from (2.22) with $m = r - k + 1$,

$$\Pr(Y_k > n) = r^{-n} \sum_{i=m}^{r-1} (-1)^{i-m} \binom{i-1}{i-m} \binom{r}{i} (r-i)^n, \qquad m = r - k + 1,$$

$$\tag{4.28}$$

for $k = 2, \ldots, r$. From this, the c.d.f. and p.m.f. can be computed. Note that, when $k = r$, this reduces to (4.25). A program to compute the c.d.f. of Y_k is given in Program Listing 4.5.

Remarks: **(a)** As with the sampling schemes introduced earlier, occupancy probabilities converge to a Poisson limit under certain conditions on the parameters, see, for example, Port (1994, p. 645(50.26)).

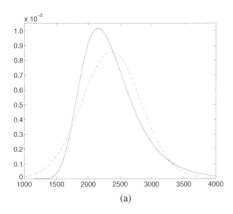

(a)

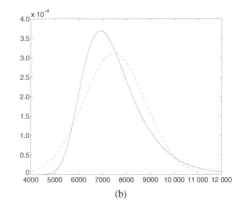

(b)

Figure 4.6 Exact mass function (4.27) (solid) and the normal approximation (dashed) for (a) $r = 365$ and (b) $r = 1000$

(b) The normal approximation to (4.27) shown in Figure 4.6 is quite poor because it cannot capture the inherent asymmetry in f_{Y_k}. However, Read (1998) has demonstrated that the *log–normal* density (Section 7.2) provides a very good approximation. (The aforementioned saddlepoint approximation, however, is superior.)
(c) With Stirling's approximation $n! \approx (2\pi)^{1/2} n^{n+1/2} e^{-n}$ from (1.42) applied to the binomial coefficient in (4.25) and simplifying,

$$\Pr(Y > n) \approx (2\pi)^{-1/2} r^m \sum_{j=1}^{r-1} (-1)^{j+1} j^{-(j+1/2)} (r-j)^{j-m}, \qquad (4.29)$$

where $m = r - n + 1/2$.
Figure 4.7 plots the relative percentage error, $100\,(\text{Approx} - \text{Exact})\,/\text{Exact}$, associated with use of (4.29) versus n for $r = 365$.
(d) Note that, for $n = r$,

$$\Pr(Y = r) = \Pr(Y > r - 1) - \Pr(Y > r)$$

$$= 1 - \Pr(Y > r)$$

$$= 1 - r^{-r} \sum_{i=1}^{r-1} (-1)^{i+1} \binom{r}{i} (r-i)^r,$$

while, from basic principles, $\Pr(Y = r) = r!/r^r$. Equating these gives the interesting combinatoric identity

$$r! = r^r - \sum_{i=1}^{r-1} (-1)^{i+1} \binom{r}{i} (r-i)^r. \qquad (4.30)$$

(e) Matters get more interesting and more difficult, but more realistic, when the urns are no longer equally likely to be chosen. This is addressed in Rosen (1970),

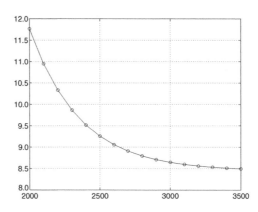

Figure 4.7 Relative percentage error incurred when using Stirling's approximation (4.29) as a function of n, for $r = 365$

Aldous (1989, pp. 108–110), Diaconis and Mosteller (1989), Levin (1992) and Nunnikhoven (1992); see also Ross (1997, pp. 123–124).

(f) There are many discussions of occupancy distributions and their myriad of applications, see Feller (1968), Johnson and Kotz (1977), Dawkins (1991), Johnson, Kotz and Kemp (1993, Section 10.4), Langford and Langford (2002), Scheutzow (2002), and the references therein. Via Poisson processes, Holst (1986) presents a unified approach to several questions occurring in the study of occupancy distributions including, for example, the number of cereal boxes required to get all r prizes, m times each. A generalization to coupon subsets is considered by Adler and Ross (2001) and the references therein. ∎

We end this section with a simple illustration which, like the derivation of the Poisson approximation to the binomial distribution, makes use of asymptotic arguments. In particular, assume as before that n indistinguishable balls are randomly distributed into r urns, *but such that all the different ways of arranging the balls are equally likely.* Let X_1 denote the number of balls in urn 1. From (2.1), there are

$$K_{n,r}^{(0)} = \binom{n+r-1}{n}$$

ways of arranging all the balls and, if x balls are in urn 1, $0 \le x \le n$, then there are

$$K_{n-x,r-1}^{(0)} = \binom{(n-x)+(r-1)-1}{n-x}$$

ways of arranging the remaining $n - x$ balls. Thus,

$$f_{X_1}(x) = \Pr(X_1 = x) = \frac{\binom{n-x+r-2}{n-x}}{\binom{n+r-1}{n}} \mathbb{I}_{\{0,1,\dots,n\}}(x).$$

Applying Stirling's approximation $n! \approx (2\pi)^{1/2} n^{n+1/2} e^{-n}$ from (1.42) to each factorial in f_{X_1} gives, after simplifying,

$$
\begin{aligned}
f_{X_1}(x) &= \frac{(n-x+r-2)! \; n! \, (r-1)!}{(n-x)! \, (r-2)! \, (n+r-1)!} \\
&\approx \frac{(n-x+r-2)^{(n-x+r-2)+1/2} \, (r-1)^{(r-1)+1/2}}{(n-x)^{(n-x)+1/2}} \frac{n^{n+1/2}}{(r-2)^{(r-2)+1/2} \, (n+r-1)^{(n+r-1)+1/2}}
\end{aligned}
$$

or, rearranging,

$$\frac{r-1}{n+r-1} \left(\frac{n-x}{n-x+r-2}\right)^x \left(\frac{r-1}{r-2}\right)^{r-3/2} \left(\frac{n-x+r-2}{n+r-1}\right)^{n+r-3/2} \left(\frac{n}{n-x}\right)^{n+1/2}.$$

Now consider taking the limit as both r and n go to infinity, but such that $n/r \to \lambda \in \mathbb{R}_{>0}$. This can be interpreted as follows. As the number of different prizes available in each cereal box (r) increases, so does the number of boxes you purchase (n), and you do so in such a way that the ratio n to r approaches a constant, nonzero value. Then, as

$$\lim_{\substack{r,n\to\infty \\ n/r\to\lambda}} \frac{r-1}{n+r-1} = \lim_{\substack{r,n\to\infty \\ n/r\to\lambda}} \frac{r/r-1/r}{n/r+r/r-1/r} = \frac{1}{1+\lambda},$$

$$\lim_{\substack{r,n\to\infty \\ n/r\to\lambda}} \left(\frac{n-x}{n-x+r-2}\right)^x = \lim_{\substack{r,n\to\infty \\ n/r\to\lambda}} \left(\frac{n/r-x/r}{n/r-x/r+r/r}\right)^x = \left(\frac{\lambda}{1+\lambda}\right)^x,$$

$$\lim_{\substack{r,n\to\infty \\ n/r\to\lambda}} \left(\frac{r-1}{r-2}\right)^{r-3/2} = \lim_{r\to\infty} \left(1+\frac{1}{r-2}\right)^r$$

$$\stackrel{u=r-2}{=} \lim_{u\to\infty} \left(1+\frac{1}{u}\right)^{u+2} = e,$$

$$\lim_{\substack{r,n\to\infty \\ n/r\to\lambda}} \left(\frac{n-x+r-2}{n+r-1}\right)^{n+r-3/2} \stackrel{s=n+r}{=} \lim_{s\to\infty} \left(\frac{s-x-2}{s-1}\right)^s$$

$$= \lim_{s\to\infty} \left[\frac{(s-1)-(x+1)}{(s-1)}\right]^s$$

$$\stackrel{m=s-1}{=} \lim_{m\to\infty} \left(1-\frac{x+1}{m}\right)^{m+1} = e^{-(x+1)},$$

and

$$\lim_{\substack{r,n\to\infty \\ n/r\to\lambda}} \left(\frac{n}{n-x}\right)^{n+1/2} = \lim_{n\to\infty} \left(1-\frac{x}{n}\right)^{-n} = e^x,$$

it follows that, with $p = 1/(1+\lambda)$,

$$\lim_{\substack{r,n\to\infty \\ n/r\to\lambda}} f_{X_1}(x) = \frac{1}{1+\lambda} \left(\frac{\lambda}{1+\lambda}\right)^x \mathbb{I}_{\{0,1,\ldots\}}(x) = p(1-p)^x \mathbb{I}_{\{0,1,\ldots\}}(x),$$

which is the same as (4.16), i.e. asymptotically, X_1 is geometrically distributed.

4.3 Transformations

There are many occasions for which interest centers less on the outcomes of a particular r.v., but rather on a certain function of it. Let X be a random variable and g a real-valued measurable function. If $Y = g(X)$ $(= g \circ X)$, then Y is itself a random variable. In the discrete case, matters are very straightforward. For some set C, $\Pr(Y \in C) = \Pr(X \in g^{-1}(C))$, where $g^{-1}(C) = \{x \in \mathcal{B} : g(x) \in C\}$. If C just contains one element y, then $\Pr(Y = c) = \Pr(X \in g^{-1}(\{y\})) = \Pr(X \in \{x \in \mathcal{B} : g(x) = y\})$. It is important to note that g^{-1} is not necessarily a function. As an illustration, if X has discrete support on -2, -1, 0, 1, 2 and $Y = g(x) = x^2$, then the image of g is $\{0, 1, 4\}$. For $C = \{y : 1 \leq y \leq 5\}$, $g^{-1}(C) = \{-1, 1, -2, 2\}$ and $\Pr(Y \in C) = \Pr(X = -1) + \Pr(X = 1) + \Pr(X = -2) + \Pr(X = 2)$. A simple example which we will make use of several times is given next.

Example 4.17 A manufacturer produces three types of solid cylindrical metal shafts of equal length ℓ and diameters $d_1 = 1$, $d_2 = 2$ and $d_3 = 3$cm. Let D be a r.v. denoting the diameter requested by a customer with

$$\Pr(D = 1) = 2 \times \Pr(D = 2) = 3 \times \Pr(D = 3),$$

i.e. $\Pr(1) = 6/11$, $\Pr(2) = 3/11$ and $\Pr(3) = 2/11$. For shipping reasons, the weight of the shafts are of primary interest which, for a single shaft, is given by $w = \ell k \pi d^2 / 4$ for constant material density $k > 0$. Then the probability mass function of the weight W is given by

w	$g^{-1}(w)$	$\Pr(W = w)$
$w_1 = \ell k \pi / 4$	1	6/11
$w_2 = \ell k \pi$	2	3/11
$w_3 = 9\ell k \pi / 4$	3	2/11

and zero otherwise. ∎

Methods for transformations of continuous r.v.s will be developed in Chapter 7.

4.4 Moments

I think of lotteries as a tax on the mathematically challenged.

(Roger Jones)

Among the most important characteristics of a random variable X are (i) its 'average behavior' or what one might 'expect' before a value of X is realized; (ii) the extent to which values of X are spread out or tend to vary around, say, its average value, and (iii) how much, and in what direction X deviates from symmetry, if at all. While several useful measures exist for quantifying these notions, the most popular are based on the *moments* of X. Table B.7 summarizes the various kinds of moments discussed in this section.

4.4.1 Expected value of *X*

For convenience, define the integral

$$\int_{-\infty}^{\infty} g(x) \, dF_X(x) := \begin{cases} \int_S g(x) f_X(x) \, dx, & \text{if } X \text{ is continuous,} \\ \sum_{i \in S} g(x_i) f_X(x_i), & \text{if } X \text{ is discrete,} \end{cases} \tag{4.31}$$

(assuming the right-hand side exists), where S is the support of X and $g(x)$ is a real-valued function. The *expected value* of random variable X is defined by[6] $\int_S x \, dF_X(x)$ and denoted by $\mathbb{E}[X]$ or μ, i.e.

$$\mu = \mathbb{E}[X] := \int_S x \, dF_X(x) \,.$$

⊙ ***Example 4.18*** Let X_0 and X_1 be geometric random variables with densities given in (4.16) and (4.18), respectively. A clever way of computing the expected value is,[7] with $q = 1 - p$,

$$\mathbb{E}[X_0] = \sum_{x=0}^{\infty} x f_X(x) = p \sum_{x=1}^{\infty} x q^x = pq \sum_{x=1}^{\infty} x q^{x-1} = pq \frac{d}{dq} \left(\sum_{x=1}^{\infty} q^x \right) \qquad (4.32)$$

$$= pq \frac{d}{dq} \left(\frac{q}{1-q} \right) = pq(1-q)^{-2} = \frac{1-p}{p} \,.$$

For X_1, the same method could be used, but note that, from its definition, it must be the case that

$$\mathbb{E}[X_1] = \mathbb{E}[X_0] + 1 = p^{-1}. \qquad (4.33)$$

Of course, both results can also be arrived at by direct use of the definition – see Problem 4.3.

An arguably more interesting, and certainly easier solution for calculating the mean of X_1 is available, but uses the notion of conditioning, which will be formally introduced and developed in Chapter 8. Following Mosteller (1965, p. 19), condition on the outcome of the first trial. If it is a success, then the average number of tosses is just $x = 1$. Otherwise, on average, we expect $x = 1 + \mathbb{E}[X]$. By weighting the two possibilities with their respective probabilities, $\mathbb{E}[X] = p \cdot 1 + (1-p) \cdot (1 + \mathbb{E}[X])$ or $\mathbb{E}[X] = p^{-1}$. ∎

⊙ ***Example 4.19*** How many times do you expect to roll a fair, k-sided die until you get a 1, i.e. the first side? These are geometric trials with success probability $1/k$, so that, from (4.33), the answer is just k. Knowing this, one could possibly think that, if you role the die k times, the probability of getting a 1 is 1/2. Is this true?

The question needs to be clarified in terms of how many 1s get rolled; in particular, *exactly* or *at least one* 1. Let $p_e(k)$ be the probability of getting *exactly* one 1. As the number of roles is determined, these are binomial trials, and the answer is

$$p_e(k) = \binom{k}{1} \left(\frac{1}{k} \right) \left(\frac{k-1}{k} \right)^{k-1} = \left(\frac{k-1}{k} \right)^{k-1}.$$

---

[6] This definition will suffice for our purposes, but there are some pathological examples when there is a need for a broader definition of the expected value; see, for example, Fristedt and Gray (1997, p. 56).

[7] With regard to the use of the derivative in (4.32) to arrive at the answer, Wilf (1994, p. 6) writes '...we use the following stunt, which seems artificial if you haven't seen it before, but after using it 4993 times, it will seem quite routine'.

Similarly, the probability $p_a(k)$ of getting a 1 *at least* one time in k rolls is just

$$p_a(k) = 1 - \Pr(\text{never get it}) = 1 - \left(\frac{k-1}{k}\right)^k.$$

Recall that $\lim_{s \to \infty}(1 + g/s)^{s-w} = e^g$ for any $g, w \in \mathbb{R}$, so that

$$\lim_{k \to \infty}\left(\frac{k-1}{k}\right)^{k-1} = e^{-1} = \lim_{k \to \infty}\left(\frac{k-1}{k}\right)^k,$$

i.e. as k grows, $p_e(k) \approx e^{-1} \approx 0.3679$ and $p_a(k) \approx 1 - e^{-1} \approx 0.6321$. Thus, the limiting average of $p_e(k)$ and $p_a(k)$ is exactly 1/2. ∎

Example 4.18 demonstrated that direct use of the definition is not always necessary for evaluating the expected value; although, in this case, the derivation in (4.32) did not save much work. It turns out that, for the other sampling distributions (binomial, negative binomial, hypergeometric and inverse hypergeometric), other ways of evaluating the mean (and higher moments – see below) also exist which are far more expedient than direct use of the definition. These will be detailed in Section 6.4. The direct calculation of the mean for $X \sim \text{Bin}(n, p)$, however, is still straightforward and yields $\mathbb{E}[X] = np$ (see Problem 4.4). For a Poisson r.v., a direct calculation is indeed preferred. If $X \sim \text{Poi}(\lambda)$, then

$$\mathbb{E}[X] = \sum_{x=0}^{\infty} \frac{x e^{-\lambda}\lambda^x}{x!} = \sum_{x=1}^{\infty} \frac{e^{-\lambda}\lambda^x}{(x-1)!} = \lambda e^{-\lambda} \sum_{x=1}^{\infty} \frac{\lambda^{x-1}}{(x-1)!} = \lambda. \qquad (4.34)$$

Recalling the relationship between the binomial and Poisson, this result should not be surprising.

Before moving on to higher-order moments, we state a fact about expectations of sums of random variables: if $X_1, X_2, \ldots, X_n$ are r.v.s which each possess a finite mean, then

$$\boxed{\mathbb{E}[X_1 + X_2 + \cdots + X_n] = \mathbb{E}[X_1] + \mathbb{E}[X_2] + \cdots + \mathbb{E}[X_n]}. \qquad (4.35)$$

This is intuitive and not difficult to prove, but we wait until the beginning of Section 6.1 to formally show it, as it requires some notions of multivariate r.v.s, which have not yet been introduced.

4.4.2 Higher-order moments

If $g(X)$ is a real-valued measurable function of r.v. X, then the expected value of $Y = g(X)$ (provided it exists) can be obtained by first finding the mass or density function corresponding to Y (as in Section 4.3 for the discrete case; see Chapter 7 for

the continuous case) and then computing $\int_{S_Y} y \, dF_Y(y)$. However, it can be proven[8]
that

$$\mathbb{E}\left[g(X)\right] = \int_{S_X} g(x) \, dF_X(x) \,, \qquad (4.36)$$

so that a direct computation is possible without having to first compute function f_Y.
In the discrete case, this result should be clear on comparing $\sum_{S_Y} y \Pr(Y = y)$ and
$\sum_{S_X} g(x) \Pr(X = x)$.

An (immensely) important special case is when g is linear, i.e. $g(X) = a + bX$ for
constants a and b; if $\mathbb{E}[X] \in \mathbb{R}$, then

$$\mathbb{E}[a + bX] = a + b \mathbb{E}[X] \,, \qquad (4.37)$$

which follows from (4.35) and basic properties of summation and integration. We will
refer to this property as *linearity of expectation*. Observe also that, if $\mathbb{E}[X] = \pm\infty$, $b \in$
$\mathbb{R}_{>0}$ and $a \in \mathbb{R}$, then $\mathbb{E}[Y] = \pm\infty$. That is, property (4.37) also holds when $\mathbb{E}[X] \in \mathbb{X}$,
where $\mathbb{X}$ is the extended real line $\mathbb{R} \cup \{-\infty, \infty\}$.

Another useful property of expectation is that it *preserves inequalities*, i.e. if X and
Y are r.v.s with finite means and $X \geq Y$, then $\mathbb{E}[X] \geq \mathbb{E}[Y]$ (but the converse need
not be true).

⊖ **Example 4.20 (Example 4.17 cont.)**   Using the tabulated values, the expected value
of weight W is given by either

$$\int_{S_W} w \, dF(w) = \sum_{i=1}^{3} w_i \Pr(W = w_i) = \frac{6}{11} \frac{\ell k \pi}{4} + \frac{3}{11} \ell k \pi + \frac{2}{11} \frac{9 \ell k \pi}{4} = \frac{9}{11} \ell k \pi$$

or

$$\int_{S_D} g(d) \, dF(d) = \sum_{i=1}^{3} g(d_i) \Pr(D = d_i) = \sum_{i=1}^{3} w_i \Pr(W = w_i) \,. \qquad ∎$$

⊖ **Example 4.21**   Let A be a continuous random variable with support $S \subset \mathbb{R}$ which
describes the value (cash payoff) of a certain financial investment, so that $F_A(x)$ is the
probability of making less than or equal to x (say) dollars. Similarly, let B be an r.v.
referring to the payoff of a different investment. If, for every $x \in S$, $F_A(x) \leq F_B(x)$,
then investment A is said to *first-order stochastically dominate* investment B, or A
FSD B, and A would be preferred by all (rational) investors. This is because, for any
$x \in S$, $\Pr(A > x) \geq \Pr(B > x)$, i.e. the probability of making more than x dollars is
higher with investment A, for all possible x.

---

[8] See, for example, Ross (1988, pp. 255-257, 313(2)) or Mittelhammer (1996, pp. 117–8) for elementary
proofs.

As a trivial example, if investments A and B are such that $A = B + k$, $k > 0$, then clearly, A is to be preferred, and indeed, A FSD B, because

$$F_B(x) = \Pr(B \le x) = \Pr(A \le x + k) = F_A(x + k) \ge F_A(x),$$

using the fact that F is a nondecreasing function. Similarly, if the support of A and B is positive and if $A = Bk$, $k > 1$, then $F_B(x) = \Pr(B \le x) = \Pr(A \le xk) = F_A(xk) \ge F_A(x)$. Use of the exponential distribution with different parameters is a special case of the latter result, recalling that the parameter in the exponential distribution is an (inverse) scale parameter. In particular, if $0 < \lambda_1 < \lambda_2 < \infty$, then it is easy to see that, $\forall x > 0$, $e^{-\lambda_1 x} > e^{-\lambda_2 x}$ or $1 - e^{-\lambda_1 x} < 1 - e^{-\lambda_2 x}$, so that, if $X_i \sim \text{Exp}(\lambda_i)$, $i = 1, 2$, then, $\forall x > 0$, $F_{X_1}(x) < F_{X_2}(x)$, and X_1 FSD X_2. A distribution with finite support might make more sense in the following context. Let $X_i \sim \text{Beta}(p, q_i)$, $i = 1, 2$, with $0 < q_1 < q_2$. A graphical analysis suggests that $F_{X_1}(x) < F_{X_2}(x)$ $\forall x \in (0, 1)$ and any $p > 0$. The reader is invited to prove this, or search for its proof in the literature.

Let $U(W)$ be a wealth utility function as in Example A.12 and let $\mathbb{E}[U(A)]$ be the expected utility of the return on investment A. If A FSD B, then one might expect that $\mathbb{E}[U(w + A)] \ge \mathbb{E}[U(w + B)]$ for *any* increasing utility function and any fixed, initial level of wealth w. This is easily proven when U is differentiable, with $U'(W) > 0$, and A and B are continuous r.v.s.[9] First note that, if A FSD B, then $(w + A)$ FSD $(w + B)$, so we can take $w = 0$ without loss of generality. Let interval (a, b) be the union of the support of A and B. Integrating by parts shows that

$$\int_a^b U(x) f_A(x)\,dx = U(b) F_A(b) - U(a) F_A(a) - \int_a^b F_A(x) U'(x)\,dx$$

$$= U(b) - \int_a^b F_A(x) U'(x)\,dx,$$

as $F_A(b) = 1$ and $F_A(a) = 0$. Similarly,

$$\int_a^b U(x) f_B(x)\,dx = U(b) - \int_a^b F_B(x) U'(x)\,dx,$$

so that, from (4.36),

$$\mathbb{E}[U(A)] - \mathbb{E}[U(B)] = \int_a^b U(x) f_A(x)\,dx - \int_a^b U(x) f_B(x)\,dx$$

$$= \int_a^b [F_B(x) - F_A(x)] U'(x)\,dx \ge 0,$$

which follows because $U'(x) > 0$ and $F_B(x) > F_A(x)$ by assumption. ∎

---

[9] If A or B (or both) are discrete, then the proof is essentially identical, but based on the Riemann–Stieltjes integral; see, for example, Ostaszewski (1990, Section 17.9) or Stoll (2001, Section 6.5) for an introduction to the Riemann–Stieltjes integral.

By taking $Y = g(X) = X^r$ in (4.36), the rth *raw moment* of random variable X is

$$\mu'_r = \mu'_r(X) := \mathbb{E}[X^r] = \int_S x^r \, dF_X(x),$$

(4.38)

while the rth *central moment* of X is

$$\mu_r = \mu_r(X) := \mathbb{E}[(X - \mu)^r] = \int_S (X - \mu)^r \, dF_X(x),$$

(4.39)

where $\mu = \mathbb{E}[X]$. Usually r is a positive integer, but it can be any real number. As part of the definition, the rth raw and central moment of X exists if and only if $\int_S |x|^r \, dF_X(x)$ is finite. If the rth moment (raw or central) exists, $r > 0$, it seems intuitive that the sth moment, $0 < s < r$, also exists. This is indeed true and is easily proven:

$$\int_S |x|^s \, dF_X(x) = \int_{|x| \leq 1} |x|^s \, dF_X(x) + \int_{|x| > 1} |x|^s \, dF_X(x)$$

$$\leq \int_{|x| \leq 1} dF_X(x) + \int_{|x| > 1} |x|^s \, dF_X(x)$$

$$\leq 1 + \int_{|x| > 1} |x|^s \, dF_X(x) \leq 1 + \int_{|x| > 1} |x|^r \, dF_X(x) < \infty.$$

The result also holds for central moments via (4.47) below.

The second central moment, μ_r, plays an important role in many statistical models and is referred to as the *variance* of X:

$$\mu_2 = \mathbb{V}(X) := \int_S (x - \mu)^2 \, dF_X(x) = \mu'_2 - \mu^2$$

(4.40)

and often denoted as σ^2.

Another simple, but immensely important result is that, if $a \in \mathbb{R}$ is a constant, then $\mathbb{V}(aX) = a^2 \mathbb{V}(X)$. This is easily verified from the following definition. As $\mathbb{E}[aX] = a\mathbb{E}[X]$ from the linearity of expectation (4.37), we have

$$\mathbb{V}(aX) = \mathbb{E}[(aX - a\mathbb{E}[X])^2]$$

$$= \mathbb{E}[a^2(X - \mathbb{E}[X])^2] = a^2 \mathbb{E}[(X - \mathbb{E}[X])^2] = a^2 \mathbb{V}(X),$$

i.e.

$$\mathbb{V}(aX) = a^2 \mathbb{V}(X).$$

(4.41)

The *standard deviation* of an r.v. is defined to be $\sigma := \sqrt{\mu_2}$.

The *skewness (coefficient)* is a moment-based measure of the extent to which the p.d.f. of an r.v. deviates from symmetry. It is given by

$$
\text{skew}(X) = \mathbb{E}\left[\left(\frac{X-\mu}{\sigma}\right)^3\right] = \frac{\mu_3}{\mu_2^{3/2}},
\tag{4.42}
$$

if the expectation exists, where X is an r.v. with mean μ and variance σ^2. It is obvious from the definition that skew(X) is invariant to location and scale transformations.

Remark: There are other measures of asymmetry which do not involve the third moment. Besides having different properties than skew(X) which might be more valuable in certain contexts, there are distributions for which third (and higher) moments do not exist (see Section 7.2). Let X be a random variable with $0 < \mathbb{V}(X) < \infty$ and mean, median and standard deviation given by μ, m and $\sigma = \sqrt{\mathbb{V}(X)}$, respectively. Then

$$
S = \frac{\mu - m}{\sigma}
\tag{4.43}
$$

is a measure of asymmetry going back at least to Hotelling and Solomons (1932). Building on their results, Majindar (1962) showed that

$$
|S| < \frac{2(pq)^{1/2}}{(p+q)^{1/2}},
$$

where $p = \Pr(X > \mu)$ and $q = \Pr(X < \mu)$. Another measure is

$$
\eta = \int_{-\infty}^{\infty} \left| f_X(\mu + y) - f_X(\mu - y) \right| dy,
\tag{4.44}
$$

proposed and studied by Li and Morris (1991). It requires only the existence of the first moment. A measure which does not even require that is

$$
\lim_{x \to \infty} \frac{P(X > x) - P(X < -x)}{P(X > x) + P(X < -x)},
$$

which is the measure of skewness associated with the *asymmetric stable Paretian distribution* (see Volume II). ∎

A measure of the 'peakedness' (how spiked the center of a unimodal p.d.f. is) or 'heaviness of the tails' of a distribution is given by the *kurtosis (coefficient)*, defined as

$$
\text{kurt}(X) = \mathbb{E}\left[\left(\frac{X-\mu}{\sigma}\right)^4\right] = \frac{\mu_4}{\mu_2^2},
\tag{4.45}
$$

if the expectation exists, where X is an r.v. with mean μ and variance σ^2. Let $Z = (X - \mu)/\sigma$, so that, by construction, $E[Z^2] = \text{Var}(Z) = 1$ and thus

$$\text{kurt}(X) = E[Z^4] = \text{Var}(Z^2) + (E[Z^2])^2 = \text{Var}(Z^2) + 1, \qquad (4.46)$$

from which we obtain the lower-bound $\text{kurt}(X) \geq 1$. Based on (4.46), Moors (1986) offers the interpretation of $\text{kurt}(X)$ as being a measure of spread, or dispersion, of X around the two points $\mu \pm \sigma$. This is quite useful, because if the distribution of X is highly dispersed around $\mu \pm \sigma$, then this implies that the density must be relatively smaller in that region, and instead has correspondingly more mass either near μ, and/or in the tails. This corresponds precisely to the interpretation of high kurtosis as being present when the p.d.f. of X is highly peaked around its mode and/or has fat tails. As an example, the Student's t distribution with $v > 4$ degrees of freedom has finite kurtosis and, indeed, as v decreases, the density becomes both more peaked near the center, and exhibits fat tails.

Remark: While the kurtosis is clearly invariant to location and scale changes, its value can be influenced by the amount of skewness. This was studied by Blest (2003), who also proposed a new measure of kurtosis which adjusts for the skewness.

The standard measures of skewness and kurtosis given above are not robust to outliers, and so can be very misleading. Kim and White (2004) have addressed this issue and proposed alternative measures which are of particular use in the field of empirical finance. ∎

Using (4.35), raw and central moments of X (if they exist) are related by

$$\mu_n = \mathbb{E}\left[(X - \mu)^n\right] = \mathbb{E}\left[\sum_{i=0}^{n} \binom{n}{i} X^{n-i} (-\mu)^i\right]$$

$$= \sum_{i=0}^{n} \binom{n}{i} \mathbb{E}\left[X^{n-i}\right] (-\mu)^i = \sum_{i=0}^{n} (-1)^i \binom{n}{i} \mu'_{n-i} \mu^i \qquad (4.47)$$

and

$$\mu'_n = \mathbb{E}\left[(X - \mu + \mu)^n\right] = \mathbb{E}\left[\sum_{i=0}^{n} \binom{n}{i} (X - \mu)^{n-i} \mu^i\right] = \sum_{i=0}^{n} \binom{n}{i} \mu_{n-i} \mu^i, \quad (4.48)$$

where $n \in \mathbb{N}$ and the fact that, if X_i, $i = 1, \ldots, n$ are r.v.s, then

$$\mathbb{E}\left[\sum_i g(X_i)\right] = \sum_i \mathbb{E}\left[g(X_i)\right],$$

which follows from (4.31) and the fact that $g(X_i)$ are r.v.s themselves.

For low-order moments, these simplify to

$$\mu_2' = \mu_2 + \mu^2, \qquad\qquad \mu_2 = \mu_2' - \mu^2,$$

$$\mu_3' = \mu_3 + 3\mu_2\mu + \mu^3, \qquad\qquad \mu_3 = \mu_3' - 3\mu_2'\mu + 2\mu^3, \qquad (4.49)$$

$$\mu_4' = \mu_4 + 4\mu_3\mu + 6\mu_2\mu^2 + \mu^4, \qquad \mu_4 = \mu_4' - 4\mu_3'\mu + 6\mu_2'\mu^2 - 3\mu^4,$$

using $\mu' = \mu$ and $\mu_1 = 0$. Not all random variables possess finite moments of all order; examples include the Pareto and Student's t distributions (see Chapter 7).

Finally, it can be shown (see, for example, Lange, 2003, p. 37) that

$$\mathbb{E}[X^n] = \sum_{k=0}^{\infty} \left[(k+1)^n - k^n\right]\left[1 - F_X(k)\right] \qquad (4.50)$$

for nonnegative discrete r.v. X. The analog of this and some extensions for continuous r.v.s are developed in Problem 7.13.

Example 4.22 Let $X \sim \text{Geo}(p)$ with mass function (4.16). From (4.17) and (4.50),

$$\mathbb{E}[X] = \sum_{x=0}^{\infty} (1 - F_X(x)) = \sum_{x=0}^{\infty} q^{x+1} = \frac{q}{1-q} = \frac{1-p}{p}$$

and, with $S_1 = \sum_{x=0}^{\infty} x q^x = q(1-q)^{-2}$ (see Problem 4.3),

$$\mathbb{E}[X^2] = \sum_{x=0}^{\infty} \left[(x+1)^2 - x^2\right]\left[1 - F_X(x)\right]$$

$$= \sum_{x=0}^{\infty} (2x+1)\, q^{x+1} = \frac{2(1-p)^2}{p^2} + \frac{1-p}{p},$$

from which (4.40) implies

$$\mathbb{V}(X) = \frac{2(1-p)^2}{p^2} + \frac{1-p}{p} - \left(\frac{1-p}{p}\right)^2 = \frac{1-p}{p^2}. \qquad (4.51)$$

The variance of X is the same whether or not the single success is counted; this can be directly verified, but it is more important to realize that the variance of an r.v. is not affected by location shifts. ∎

We now consider some examples using mass functions which were previously introduced.

Example 4.23 Let Y_k be the number of cards drawn without replacement from a well-shuffled deck of 52 playing cards until the kth king occurs, $k = 1, 2, 3, 4$. Random variable Y_k is inverse hypergeometric with $w = 4$ and $b = 48$. From the mass function, the mean and standard deviation of Y_k can be computed. However, Section 6.4.3 will give simple, closed-form solutions for the mean and variance; in particular, from (6.25) and (6.29), $\mathbb{E}[Y_k] = 53k/5$ and $\mathbb{V}(Y_k) = 424k(5-k)/25$, i.e.

k	$\mathbb{E}[Y_k]$	$\sqrt{\mathbb{V}(Y_k)}$
1	10.6	8.24
2	21.2	10.1
3	31.8	10.1
4	42.4	8.24

Calculation without the closed-form solutions would be tedious in this case. For example, with $k = 4$, direct use of (4.23) gives

$$\mathbb{E}[Y_4] = \sum_{y=4}^{52} y \binom{y-1}{4-1} \frac{\binom{52-y}{4-4}}{\binom{52}{4}} = \frac{1}{\binom{52}{4}} \sum_{y=4}^{52} y \binom{y-1}{3}$$

$$= \frac{1}{270\,725} \sum_{y=4}^{52} \frac{(y-3)(y-2)(y-1)y}{6}$$

$$= \frac{1/6}{270\,725} \sum_{y=4}^{52} (y^4 - 6y^3 + 11y^2 - 6y)$$

$$= \frac{212}{5} = 42.4,$$

using the rules for summation detailed in Example 1.18.[10] The variance could be similarly obtained but at even more expense.   ∎

⊙ **Example 4.24** The expected number of matches, M, in the problem of coincidences is, using (2.18) and letting the sum over i start from zero instead of two for numerical evaluation purposes,

$$\mathbb{E}[M] = \sum_{m=0}^{n} m\, c_{m,n} = \sum_{m=1}^{n} \frac{1}{(m-1)!} \sum_{i=0}^{n-m} \frac{(-1)^i}{i!}. \tag{4.52}$$

For $n = 1$, $\mathbb{E}[M]$ has to be one. For $n = 2$, either the first round is a match, in which case the second round also results in a match, or the first round is not a match, in which case the second is not either. Because these two scenarios have equal probability, the expected number of matches is $(0+2)/2 = 1$. For $n = 3$, (4.52) simplifies to

$$\mathbb{E}[M] = \frac{1}{(1-1)!} \sum_{i=0}^{2} \frac{(-1)^i}{i!} + \frac{1}{(2-1)!} \sum_{i=0}^{1} \frac{(-1)^i}{i!} + \frac{1}{(3-1)!} \sum_{i=0}^{0} \frac{(-1)^i}{i!}$$

$$= \left(1 - 1 + \frac{1}{2}\right) + (1 - 1) + \frac{1}{2}(1) = 1.$$

---

[10] Or, far faster, slicker, modern, and more reliable, is just to use a symbolic computing package.

It turns out that $\mathbb{E}[M]$ is one for *all* values of n; Problem 4.14 verifies this induction, while Example 6.3 proves it in another way. ∎

4.4.3 Jensen's inequality

Settling for the market averages strikes many investors as rather dull. These folks enjoy picking stocks and funds in an effort to overcome their investment costs and thereby beat the market. But my goal is a little different. I am not looking for the thrill of victory. Instead, I am trying to avoid the agony of defeat.

(Jonathan Clements)[11]

Reproduced by permission of Dow Jones, Wall Street Journal Europe

Jensen's inequality is a simple and important result which states that, for any r.v. X with finite mean,

$$
\begin{array}{ll}
\mathbb{E}[g(X)] \geq g(\mathbb{E}[X]), & g(\cdot) \text{ convex}, \\
\mathbb{E}[g(X)] \leq g(\mathbb{E}[X]), & g(\cdot) \text{ concave}.
\end{array}
\tag{4.53}
$$

To verify this, assume that $g''(x)$ exists (which is usually fulfilled in applications), so that g is convex if $g''(x) \geq 0$ for all x. Let X be an r.v. with finite mean μ. Then, for g a twice differentiable, convex function, there exists a value ξ such that (see Appendix A.2.4.5)

$$
g(x) = g(\mu) + g'(\mu)(x - \mu) + \frac{1}{2}g''(\xi)(x - \xi)^2,
$$

i.e. $g(x) \geq g(\mu) + g'(\mu)(x - \mu)$ for all x. Thus, $g(X) \geq g(\mu) + g'(\mu)(X - \mu)$; taking expectations of both sides and using the preservation of inequality and linearity properties of expectation, it follows that $\mathbb{E}[g(X)] \geq g(\mathbb{E}[X])$. If $g''(x) \leq 0$, then g is *concave* and the proof of (4.53) is similar.[12]

In the following simple examples, assume X is an r.v. with finite mean μ.

⊙ ***Example 4.25*** Let $g(x) = x^2$ with $g''(x) = 2$, so that g is convex. Then $\mathbb{E}[X^2] \geq \mu^2$. Of course, from (4.40), if $\mathbb{E}[X^2] < \infty$, then $\mathbb{E}[X^2] = \mathbb{V}(X) + \mu^2$, which, as $\mathbb{V}(X) \geq 0$, shows the result immediately. ∎

⊙ ***Example 4.26*** Let X be a nonnegative r.v. and take $g(x) = \sqrt{x}$. It is easy to see that $g''(x)$ is negative for $x > 0$, so that g is concave and $\mathbb{E}[\sqrt{X}] \leq \sqrt{\mu}$. However, as $0 \leq \mathbb{V}(\sqrt{X}) = \mathbb{E}[X] - (\mathbb{E}[\sqrt{X}])^2$, the result also follows immediately. ∎

---

[11] Quoted from Clements' personal finance article in the *Wall Street Journal Europe*, January 7–9th, 2005, page P4.

[12] The definition of convexity and the statement of Jensen's inequality can be made somewhat more precise as follows. A subset S of $\mathbb{R}^n$ is convex if all its elements are connected, i.e. for any $s_1, s_2 \in S$ and all $\lambda \in (0, 1)$, $\lambda s_1 + (1 - \lambda) s_2 \in S$. If S is convex, then function $g : S \mapsto \mathbb{R}$ is convex if $g(v s_1 + w s_1) \leq g(v s_1) + g(w s_1)$ for all $s_1, s_2 \in S$, $v \in (0, 1)$ and $w = 1 - v$. If $-g$ is convex, then g is said to be concave.

⊚ ***Example 4.27***   Let $g(x) = x^{-1}$ for $x > 0$. Then g is convex because $g''(x) = 2/x^3 \geq 0$ and $\mathbb{E}\left[X^{-1}\right] \geq \mu^{-1}$.   ∎

⊚ ***Example 4.28***   Let $g(x) = e^{ax}$ with $g''(x) = a^2 e^{ax} \geq 0$ for all a and x, so that g is convex. Then $\mathbb{E}\left[e^{aX}\right] \geq e^{a\mu}$.   ∎

⊚ ***Example 4.29***   Let $g(x) = \ln(x)$ for $x > 0$. Then g is concave because $g''(x) = -x^{-2}$ and $\mathbb{E}[\ln X] \leq \ln\mu$.   ∎

⊖ ***Example 4.30***   The notion of a wealth utility function is introduced in Example A.12; in particular, let $U(W)$ be a twice-differentiable, concave function of wealth W, i.e. $U'(W) > 0$ and $U''(W) < 0$. Letting A be a random variable associated with the payoff of a financial investment, (4.53) implies that $\mathbb{E}[U(A)] \leq U(\mathbb{E}[A])$.

The intuition behind this result is that a risk-averse person (one for whom $U''(W) < 0$) prefers a sure gain of zero (the utility of the expected value of zero) to taking a fair bet (win or lose x dollars with equal probability). This also follows geometrically from the plot of U, as shown in most every microeconomics textbook; see, for example, Nicholson (2002, p. 205).   ∎

We now discuss another measure of central tendency, the *geometric expectation*, and use Jensen's inequality to determine its relation to the usual expected value. We illustrate it in the context of the returns on a financial asset, which should be of general interest.

The *simple return*, R_t, of a financial investment at time t with respect to the price from the previous period, $t - 1$, is (neglecting dividend payments) $R_t = (P_t - P_{t-1}) / P_{t-1}$, so that

$$P_t = \frac{P_t}{P_{t-1}} P_{t-1} = (1 + R_t) P_{t-1}. \tag{4.54}$$

From the Taylor series expansion

$$\ln(1 + a) = \sum_{i=1}^{\infty} (-1)^{i+1} a^i / i$$

and the fact that R_t is small compared to 1,

$$\ln\left(\frac{P_t}{P_{t-1}}\right) = \ln(1 + R_t) \approx R_t,$$

which explains the common use of *continuous (compounded) (percentage) returns* which, instead of $100\,R_t$, are defined as $r_t := 100\ln(P_t/P_{t-1})$.

Notice the following difference. If $P_1 = 100$ and $P_2 = 105$, then $R_2 = 0.05 = 5\%$, and $r_2 = 100\ln(1.05) = 4.879\%$. If $P_3 = P_1 = 100$, then $R_3 = -4.762\%$ and $r_3 = -4.879\% = -r_2$. The (arithmetic) average continuous percentage return, $(r_2 + r_3)/2$, is zero, which is desirable, as the last period's price is the same as the first period's price. In contrast, the (arithmetic) average discrete return is positive! However, as $P_3 = P_1$, the *geometric average* based on the discrete returns R_t given by

$$[(1 + R_2)(1 + R_3)]^{1/2} = \left[\frac{P_2}{P_1}\frac{P_3}{P_2}\right]^{1/2} = 1 = 1 + 0,$$

also gives an average return of zero. As another illustration, if $P_t = 100$ and $P_{t+1} = 90$, so that $R_{t+1} = -10\%$, then the value of R_{t+2} such that $P_{t+2} = P_t$ is, from (4.54), $P_t \equiv P_{t+2} = (1 + R_{t+2})P_{t+1}$ or

$$R_{t+2} = P_t/P_{t+1} - 1 = 100/90 - 1 = 0.11\overline{1} > -R_{t+1}.$$

In general with $P_t \equiv P_{t+2}$,

$$R_{t+2} = \frac{P_{t+2} - P_{t+1}}{P_{t+1}} \equiv \frac{P_t - P_{t+1}}{P_{t+1}} = \frac{1}{R_{t+1} + 1} - 1 = -R_{t+1} + R_{t+1}^2 + \cdots. \quad (4.55)$$

These examples illustrate that the arithmetic average of the R_t (or $1 + R_t$) over a period of time is too optimistic (i.e. upwards biased) as a measure of 'average performance' (but provides a nice trick for unscrupulous sellers of financial products).

Thus, the geometric average of the $1 + R_t$ is more suitable, because it takes compounding into account. Let $Y = 1 + R$ be a discrete random variable (i.e. assume for now that the return can take on one of a finite number of values) with p.m.f. f_Y and support $\mathcal{S}$, the elements of which are all nonnegative. The geometric expected value for discrete r.v.s is then denoted and defined as

$$\mathbb{G}[Y] = \prod_{y \in \mathcal{S}} y^{f_Y(y)}. \quad (4.56)$$

For example, based on (4.55), if the return is R with probability 0.5 and $(R + 1)^{-1} - 1$ with probability 0.5, then $\mathbb{G}[Y] = (1 + R)^{1/2}(1 + R)^{-1/2} = 1$. Rewriting (4.56) as

$$\mathbb{G}[Y] = \exp\{\ln(\mathbb{G}[Y])\}$$

shows that

$$\mathbb{G}[Y] = \exp\left[\sum_{y \in \mathcal{S}} f_Y(y)\ln y\right] = \exp(\mathbb{E}[\ln Y]),$$

which is the definition in the more general case for which Y is discrete or continuous. Writing this as $\ln \mathbb{G}[Y] = \mathbb{E}[\ln Y]$ and using Example 4.29, i.e. that $\mathbb{E}[\ln Y] \leq \ln \mathbb{E}[Y]$, we obtain

$$\mathbb{G}[Y] \leq \mathbb{E}[Y],$$

with equality holding if $\mathbb{V}(Y) = 0$.

4.5 Poisson processes

Let t denote a measure of time and $N(t)$ a sequence of random variables indexed by $t > 0$. For example, $N(t)$ could represent your blood pressure at any given moment, the speed of a moving object, monetary value of a financial asset, etc. If, less generally, $N(t)$ measures the total number of discrete occurrences of some well-defined event up to time t, then $N(t)$ is called a *counting process*. For example, $N(t)$ could measure the total number of credit card transactions in a specific region, or the number of cars entering a gas station, etc. More formally, for $N(t)$ to be a counting process, we require (i) that $N(t) \in \{0, 1, 2, \ldots\}$, (ii) if $t_1 < t_2$, then $N(t_1) \le N(t_2)$, and (iii) for $t_1 < t_2$, $N(t_2) - N(t_1)$ is the number of events which occurred in the interval (t_1, t_2).

If the numbers of events which occur in disjoint time intervals are independent, then $N(t)$ is said to have *independent increments*, e.g. if $t_1 < t_2 \le t_3 < t_4$, then $N(t_2) - N(t_1)$ is independent of $N(t_4) - N(t_3)$. In the credit card example, the independent increments assumption appears tenable. For the gas station example, if $N(t+1) - N(t)$ is large, then $N(t+2) - N(t+1)$ might be small (because potential customers would have to wait for service, and thus go to the competitor next door) and vice versa, so that the increments are not independent but *negatively correlated* instead.

If the probability distribution of the number of events which occur in any interval of time depends only on the length of the time interval, then $N(t)$ is said to have *stationary increments*, e.g. for any $t_1 < t_2$ and any $s > 0$, if $N(t_2 + s) - N(t_1 + s)$ and $N(t_2) - N(t_1)$ have the same distribution. In both the credit card and gas station examples, the extent to which stationary increments can be assumed depends on the timescale considered. Certainly, the 'intensity' of purchases with credit cards or the demand for gasoline will depend on the time of day, the day of the week and the season, as well as other microeconomic factors which can change over time such as the price elasticity of demand. Over a small enough timeframe, the assumption of stationary increments might well be tenable.

We can now introduce the *Poisson process with rate* λ. This is a very specific, yet popular and useful counting process, such that $N(0) = 0$ and

$$\Pr\left[N(s+t) - N(s) = x\right] = e^{-\lambda t} \frac{(\lambda t)^x}{x!} \mathbb{I}_{\{0,1,2,\ldots\}}(x)$$

for all $t > 0$ and $s > 0$. Note that $N(t)$ has independent and stationary increments. Recalling the expected value of a Poisson r.v., the expected number of events in an interval $(s+t, t)$ increases by either increasing t, increasing λ, or both. Just as the Poisson served as a limiting distribution for the binomial and negative binomial under certain assumptions, the Poisson process arises by specifying particular assumptions for a counting process. In particular, if $N(t)$ is a counting process for $t \ge 0$ with stationary and independent increments such that $N(0) = 0$,

$$\Pr\left[N(s) = 1\right] = \lambda s + o(s) \quad \text{and} \quad \Pr\left[N(s) \ge 2\right] = o(s), \tag{4.57}$$

then $N(t)$ is a Poisson process with rate λ.[13]

---

[13] For some $a \in \mathbb{R}$, we write $f(x) = o(g(x))$ ('little oh') if, for *every* $C > 0$, there exists a $d > 0$ such that $|f(x)| < C|g(x)|$ for $|x - a| < d$. Informally, this says that $f(x)/g(x)$ gets arbitrarily close to zero as x

The arguably least sophisticated way to show this is to divide the interval $[0, t]$ into n nonoverlapping equal-length subintervals and let $n \to \infty$. That is, following Ross (1988, pp. 133–5), let $n \geq k$ and write

$$\Pr[N(t) = k] = \Pr(A) + \Pr(B),$$

where event $A = \{k$ of the n subintervals contain exactly one occurrence, the other $n - k$ contain none}, event $B = \{$there are exactly k occurrences among the n subintervals, and at least one of them contains two or more} and A and B are mutually exclusive. Then, with S_i denoting the number of occurrences in the ith subinterval, $i = 1, \ldots, n$,

$$\Pr(B) \leq \Pr(\text{at least one } S_i \geq 2)$$
$$= \Pr\left(\cup_{i=1}^{n} \{S_i \geq 2\}\right)$$
$$\leq \sum_{i=1}^{n} \Pr(S_i \geq 2) \quad \text{(Boole's inequality, Section 2.3.1)}$$
$$= \sum_{i=1}^{n} o(t/n)$$
$$= no(t/n) = t\left[\frac{o(t/n)}{t/n}\right]. \tag{4.58}$$

However, the term in brackets goes to zero as $n \to \infty$ so that $\Pr(B) \to 0$. For event A, the assumption of stationary and independent increments implies that

$$\Pr(A) = \binom{n}{k} p^k q^{n-k},$$

where, from (4.57), $p = \Pr(S_i = 1) = \lambda t/n + o(t/n)$ and

$$q = \Pr(S_i = 0) = 1 - \Pr(S_i > 0) = 1 - \Pr(S_i = 1) - \Pr(S_i > 1)$$
$$= 1 - p - o(t/n),$$

so that q and $1 - p$ are arbitrarily close as $n \to \infty$. Thus, we have a binomial model whereby $n \to \infty$ and $p \to 0$ and such that

$$np = \lambda t + no(t/n) \to \lambda t$$

from (4.58). It then follows from the results at the beginning of this subsection and letting $n \to \infty$ that

$$\Pr[N(t) = k] = \frac{e^{-\lambda t}(\lambda t)^k}{k!} \mathbb{I}_{\{0,1,2,\ldots\}}(k). \tag{4.59}$$

---

approaches a, i.e. that f converges to zero faster than g. In the case discussed here, take $f(s) = \Pr(N(s) \geq 2)$, $g(s) = s$ and $a = 0$. 'Big oh' is defined similarly, but changing 'for every $C > 0$' to 'there exists a $C > 0$', i.e. $f(x) = O(g(x))$ if $|f(x)|/g(x)$ stays bounded by a positive constant C for x in a specified range. The letter O for Landau's symbol is used because the rate of growth of a function is also called its *order*. The name is used in honor of the inventor of the notation, the German number theorist Edmund Georg Hermann Landau, 1877–1938 see, for example, Lang (1997, p. 117) for further details on the order of magnitude.

Other methods of proof of this fundamental result can be found in virtually all books on stochastic processes; very readable sources which also develop the theory of Poisson processes further than our very basic presentation include Gut (1995, Chapter 7), Ross (1997, Section 5.3), and Jones and Smith (2001, Chapter 5).

Finally, if $N(t)$ is a Poisson process with rate λ and T_1 denotes the time at which the first event occurs, then $\{T_1 > t\} = \{N(t) = 0\}$ and the probability of this is, from (4.59), just $\exp(-\lambda t)$, i.e. $\Pr(T_1 \leq t) = 1 - \exp(-\lambda t)$. Thus, recalling Example 4.6, random variable T_1 follows an exponential distribution with rate λ. See the discussion of the exponential and gamma distributions in Section 7.1 for more details.

4.6 Problems

It is not enough to do your best; you must know what to do, and THEN do your best.   (William Edwards Deming)

It is no use saying, 'We are doing our best.' You have got to succeed in doing what is necessary.   (Sir Winston Churchill)

4.1. Assume that X is a positive, integer-valued random variable such that

$$\Pr(X = 1) = 0.3 \quad \text{and} \quad \Pr(X = 2 \mid X > 1) = 0.8.$$

Compute $\Pr(X > 2)$.

4.2. Let $f_X(x) = p(1-p)^x \mathbb{I}_{\{0,1,\dots\}}(x)$ whereby $\Pr(X \leq 1) = \Pr(X > 1)$. Calculate $\mathbb{E}[X]$.

4.3. ★ Let $X \sim \text{Geo}(p)$ (with either density (4.16) or (4.18)). Calculate $\mathbb{E}[X]$ and $\mathbb{E}[X^2]$ directly from the definition of expected value and, from these, show (4.51), i.e. that $\mathbb{V}(X) = (1-p)/p^2$.

4.4. Calculate the expected value of $X \sim \text{Bin}(n, p)$ directly from the definition.

4.5. Verify that the kurtosis coefficient (4.45) is invariant to scale transformations.

4.6. ★ ★ ★ The following illustrates a random variable which generalizes the geometric, and whose expectation is tedious to evaluate directly, but can be accomplished easily by recursion. Simultaneously roll d fair, six-sided dice. Those showing a 6 are set aside and the remaining ones are again simultaneously thrown. This is continued until all d dice show a 6. Let N be the required number of throws and denote its expected value as $\mathbb{E}_d[N]$, using the subscript d to indicate its dependence on the number of dice, d. (Notice that N is NOT a sum of d geometrically distributed r.v.s.)

(a) Show that

$$\mathbb{E}_d[N] = \frac{6^d + \sum_{k=1}^{d-1} 5^k \binom{d}{k} \mathbb{E}_k[N]}{6^d - 5^d}$$

for $d = 2, 3, \ldots$, and $\mathbb{E}_1[N] = 6$.   (Epstein, 1977, p. 141)

(b) Write a program to compute $\mathbb{E}_d[N]$ and plot it for a grid of values from 1 to 200.

(c) For large d, the recursive program can take a long time to evaluate. Using the previously computed values of $\mathbb{E}_{150}[N]$ to $\mathbb{E}_{200}[N]$, approximate the expectation as a (nonlinear) function of d (using, say, least squares).

(d) While a computable formula for the expectation was obtainable without having to compute the mass function of N, we may not be so lucky if interest centers on higher moments of N. Instead of trying to determine the mass function algebraically, we can easily simulate it. Construct a program to do so and plot a histogram of the resulting values. Example output is shown in Figure 4.8 for $d = 10$ and $d = 50$.

(e) It turns out after all that the mass function of N is not too difficult to determine! Define event A to be {after x throws, all d dice show 6}, and event B to be {after $x - 1$ throws, at least one dice does not show a 6}. Then

$$f_N(x; d) = \Pr(N = x ; d) = \Pr(A \cap B) = \Pr(A) + \Pr(B) - \Pr(A \cup B).$$

(i) Compute $\Pr(A^c \cap B^c)$ and $\Pr(A \cup B)$, which are easy.
(ii) Specify the relation between A^c and B, and the relation between B^c and A.
(iii) Draw a Venn diagram with the events.
(iv) Compute $\Pr(A)$ and $\Pr(B)$, from which $\Pr(N = x ; d)$ follows. (Hint: first consider only one die ($d = 1$) and work with the complement.)
(v) Graphically compare your result with simulated values of the mass function.

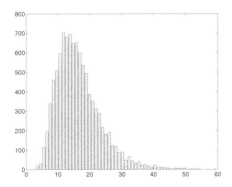
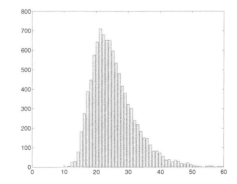

Figure 4.8 Histogram of 10 000 simulated values of N for (a) $d = 10$ and (b) $d = 50$

(f) Using the mass function, a closed-form nonrecursive expression for the expected value can be obtained. In particular, simplify the definition

$$\mathbb{E}_d\,[N] = \sum_{x=1}^{\infty} x f_N(x;d)$$

to obtain the pleasingly simple expression

$$\mathbb{E}_d\,[N] = \sum_{j=1}^{d} \binom{d}{j} \frac{(-1)^{j+1}}{1 - (5/6)^j}.$$

4.7. ★ Consider repeated independent trials consisting of rolling two (six-sided) dice until the sum of the two, say random variable X, is either 5 or 7. Let E be the event that $X = 5$ occurs before $X = 7$.

(a) How many trials do you expect to have to perform until either a 5 or 7 appears?

(b) What is $\Pr(E)$?
(Ross, 1988, p. 77; Rohatgi, 1984, p. 388(13); Roussas, 1997, pp. 50–51)

4.8. Two fair, six-sided dice are repeatedly thrown until either $S = 7$ or $S \leq 4$, where r.v. S is the sum of the two dice. Let r.v. N be the number of throws required.

(a) Compute $\mathbb{E}\,[N]$ and $\mathbb{V}\,(N)$.

(b) Compute the probability that the event $S = 7$ occurs before $S \leq 4$.

4.9. ★ Let random variable X follow the zero-truncated Poisson (λ) distribution. This distribution arises, for example, as a prior for the number of components in mixture models. Calculate the p.m.f. of X, $\mathbb{E}\,[X]$ and $\mathbb{V}\,(X)$.

4.10. An urn contains three white and two black balls. Compute the expected value of X, i.e. the number balls that have to be drawn *without replacement*, until the first white ball appears, using the definition of $\mathbb{E}\,[X]$ and the fact that x is either 1,2 or 3.

4.11. ★ ★ ★ A set of N electrical components is known to contain exactly $D = 2$ defectives. They are to be tested, one at a time, until both defectives have been *identified*. Let T be the number of tests required.

(a) Calculate $f_T(t)$ and $\mathbb{E}\,[T]$ for $N = 5$.

(b) Calculate $f_T(t)$ for general N.

(c) Show that, for general N,

$$\mathbb{E}\,[T] = \frac{2\,(N-2)\,(N+2)}{3\,(N-1)}$$

and

$$V(T) = \frac{N^5 - 3N^4 - 23N^3 + 111N^2 - 86N - 72}{18N(N-1)^2}.$$

(d) What is $\mathbb{E}[T]$ as $N \to \infty$? Can you justify this intuitively?

(e) Write a program which simulates the number of tests required. Plot the resulting values as a histogram and compare with f_T. For example, Figure 4.9(a) shows the case with $N = 50$ (and $D = 2$) based on 10 000 replications. From these values, the sample mean was 34.04 and the sample variance was 132.2. The exact values from the above formulae give 34.00 and 134.7, respectively, which are in close agreement with the values obtained via simulation.

(f) An obvious extension is to lift the restriction on D, i.e. allow it to be any integer value in $[1, N - 1]$. (Note that the mass functions of $f_T(t; N, D)$ and $f_T(t; N, N - D)$ will be the same.) What is $\Pr(T = N - 1)$? Instead of working out the algebraic expression for the entire mass function, use simulation to approximate the mass function and moments for given N and D. Figure 4.9 also shows the (b) $D = 3$, (c) $D = 15$ and (d) $D = 25$ cases using 10 000 replications. Let $X = N - T$. From visual inspection of the plots, particularly for $D \gg 1$, what might be a reasonable approximation to the p.m.f.?

4.12. ★ ★ Recall Banach's matchbox problem from Example 4.13.

(a) One natural generalization is to allow for different numbers of matches in the left and right pockets, say N_1 and N_2, and probability p not necessarily 1/2 of drawing from the left side.[14] Derive the mass function $f(k; N_1, N_2, p)$ and construct a program, say `banach(n1,n2,p,sim)` which computes it, simulates the process `sim` times, and finally plots the true and simulated mass functions overlaid. As an example, Figure 4.10(a) was produced with

```
vec=banach(30,10,1/2,10000);
text(3,0.07,'N_1=30, N_2=10, p=1/2','fontsize',14)
```

(b) Further extensions could allow for more than two matchboxes; see, for example, Cacoullos (1989, p. 80(317,318)) for some analytic results. For now, assume that Banach has $n \geq 2$ matchboxes distributed throughout his many pockets, each of which initially contains $N_i > 0$ matches, $i = 1, \ldots, n$. Associated with each box is a probability p_i, with $\sum_i p_i = 1$. Construct a program which simulates the process and, once an empty box is discovered, reports the minimum and maximum number of matches in the remaining matchboxes. Simulate a large number of times and plot the marginal mass functions together. For example, Figure 4.11 shows the result with $N = [20, 50, 100]$, $\mathbf{p} = [0.1, 0.4, 0.5]$ and 20 000 replications.

---

[14] More general setups and asymptotic analysis were considered by Cacoullos (1967).

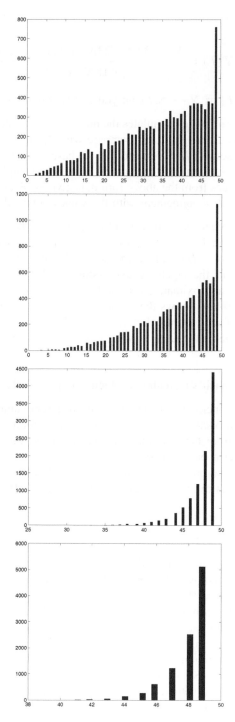

Figure 4.9   Simulated mass function of the number of tests required to identify D out of $N = 50$ defectives using (a) $D = 2$, (b) $D = 3$, (c) $D = 15$ and (d) $D = 25$

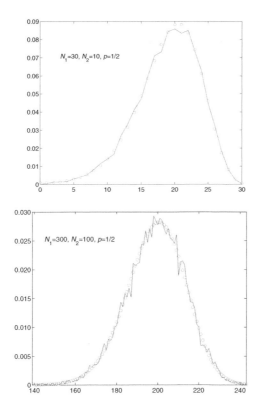

Figure 4.10 True (circles) and simulated (lines) mass function for the generalized Banach match-box problem

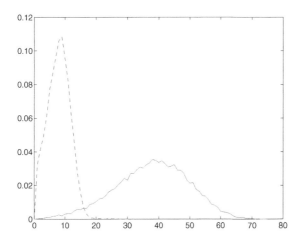

Figure 4.11 Mass function of minimum and maximum in the generalized Banach matchbox problem.

4.13. ★ ★   Recall Banach's matchbox problem from Example 4.13. With $N = 1$, it is clear that $\mathbb{E}[K] = 1/2$. In general, the expression $\sum k f(k; N)$ does not appear to admit a closed-form expression. However, with $\mathbb{E}_N[K]$ denoting the expected value of K given N, letting $k = j - 1$ and using (1.4), $\mathbb{E}_N[K]$ is given by

$$\sum_{j=0}^{N} j \binom{2N-j}{N} \left(\frac{1}{2}\right)^{2N-j} = \sum_{k=0}^{N-1} (k+1) \binom{2N-k-1}{N} \left(\frac{1}{2}\right)^{2N-k-1}$$

$$= \sum_{k=0}^{N-1} (k+1) \binom{2N-k-2}{N-1} \left(\frac{1}{2}\right)^{2N-k-1}$$

$$+ \sum_{k=0}^{N-1} (k+1) \binom{2N-k-2}{N} \left(\frac{1}{2}\right)^{2N-k-1}$$

$$=: G + H.$$

Show that

$$G = \frac{1}{2}\mathbb{E}_{N-1}[K] + \frac{1}{2}, \quad H = \frac{1}{2}\mathbb{E}_N[K] - \frac{1}{2} + \binom{2N}{N} \left(\frac{1}{2}\right)^{2N+1},$$

and, simplifying $G + H$,

$$\mathbb{E}_N[K] = \mathbb{E}_{N-1}[K] + \binom{2N}{N} \left(\frac{1}{2}\right)^{2N}, \quad \mathbb{E}_1[K] = \frac{1}{2}.$$

By solving this equation, show that

$$\mathbb{E}_N[K] = \binom{N+1/2}{N} - 1,$$

as was found by Feller (1957, p. 212) using a different method. Finally, show that, via (A.95),

$$\mathbb{E}_N[K] \approx -1 + 2\sqrt{N/\pi}.$$

These results can be extended to expressions for all moments of K. For example,

$$\mathbb{V}(K) = [N - \mathbb{E}_N(K)] - \mathbb{E}_N(K)[1 + \mathbb{E}_N(K)] \approx \left(2 - \frac{4}{\pi}\right) N - \frac{2}{\sqrt{\pi}}\sqrt{N} + 2 \tag{4.60}$$

and

$$\mathbb{E}_N(K^3) = 6\binom{N+\frac{3}{2}}{N} + 7\binom{N+\frac{1}{2}}{N} - 12N - 13,$$

with higher moments similarly obtained.   (Contributed by Markus Haas)

4.14. ★ ★ Recall Example 4.24, in which

$$\mathbb{E}[M \mid N] = \sum_{m=0}^{n} m \, c_{m,n} = \sum_{m=1}^{n} \frac{1}{(m-1)!} \sum_{i=0}^{n-m} \frac{(-1)^i}{i!}.$$

Show that $\mathbb{E}[M \mid N] = 1$ by induction.

4.15. Let $X \sim \text{NBin}(r, p)$ with mass function (4.19) and define $\lambda = r(1-p)$. Let $r \to \infty$ and $p \to 1$ such that λ stays constant. Show that $X \stackrel{\text{asy}}{\sim} \text{Poi}(\lambda)$.

5

Multivariate random variables

A lecture is a process by which the notes of the professor become the notes of the students without passing through the minds of either.　　(R. K. Rathbun)

A graduation ceremony is an event where the commencement speaker tells thousands of students dressed in identical caps and gowns that individuality is the key to success.　　(Robert Orben)

Many applications involve several random variables, either because (i) more than one random quantity is associated or observed in conjunction with the process of interest, or (ii) the variable(s) of interest can be expressed as functions of two or more (possibly unobserved) random variables. This chapter is concerned with the former, and the next chapter with the latter.

5.1 Multivariate density and distribution

Similar to the univariate case, the n-variate vector function

$$\mathbf{X} = (X_1, X_2, \ldots, X_n) = (X_1(\omega), X_2(\omega), \ldots, X_n(\omega)) = \mathbf{X}(\omega)$$

is defined to be a (*multivariate* or *vector*) *random variable* relative to the collection of events $\mathcal{A}$ from the probability space $\{\Omega, \mathcal{A}, \Pr(\cdot)\}$ if and only if it is a function with domain Ω and range $\mathbb{R}^n$ and such that

for every $\mathbf{x} = (x_1, \ldots, x_n) \in \mathbb{R}^n$, $\quad \{\omega \in \Omega \mid X_i(\omega) \leq x_i, \ i = 1, \ldots n\} \in \mathcal{A}$.

Thus, the random variable $\mathbf{X}$ maps the sample space Ω to $\mathbb{R}^n$ such that, for any real n-dimensional vector $\mathbf{x}$, the set of elements in Ω which are associated with real n-vectors which are elementwise less than or equal to $\mathbf{x}$ belong to $\mathcal{A}$ and are measurable.

Similar to the univariate case, the multivariate random variable $\mathbf{X}$ induces the probability space $\{\mathbb{R}^n, \mathcal{B}^n, \Pr(\cdot)\}$, where $\mathcal{B}^n$ is the Borel σ-field of $\mathbb{R}^n$. For $\mathbf{a} = (a_1, \ldots, a_n)$

Fundamental Probability: A Computational Approach　M.S. Paolella
© 2006 John Wiley & Sons, Ltd

and $\mathbf{b} = (b_1, \ldots, b_n)$, the subset of $\mathbb{R}^n$ defined by $\{\mathbf{x} : \mathbf{a} < \mathbf{x} \le \mathbf{b}\}$ is called a *hyper-rectangle* $\mathbb{R}^n$. Similar to the univariate case, $\mathcal{B}^n$ is generated by the collection of hyper-rectangles in $\mathbb{R}^n$.

The support $\mathcal{S}$ of $\mathbf{X}$ is the smallest closed subset of $\mathbb{R}^n$ such that $\Pr(\mathcal{S}) = 1$; if there exists a closed subset $\mathcal{P}$ in $\mathbb{R}^n$ such that $\mathbf{X}(\omega) \in \mathcal{P} \ \forall \ \omega \in \mathcal{S}$, then $\mathbf{X}$ has bounded support, which can usually be taken to be the subset of $\mathbb{R}^n$ such that $f_{\mathbf{X}}$ (see below) is positive. Similar to the univariate case, $\mathbf{X}$ is discrete if it has a finite or countably infinite support, and is continuous otherwise (we do not consider any 'mixed' r.v.s).

5.1.1   Joint cumulative distribution functions

The (*multivariate* or *vector* or *joint*) cumulative distribution function of $\mathbf{X}$, denoted $F_{\mathbf{X}}(\cdot)$ is defined to be the function with domain $\mathbb{R}^n$ and range $[0, 1]$ given by

$$F_{\mathbf{X}}(\mathbf{x}) = \Pr(\mathbf{X} \le \mathbf{x}) := \Pr(-\infty < X_i \le x_i, \ i = 1, \ldots, n)$$

for any $\mathbf{x} \in \mathbb{R}^n$, where vector inequalities are defined to operate elementwise on the components. It is sometimes useful to write the individual elements of the vector $\mathbf{x} = [x_1, \ldots, x_n]$, in which case we omit one set of parentheses and use the shorthand notation

$$F_{\mathbf{X}}(\mathbf{x}) = F_{\mathbf{X}}(x_1, \ldots, x_n) := F_{\mathbf{X}}([x_1, \ldots, x_n]).$$

Recall that, for the univariate r.v. X, $\Pr(a < X \le b)$ can be expressed in terms of the c.d.f. as $F_X(b) - F_X(a)$. Now consider the bivariate r.v. $\mathbf{X} = (X_1, X_2)$. To show that the probability of any rectangle event

$$I := \mathbb{I}_{(a_1,a_2),(b_1,b_2)} := \{(x_1, x_2) : a_1 < x_1 \le b_1, \ a_2 < x_2 \le b_2\} \in \mathcal{B}^2$$

can be expressed in terms of the joint c.d.f., let (measurable) event $E_{(b_1,b_2)} := \mathbb{I}_{(-\infty,-\infty),(b_1,b_2)}$ and note that I can be expressed as the intersection of various $E_{(\cdot,\cdot)}$ and their complements, i.e.

$$I = E_{(b_1,b_2)} E^c_{(a_1,b_2)} E^c_{(b_1,a_2)} = E_{(b_1,b_2)} \setminus \left(E_{(a_1,b_2)} \cup E_{(b_1,a_2)} \right) \qquad (5.1)$$

from De Morgan's law. Thus, $0 \le \Pr(I) \le 1$. Also, from (2.8),

$$\begin{aligned} \Pr(I) &= \Pr\left(E_{(b_1,b_2)}\right) - \Pr\left(E_{(a_1,b_2)} \cup E_{(b_1,a_2)}\right) \\ &= \Pr\left(E_{(b_1,b_2)}\right) - \left(\Pr\left(E_{(a_1,b_2)}\right) + \Pr\left(E_{(b_1,a_2)}\right) - \Pr\left(E_{(a_1,a_2)}\right)\right) \\ &= F(b_1, b_2) - F(a_1, b_2) - F(b_1, a_2) + F(a_1, a_2), \end{aligned}$$

where $E_{(a_1,b_2)} E_{(b_1,a_2)} = E_{(a_1,a_2)}$. That is, for $a_1 \le b_1, \ a_2 \le b_2$,

$$\Pr(a_1 < X_1 \le b_1, a_2 < X_2 \le b_2) = F(b_1, b_2) - F(a_1, b_2) - F(b_1, a_2) + F(a_1, a_2). \qquad (5.2)$$

This becomes intuitively obvious when using a graphical depiction as in Figure 5.1.

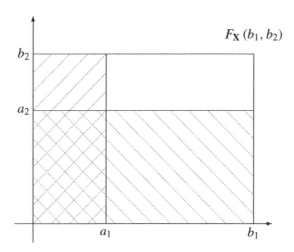

Figure 5.1 $\Pr(a_1 < X_1 < b_1, a_2 < X_2 < b_2)$ is equivalent to $F(b_1, b_2) - F(a_1, b_2) - F(b_1, a_2) + F(a_1, a_2)$

⊖ **Example 5.1** Let X and Y have a joint distribution function $F_{X,Y}$. Depending on the distribution of X and Y, it might be easier to compute $\overline{F}_{X,Y}(x, y) = \Pr(X > x, Y > y)$ instead of $F_{X,Y}$, or it could be the case that only $\overline{F}$ is available. (For example, when X and Y are bivariate normal, tables of $\overline{F}_{X,Y}$ are more common than those of $F_{X,Y}$.) From (5.2) with $b_1 = b_2 = \infty$, $\overline{F}_{X,Y}$ and $F_{X,Y}$ are related by

$$\overline{F}_{X,Y}(x, y) = \Pr(x < X < \infty, \ y < Y < \infty)$$

$$= F_{X,Y}(\infty, \infty) - F_{X,Y}(x, \infty) - F_{X,Y}(\infty, y) + F_{X,Y}(x, y)$$

$$= 1 - F_X(x) - F_Y(y) + F_{X,Y}(x, y). \tag{5.3}$$

Thus, computation of the marginal c.d.f.s is necessary. ∎

Expression (5.2) implies that, in order for a function $F(x_1, x_2)$ to be a c.d.f. of a bivariate r.v. it must, among other things, satisfy

$$F(b_1, b_2) - F(a_1, b_2) - F(b_1, a_2) + F(a_1, a_2) \geq 0. \tag{5.4}$$

For the trivariate case with r.v. $\mathbf{X} = (X_1, X_2, X_3)$ and defining $E_{(b_1, b_2, b_3)}$ in an obvious way as $\mathbb{I}_{(-\infty, -\infty, -\infty), (b_1, b_2, b_3)}$, then, with $a_i \leq b_i$, $i = 1, 2, 3$,

$$I := \mathbb{I}_{(a_1, a_2, a_3), (b_1, b_2, b_3)} = E_{(b_1, b_2, b_3)} E^c_{(a_1, b_2, b_3)} E^c_{(b_1, a_2, b_3)} E^c_{(b_1, b_2, a_3)}$$

$$= E_{(b_1, b_2, b_3)} \setminus \left(E_{(a_1, b_2, b_3)} \cup E_{(b_1, a_2, b_3)} \cup E_{(b_1, b_2, a_3)} \right)$$

and, from (2.8) and Poincaré's theorem (2.11), $\Pr(I)$ is given by

$$\Pr\left(E_{(b_1, b_2, b_3)}\right) - \Pr\left(E_{(a_1, b_2, b_3)} \cup E_{(b_1, a_2, b_3)} \cup E_{(b_1, b_2, a_3)}\right)$$

$$= \Pr\left(E_{(b_1, b_2, b_3)}\right)$$

$$
\begin{aligned}
&-\left(
\begin{array}{c}
\Pr\left(E_{(a_1,b_2,b_3)}\right) + \Pr\left(E_{(b_1,a_2,b_3)}\right) + \Pr\left(E_{(b_1,b_2,a_3)}\right) \\
- \Pr\left(E_{(a_1,b_2,b_3)} E_{(b_1,a_2,b_3)}\right) - \Pr\left(E_{(a_1,b_2,b_3)} E_{(b_1,b_2,a_3)}\right) \\
- \Pr\left(E_{(b_1,a_2,b_3)} E_{(b_1,b_2,a_3)}\right) + \Pr\left(E_{(a_1,b_2,b_3)} E_{(b_1,a_2,b_3)} E_{(b_1,b_2,a_3)}\right)
\end{array}
\right) \\
&= F\left(b_1, b_2, b_3\right) - F\left(a_1, b_2, b_3\right) - F\left(b_1, a_2, b_3\right) - F\left(b_1, b_2, a_3\right) \\
&\quad + F\left(a_1, a_2, b_3\right) + F\left(a_1, b_2, a_3\right) + F\left(b_1, a_2, a_3\right) - F\left(a_1, a_2, a_3\right). \quad (5.5)
\end{aligned}
$$

Expression (5.5) must be nonnegative in order for $F(x_1, x_2, x_3)$ to be a valid c.d.f. The constraint in the general case for the joint c.d.f. of n r.v.s can be determined by generalizing the above expressions via Poincaré's theorem.

For the n-variate vector random variable $\mathbf{X}$, the c.d.f. $F_{\mathbf{X}}$ has the properties:

(i) $0 \leq F_{\mathbf{X}}(\mathbf{x}) \leq 1$ for all $\mathbf{x} \in \mathbb{R}^n$;

(ii) $F_{\mathbf{X}}(\cdot)$ is nondecreasing, i.e. if $\mathbf{x}_1 \leq \mathbf{x}_2$, then $F_{\mathbf{X}}(\mathbf{x}_1) \leq F_{\mathbf{X}}(\mathbf{x}_2)$ – in particular, $F_{\mathbf{X}}(\mathbf{x})$ is nondecreasing in each element of $\mathbf{x}$;

(iii) $F_{\mathbf{X}}(\cdot)$ is right continuous in each element of $\mathbf{x}$, i.e. $\lim_{x_i \to x_{i,0}^+} F_{\mathbf{X}}(\mathbf{x}) = F_{\mathbf{X}}(\mathbf{x}_0)$ $\forall x_{i,0} \in \mathbb{R}$, $i = 1, \ldots, n$;

(iv) $\lim_{x_i \to -\infty} F_{\mathbf{X}}(\mathbf{x}) = 0$, $i = 1, \ldots, n$, and $\lim_{\mathbf{x} \to \infty} F_{\mathbf{X}}(\mathbf{x}) = 1$;

(v) in the bivariate case, (5.4) holds, with appropriate generalization to the n-variate case via Poincaré's theorem.

Proofs of (i)–(iv) are similar to those for univariate F_X.

⊖ ***Example 5.2*** (Rohatgi, 1984, p. 116) It is straightforward to verify that the function

$$
F_{X,Y}(x, y) = 1 - \exp\{-x - y\}\, \mathbb{I}_{(0,\infty)}(x)\, \mathbb{I}_{(0,\infty)}(y)
$$

satisfies conditions (i) to (iv) above. However, for $0 < x_1 < x_2$ and $0 < y_1 < y_2$, condition (v) fails, because the l.h.s. of (5.4) simplifies to $e^{-x_2-y_1} - e^{-x_2-y_2} < 0$. Moreover, from (5.8) below, $f_{X,Y}$ is given by $-e^{-x-y}$, confirming that $F_{X,Y}$ is not a valid c.d.f. ∎

5.1.2 Joint probability mass and density functions

Let $\mathbf{X} = (X_1, \ldots, X_n)$ be a discrete vector random variable. The *multivariate probability mass function*, or just p.m.f., of $\mathbf{X}$ is given by

$$
f_{\mathbf{X}}(\mathbf{x}) = \Pr(\{\omega\}) = \Pr(\omega : \omega \in \Omega \mid X_i(\omega) = x_i \quad i = 1, \ldots, n), \quad (5.6)
$$

where $\mathbf{x} = (x_1, \ldots, x_n) \in \mathbb{R}^n$. The p.m.f. has the properties that $f_\mathbf{X}(\mathbf{x}) = 0$ for any $\mathbf{X}(\omega) \notin \mathcal{S}_\mathbf{X}$ and

$$\sum_{\omega \in \Omega} \Pr(\{\omega\}) = \sum_{\mathbf{x} \in \mathcal{S}_\mathbf{X}} f_\mathbf{X}(\mathbf{x}) = \sum_{j_1=-\infty}^{\infty} \sum_{j_2=-\infty}^{\infty} \cdots \sum_{j_n=-\infty}^{\infty} \Pr(X_1 = j_1, \ldots, X_n = j_n) = 1.$$

For continuous $\mathbf{X}$, the function $f_\mathbf{X}(\mathbf{x})$ is the *multivariate probability density function*, or just p.d.f., if, for all $A \in \mathcal{B}^n$,

$$\Pr(\mathbf{X} \in A) = \int \cdots \int_A f_\mathbf{X}(\mathbf{x}) \, d\mathbf{x}. \tag{5.7}$$

The p.d.f. has the property that $\int_{\mathbb{R}^n} f_\mathbf{X}(\mathbf{x}) \, d\mathbf{x} = 1$. The integral in (5.7) refers to an area or volume in $\mathbb{R}^n$ and, for our purposes, will always be equal to any of the $n!$ iterated integrals (see Section A.3.4).

We illustrate this with the event for which A is a hyper-rectangle in $\mathbb{R}^n$, i.e. $A = \{\mathbf{x} : a_i < x_i \leq b_i, i = 1, \ldots, n\}$, so that

$$\Pr(\mathbf{X} \in A) = \int_{a_1}^{b_1} \cdots \int_{a_n}^{b_n} f_\mathbf{X}(\mathbf{x}) \, dx_n \cdots dx_1.$$

As an iterated integral, it is to be evaluated from the innermost univariate integral outwards, holding $x_1, \ldots, x_{n-i}$ constant when evaluating the ith one. Another common event is when

$$A = \{\mathbf{x} : a_i(\mathbf{x}_{i-1}) < x_i \leq b_i(\mathbf{x}_{i-1}), \ i = 1, \ldots, n\},$$

where $\mathbf{x}_j := (x_1, \ldots, x_j)$, i.e. bounds a_i and b_i are functions of $x_1, x_2, \ldots, x_{i-1}$.

Example 5.3 Let $f_{X,Y}(x, y) = e^{-y} \mathbb{I}_{(0,\infty)}(x) \mathbb{I}_{(x,\infty)}(y)$, i.e. $f_{X,Y}(x, y) > 0$ if and only if $0 < x < y < \infty$. Then, for value $a > 0$, $\Pr(aX < Y) = 1$ for $a \leq 1$ and, for $a > 1$,

$$\Pr(aX < Y) = \iint_{ax<y} f_{X,Y}(x, y) \, dy dx = \int_0^\infty \int_{ax}^\infty e^{-y} dy dx$$

$$= \int_0^\infty e^{-ax} dx = \frac{1}{a}, \quad a > 1.$$

See also Example 8.12. ∎

Example 5.4 Let (X, Y, Z) be jointly distributed with density

$$f_{X,Y,Z}(x, y, z) = k \, xy \exp\left(-\frac{x+y+z}{3}\right) \mathbb{I}_{(0,\infty)}(x) \mathbb{I}_{(0,\infty)}(y) \mathbb{I}_{(0,\infty)}(z).$$

Constant k is given by

$$1 = k \int_0^\infty \int_0^\infty \int_0^\infty xy \exp\left(-\frac{x+y+z}{3}\right) dz\,dy\,dx = 243k,$$

i.e. $k = 1/243$. There are 3! ways of expressing $\Pr(X < Y < Z) = 7/108$, given by

$$\int_0^\infty \int_0^z \int_0^y f\,dx\,dy\,dz, \qquad \int_0^\infty \int_y^\infty \int_0^y f\,dx\,dz\,dy, \qquad \int_0^\infty \int_0^z \int_x^z f\,dy\,dx\,dz,$$

and

$$\int_0^\infty \int_x^\infty \int_x^z f\,dy\,dz\,dx, \qquad \int_0^\infty \int_0^y \int_y^\infty f\,dz\,dx\,dy, \qquad \int_0^\infty \int_x^\infty \int_y^\infty f\,dz\,dy\,dx,$$

where f stands for $f_{X,Y,Z}(x, y, z)$. ∎

Remark: Students less adept at multivariate calculus might profit from a 'procedure' to construct integrals common in this context. First using Example 5.3 to illustrate, the desired probability $\Pr(aX < Y)$ involves *two* r.v.s, so that a *double* integral is required.[1] Starting from

$$\int_?^? \int_?^? e^{-y} dy\,dx,$$

the bounds on the inner integral are obtained from the constraint that $ax < y$, i.e. $\int_{ax}^\infty$. For the outer integral, one might be tempted to use this constraint again, producing $\int_0^{y/a}$. However, this involves y, *which has already been integrated out*. The limits on x are, using the indicator function and the constraint, $0 < x < y/a < \infty$. As the y/a upper limit is no longer applicable, the next one is used, which is ∞. Similarly, if we started with

$$\int_?^? \int_?^? e^{-y} dx\,dy,$$

the inner integral is $\int_0^{y/a}$, and for the outer integral in y, use the tightest bounds on x from the constraint $0 < x < ya < \infty$ *which do not involve variables already integrated out*, yielding $\int_0^\infty$.

Similarly with Example 5.4, as $\Pr(X < Y < Z)$ involves three r.v.s, we need a triple integral. Write

$$\int_?^? \int_?^? \int_?^? f\,dy\,dz\,dx$$

---

[1] While this is true in general, note that $\Pr(aX < Y) = \Pr(aX - Y < 0)$, so that, if the distribution of $aX - Y$ can be determined, then only (at most) one integral is necessary. Chapter 8 will show this in more detail.

and note the constraint on y for the innermost integral is $0 < x < y < z < \infty$. The tightest bound on y is $x < y < z$, which can be used because x and z 'are still available'. For the middle integral over z, bound $y < z < \infty$ is invalid because y has been integrated out. The next is $x < z < \infty$. The outermost integral uses $0 < x < \infty$ because both y and z have been integrated out. ∎

The relation between the p.d.f. and c.d.f. in the continuous case is a natural extension of the univariate case. In particular, for the continuous bivariate case,

$$f_{X,Y}(x, y) = \frac{\partial^2 F_{X,Y}(x, y)}{\partial x \partial y}, \tag{5.8}$$

with the extension to n random continuous variables $X_1, \ldots, X_n$ analogously given. Also, similar to the interpretation of $\Pr(X \in (x, x + \triangle x))$ as being approximately equal to $f_X(x) \triangle x$, we have, for $n = 2$,

$$\Pr(a < X \le a + \triangle a, \, b < Y \le b + \triangle b) = \int_b^{b + \triangle b} \int_a^{a + \triangle a} f_{X,Y}(x, y) \, dx \, dy$$

$$\approx f_{X,Y}(a, b) \triangle a \triangle b,$$

with similar results holding for general n.

5.2 Fundamental properties of multivariate random variables

5.2.1 Marginal distributions

For a given n-variate c.d.f. $F_{\mathbf{X}}$, interest often centers on only a subset of $\{X_i, i = 1, \ldots, n\}$. The p.d.f. and c.d.f. of this subset can be derived from $F_{\mathbf{X}}$ and are referred to as the *marginal* density and distribution for the chosen subset, respectively, of which there are a total of $2^n - 2$. In the bivariate continuous case, the marginal densities of X and Y are given by

$$\boxed{f_X(x) = \int_{-\infty}^{\infty} f_{X,Y}(x, y) \, dy, \quad f_Y(y) = \int_{-\infty}^{\infty} f_{X,Y}(x, y) \, dx}$$,

with marginal c.d.f.s given by

$$F_X(x) = \int_{-\infty}^{x} \int_{-\infty}^{\infty} f_{X,Y}(t, y) \, dy \, dt = \int_{-\infty}^{x} f_X(z) \, dz$$

$$F_Y(y) = \int_{-\infty}^{y} \int_{-\infty}^{\infty} f_{X,Y}(x, t) \, dx \, dt = \int_{-\infty}^{y} f_X(z) \, dz.$$

Similar expressions hold in the discrete case with summations replacing integrals. In the general case for joint c.d.f. $F_X = F_{(X_1, \dots, X_n)}$ with joint density

$$f_X(x) = \frac{\partial^n F_X(x)}{\partial x_1 \cdots \partial x_n},$$

the marginal density of subset $X_m := (X_{i_1}, \dots, X_{i_m})$, $1 \leq m \leq n$, is obtained by *integrating out* (or *summing out*) all of the remaining X_j, i.e.

$$X_{\overline{m}} := \{X_j : j \in \{1, \dots, n\} \setminus \{i_1, \dots, i_m\}\}.$$

That is,

$$f_{X_m}(x_m) = \int_{x_{\overline{m}} \in \mathbb{R}^{n-m}} dF_{X_{\overline{m}}}(x_{\overline{m}}) = \begin{cases} \sum_{x_{\overline{m}} \in \mathbb{Z}^{n-m}} f_X(x), & \text{if } X \text{ is discrete,} \\ \int_{x_{\overline{m}} \in \mathbb{R}^{n-m}} f_X(x) \, dx_{\overline{m}}, & \text{if } X \text{ is continuous.} \end{cases}$$

Remark: As often arises with such general math expressions, things look complicated while the underlying concept is quite simple. For example, the bivariate marginal density $f_{(X_1, X_3)}$ of the trivariate continuous density $f_X = f_{(X_1, X_2, X_3)}$ is given by $\int_{\mathbb{R}} f_X(x_1, x_2, x_3) \, dx_2$. ∎

Some more insight into a marginal distribution can be gained by examining a limiting case of the joint c.d.f. For example, in the bivariate case with c.d.f. F_{X_1, X_2} and monotone increasing events $E_{(b_1, b_2)}$, $b_2 = 1, 2, \dots$,

$$\lim_{b_2 \to \infty} E_{(b_1, b_2)} = E_{(b_1)} = \{x : -\infty < x \leq b_1\},$$

so that, from (2.23),

$$\lim_{b_2 \to \infty} F(b_1, b_2) = \lim_{b_2 \to \infty} \Pr\left(E_{(b_1, b_2)}\right) = \Pr\left(\lim_{b_2 \to \infty} E_{(b_1, b_2)}\right) = \Pr\left(E_{(b_1)}\right) = F_{X_1}(b_1).$$

⊖ **Example 5.5** (Example 4.17 cont.) Envisage the (slightly) more realistic situation in which the length ℓ is not fixed but specified by the customer as one of four possible values – 10, 20, 30 or 40 cm; and the density k can be chosen as one of two values, 40 or 80 g/cm³. Past sales records were used to obtain the probability of a customer choosing one of the 24 possible combinations, $f_{d,\ell,k}$, and given as follows:

$360 \, f_{d,\ell,k}$	$k = 40$			$k = 80$		
ℓ \ d	1	2	3	1	2	3
10	30	15	5	30	25	15
20	25	20	5	20	20	10
30	20	10	10	20	10	5
40	10	10	20	10	10	5

This is an example of a discrete trivariate mass function (5.6). For instance,

$$\Pr(d = 2, \ell = 10, k = 80) = 25/360 = \frac{5}{72}$$

and

$$\Pr(d = 2, \ell \le 20) = \frac{15 + 20 + 25 + 20}{360} = \frac{2}{9}.$$

The marginals are computed to be

$360\, f_{d,\ell}$	d		
ℓ	1	2	3
10	60	40	20
20	45	40	15
30	40	20	15
40	20	20	25

$360\, f_{d,k}$	k	
d	40	80
1	85	80
2	55	65
3	40	35

$360\, f_{\ell,k}$	k	
ℓ	40	80
10	50	70
20	50	50
30	40	35
40	40	25

and

d	1	2	3
$360\, f_d$	165	120	75

ℓ	10	20	30	40
$360\, f_\ell$	120	100	75	65

and

k	40	80
$360\, f_d$	180	180

∎

5.2.2 Independence

It's déjà vu all over again.   (Yogi Berra)
Reproduced by permission of Workman Publishing Company

We have already informally encountered the notion of *independence* several times above in the context of random sampling. Independence is an extremely important characteristic of a set of random variables and informally means that they have 'nothing whatsoever to do with one another'. For example, if X and Y are r.v.s modeling how much evening television you watch and weekly sales from a certain car dealership in Belgium, respectively, one might assume a priori that X and Y are independent, as it seems clear that neither can influence the other in any possible way (unless you happen to work for that car dealership!). More formally, r.v.s $X_1, \ldots, X_n$ are said to be *mutually independent* or just *independent* if, for all rectangles $I_{\mathbf{a},\mathbf{b}} = \mathbb{I}_{(a_1,a_2,\ldots,a_n),(b_1,b_2,\ldots,b_n)}$ for which $a_i \le b_i$, $i = 1, \ldots, n$,

$$\Pr(\mathbf{X} \in I) = \Pr\left(X_1 \in I_{a_1,b_1}, \ldots, X_n \in I_{a_n,b_n}\right) = \prod_{i=1}^{n} \Pr\left(X_i \in I_{a_i,b_i}\right).$$

As a special case, this implies that the joint c.d.f. can be expressed as $F_{\mathbf{X}}(\mathbf{x}) = \prod_{i=1}^{n} F_{X_i}(x_i)$. It is not hard to show[2] that this definition is equivalent with one in terms of the p.d.f. of $\mathbf{X}$: The r.v.s $X_1, \ldots, X_n$ are independent if and only if their joint density can be factored as

$$f_{\mathbf{X}}(\mathbf{x}) = \prod_{i=1}^{n} f_{X_i}(x_i) .$$
(5.9)

This can be further generalized to the case when subsets of $X_1, \ldots, X_n$ are independent of other (nonoverlapping) subsets of $X_1, \ldots, X_n$ but such that the r.v.s of a given subset are not necessarily independent of one another. Instead of struggling with the notation necessary to show this in general, we give the case for a six-dimensional random variable and three subsets. If $\mathbf{X} = (X_1, \ldots, X_6)$ and

$$f_{\mathbf{X}}(\mathbf{x}) = f_{(X_1, X_2)}(x_1, x_2) \, f_{(X_3, X_4, X_5)}(x_3, x_4, x_5) \, f_{X_6}(x_6) ,$$
(5.10)

then the subsets of r.v.s $\{X_1, X_2\}$, $\{X_3, X_4, X_5\}$ and $\{X_6\}$ are independent. Note that X_1 and X_2 need not be independent of one another, etc.

In Example 5.5, observe that

$$\frac{1}{6} = \Pr(d = 1, \ell = 10) \neq \Pr(d = 1) \cdot \Pr(\ell = 10) = \frac{165}{360} \frac{120}{360} = \frac{11}{72},$$

which, from (5.9), implies that ℓ and d are not independent. A similar calculation reveals that k is not independent of $\{\ell, d\}$.

If n random variables $X_1, \ldots, X_n$ are not only independent but each follow the same distribution, i.e. $F_{X_i} = F_X$, $i = 1, \ldots, n$, for some distribution F_X, then $X_1, \ldots, X_n$ are said to be *independently and identically distributed*, which is often abbreviated as *i.i.d.*, spoken 'eye eye dee', and expressed as $X_i \overset{\text{i.i.d.}}{\sim} f_X$. If the X_i are independent and described or *indexed* by the same family of distributions but with different parameters, say θ_i, then we write $X_i \overset{\text{ind}}{\sim} f_{X_i}(x_i; \theta_i)$ or, to emphasize that the functional form of f is the same, use the distributional name, e.g. $X_i \overset{\text{ind}}{\sim} \text{Ber}(p_i)$.

5.2.3 Exchangeability

A set of n random variables is said to be *exchangeable* if every permutation of them has the same joint distribution. In most situations, a symmetric structure will be apparent and so it will not be necessary to actually verify exchangeability directly from the definition. Exchangeability is a weaker assumption than i.i.d. In particular, a set of

---

[2] See, for example, Mittelhammer (1996, p. 91).

i.i.d. r.v.s are exchangeable, but an exchangeable set may or may not be i.i.d. If the events A_i, $i = 1, \ldots, n$, are exchangeable, then

$$\Pr\left(A_{i_1} A_{i_2} \cdots A_{i_j}\right) = \Pr\left(A_1 A_2 \cdots A_j\right), \quad 1 \le j \le n,$$

so that Poincaré's and De Moivre–Jordan's theorems simplify to

$$\Pr\left(\bigcup_{i=1}^{n} A_i\right) = \sum_{i=1}^{n} (-1)^{i+1} \binom{n}{i} \Pr\left(A_1 \cdots A_i\right), \tag{5.11}$$

$$p_{m,n}\left(\{A_i\}\right) = \sum_{i=0}^{n-m} (-1)^i \binom{m+i}{m} \binom{n}{m+i} \Pr\left(A_1 \cdots A_{m+i}\right) \tag{5.12}$$

and

$$P_{m,n}\left(\{A_i\}\right) = \sum_{i=0}^{n-m} (-1)^i \binom{m+i}{m} \binom{n}{m+i} \Pr\left(A_1 \cdots A_{m+i}\right), \tag{5.13}$$

respectively. Note that, in all the examples in Chapters 2 and 3 which used these theorems, the A_i were always exchangeable.

A nice discussion of exchangeability is given by Johnson and Kotz (1977, Section 2.9). Other good sources include Lindley and Novick (1981); Koch (1982); Port (1994, Chapter 15 and p. 200(15.3)); Bernardo and Smith (1994, Chapter 4) and Gelman, Carlin, Stern and Rubin (2003).

5.2.4 Transformations

Generalizing the univariate p.m.f. case, interest might center on one or more functions of the r.v.s $X_1, \ldots, X_n$ from a particular n-variate p.m.f. The resulting functions, say $Y_1 = g_1(X_1, \ldots, X_n)$, $Y_2 = g_2(X_1, \ldots, X_n)$, $\ldots$, $Y_m = g_m(X_1, \ldots, X_n)$ are themselves r.v.s and give rise to an m-variate p.m.f. with probabilities $\Pr(Y_1 = y_1, \ldots, Y_m = y_m)$ calculated from

$$\sum_{j_1=-\infty}^{\infty} \sum_{j_2=-\infty}^{\infty} \cdots \sum_{j_n=-\infty}^{\infty} \Pr(X_1 = j_1, \ldots, X_n = j_n) \prod_{i=1}^{m} \mathbb{I}\left(g_i(X_1, \ldots, X_n) = y_i\right).$$

Example 5.6 (Example 5.5 cont.) The weight W of the shaft is a function of the r.v.s d, ℓ and k, namely $w = \ell k \pi d^2 / 4$, so that the p.m.f. is as follows.

$4w/\pi = \ell k d^2$	$360 \Pr(W = w)$	$\{d, \ell, k\}$
400	30	$\{1, 10, 40\}$
800	$30 + 25$	$\{1, 10, 80\}, \{1, 20, 40\}$
1600	$20 + 10 + 15$	$\{1, 20, 80\}, \{1, 40, 40\}, \{2, 10, 40\}$
1200	20	$\{1, 30, 40\}$
2400	20	$\{1, 30, 80\}$
3200	$10 + 25 + 20$	$\{1, 40, 80\}, \{2, 10, 80\}, \{2, 20, 40\}$
6400	$20 + 10$	$\{2, 20, 80\}, \{2, 40, 40\}$
4800	10	$\{2, 30, 40\}$
9600	10	$\{2, 30, 80\}$
12800	10	$\{2, 40, 80\}$
3600	5	$\{3, 10, 40\}$
7200	$15 + 5$	$\{3, 10, 80\}, \{3, 20, 40\}$
14400	$10 + 20$	$\{3, 20, 80\}, \{3, 40, 40\}$
10800	10	$\{3, 30, 40\}$
21600	5	$\{3, 30, 80\}$
28800	5	$\{3, 40, 80\}$

From this, any quantity of interest can be calculated, such as $\Pr(W \leq 10\,000)$ and $\mathbb{E}[W]$. ∎

Transformations involving continuous random variables will be discussed at length in Chapters 7 and 8.

5.2.5 Moments

The invalid assumption that correlation implies cause is probably among the two or three most serious and common errors of human reasoning.

(Stephen Jay Gould, The Mismeasure of Man)

The moments of a single random variable were introduced in Section 4.4, whereby the rth raw moment of random variable X is $\mu'_r = \mathbb{E}[X^r] = \int x^r \mathrm{d}F_X(x)$ and, with $\mu = \mathbb{E}[X]$, the rth central moment of X is $\mu_r = \mathbb{E}[(X - \mu)^r] = \int (X - \mu)^r \mathrm{d}F_X(x)$. The most important central moment is for $r = 2$, which is the variance of X, denoted by σ_X^2. The extension to the multivariate case will be seen to follow along very similar lines.

However, before doing so, we return to the calculation of σ_X^2, now using some basic concepts of multivariate random variables introduced so far in this chapter. In particular, let X and Y be i.i.d. continuous random variables with p.d.f. f and c.d.f. F. Then the variance of X can be interpreted (and expressed) as half the expected squared difference between X and Y, i.e.

$$\sigma^2 = \frac{1}{2}\mathbb{E}[(X - Y)^2] = \frac{1}{2} \int_{-\infty}^{\infty} \int_{-\infty}^{\infty} (x - y)^2 f(x) f(y) \, \mathrm{d}x \, \mathrm{d}y. \qquad (5.14)$$

This is easily verified by directly computing the iterated integrals:

$$\int_{-\infty}^{\infty} \left(x^2 - 2xy + y^2\right) f(x) \, dx = \mathbb{E}\left[X^2\right] - 2y\mathbb{E}\left[X\right] + y^2$$

and

$$\int_{-\infty}^{\infty} \left(\mathbb{E}\left[X^2\right] - 2y\mathbb{E}\left[X\right] + y^2\right) f(y) \, dy = \mathbb{E}\left[X^2\right] - 2\mathbb{E}\left[X\right]\mathbb{E}\left[Y\right] + \mathbb{E}\left[Y^2\right]$$

$$= 2\mathbb{E}\left[X^2\right] - 2\left(\mathbb{E}\left[X\right]\right)^2 = 2\mathbb{V}(X).$$

Another expression for the variance is

$$\sigma^2 = 2 \iint_{-\infty < x < y < \infty} F(x)(1 - F(y)) \, dx \, dy, \tag{5.15}$$

which is attractive because it only involves the c.d.f. in the integrand. Expression (5.15) 'does not seem to be very familiar, but was known a long time ago' (Jones and Balakrishnan, 2002). It is verified in Problem 5.3.

Example 5.7 Recall the exponential distribution introduced in Example 4.6. Using the new notation introduced in Section 5.2.2, let $X, Y \overset{\text{i.i.d.}}{\sim} \text{Exp}(1)$, so that $f(x) = e^{-x}$ and $F(x) = 1 - e^{-x}$ for $x \geq 0$. Then (5.15) implies

$$\sigma^2 = 2 \int_0^{\infty} e^{-y} \int_0^y \left(1 - e^{-x}\right) dx \, dy = 2 \int_0^{\infty} e^{-y} \left[y + e^{-y} - 1\right] dy \tag{5.16}$$

$$= 2 \int_0^{\infty} \left(ye^{-y} + e^{-2y} - e^{-y}\right) dy = 1. \tag{5.17}$$

A similar calculation will reveal that, if $U, V \overset{\text{i.i.d.}}{\sim} \text{Exp}(\lambda)$, then $\sigma^2 = \lambda^{-2}$. (This latter result can be obtained much more easily by using (5.17), (4.41) and the fact that $V = X/\lambda$, which follows from expression (7.1) derived later.) ■

Remark: Jones and Balakrishnan (2002) generalized (5.15) to the third and fourth central moments to get the impressive looking

$$\mu_3 = \mathbb{E}\left[(X - \mu)^3\right] = 6 \iiint_{-\infty < x < y < z < \infty} F(x)\left[2F(y) - 1\right]\left[1 - F(z)\right] dx \, dy \, dz$$

and

$$\mu_4 = \mathbb{E}\left[(X - \mu)^4\right] = 24 \iiiint_{-\infty < w < x < y < z < \infty} F(w)\left[3F(x)F(y) - 2F(x) - F(y) + 1\right] \times \left[1 - F(z)\right] dw \, dx \, dy \, dz,$$

as well as other related formulae. ■

Moving to the multivariate case, the expected value of a function $g(\mathbf{X})$ for $g:$ $\mathbb{R}^n \to \mathbb{R}$, with respect to the n-length vector random variable $\mathbf{X}$ with p.m.f. or p.d.f. $f_{\mathbf{X}}$, is defined by

$$
\mathbb{E}\left[g(\mathbf{X})\right] = \int_{\mathbf{x} \in \mathbb{R}^n} g(\mathbf{X}) \, dF_{\mathbf{X}}(\mathbf{x}) =
\begin{cases}
\sum_{\mathbf{x} \in \mathbb{Z}^n} g(\mathbf{x}) f_{\mathbf{X}}(\mathbf{x}), & \text{if } \mathbf{X} \text{ is discrete,} \\
\int_{\mathbf{x} \in \mathbb{R}^n} g(\mathbf{x}) f_{\mathbf{X}}(\mathbf{x}) \, d\mathbf{x}, & \text{if } \mathbf{X} \text{ is continuous.}
\end{cases}
$$

(5.18)

For example, if

$$
f_{X,Y}(x, y) = ab\,e^{-ax} e^{-by} \mathbb{I}_{(0,\infty)}(x)\, \mathbb{I}_{(0,\infty)}(y)
$$

(5.19)

for $a, b \in \mathbb{R}_{>0}$, then

$$
\mathbb{E}[XY] = ab \int_0^\infty \int_0^\infty xy\,e^{-ax} e^{-by} \, dx\,dy = ab \int_0^\infty x e^{-ax} \, dx \int_0^\infty y e^{-by} \, dy = \frac{1}{ab},
$$

an easy calculation because X and Y are independent.

Often only a subset of the $\mathbf{X}$ are used in g, say $g(\mathbf{X}_m) = g(X_1, \ldots, X_m)$, $m < n$, in which case $X_{m+1}, \ldots, X_n$ are integrated out in (5.18) so that

$$
\mathbb{E}\left[g(\mathbf{X}_m)\right] = \int_{\mathbf{x} \in \mathbb{R}^m} g(\mathbf{X}) \, dF_{\mathbf{X}_m}(\mathbf{x}),
$$

where $F_{\mathbf{X}_m}$ denotes the marginal c.d.f. of $(X_1, \ldots, X_m)$. Continuing with $f_{X,Y}$ given in (5.19) and f_X the marginal p.d.f. of X,

$$
\mathbb{E}\left[X^2\right] = \int_0^\infty \int_0^\infty x^2 f_{X,Y}(x, y) \, dy\,dx
$$

$$
= \int_0^\infty x^2 \int_0^\infty f_{X,Y}(x, y) \, dy\,dx = \int_0^\infty x^2 f_X(x) \, dx = \frac{2}{a^2}.
$$

The computation of $\mathbb{E}[XY]$ above is a special case of the following. If the n components of r.v. $\mathbf{X} = (X_1, \ldots, X_n)$ are independent and function $g(\mathbf{X})$ can be partitioned as, say, $g_1(X_1) g_2(X_2) \cdots g_n(X_n)$, then

$$
\mathbb{E}_{\mathbf{X}}\left[g(\mathbf{X})\right] = \prod_{i=1}^n \int_{x_i \in \mathbb{R}} g_i(X_i) \, dF_{X_i}(x_i) = \prod_{i=1}^n \mathbb{E}_{X_i}\left[g_i(X_i)\right],
$$

(5.20)

where the notation $\mathbb{E}_Y$ denotes taking the expectation with respect to the distribution of r.v. Y. The result follows because $f_{\mathbf{X}}$ can be expressed as the product of the n marginals. For example, in the bivariate case,

$$
\mathbb{E}\left[g(X) h(Y)\right] = \mathbb{E}\left[g(X)\right] \mathbb{E}\left[h(Y)\right], \quad X \perp Y,
$$

or, with g and h the identity functions, $\mathbb{E}[XY] = \mathbb{E}[X]\mathbb{E}[Y]$. Similar to the more general subsets in (5.10), this could be generalized in a natural way.

A simple inequality of occasional use is that

$$|\mathbb{E}[UV]| \leq \mathbb{E}[|UV|]. \tag{5.21}$$

To see this, let $A = UV$. Then, $\forall A \in \mathbb{R}$, $A \leq |A|$, so that $\mathbb{E}[A] \leq \mathbb{E}[|A|]$. Similarly, as $-|A| \leq A$, $-\mathbb{E}[|A|] = \mathbb{E}[-|A|] \leq \mathbb{E}[A]$, so that $-\mathbb{E}[|A|] \leq \mathbb{E}[A] \leq \mathbb{E}[|A|]$, or $|\mathbb{E}[A]| \leq \mathbb{E}[|A|]$, which is (5.21).

A much more important result is the *Cauchy–Schwarz inequality (for random variables)*, given by

$$\boxed{|\mathbb{E}[UV]| \leq +\sqrt{\mathbb{E}[U^2]\mathbb{E}[V^2]}}, \tag{5.22}$$

for any two r.v.s U and V. Similar to the derivation of the Cauchy–Schwarz inequality in (A.7), the standard trick to proving (5.22) is to consider $\mathbb{E}[(rU + V)^2]$. Then, $\forall r \in \mathbb{R}$,

$$0 \leq \mathbb{E}[(rU + V)^2] = \mathbb{E}[U^2]r^2 + 2\mathbb{E}[UV]r + \mathbb{E}[V^2] = ar^2 + br + c, \tag{5.23}$$

where $a := \mathbb{E}[U^2]$, $b := 2\mathbb{E}[UV]$ and $c = \mathbb{E}[V^2]$, and recalling the linearity property (4.35). First let $V = -rU$, so that $0 = \mathbb{E}[0] = \mathbb{E}[(rU + V)^2]$, and the l.h.s. of (5.22) is $|\mathbb{E}[UV]| = |-r\mathbb{E}[U^2]| = |r|\mathbb{E}[U^2]$. Likewise, the r.h.s. of (5.22) is

$$\sqrt{\mathbb{E}[U^2]\mathbb{E}[V^2]} = \sqrt{\mathbb{E}[U^2]\mathbb{E}[(-rU)^2]} = \sqrt{r^2\mathbb{E}[U^2]\mathbb{E}[U^2]} = |r|\mathbb{E}[U^2],$$

and (5.22) holds with equality. Now assume $V \neq -rU$, so that the inequality in (5.23) is strict. This implies that the quadratic $ar^2 + br + c$ has no real roots, or that its discriminant $b^2 - 4ac < 0$. Substituting gives

$$\sqrt{4(\mathbb{E}[UV])^2 - 4\mathbb{E}[U^2]\mathbb{E}[V^2]} < 0$$

or $(\mathbb{E}[UV])^2 - \mathbb{E}[U^2]\mathbb{E}[V^2] < 0$, i.e. $(\mathbb{E}[UV])^2 < \mathbb{E}[U^2]\mathbb{E}[V^2]$, which gives (5.22) after taking square roots of both sides.

Interest often centers on the mean $\mu_i = \mathbb{E}[X_i]$ and variance $\sigma_i^2 = \mathbb{E}[(X_i - \mu_i)^2]$ (or possibly the standard deviation, $\sigma_i = +\sqrt{\sigma_i^2}$) of the individual components of **X**. A generalization of the variance is the *covariance*: for any two X_i, the covariance is given by

$$\boxed{\sigma_{ij} := \text{Cov}(X_i, X_j) = \mathbb{E}[(X_i - \mu_i)(X_j - \mu_j)] = \mathbb{E}[X_iX_j] - \mu_i\mu_j}, \tag{5.24}$$

where $\mu_i = \mathbb{E}[X_i]$, and is a measure of the *linear association* between the two variables. If σ_{ij} is positive then, generally speaking, relatively large (small) values of X_1

tend to occur with relatively large (small) values of X_2, while if $\sigma_{ij} < 0$, then relatively small (large) values of X_1 tend to occur with relatively large (small) values of X_2. From symmetry, $\mathrm{Cov}\left(X_i, X_j\right) = \mathrm{Cov}\left(X_j, X_i\right)$. Thus, for $i \neq j$, there are $(n^2 - n)/2$ unique covariance terms among n random variables. If X_i and X_j are independent, then from (5.20) for $i \neq j$, $\mathrm{Cov}\left(X_i, X_j\right) = \mathbb{E}\left[X_i - \mu_i\right] \mathbb{E}\left[X_j - \mu_j\right] = 0$, i.e.

$$X_i \perp X_j \quad \Rightarrow \quad \mathrm{Cov}\left(X_i, X_j\right) = 0. \tag{5.25}$$

Note that, if $i = j$, $\sigma_{ii} = \mathrm{Cov}\,(X_i, X_i) = \mathbb{V}\,(X_i) = \sigma_i^2$.

The *correlation* of two r.v.s is defined to be

$$\mathrm{Corr}\left(X_i, X_j\right) = \frac{\mathrm{Cov}\left(X_i, X_j\right)}{\sqrt{\mathbb{V}\left(X_i\right)\mathbb{V}\left(X_j\right)}} = \frac{\sigma_{ij}}{\sigma_i \sigma_j}. \tag{5.26}$$

While the covariance can, in general, be any value in $\mathbb{R}$, depending on the scaling and range of the r.v.s of interest,

$$-1 \leq \mathrm{Corr}\left(X_i, X_j\right) \leq 1, \tag{5.27}$$

with high positive (negative) correlation associated with values near 1 (-1). Bound (5.27) follows by squaring the left and right sides of the Cauchy–Schwarz inequality (5.22) and setting $U = X_1 - \mu_1$ and $V = X_2 - \mu_2$, to give

$$(\mathbb{E}\left[(X_1 - \mu_1)\,(X_2 - \mu_2)\right])^2 \leq \mathbb{E}\left[(X_1 - \mu_1)^2\right]\mathbb{E}\left[(X_2 - \mu_2)^2\right] = \sigma_1^2 \sigma_2^2$$

or

$$\left|\mathbb{E}\left[(X_1 - \mu_1)\,(X_2 - \mu_2)\right]\right| \leq \sigma_1 \sigma_2. \tag{5.28}$$

Equality holds in (5.28) when X_1 and X_2 are linearly related, i.e. say $X_2 = bX_1 + c$ so that $\mu_2 = b\mu_1 + c$, $\sigma_2^2 = b^2 \sigma_1^2$ and the l.h.s. of (5.28) becomes

$$\left|\mathbb{E}\left[(X_1 - \mu_1)\,(bX_1 - b\mu_1)\right]\right| = |b|\sigma_1^2,$$

while the r.h.s. of (5.28) is $\sigma_1 \sigma_2 = |b|\sigma_1^2$. Problem 5.2 illustrates another way of showing (5.27).

⊖ *Example 5.8* (Example 5.6 cont.) The means and variances of d, ℓ and k can be most easily computed from the marginals and are given by

$$\mu_d = \mathbb{E}\,[d] = 1 \cdot \frac{165}{360} + 2 \cdot \frac{120}{360} + 3 \cdot \frac{75}{360} = \frac{7}{4},$$

$$\sigma_d^2 = \mathbb{V}\,(d) = \left(1 - \frac{7}{4}\right)^2 \cdot \frac{165}{360} + \left(2 - \frac{7}{4}\right)^2 \cdot \frac{120}{360} + \left(3 - \frac{7}{4}\right)^2 \cdot \frac{75}{360} = \frac{29}{48}$$

and, similarly,

$$\mu_\ell = \frac{805}{36} = 22.36\overline{11}, \quad \sigma_\ell^2 = 120.814, \quad \mu_k = 60, \quad \sigma_k^2 = 400.$$

From the bivariate marginals, the covariances among the three r.v.s are computed to be

$$\text{Cov}\,(d, \ell) = 1 \cdot 10 \cdot \frac{60}{360} + 1 \cdot 20 \cdot \frac{45}{360} + \cdots + 3 \cdot 40 \cdot \frac{25}{360} - \mu_d \mu_\ell = \frac{185}{144}$$

and, similarly,

$$\text{Cov}\,(d, k) = 0, \quad \text{Cov}\,(\ell, k) = -\frac{275}{9}.$$

The correlations are thus $\text{Corr}\,(d, k) = 0$,

$$\text{Corr}\,(d, \ell) = \frac{\frac{185}{144}}{\sqrt{\frac{29}{48} \cdot 120.814}} \approx 0.150, \quad \text{Corr}\,(\ell, k) = \frac{-\frac{275}{9}}{\sqrt{120.814 \cdot 400}} \approx -0.139.\blacksquare$$

Clearly, $\text{Corr}\,(X_i, X_j) = 0$ if X_i and X_j are independent. The converse does not hold in general; a correlation (or covariance) of zero does not necessarily imply that the two r.v.s are independent.

↻ **Example 5.9**   Let X and Y be discrete random variables such that

$$\Pr\,(X = Y = 0) = \Pr\,(X = Y = m) = \Pr\,(X = -m, \, Y = m) = \frac{1}{3}$$

for any $m \neq 0$. As

$$0 = \Pr\,(X = m \mid Y = 0) \neq \Pr\,(X = m \mid Y = m) = 1/2,$$

X and Y are not independent. The marginal distribution of X is given by

$$\Pr\,(X = 0) = \Pr\,(X = m) = \Pr\,(X = -m) = 1/3,$$

while that for Y is $\Pr\,(Y = 0) = 1/3$ and $\Pr\,(Y = m) = 2/3$. Thus, $\mathbb{E}\,[X] = 0$, $\mathbb{E}\,[Y] = 2m/3$ and $\mathbb{E}\,[XY] = (0 + m^2 - m^2)/3 = 0$, so that, from (5.24), $\text{Cov}\,(X, Y) = 0$. See also Billingsley (1995, p. 281 (21.11)) for a similar example. $\blacksquare$

Remark: Before moving on to the next section, it is worth commenting on the quotation above from S. Gould regarding the misunderstanding between cause and correlation. While we all intuitively know what 'causality' means, it turns out to be a rather elusive concept which plays a central role in the philosophy of science, and so commands quite some attention and academic contemplation. Correlation, on the other hand, is a simplistic, mathematical measure of linear dependence, which has nothing, per se, to do with causality, though of course there are links between the two.

To avoid getting sidetracked, we just mention here the notion of *spurious correlation*, which describes the situation when two occurrences are statistically correlated, but there is no causation. As an example, Fair (2002) found that high grades in an introductory economics class at university are *not* related to hours of study or quality of preparation in high school, but *are* significantly positively correlated with the frequency of simply showing up to class. Taken by itself, one might conclude that going to class thus helps to secure a higher grade. A minor critique of this conclusion is simply this – we don't need statistical evidence to tell us this because, at least on average, of course going to lectures helps! The major criticism is this. The conclusion that going to lectures *is responsible for*, i.e. *is causal for* getting high grades, is not supported. The relationship is partially *spurious*, because getting high grades and going to class are both correlated with a third variable which was not controlled for: ability and desire of the student. 'Good', motivated students tend to go to class. Even if they didn't go to class, they would still get better grades than 'bad', disinterested students, who tend not to go to class in the first place!

There are many such examples. Consider the often-heard idea that children who learn to play a musical instrument do better on average in mathematics and other subjects than children who do not play music. This could well be a fact, but it is not necessarily causal. Maybe children who take part in a musical education are simply the benefactors of more guided and skilled upbringing from their parents, or maybe children who develop and keep an interest in music just have longer attention spans and/or more patience.

The third and last example is a classic one, and is more clear cut than the previous two. The number of violent crimes committed against women is significantly correlated with ice cream consumption. Should ice cream be banned? To what missing factor are both occurrences correlated? ∎

5.3 Discrete sampling schemes

Recall the four basic random sampling methods given in Section 4.2, which arise from choosing between (i) either a fixed number of random trials or sampling until a specified number of 'successes' occur, and (ii) sampling with or without replacement. Each of these four cases has a multivariate generalization, which we discuss in turn. We also discussed the Poisson and occupancy distribution in Section 4.2. Of course, these also have multivariate analogs, but we do not examine them here – see Johnson, Kotz and Balakrishnan (1997) and Johnson and Kotz (1977) for a wealth of detail on these and other discrete multivariate random variables.

5.3.1 Multinomial

Recall the binomial distribution

$$\Pr(X = x) = \binom{n}{x} p^x (1 - p)^{n-x} = \frac{n!}{x! \, (n - x)!} p^x (1 - p)^{n-x}, \quad x = 0, 1, \ldots, n,$$

where p is the 'success' probability on a given trial. By defining

$$
\begin{aligned}
p_1 &= p, & x_1 &= x, & X_1 &= X, \\
p_2 &= 1 - p, & x_2 &= n - x, & X_2 &= n - X,
\end{aligned}
$$

$\Pr(X = x)$ can be written as

$$
\Pr(X = x) = \Pr(X_1 = x_1, X_2 = x_2) = \frac{n!}{x_1! x_2!} p_1^{x_1} p_2^{x_2},
$$

where $x_1 = 0, 1, \ldots, n$, $x_2 = 0, 1, \ldots, n$ and $x_1 + x_2 = n$. Generalizing this to k possible outcomes instead of just two, we obtain the *multinomial* mass function

$$
\Pr(\mathbf{X} = \mathbf{x}) = \Pr(X_1 = x_1, X_2 = x_2, \ldots, X_k = x_k)
$$

$$
= \frac{n!}{x_1! x_2! \cdots x_k!} p_1^{x_1} p_2^{x_2} \cdots p_k^{x_k} \tag{5.29}
$$

$$
= n! \frac{\prod_{i=1}^{k} p_i^{x_i}}{\prod_{i=1}^{k} x_i!},
$$

for $0 \le x_i \le n$, $i = 1, \ldots, k$ and $\sum_{i=1}^{k} x_i = n$. The probabilities p_i sum to one as in the binomial model, i.e. $\sum_{i=1}^{k} p_i = 1$. We will use the notation $\mathbf{X} = (X_1, X_2, \ldots, X_k) \sim \text{Multinom}(n, p_1, \ldots, p_k)$ or, with $\mathbf{p} = (p_1, \ldots, p_k)$, just $\mathbf{X} \sim \text{Multinom}(n, \mathbf{p})$, to denote a vector random variable which follows the multinomial distribution.

Like the binomial, it is used to model outcomes sampled with replacement. For example, if a six-sided die is flipped n times and X_1 ($X_2, \ldots, X_6$) denotes the total number of ones (twos, ..., sixes) which were facing up, then $\mathbf{X} = (X_1, X_2, \ldots, X_6) \sim \text{Multinom}(n, 1/6, \ldots, 1/6)$. If $k = 3$, we say $\mathbf{X}$ is *trinomially* distributed, or $\mathbf{X} \sim \text{trinom}(\mathbf{x}; n, p_1, p_2)$. The usual probabilistic model is sampling from an urn with, say, r red marbles, w white marbles, etc., and X_1 is the number of red marbles drawn, X_2 is the number of white marbles drawn, etc. The k-dimensional vector $\mathbf{X}$ has only a $(k-1)$-dimensional distribution, i.e. the p.m.f. of $\mathbf{X}$ is *degenerate* (in one dimension), because $\sum X_k = n$; usually the last element is redundant, i.e. X_k is taken to be $n - \sum_{i=1}^{k-1} X_i$.

Example 5.10 (Example 4.12 cont.) The probability that A wins g rounds before B wins g rounds can also be determined using the multinomial distribution.

Let $n + 1$ be the total number of games played, t the number of ties, and b be the number of games B wins, $0 \le b \le 3$. Note that A must win the last game and three out of the n, while B can have no more than three wins in the n games. Thus, given b and t and $n = 3 + b + t$,

$$
\Pr(A) = 0.3 \times \binom{n}{3, b, t} (0.3)^3 (0.2)^b (0.5)^t .
$$

As $0 \leq b \leq 3$, the possible combinations for various n such that A wins on the $(n+1)$th trial are as follows.

$$n = 3 : \begin{pmatrix} 3 \\ 3, 0, 0 \end{pmatrix}$$

$$n = 4 : \begin{pmatrix} 4 \\ 3, 0, 1 \end{pmatrix}, \begin{pmatrix} 4 \\ 3, 1, 0 \end{pmatrix}$$

$$n = 5 : \begin{pmatrix} 5 \\ 3, 0, 2 \end{pmatrix}, \begin{pmatrix} 5 \\ 3, 1, 1 \end{pmatrix}, \begin{pmatrix} 5 \\ 3, 2, 0 \end{pmatrix}$$

$$n = 6 : \begin{pmatrix} 6 \\ 3, 0, 3 \end{pmatrix}, \begin{pmatrix} 6 \\ 3, 1, 2 \end{pmatrix}, \begin{pmatrix} 6 \\ 3, 2, 1 \end{pmatrix}, \begin{pmatrix} 6 \\ 3, 3, 0 \end{pmatrix}$$

$$n = 7 : \begin{pmatrix} 7 \\ 3, 0, 4 \end{pmatrix}, \begin{pmatrix} 7 \\ 3, 1, 3 \end{pmatrix}, \begin{pmatrix} 7 \\ 3, 2, 2 \end{pmatrix}, \begin{pmatrix} 7 \\ 3, 3, 1 \end{pmatrix}$$

$$\vdots$$

$$n \geq 6 : \begin{pmatrix} n \\ 3, 0, n-3 \end{pmatrix}, \begin{pmatrix} n \\ 3, 1, n-4 \end{pmatrix}, \begin{pmatrix} n \\ 3, 2, n-5 \end{pmatrix}, \begin{pmatrix} n \\ 3, 3, n-6 \end{pmatrix}.$$

Some thought shows that we can express this compactly as

$$\Pr(A) = 0.3 \lim_{U \to \infty} \sum_{n=3}^{U} \sum_{i=0}^{\min(3, n-3)} \begin{pmatrix} n \\ 3, i, n-3-i \end{pmatrix} (0.3)^3 (0.2)^i (0.5)^{n-3-i},$$

and, for various values of U, is numerically evaluated to be

U	Pr (A)
10	0.414
25	0.70911
50	0.710207998212
100	0.710207999991

which approaches the answer 0.710208. ∎

An interesting special case of the multinomial is if all the p_i are equal, i.e. $p_i = 1/k$, so that

$$\Pr(\mathbf{X} = \mathbf{x}) = \frac{n!}{x_1! x_2! \cdots x_k! \, k^n},$$

as $\sum_{i=1}^{k} x_i = n$. If also $k = n$ and each $x_i = 1$,

$$\Pr(\mathbf{X} = \mathbf{1}) = \Pr(X_1 = 1, X_2 = 1, \ldots, X_r = 1) = \frac{n!}{n^n} = \frac{(n-1)!}{n^{n-1}}, \tag{5.30}$$

because $1! = 1$.

⊖ ***Example 5.11*** Four indistinguishable balls are, independently of one another, randomly placed in one of four large urns. The probability that each urn contains one ball is given by (5.30) for $n = 4$, i.e. $4! / 4^4 = 3/32 \approx 0.094$. This could, of course, be computed directly as follows. The first ball can go into any urn. The next ball can go into one of the remaining three; this occurs with probability $3/4$. The third ball enters one of the two remaining empty urns with probability $2/4$ while the last enters the remaining empty urn with probability $1/4$. Multiplying these gives $(n - 1)!/n^{n-1}$ for $n = 4$.

Now consider the probability that one of the urns contains three balls. There are several ways for this to occur, which are outlined in the following table.

#	x_1	x_2	x_3	x_4
1	3	1	0	0
2	3	0	1	0
3	3	0	0	1
4	1	3	0	0
5	0	3	1	0
6	0	3	0	1
7	1	0	3	0
8	0	1	3	0
9	0	0	3	1
10	1	0	0	3
11	0	1	0	3
12	0	0	1	3

From the multinomial distribution with $n = r = 4$,

$$\Pr(X_1 = x_1, \ X_2 = x_2, \ X_3 = x_3, \ X_4 = x_4) = \frac{4!}{x_1! x_2! x_3! x_4!} \frac{1}{4^4},$$

as $p_i = 1/4$ for $i = 1, 2, 3, 4$.

Each of the 12 possibilities has the same probability; for example, for case #1 in the above table,

$$\Pr(X_1 = 3, \ X_2 = 1, \ X_3 = 0, \ X_4 = 0) = \frac{4!}{3! 1! 0! 0!} \times \frac{1}{4^4} = \frac{1}{64} \approx 0.016,$$

so that the probability that one urn contains three balls is $12/64 \approx 0.1875$.

Next, for the probability that one of the urns contains (all) four balls, an analogous calculation is

$$4 \times \Pr(X_1 = 4, \ X_2 = 0, \ X_3 = 0, \ X_4 = 0) = 4 \times \frac{4!}{4! 0! 0! 0!} \times \frac{1}{4^4} = \frac{1}{64}.$$

Lastly, the probability that no urn contains more than two balls is, using the complement, just

$$1 - \Pr(\text{one urn has } 3) - \Pr(\text{one urn has } 4) = 1 - \frac{12}{64} - \frac{1}{64} = \frac{51}{64} \approx 0.80$$

using the above results. ∎

If we randomly place n balls into k urns (multinomial setup with equal p_i), but fix attention on one particular urn, the number of balls that it ultimately holds is binomially distributed with n trials and probability $1/k$. The other $k-1$ urns can be viewed as 'failures', with probability of failure $(k-1)/k$. More generally, if $X_1, X_2, \ldots, X_k$ is multinomially distributed, based on n trials and probabilities $p_1, p_2, \ldots, p_k$, then, for a particular i, $X_i \sim \text{Bin}(n, p_i)$. This is algebraically demonstrated next for $k=3$.

$\ominus$ **Example 5.12**   Let $\mathbf{X} \sim \text{trinom}(\mathbf{x}; n, p_1, p_2)$. To find the marginal distribution of X_1, we need to sum over all possible x_2 and x_3; but x_3 is redundant so, if $X_1 = x$, then x_2 can take on values $0, 1, \ldots, n-x$, or

$$\Pr(X_1 = x) = \sum_{x_2=0}^{n-x} \frac{n!}{x! x_2! x_3!} p_1^x p_2^{x_2} p_3^{n-x-x_2}$$

$$= \frac{n!}{x!(n-x)!} p_1^x \sum_{x_2=0}^{n-x} \frac{(n-x)!}{x_2!(n-x-x_2)!} p_2^{x_2} p_3^{n-x-x_2}$$

$$= \binom{n}{x} p_1^x \sum_{x_2=0}^{n-x} \binom{n-x}{x_2} p_2^{x_2} p_3^{(n-x)-x_2} = \binom{n}{x} p_1^x (p_2 + p_3)^{n-x}$$

from the binomial theorem; but $p_3 = 1 - p_1 - p_2$ so that

$$\Pr(X_1 = x) = \binom{n}{x} p_1^x (1 - p_1)^{n-x}$$

and X_1 is marginally binomial. This is intuitive; the number of red marbles in a sample of size n drawn with replacement from an urn with r red, w white and b blue marbles is binomial with parameters n and $p = r/(r+w+b)$. Similar reasoning shows that this holds in general, i.e. the marginal of any set of X_i from a multinomial is itself multinomial. ∎

The previous result on the marginals of a multinomial random vector implies that, if $\mathbf{X} \sim \text{Multinom}(n, \mathbf{p})$, then $\mathbb{E}[X_i] = np_i$, $i = 1, \ldots, k$ (see Problem 4.4). Example 6.7 will show that $\mathbb{V}(X_i) = np_i(1 - p_i)$ and that the covariance between the ith and jth elements of $\mathbf{X}$ is given by $\text{Cov}(X_i, X_j) = -np_i p_j$.

It is instructive to see how easy it is to simulate from the multinomial distribution (and we will need it in Section 5.3.4 below). As trials in the corresponding sampling scheme are independent, it suffices to consider the $n = 1$ case which, once programmed, is just repeated n times and the results tallied. First recall the binomial case (Bernoulli when $n = 1$) with success probability p, and label the possible outcomes as zero (failure) and one (success). A Bernoulli realization, X, is obtained by generating a uniform r.v., U, and setting X to one if $U < p$, and zero otherwise.

Generalizing this, let $\mathbf{X} \sim \text{trinom}(\mathbf{x}; 1, p_1, p_2)$, and label the possible outcomes as 1,2,3. Again with U a uniform r.v., set X to one if $U < p_1$ (which occurs with probability p_1); set X to two if $p_1 < U < p_1 + p_2$ (which occurs with probability $(p_1 + p_2) - p_1 = p_2$); and set X to three if $U > p_2$. Repeating this n times, it is clear that the number of occurrences of $X = i$, $i = 1, 2, 3$, will be proportional to their respective probabilities. The only drawback with this formulation is that a computer program will need a series of nested IF... THEN structures, which becomes awkward when k is large. The more elegant CASE structure could be used, but either way, this works for a predetermined value of k, and it is not clear how to support general k. However, all this is unnecessary, seen best by moving to the general case with k outcomes: X is set to two if $p_1 < U < p_1 + p_2$; three, if $p_1 + p_2 < U < p_1 + p_2 + p_3$, etc., so that the cumulative sum, abbreviated *cumsum* or *cusum*, of the p_i vector is clearly the essential structure. It is given by, say, $pp = (p_1, p_1 + p_2, \ldots, 1)$, and X is set to $1 + \text{sum}(U > pp)$. This is shown in Program Listing 5.1.

```
function X=randmultinomial(p)
pp=cumsum(p); u=rand; X=1+sum(u>pp);
```

Program Listing 5.1 Simulates one draw from a multinomial distribution with vector of probabilities $p = (p_1, \ldots, p_k)$

The program is only for $n = 1$. For general n, just call `randmultinomial` n times, and tally the results. An example of how to do this is given in the following code:

```
plen=length(p); A=zeros(plen,1);
for j=1:n,   X=randmultinomial(p); A(X)=A(X)+1; end
```

⊖ *Example 5.13* Let $\mathbf{X} \sim \text{Multinom}(10, \mathbf{p})$ with $k = 4$ and $p_i = i/10$, $i = 1, 2, 3, 4$. The mass function of $\mathbf{X}$ is given in (5.29), from which that of, say, random variable $S = X_1 X_2 + X_3 X_4$, could be worked out. However, simulation offers a fast and easy way to approximate f_S, with code for doing so given in Program Listing 5.2. Figure 5.2 shows the simulated mass function of S based on 50 000 replications. We see that the mass function is not at all 'smooth', though does have a basic resemblance to the bell curve. ∎

```
p=[0.1 0.2 0.3 0.4]; n=10; sim=50000;
plen=length(p); S=zeros(sim,1);
for i=1:sim
  A=zeros(plen,1);
  for j=1:n
    X=randmultinomial(p);
    A(X)=A(X)+1;
  end
  S(i) = A(1)*A(2)+A(3)*A(4);
end
[hcnt, hgrd] = hist(S,max(S)-min(S)+1);
h1=bar(hgrd,hcnt/sim);
```

Program Listing 5.2 Simulates a particular function of the components of a multinomial random variable

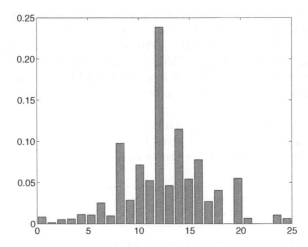

Figure 5.2   Simulated mass function of $S = X_1 X_2 + X_3 X_4$, using 50 000 replications, where $\mathbf{X} \sim \text{Multinom}\,(10, \mathbf{p})$ with $k = 4$ and $p_i = i/10$, $i = 1, 2, 3, 4$

5.3.2   Multivariate hypergeometric

> Whenever I want to learn a new subject, I announce a graduate course in it, since the best way to learn is by teaching. But even better than teaching humans is teaching computers, i.e. program! Since computers will not let you wave your hands and wing it.   (Doron Zeilberger)

The *multivariate* or *k-variate hypergeometric* distribution generalizes the (bivariate) hypergeometric; sampling is conducted without replacement from the proverbial urn which initially contains N marbles, each of which is one of k colors, $1 \leq k \leq N$. In particular, if the urn contains $n_i > 0$ marbles of color i, $i = 1, \ldots, k$, and n marbles are drawn without replacement, then we write

$$\mathbf{X} = (X_1, \ldots, X_k) \sim \text{MultiHGeo}\,(\mathbf{x}; n, n_1, n_2, \ldots, n_k)\,,$$

where X_i is the number of marbles with color i drawn, $i = 1, \ldots, k$, and

$$\Pr\,(\mathbf{X} = \mathbf{x}) = \frac{\prod_{i=1}^{k} \binom{n_i}{x_i}}{\binom{N}{n}}, \quad \sum_{i=1}^{k} n_i = N, \quad \sum_{i=1}^{k} x_i = n, \tag{5.31}$$

with X_k being redundant, i.e. $X_k = n - \sum_{i=1}^{k-1} X_i$.

For simulating from the multivariate hypergeometric distribution, we first construct an urn in the computer as an array with n_1 ones, n_2 twos, . . . , and n_k ks. This is done in the following code segment:

```
nvec=[4 1 3]; % Example with n1=4, n2=1 and n3=3
N=sum(nvec); k=length(nvec); urn=[];
```

```
for i=1:k
  urn=[urn  i*ones(1,nvec(i))];
end
```

which, for these values of n_1, n_2 and n_3, yields vector urn to be:

| 1 | 1 | 1 | 1 | 2 | 3 | 3 | 3 |

With our 'computer urn', simulation is easily done by choosing a random integer between one and N (with Matlab's command unidrnd), say r, and removing the rth value from the urn vector. This is then repeated n times, tallying the results along the way. The program is given in Program Listing 5.3.

```
function X=hypergeometricsim(n,nvec)
N=sum(nvec); k=length(nvec);
urn=[]; for i=1:k, urn=[urn  i*ones(1,nvec(i))]; end
marbles=length(urn); X=zeros(k,1);
for i=1:n
  r=unidrnd(marbles); % random index into the urn
  c=urn(r); % c is the 'color' of the marble, 1,2,...,k
  X(c)=X(c)+1; % tally it
  urn=[urn(1:(r-1)) urn((r+1):end)]; % remove it from the urn
  marbles=marbles-1; % number of marbles remaining
end
```

Program Listing 5.3 Simulates a single vector r.v. from the hypergeometric distribution, with n the number of draws, and nvec $= (n_1, \ldots, n_k)$ giving the contents of the urn

⊖ *Example 5.14* An urn contains eight red, six green and six blue marbles, from which n marbles are randomly drawn without replacement. We are interested in the probability, p_n, that the number of red marbles drawn, R, exceeds the *sum* of the drawn green marbles, G, and drawn blue marbles, B.

Doing this via simulation is easy with the previous program hypergeometric-sim, as the following code shows.

```
nvec=[8 6 6]; sim=50000;
p=[];
for n=1:15
  exceed=zeros(sim,1);
  for i=1:sim
    X=hypergeometricsim(n,nvec);
    if X(1)>(X(2)+X(3))
      exceed(i)=1;
    end
  end
  p=[p mean(exceed)];
end
```

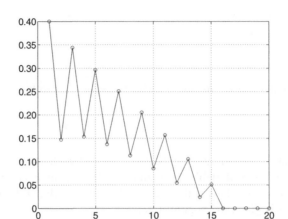

Figure 5.3 p_n versus n in Example 5.14

For example, running this gave $p_1 = 0.3998$, $p_2 = 0.1498$, $p_3 = 0.3450$ and $p_4 = 0.1507$. For this situation, however, the answer is readily obtained algebraically. A bit of thought shows that

$$p_n = \Pr\left(R > G + B\right) = \sum_{r=\lfloor n/2 \rfloor+1}^{n} \sum_{g=0}^{n-r} \frac{\binom{8}{r}\binom{6}{g}\binom{6}{n-r-g}}{\binom{20}{n}}$$

for $n = 1, \ldots, 20$ and with $\binom{a}{b} = 0$ for $a < b$. Clearly, for $16 \leq n \leq 20$, $p_n = 0$. Because only the sum of G and B is relevant, p_n can be expressed in terms of a (bivariate) multinomial distribution with eight red and 12 'non red' marbles, i.e.

$$p_n = \sum_{r=\lfloor n/2 \rfloor+1}^{n} \frac{\binom{8}{r}\binom{12}{n-r}}{\binom{20}{n}}.$$

Figure 5.3 plots p_n versus n. ∎

Remark: The computation of the c.d.f. or rectangular probabilities of the form $\Pr(\mathbf{a} \leq \mathbf{X} \leq \mathbf{b})$ can be burdensome for large k. Highly accurate and easily computed approximations for these probabilities in the multinomial and multivariate hypergeometric cases (and for two other discrete distributions) are detailed in Butler and Sutton (1998). ∎

5.3.3 Multivariate negative binomial

> Now, Hear Yea, all you self-righteous defenders of insight and understanding. A good proof may give the appearance of understanding, but in order to understand something really deeply, you should program it. (Doron Zeilberger)

Let our tried-and-true urn again contain $n_i > 0$ marbles of color i, $i = 1, \ldots, k$, with $N = n_1 + \cdots + n_k$ and $k \leq N$. Marbles are drawn, with replacement, until at least a_1 marbles of color 1, at least a_2 marbles of color 2, $\ldots$, and at least a_k marbles of color k are drawn, with $0 \leq a_i$, $i = 1, \ldots, k$. In particular, note that some of the a_i can be zero. Let X_j be the number of marbles of color j drawn, $j = 1, \ldots, k$. Interest might center on $S = \sum_{i=1}^{k} X_i$. For example, let S designate the total number of tosses of a fair, six-sided die until you get at least one 1, at least two 2s, $\ldots$, and at least six 6s. Interest might also center on the number of 'extras', say $E = S - A$, where $A = a_1 + \cdots + a_k$; or the joint distribution of a subset of the X_i, or some function thereof, such as the total number of marbles from the 'undesirable' categories, i.e. $U = \sum_{i=1}^{k} X_i \mathbb{I}(a_i = 0)$.

It is worth examining how each of the aforementioned possibilities reduces in the univariate, i.e. negative binomial, case. As an example of a random variable in the multivariate setting which has no univariate analog, let $p_{2,4;3,7,10}$ be the probability that the desired quota of marbles of colors 2 and 4 are obtained before those of colors 3, 7 and 10. There are obviously many random variables which could be of interest.

⊖ **Example 5.15** Motivated by use in medical experiments, Zhang, Burtness and Zelterman (2000) consider the $k = 2$ case with $a_1 = a_2$ in great detail, deriving various nontrivial quantities of interest such as moments and modes. To illustrate, assume that you want to have enough children so that you have at least c boys and at least c girls, i.e. at least c of each, where the probability of getting a boy on any trial is p, $0 < p < 1$. Let T be the total number of children required, and S be the number of 'superfluous', or extra children, i.e. $T = S + 2c$. Then

$$f_S(s) = \Pr(S = s) = \binom{s + 2c - 1}{c - 1} (p^s + q^s)(pq)^c \, \mathbb{I}_{\{0,1,2,\ldots\}}(s), \qquad (5.32)$$

where $q = 1 - p$. To see this, let $A_t = \{t \text{ trials are required to get } c \text{ of each}\}$ and $B = \{t\text{th trial is a boy}\}$, so that, among the first $t - 1$ trials, there are $c - 1$ boys and $t - 1 - (c - 1) = t - c$ girls, which can occur in $\binom{t-1}{c-1}$ ways. Thus,

$$\Pr(A_t \cap B) = \binom{t - 1}{c - 1} p^c q^{t-c} \quad \text{and} \quad \Pr(A_t \cap G) = \binom{t - 1}{c - 1} q^c p^{t-c},$$

where $G = B^c = \{t\text{th trial is a girl}\}$ and $\Pr(A_t \cap G)$ follows from symmetry. As these two events are disjoint, the probability of having a total of T children is

$$\Pr(T = t) = \binom{t - 1}{c - 1} (p^c q^{t-c} + q^c p^{t-c})$$

$$= \binom{t - 1}{c - 1} (q^{t-2c} + p^{t-2c})(pq)^c \, \mathbb{I}_{\{2c,2c+1,\ldots\}}(t),$$

and, as $S = T - 2c$,

$$\Pr(S = s) = \Pr(T = s + 2c) = \binom{s + 2c - 1}{c - 1} (q^s + p^s)(pq)^c \, \mathbb{I}_{\{0,1,\ldots\}}(s),$$

which is (5.32).

Because the mass function (5.32) is so easy to evaluate numerically, we can compute, say, the expected value as $\mathbb{E}[S] \approx \sum_{s=0}^{L} s\, f_S(s)$ for a large upper limit L. For $c = 7$ and $p = 0.3$ (for which the p.m.f. is bimodal), the mean of X is 9.5319. ∎

The next example shows another way of calculating the expected value in a more general setting.

⊖ ***Example 5.16*** (Lange, 2003, p. 30) Upon hearing about your interesting family planning strategy, your cousin and his wife decide that they want at least s sons and d daughters. Under the usual assumptions, and with p denoting the probability of getting a son at each and every birth, what is the expected number of children they will have?

Let N_{sd} be the random variable denoting the total number of children they have. First note that, if either $s = 0$ or $d = 0$, then N_{sd} follows a negative binomial distribution with mass function (4.20) and, as will be shown below in Section 6.4.1, $\mathbb{E}[N_{0d}] = d/q$, where $q = 1 - p$, and $\mathbb{E}[N_{s0}] = s/p$.

Now assume both s and d are positive. Instead of trying to work out the distribution of N_{sd}, consider setting up a difference equation. Conditioning on the gender of the first child, it follows from the independence of the children that

$$\mathbb{E}[N_{sd}] = p\left(1 + \mathbb{E}[N_{s-1,d}]\right) + q\left(1 + \mathbb{E}[N_{s,d-1}]\right)$$
$$= 1 + p\,\mathbb{E}[N_{s-1,d}] + q\,\mathbb{E}[N_{s,d-1}].$$

This is easily programmed as a recursion, as shown in Program Listing 5.4. Use the program to plot $\mathbb{E}[N_{sd}]$ as a function of d, for a given value of s, say $s = 4$. Before doing so, think about how the graph should look. ∎

```
function e=esd(s,d,p)
if s==0, e=d/(1-p);
elseif d==0, e=s/p;
else, e=1 + p*esd(s-1,d,p) + (1-p)*esd(s,d-1,p);
end
```

Program Listing 5.4 Expected number of children if you want s sons and d daughters, with probability p at each 'trial' of getting a son

Further discussion of the multivariate negative binomial can be found in Panaretos and Xekalaki (1986), Port (1994), Johnson, Kotz and Balakrishnan (1997) and the references therein.

5.3.4 Multivariate inverse hypergeometric

Most importantly, teach your children how to program! This will develop their minds much faster than any of the traditional curriculum, including, memorizing proofs in Euclidean Geometry, and being able to make minor variations, without really understanding the notion of proof.   (Doron Zeilberger)

This sampling scheme is the same as that for the multivariate negative binomial, but sampling is done *without* replacement. While some analytic results are possible, we skip right to simulation, as it is easy, instructive and versatile.

Simulating the total number of draws required until all the quotas are satisfied is straightforward. Let U be an 'urn vector' with $U(1)$ the initial number of white marbles in the urn, $U(2)$ be the initial number of black balls, etc. (Note how this differs from the 'urn structure' used in Listing 5.3 for the multivariate hypergeometric distribution.) Also let want be the vector giving the quotas to be drawn on white balls, black balls, etc. For example, for the inverse hypergeometric we studied in Section 4.2.4, if we want to draw at least $k = 20$ white balls from an urn originally containing $w = 50$ white and $b = 40$ black balls, then we would set want=[k 0] and U=[w b]. Program MIH in Program Listing 5.5 implements the general case. Note that each draw is the same as a draw based on a multinomial trial, *but the contents of the urn changes after each draw*. Observe also what happens if a certain color is exhausted from the urn, namely that the program still works.

```
function n=MIH(want,U)
m=length(want); want=reshape(want,1,m); U=reshape(U,1,m);
n=0; done=0;
h=zeros(1,m); % h (have) is the vector of what we have accumulated
while ~done
    p=U/sum(U);   d=randmultinomial(p);   n=n+1;
    h(d)=h(d)+1;  U(d)=U(d)-1;  done = all(h>=want);
end
```

Program Listing 5.5   Simulates the multivariate inverse hypergeometric distribution, when interest centers on the total number of marbles drawn; this uses program randmultinomial in Listing 5.1

To illustrate, consider the bivariate case, as in Section 4.2.4, with $k = 20$, 50 white and 40 black balls. Simulating the mass function for this case would be done with the following code.

```
sim=50000; k=20; white=50; black=40;
want=[k 0]; U=[white black]; n=zeros(sim,1);
for i=1:sim
    n(i)=mih(want, U);
end
[hcnt, hgrd] = hist(n-0.3,max(n)-min(n)+1); h1=bar(hgrd,hcnt/sim);
a=axis; axis([a(1) a(2) 0 1.05*max(hcnt)/sim])
```

The graphical output is shown in Figure 5.4, as one of four simulations. Of course, in this case, we have a simple expression for the mass function (which serves to corroborate the validity of the mass function and the simulation program), though in the general case, or for different functions of the multivariate inverse hypergeometric random variable, use of program MIH will probably be the fastest way to arrive at a numerical solution.

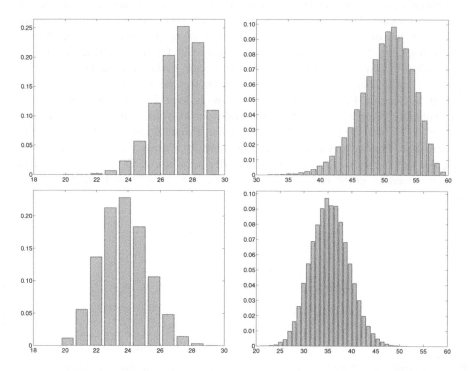

Figure 5.4 The inverse hypergeometric p.m.f. based on simulation with 50 000 replications, using the program in Program Listing 5.5; based on $k = 20$, with $w = 25$ (top) and $w = 50$ (bottom), and $b = 10$ (left) and $b = 40$ (right) – compare with the exact values shown in Figure 4.2 on page 130.

5.4 Problems

Hold yourself responsible for a higher standard than anybody expects of you. Never excuse yourself.   (Henry Ward Beecher)

If you can describe a job precisely, or write rules for doing it, it's unlikely to survive. Either we'll program a computer to do it, or we'll teach a foreigner to do it.   (Frank Levy, MIT economist)[3]
Reproduced by permission of Dow Jones, Wall Street Journal Europe

---

[3] Quoted in David Wessel's lead article of 2 April, 2004, *Wall Street Journal Europe* dealing with the outsourcing of jobs in the United States.

5.1. From a sketch, verify (5.2) and express $\Pr(X_1 > b_1, X_2 > b_2)$ in terms of the quantities $F_X(\cdot)$, $F_Y(\cdot)$ and $F_{X,Y}(\cdot, \cdot)$.

5.2. Prove $|\text{Corr}(X_1, X_2)| \leq 1$ by computing $\mathbb{V}(Y_1 + Y_2)$ and $\mathbb{V}(Y_1 - Y_2)$, where $Y_i = X_i/\sigma_i$ and $\sigma_i^2 := \mathbb{V}(X_i)$, $i = 1, 2$.

5.3. ★ Verify the equivalence of expressions (5.14) and (5.15) by performing successive integration by parts on the latter expression.

(Contributed by Chris Jones)

5.4. For events A_i, let $B_i = 1$ if A_i occurs and zero otherwise, $i = 1, \ldots, n$, and define $P = 1 - \prod_{i=1}^{n}(1 - B_i)$. Determine the relation between $\mathbb{E}[P]$ and events A_i and use the fact that

$$\prod_{i=1}^{n}(1 - x_i) = 1 - \sum_{i=1}^{n} x_i + \sum_{i<j} x_i x_j - \cdots + (-1)^n x_1 \cdots x_n$$

to prove (2.11).

5.5. Recall Example 5.16 about having s sons and d daughters. Construct a difference equation yielding the probability, say R_{sd}, that the quota of s sons is reached before the quota of d daughters. (Lange, 2003, p. 30)

5.6. ★ An 'Überraschungsei' is a hollow chocolate egg with a hidden plastic toy inside. Assume there are r different types of toys and that each is equally likely to be found in an egg. Further assume that the eggs come from a population so large that your individual purchases make no difference, so that sampling can be thought of as being 'with replacement'. Write a program which simulates and shows a histogram of how many eggs need to be purchased to amass k different toys out of y, $1 \leq k \leq y$. If $k = y$, also plot the discrepancy between the scaled simulated values and the true mass function.

6

Sums of random variables

Look around when you have got your first mushroom or made your first discovery: they grow in clusters.
(George Pólya)

Teaching is not a science; it is an art. If teaching were a science there would be a best way of teaching and everyone would have to teach like that. ...Let me tell you what my idea of teaching is. Perhaps the first point, which is widely accepted, is that teaching must be active, or rather active learning. ...The main point in mathematics teaching is to develop the tactics of problem solving.
(George Pólya)

Interest often centers not on the particular outcomes of several random variables individually, but rather on their sums. For instance, a system of components might be set up in such a way that a single component is used at a time; when it expires, the next one takes over, etc., so that the sum of their individual lifetimes is of primary interest. Another example is the profit of a company with many individual stores – the individual profits are of less interest than their sum.

Going in the other direction, in many interesting situations, the calculation of certain quantities of interest (such as expected values) may be dramatically simplified by artificially introducing (a judiciously chosen set of) simple random variables associated with the actual variables of interest. Several examples of this powerful concept are provided below.

6.1 Mean and variance

Let $Y = \sum_{i=1}^{n} X_i$, where the X_i are random variables. Then

$$\mathbb{E}\left[\sum_{i=1}^{n} X_i\right] = \sum_{i=1}^{n} \mathbb{E}[X_i],$$ (6.1)

Fundamental Probability: A Computational Approach   M.S. Paolella
© 2006 John Wiley & Sons, Ltd

if the expected value exists for each X_i. To show (6.1) for $n = 2$, let $g(\mathbf{X}) = X_1 + X_2$ so that, from (5.18),

$$
\begin{aligned}
\mathbb{E}[X_1 + X_2] &= \int_{-\infty}^{\infty} \int_{-\infty}^{\infty} (x_1 + x_2) f_{X_1, X_2}(x_1, x_2) \, dx_2 dx_1 \\
&= \int_{-\infty}^{\infty} \int_{-\infty}^{\infty} x_1 f_{X_1, X_2}(x_1, x_2) \, dx_2 dx_1 + \int_{-\infty}^{\infty} \int_{-\infty}^{\infty} x_2 f_{X_1, X_2}(x_1, x_2) \, dx_1 dx_2 \\
&= \int_{-\infty}^{\infty} x_1 \int_{-\infty}^{\infty} f_{X_1, X_2}(x_1, x_2) \, dx_2 dx_1 + \int_{-\infty}^{\infty} x_2 \int_{-\infty}^{\infty} f_{X_1, X_2}(x_1, x_2) \, dx_1 dx_2 \\
&= \int_{-\infty}^{\infty} x_1 f_{X_1}(x_1) \, dx_1 + \int_{-\infty}^{\infty} x_2 f_{X_2}(x_2) \, dx_2 \\
&= \mathbb{E}[X_1] + \mathbb{E}[X_2].
\end{aligned}
$$

The result for $n > 2$ can be similarly derived; it also follows directly from the $n = 2$ case by induction.

If the variance for each X_i exists, then $\mathbb{V}(Y)$ is given by

$$
\mathbb{V}(Y) = \sum_{i=1}^{n} \mathbb{V}(X_i) + \sum \sum_{i \neq j} \mathrm{Cov}(X_i, X_j) \tag{6.2}
$$

with special case

$$
\boxed{\mathbb{V}(X_i \pm X_j) = \mathbb{V}(X_i) + \mathbb{V}(X_j) \pm 2 \, \mathrm{Cov}(X_i, X_j)} \tag{6.3}
$$

(and the extension to $X_i - X_j$ following from (6.4) below). Furthermore, if X_i and X_j are independent, then from (5.25), $\mathbb{V}(X_i + X_j) = \mathbb{V}(X_i) + \mathbb{V}(X_j)$. Formula (6.2) is most easily proven using matrix algebra and is given in Chapter 8.

Results (6.1) and (6.2) can be extended to the case when Y is a weighted sum of the X_i, i.e. $X = \sum_{i=1}^{n} a_i X_i$. Using properties of sums and integrals,

$$
\boxed{\mathbb{E}\left[\sum_{i=1}^{n} a_i X_i\right] = \sum_{i=1}^{n} \mathbb{E}[a_i X_i] = \sum_{i=1}^{n} a_i \mathbb{E}[X_i] = \sum_{i=1}^{n} a_i \mu_i}
$$

and

$$
\boxed{\mathbb{V}\left(\sum_{i=1}^{n} a_i X_i\right) = \sum_{i=1}^{n} a_i^2 \mathbb{V}(X_i) + \sum \sum_{i \neq j} a_i a_j \, \mathrm{Cov}(X_i, X_j)}, \tag{6.4}
$$

which generalize (4.37) and (4.41), respectively.

An important special case is when X_1 and X_2 are uncorrelated and $a_1 = -a_2 = 1$, giving

$$\mathbb{V}(X_1 - X_2) = \mathbb{V}(X_1) + \mathbb{V}(X_2).$$

Finally, the covariance between two r.v.s $X = \sum_{i=1}^{n} a_i X_i$ and $Y = \sum_{i=1}^{m} b_i Y_i$ is given by

$$\boxed{\text{Cov}(X, Y) = \sum_{i=1}^{n} \sum_{j=1}^{m} a_i b_j \, \text{Cov}(X_i, Y_j)}, \tag{6.5}$$

of which (6.4) is a special case, i.e. $\mathbb{V}(X) = \text{Cov}(X, X)$. Formulae (6.4) and (6.5) are most easily proven using matrix algebra (see Chapter 8).

Example 6.1 In animal breeding experiments, n offspring from the same 'father' are expected to be *equicorrelated* with one another; i.e. denoting a particular characteristic of them (such as weight, etc.) with the r.v. X_i, $i = 1, \ldots, n$, we assume $\text{Corr}(X_i, X_j) = \rho > 0, i \neq j$. Assuming $\mathbb{V}(X_i) = \sigma^2 \, \forall i$, the variance of $S = \sum_{i=1}^{n} X_i$ is then given by (6.2)

$$\mathbb{V}(S) = n\mathbb{V}(X_i) + n(n-1)\text{Cov}(X_i, X_j)$$

$$= n\sigma^2 + n(n-1)\rho\sigma^2 = n\sigma^2(1 + (n-1)\rho),$$

using (5.26). From this, we see that, as $\rho \to 1$, $\mathbb{V}(S) \to n^2\sigma^2$, which makes sense; if $\rho = 1$, the X_i are identical, so that $S = nX_1$ and $\mathbb{V}(S) = n^2\sigma^2$. ∎

6.2 Use of exchangeable Bernoulli random variables

Many seemingly difficult moment calculations can be transformed into almost trivial problems using a clever choice of exchangeable Bernoulli r.v.s. The following examples provide ample illustration.

Example 6.2 Recall Example 3.4 in which $2n$ people consisting of n married couples, $n > 1$, are randomly seated around a table, and random variable K is defined to be the number of couples which happen to be sitting together. The p.m.f. of K was shown to be

$$\Pr(K = k) = \sum_{i=k}^{n} (-1)^{i-k} \binom{i}{k} \binom{n}{i} 2^i \frac{(2n-i-1)!}{(2n-1)!},$$

from which the moments could be computed using their definition, i.e.

$$\mathbb{E}[K^m] = \sum_{k=0}^{n} k^m \Pr(K = k) = \sum_{k=1}^{n} \sum_{i=k}^{n} k^m (-1)^{i-k} \binom{i}{k} \binom{n}{i} 2^i \frac{(2n-i-1)!}{(2n-1)!}.$$

A dedicated amount of work simplifying this expression would eventually yield the results $\mathbb{E}[K] = 2n/(2n-1)$ and $\mathbb{E}[K^2] = 4n/(2n-1)$. However, there is a much easier way of obtaining the first two moments. Define the Boolean r.v.s

$$X_i = \begin{cases} 1, & \text{if the } i\text{th couple is seated together,} \\ 0, & \text{otherwise,} \end{cases}$$

$i = 1, \ldots n$. From the nature of the problem, we can surmise that the X_i are exchangeable. Now consider how many couples are expected to be seated together. From (6.1),

$$\mathbb{E}\left[\sum_{i=1}^{n} X_i\right] = \sum_{i=1}^{n} \mathbb{E}[X_i] = \sum_{i=1}^{n} \Pr(X_i = 1) = n \Pr(X_1 = 1),$$

because the probability is the same for each couple. To compute $\Pr(X_1 = 1)$, consider one woman. She has two (different, because $n > 1$) people next to her, out of $2n-1$ different possible people, and only one of them is her husband, so that $\Pr(X_1 = 1) = 2/(2n-1)$. Thus, the expected number of couples seated together is $2n/(2n-1) \approx 1$.

Next observe that, for two couples i and j,

$$\Pr(X_i = 1 \text{ and } X_j = 1) = \Pr(X_j = 1 \mid X_i = 1)\Pr(X_i = 1) = \frac{2}{2n-2}\frac{2}{2n-1},$$

which follows because, if couple i is together, woman j cannot sit between them, so we can think of couple i as a single entity at the table. Using (6.4) with all the weights equal to one and the fact that the X_i are Bernoulli,

$$\mathbb{V}(X) = \mathbb{V}\left(\sum_{i=1}^{n} X_i\right)$$

$$= n\frac{2}{2n-1}\left(1 - \frac{2}{2n-1}\right) + n(n-1)\left(\frac{2}{2n-2}\frac{2}{2n-1} - \left(\frac{2}{2n-1}\right)^2\right)$$

$$= \frac{4n(n-1)}{(2n-1)^2} = \frac{4n^2 - 4n}{4n^2 - 4n + 1} \approx 1,$$

where, for $i \neq j$, $\mathrm{Cov}(X_i X_j) = \mathbb{E}[X_i X_j] - \mathbb{E}[X_i]\mathbb{E}[X_j]$ from (5.24). ■

⊚ **Example 6.3** Example 4.24 demonstrated via the definition of expected value that $\mathbb{E}[M]$, the expected number of matches in the problem of coincidences, is one. Now let M_i be the Boolean event that is one if the ith element is matched and zero otherwise, $i = 1, \ldots, n$. From symmetry, the M_i are exchangeable. As the total number of matches is $M = \sum M_i$, $\mathbb{E}[M] = \sum \mathbb{E}[M_i] = 1$, because $\mathbb{E}[M_i] = 1/n$. Furthermore, as each M_i is Bernoulli distributed, $\mathbb{V}(M_i) = n^{-1}(1 - n^{-1})$ and, for $i \neq j$,

$$\mathbb{E}[M_i M_j] = \Pr(M_i = 1, M_j = 1) = \Pr(M_i = 1 \mid M_j = 1)\Pr(M_j = 1)$$

$$= (n-1)^{-1} n^{-1}$$

from (3.7), so that, from (5.24),

$$\text{Cov}\left(M_i, M_j\right) = \mathbb{E}\left[M_i M_j\right] - \mathbb{E}\left[M_i\right]\mathbb{E}\left[M_j\right] = n^{-2}(n-1)^{-1}.$$

Then, from (6.4) with all the weights equal to one,

$$\mathbb{V}(M) = n\mathbb{V}(M_1) + n(n-1)\text{Cov}(M_1, M_2) = 1,$$

i.e. both the expected value and variance of M are unity. ∎

Example 6.4 In Section 4.2.6, r.v. X represented the number of unique prizes (out of r) obtained when n cereal boxes are purchased (or, in more general terms with respect to placing n balls randomly into one of r cells, X is the number of nonempty cells). To calculate the first two moments of X, express X as $\sum_{i=1}^{r} X_i$, where $X_i = 1$ if at least one of the n toys is of the ith kind and zero otherwise, $i = 1, \ldots, r$. Then, for each i,

$$\mathbb{E}[X_i] = \Pr(X_i = 1) = 1 - \Pr(X_i = 0) = 1 - \left(\frac{r-1}{r}\right)^n =: 1 - t_1$$

so that, from (6.1),

$$\mathbb{E}[X] = r\mathbb{E}[X_i] = r(1 - t_1) = r - r\left(\frac{r-1}{r}\right)^n.$$

Likewise, $\mathbb{V}(X)$ is given by (6.2) with $\mathbb{V}(X_i) = t_1(1-t_1)$ for each i. Next, from (5.24), for $i \neq j$, $\text{Cov}(X_i, X_j) = \mathbb{E}[X_i X_j] - \mathbb{E}[X_i]\mathbb{E}[X_j]$, where

$$\mathbb{E}[X_i X_j] = \Pr\left(\begin{array}{c}\text{at least one of the } n \text{ toys is of the } i\text{th kind} \\ and \text{ at least one of the } n \text{ toys is of the } j\text{th kind}\end{array}\right)$$

$$=: \Pr(A \cap B) = 1 - \Pr(A^c \cup B^c) = 1 - \Pr(A^c) - \Pr(B^c) + \Pr(A^c B^c)$$

$$= 1 - 2\left(\frac{r-1}{r}\right)^n + \left(\frac{r-2}{r}\right)^n =: 1 - 2t_1 + t_2$$

so that, putting everything together,

$$\mathbb{V}(X) = rt_1(1-t_1) + r(r-1)\left(1 - 2t_1 + t_2 - (1-t_1)^2\right)$$

$$= rt_1(1 - rt_1) + rt_2(r-1).$$

Note that, for $n = 1$, $\mathbb{E}[X] = 1$ and $\mathbb{V}(X) = 0$ for all $r \geq 1$, which is intuitively clear. Figure 6.1 plots the mean and variance as a function of n for $r = 100$. ∎

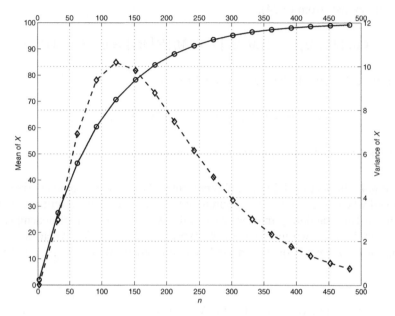

Figure 6.1 Mean (solid) and variance (dashed) of random variable X for $r = 100$

6.2.1 Examples with birthdays

This section provides a specific application of balls being randomly placed into cells, and makes use of some of the tools developed so far, such as exchangeable r.v.s. Assume that there are only 365 days a year, and the chances of being born on a particular day are equal, i.e. 1/365. Two or more people are said to have the same birthday when the month and date are the same, not necessarily the year. Consider a group of n randomly chosen people.

(1) *What is the probability that, for $n = 3$ people, all have a different birthday?*
From a direct combinatoric argument, this is just $\frac{365}{365} \cdot \frac{364}{365} \cdot \frac{363}{365} = 0.9918$.

(2) *For general n, what is the probability that all of them have unique birthdays?*
This is just

$$\frac{365 \cdot 364 \cdots (365 - n + 1)}{365^n} = \frac{365_{[n]}}{365^n}.$$

(3) *What is the probability that $2 \leq m \leq n - 1$ of them have the same birthday and the remaining $n - m$ have no birthdays in common? What about $m = 0$, $m = 1$ and $m = n$?* There are $\binom{n}{m}$ ways of choosing the m people with the same birthday and 365 ways of picking that birthday. For the remaining $n - m$ people, there are

$$(365 - 1)_{[n-m]} = 364 \cdot 363 \cdots (364 - (n - m) + 1)$$

ways so that they all have different birthdays. Multiplying these three factors and dividing by the total number of ways n people can have birthdays gives

$$\frac{\binom{n}{m}365 \cdot 364_{[n-m]}}{365^n} = \frac{\binom{n}{m}365_{[n-m+1]}}{365^n} = \frac{\binom{n}{m}}{365^n} \frac{365!}{[365 - (n-m+1)]!}. \quad (6.6)$$

Note that $m = 1$ makes no sense, but $m = 0$ does, and refers to the previous question. For $m = n$, (6.6) still holds, and means that all n people have the same birthday, which occurs with probability

$$\left(\frac{1}{365}\right)^{n-1} = \frac{365}{365^n}.$$

Note that (6.6) does not hold for $m = 0$.

(4) *For $n = 3$, $n = 4$ and $n = 5$, detail all cases which can arise with respect to the n birthdays and, for each of the three n-values, show that the probabilities sum to one.*

(a) For $n = 3$, the above analysis is applicable – either $m = 0$, $m = 2$ or $m = 3$ – this gives

$$\frac{365!}{365^3(365-3)!} + \sum_{m=2}^{3} \frac{\binom{3}{m}365!}{365^3[365-(3-m)-1]!} = 1.$$

(b) For $n = 4$, the above analysis is still useful, but the case in which two people have the same birthday and the other two also share a birthday (different from that of the first pair) needs to be taken into account. This latter event occurs with probability

$$\frac{\binom{4}{2}\binom{365}{2}}{365^4},$$

so that

$$\frac{365!}{365^4(365-4)!} + \sum_{m=2}^{4} \frac{\binom{4}{m}365!}{365^4[365-(4-m)-1]!} + \frac{\binom{4}{2}\binom{365}{2}}{365^4} = 1.$$

(c) For $n = 5$, two further possible events need to be taken into account.

(i) That two people share a birthday, two other people share a (different) birthday, and the last person has a unique birthday. This occurs with probability

$$p_{2,2,1} = \frac{\binom{n}{2,2,1}\frac{365 \cdot 364 \cdot 363}{2!}}{365^n},$$

where the factor 2! takes into account the fact that the two birthdays belonging to the two pairs of birthday people was counted twice in the expression $365 \cdot 364$.

(ii) That two people have the same birthday, while the other three also share a (different) birthday. This occurs with probability

$$p_{2,3} = \frac{\binom{n}{2,3} \cdot 365 \cdot 364}{365^n}.$$

Notice that, because the groups have different numbers of people (two and three), $365 \cdot 364$ does not repeat any possibilities and so does not have to be adjusted.

Adding all the terms yields

$$\frac{365!}{365^5 (365-5)!} + \sum_{m=2}^{5} \frac{\binom{5}{m}365!}{365^5 [365-(5-m)-1]!} + p_{2,2,1} + p_{2,3} = 1.$$

(5) *A party is thrown on each day for which at least one of the n people has a birthday. Let Z be the number of days of the year without parties. Calculate* $\Pr(Z = z)$. An expression for $\Pr(Z = z)$ is obtained directly from (2.21) with $m = z$ and $r = 365$. Using the fact that

$$\binom{i}{m}\binom{r}{i} = \binom{r}{m}\binom{r-m}{i-m}$$

and substituting $j = i - m$, this can be written as

$$\Pr(Z = z) = \binom{365}{z} \sum_{j=0}^{365-z-1} (-1)^i \binom{365-z}{j} \left(\frac{365-z-j}{365}\right)^n.$$

(6) *How would you find the minimum number of required people such that, with given probability p, each day has at least one birthday?* This is the smallest value n such that $F_Y(n) \leq 1 - p$, where $F_Y(n)$ is given in (4.25) with $r = 365$, i.e.

$$\Pr(Y > n) = \sum_{i=1}^{364} (-1)^{i+1} \binom{365}{i} \left(\frac{365-i}{365}\right)^n,$$

and would need to be numerically determined. The Figure 4.4(b) on page 135 plots $\Pr(Y > n)$ for $r = 365$, so that the required value of n for a particular probability p can be read off the graph. If 'birthday' is replaced by 'birthmonth', then $r = 12$, to which Figure 4.4(a) corresponds.

(7) *Let p_n be the probability that, of the n people, at least two have the same birthday. Derive an expression for p_n.* Consider working with the complement. We have

$$p_n = 1 - \frac{365}{365} \cdot \frac{364}{365} \cdot \frac{363}{365} \cdots \cdots \frac{365 - n + 1}{365} = 1 - \frac{365_{[n]}}{365^n}. \qquad (6.7)$$

(8) *Compute the expected number of people who celebrate their birthday alone.* Define

$$W_i = \begin{cases} 1, & \text{if person } i \text{ has a unique birthday,} \\ 0, & \text{otherwise,} \end{cases}$$

$i = 1, \ldots, n$, and observe that the W_i are exchangeable with $\mathbb{E}[W_1] = \left(\frac{364}{365}\right)^{n-1}$. Thus, the solution is given by

$$\sum_{i=1}^{n} \mathbb{E}[W_i] = n \left(\frac{364}{365}\right)^{n-1}.$$

(9) *Generalizing the previous question, compute the expected number of days of the year that exactly k people, $1 \le k \le n$, have the same birthday.* Similar to the previous question, define

$$X_i = \begin{cases} 1, & \text{if exactly } k \text{ people have a birthday on day } i, \\ 0, & \text{otherwise,} \end{cases}$$

$i = 1, \ldots, 365$. The X_i are exchangeable with $\mathbb{E}[X_1] = \binom{n}{k}\left(\frac{1}{365}\right)^k \left(\frac{364}{365}\right)^{n-k}$ so that the solution is

$$\sum_{i=1}^{365} \mathbb{E}[X_i] = 365 \binom{n}{k}\left(\frac{1}{365}\right)^k \left(\frac{364}{365}\right)^{n-k}.$$

The X_i could also have been used to answer the previous question as well.

(10) *How many days in the year do we expect to have at least one birthday among the n people?* Define

$$Y_i = \begin{cases} 1, & \text{if at least one person has a birthday on day } i, \\ 0, & \text{otherwise,} \end{cases}$$

$i = 1, \ldots, 365$. The Y_i are exchangeable with $\mathbb{E}[Y_1] = 1 - \left(\frac{364}{365}\right)^n$ so that the solution is

$$\sum_{i=1}^{365} \mathbb{E}[Y_i] = 365 \left[1 - \left(\frac{364}{365}\right)^n\right].$$

(11) *How many days in the year do we expect to have no birthdays among the n people?*
This must be 365 minus the solution to the previous question, or

$$365 - 365\left[1 - \left(\frac{364}{365}\right)^n\right] = 365\left(\frac{364}{365}\right)^n.$$

(12) (Mosteller, 1965, p. 47) *Let p_n be the probability that, of the n people, at least two have the same birthday. Derive an approximate expression for the smallest integer n such that $p_n \geq p^*$ and solve for the n such that $p^* = 0.5$.*
From the Taylor series expansion $e^{-x} = 1 - x + \frac{x^2}{2} - \frac{x^3}{3!} + \cdots$, $e^{-x} \approx 1 - x$ for $x \approx 0$. Each term in (6.7) is of the form

$$\frac{365 - k}{365} = 1 - \frac{k}{365} \approx e^{-k/365}$$

so that

$$p_n = 1 - \prod_{k=0}^{n-1}\left(1 - \frac{k}{365}\right) \approx 1 - \prod_{k=0}^{n-1} e^{-k/365} = 1 - \exp\left[-\frac{0 + 1 + \cdots + (n-1)}{365}\right]$$

$$= 1 - \exp\left[-\frac{n(n-1)}{2 \cdot 365}\right]$$

or

$$n = \left\lceil 0.5 + 0.5\sqrt{1 - 2920 \ln(1 - p^*)}\right\rceil,$$

where $\lceil \cdot \rceil$ denotes the ceiling function, or rounding upward to the nearest integer. For $p^* = 0.5$, this yields $n = 23$. Checking with the exact values in (6.7), $p_{23} = 0.5073$ and $p_{22} = 0.4757$, so that $n = 23$ is correct.

For more birthday problems, see Example 6.6 and Problem 6.3 below.

6.3   Runs distributions*

When I was young, people called me a gambler. As the scale of my operations increased, I became known as a speculator. Now I am called a banker. But I have been doing the same thing all the time.

(Sir Ernest Cassell, Banker to Edward VII)

Example 2.5 implicitly introduced the notion of *runs*. For any sequence of trials with binary outcomes, say 0 and 1, a run is a consecutive sequence of equal outcomes, i.e. a streak of either zeros or ones. By examining the frequency of runs, one can determine how plausible a series of long 'winning streaks' is under the assumption that the outcome of the trials are not influenced by one another. Besides being of theoretical interest, the examination of runs data is useful in *nonparametric hypothesis testing.* For arguments' sake, imagine you make a series of stock trades throughout some time period and you record whether the decision led to loss (0) or profit (1). Assume a series consists of Z zeros and N ones; for example, with $Z = 6$ and $N = 9$, the series might be 011101011001101, giving rise to five runs of zeros and five runs of ones. Interest centers not on assessing the 'long-term probability' of winning, e.g. $N/(Z + N)$, but rather, *given a particular sequence*, how probable is the number of runs obtained.

To calculate the probability of r runs of ones, associate variable w_i with the ith run of wins, $i = 1, 2, \ldots, r$ and ℓ_j with the jth run of losses, $j = 1, 2, \ldots, r + 1$. In the above series, $r = 5$ and

$$\underbrace{0}_{\ell_1}\ \underbrace{1\ 1\ 1}_{w_1}\ \underbrace{0}_{\ell_2}\ \underbrace{1}_{w_2}\ \underbrace{0}_{\ell_3}\ \underbrace{1\ 1}_{w_3}\ \underbrace{0\ 0}_{\ell_4}\ \underbrace{1\ 1}_{w_4}\ \underbrace{0}_{\ell_5}\ \underbrace{1}_{w_5}\ \underbrace{}_{\ell_6}$$

with $\ell_1 = 1$, $w_1 = 3$, etc., and $\ell_6 = 0$. The number of possible arrangements with r runs of ones is the same as the number of solutions to $\sum_{i=1}^{r} w_i = N$, $w_i \geq 1$, times the number of solutions to $\sum_{j=1}^{r+1} \ell_j = Z$, $\ell_1 \geq 0$, $\ell_{r+1} \geq 0$, $\ell_j \geq 1$, $j = 2, 3, \ldots, r$. From the discussion of occupancy in Section 2.1, this is given by

$$K_{N,r}^{(1)} \cdot K_{Z,r+1}^{(0,1,1,\ldots,1,1,0)} = \binom{N-1}{r-1}\binom{Z+1}{r}.$$

Supposing that all possible arrangements of the Z zeros and N ones were equally likely to have occurred, it follows that

$$\Pr(r \text{ runs of ones}) = \frac{\binom{N-1}{r-1}\binom{Z+1}{r}}{\binom{Z+N}{N}}, \quad r = 1, \ldots, N. \tag{6.8}$$

The fact that this sums to one follows from (1.55). To illustrate, Figure 6.2 plots this probability versus r for several combinations of Z and R. For example, for the lower-right panel, with $Z = 36$ and $N = 72$, it appears unlikely that $r < 20$ or $r > 30$.

Similarly, to calculate the probability of a total of t runs among both the zeros and ones, first note that there are four possible sequences, either $w_1, \ell_1, \ldots, w_r, \ell_r$ or $\ell_1, w_1, \ldots, \ell_r, w_r$ or $w_1, \ell_1, \ldots, w_r, \ell_r, w_{r+1}$ or $\ell_1, w_1, \ldots, \ell_r, w_r, \ell_{r+1}$, each with $w_i \geq 1$, $\sum w_i = N$, $\sum \ell_j = Z$, $\ell_i \geq 1$. The first two have $K_{N,r}^{(1)} \cdot K_{Z,r}^{(1)}$ total arrangements; the latter two have $K_{N,r+1}^{(1)} \cdot K_{Z,r}^{(1)}$ and $K_{Z,r+1}^{(1)} \cdot K_{N,r}^{(1)}$, respectively. Thus,

$$\Pr(t = 2r \text{ runs}) = 2\binom{N-1}{r-1}\binom{Z-1}{r-1} \Big/ \binom{Z+N}{N}$$

and

$$\Pr(t = 2r + 1 \text{ runs}) = \left\{\binom{N-1}{r}\binom{Z-1}{r-1} + \binom{Z-1}{r}\binom{N-1}{r-1}\right\} \Big/ \binom{Z+N}{N}.$$

These expressions were first obtained by Stevens (1939); see Wilks (1963, Section 6.6) and Mood (1940) for derivations of other quantities of related interest. David (1947)

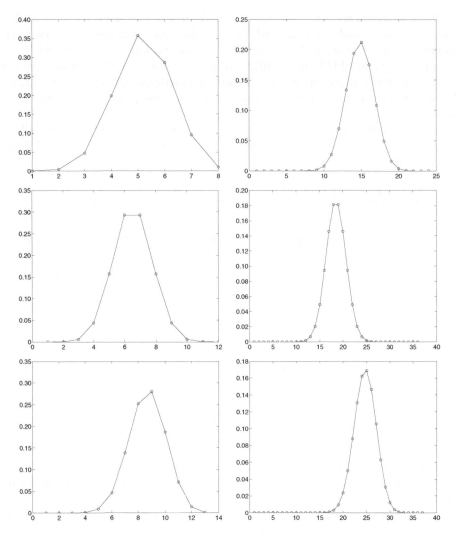

Figure 6.2 Pr (r runs of ones) versus r for $Z = 12$ (left) and $Z = 36$ (right), and $N = 2Z/3$ (top), $N = Z$ (middle) and $N = 2Z$ (bottom)

derives the distribution of the number of runs when the outcomes exhibit (first-order) dependence, which is useful for computing the power when testing the null hypothesis that the outcomes are independent.

Let random variable R be the number of runs of ones, among Z zeros and N ones, with mass function (6.8). As is often the case with discrete r.v.s, the gth *factorial moment* of R, defined by

$$\mu'_{[g]}(R) = \mathbb{E}\left[R\left(R-1\right)\cdots\left(R-g+1\right)\right],$$

is easier to calculate than the usual raw moments. The (lower-order) raw moments can then be obtained from the $\mu'_{[g]}(R)$. Problem 5.5 shows that, for $g \leq r$,

$$\mu'_{[g]}(R) = \frac{(Z+1)_{[g]}}{\binom{Z+N}{Z}}\binom{N+Z-g}{N-g}. \tag{6.9}$$

From this, the mean is

$$\mu'_{[1]}(R) = \mathbb{E}[R] = \frac{Z+1}{\binom{Z+N}{Z}}\binom{N+Z-1}{N-1} = \frac{N(Z+1)}{N+Z} \tag{6.10}$$

and, with

$$\mu'_{[2]}(R) = \mathbb{E}[R^2 - R] = \frac{(Z+1)Z}{\binom{Z+N}{Z}}\binom{N+Z-2}{N-2} = \frac{Z(Z+1)N(N-1)}{(N+Z)(N+Z-1)},$$

the variance is

$$\mathbb{V}(R) = \mathbb{E}[R^2 - R] + \mathbb{E}[R] - (\mathbb{E}[R])^2 = \frac{Z(Z+1)N(N-1)}{(N+Z)^2(N+Z-1)}. \tag{6.11}$$

The *joint* mass function of R_0, the number of runs of zeros, and R_1, the number of runs of ones, is now considered. For $R_0 = r_0$ and $R_1 = r_1$, let ℓ_i (w_i) denote the number of zeros (ones) in the ith run of zeros (ones) and observe that there are only four possible cases for which $\ell_i > 0$, $i = 1, \ldots r_0$ and $w_i > 0$, $i = 1, \ldots r_1$: either (i) $r_0 = r_1$ and the series starts with zero; (ii) $r_0 = r_1$ and the series starts with one; (iii) $r_0 = r_1 + 1$ (in which case the series must start with zero); or $r_0 = r_1 - 1$ (in which case the series must start with one). In each case, there are a total of $K_{Z,r_0}^{(1)} \times K_{N,r_1}^{(1)}$ (equally likely) arrangements. For $r_0 = r_1$, the first two cases are indistinguishable (as far as the number of runs is concerned), so that the mass function can be expressed as

$$f_{R_0,R_1}(r_0, r_1) = \gamma(r_0, r_1)\binom{Z-1}{r_0-1}\binom{N-1}{r_1-1}\Big/\binom{Z+N}{Z}, \tag{6.12}$$

where $\gamma(r_0, r_1) = \mathbb{I}(|r_0 - r_1| = 1) + 2\mathbb{I}(r_0 = r_1)$. Because $|r_0 - r_1| \leq 1$, the mass function takes on a 'tridiagonal' appearance over a grid of (r_0, r_1) values. Figure 6.3 illustrates the mass function for $Z = N = 30$. Figure 6.3(a) shows a contour plot of f_{R_0,R_1} for $Z = N = 30$; Figure 6.3(b) shows the mass function values for which $r_0 = r_1$ (solid) overlaid with f_{R_0,R_1} for $|r_0 - r_1| = 1$ (dashed). With $Z = N$, note that $f_{R_0,R_1+1} = f_{R_0+1,R_1}$.

The marginal distribution of R_0 can be obtained from (6.12) by summing out r_1. In particular, recalling the definition of $\gamma(r_0, r_1)$ and repeated usage of (1.4), quantity $\binom{Z+N}{Z}f_{R_0}(r_0)$ is given by

$$\binom{Z+N}{Z}\sum_{r_1=1}^{N} f_{r_0,r_1}(r_0, r_1)$$

$$= 2\binom{Z-1}{r_0-1}\binom{N-1}{r_0-1} + \binom{Z-1}{r_0-1}\binom{N-1}{r_0-2} + \binom{Z-1}{r_0-1}\binom{N-1}{r_0}$$

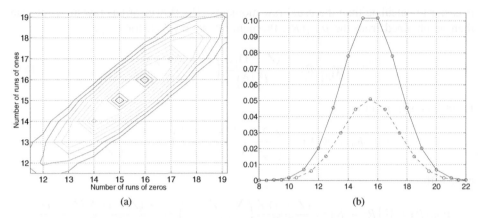

Figure 6.3   The bivariate mass function f_{R_0,R_1} for the number of runs of zeros, R_0, and the number of runs of ones, R_1, among a total of $Z = 30$ zeros and $N = 30$ ones: (a) shows a contour plot of f_{R_0,R_1}; (b) the solid plot is $f_{R_0,R_1}(r_0, r_0)$ versus r_0, the dashed plot is $f_{R_0,R_1}(r_0, r_0 + 1)$ versus $r_0 + 1/2$, with the offset $+1/2$ used to enhance readability

$$= \binom{Z-1}{r_0-1}\binom{N-1}{r_0-1} + \binom{Z-1}{r_0-1}\binom{N-1}{r_0-2} + \binom{Z-1}{r_0-1}\binom{N-1}{r_0-1}$$
$$+ \binom{Z-1}{r_0-1}\binom{N-1}{r_0}$$

$$= \binom{Z-1}{r_0-1}\left(\binom{N-1}{r_0-1} + \binom{N-1}{r_0-2}\right) + \binom{Z-1}{r_0-1}\left(\binom{N-1}{r_0-1} + \binom{N-1}{r_0}\right)$$

$$= \binom{Z-1}{r_0-1}\binom{N}{r_0-1} + \binom{Z-1}{r_0-1}\binom{N}{r_0}$$

$$= \binom{Z-1}{r_0-1}\binom{N+1}{r_0},$$

i.e.

$$f_{R_0}(r_0) = \binom{Z-1}{r_0-1}\binom{N+1}{r_0} \Big/ \binom{Z+N}{Z}.$$

Similarly,

$$f_{R_1}(r_1) = \binom{N-1}{r_1-1}\binom{Z+1}{r_1} \Big/ \binom{Z+N}{Z},$$

which agrees with (6.8).

The expected total number of runs among Z zeros and N ones, $T = R_0 + R_1$, where r_0 is the number of runs of zeros and r_1 is the number of runs of ones, is most easily computed by using (6.1), i.e. $\mathbb{E}[T] = \mathbb{E}[R_0] + \mathbb{E}[R_1]$, with $\mathbb{E}[R_1]$ given in (6.10) and $\mathbb{E}[R_0]$ obtained by symmetry, i.e. switch Z and N. That is,

$$\mathbb{E}[T] = \frac{Z(N+1)}{Z+N} + \frac{N(Z+1)}{N+Z} = 1 + 2\frac{NZ}{N+Z}. \tag{6.13}$$

From (6.3),

$$V(T) = V(R_0) + V(R_1) + 2\operatorname{Cov}(R_0, R_1),$$

with $V(R_1)$ (and $V(R_0)$ from symmetry) available from (6.11).

The covariance term needs to be obtained and can be computed from the *joint factorial moment* of $(R_0 - 1, R_1 - 1)$,

$$\mu'_{[g_0, g_1]} = \mathbb{E}\left[(R_0 - 1)_{[g_0]}(R_1 - 1)_{[g_1]}\right],$$

which is given by

$$\mu'_{[g_0, g_1]} = \frac{(Z-1)_{[g_0]}(N-1)_{[g_1]}}{\binom{Z+N}{N}}\binom{Z+N-g_0-g_1}{Z-g_1} \tag{6.14}$$

(see Problem 5.6). From (6.14) with $g_0 = g_1 = 1$,

$$\mu'_{[1,1]} = \mathbb{E}\left[(R_0 - 1)(R_1 - 1)\right] = \frac{(Z-1)(N-1)}{\binom{Z+N}{N}}\binom{Z+N-2}{Z-1}$$

$$= \frac{(Z-1)(N-1)NZ}{(Z+N)(Z+N-1)}$$

so that, from (5.24),

$$\operatorname{Cov}(R_0, R_1) = \mathbb{E}[R_0 R_1] - \mathbb{E}[R_0]\mathbb{E}[R_1]$$

$$= \mu'_{[1,1]} + \mathbb{E}[R_0] + \mathbb{E}[R_1] - 1 - \mathbb{E}[R_0]\mathbb{E}[R_1]$$

$$= ZN\frac{ZN - Z - N + 1}{(Z+N)^2(N+Z-1)},$$

which yields

$$V(T) = 2ZN\frac{2ZN - (Z+N)}{(Z+N)^2(N+Z-1)}$$

after simplification. Figure 6.4 plots

$$f_T(t; N, Z) = \binom{Z+N}{N}^{-1}\begin{cases} 2\binom{N-1}{r-1}\binom{Z-1}{r-1}, & \text{if } t = 2r, \\ \binom{N-1}{r}\binom{Z-1}{r-1} + \binom{Z-1}{r}\binom{N-1}{r-1}, & \text{if } t = 2r+1, \end{cases}$$

$r = 1, 2, \ldots$, versus t for $N = 14$ and several Z values (solid) and overlaid with the normal approximation with $\mu = \mathbb{E}[T]$ and $\sigma^2 = V(T)$. In general, for large $N \approx Z$, the normal approximation is excellent.

The expected value of T, the total number of runs, $T = R_0 + R_1$, from a sequence of N ones and Z zeros was found via direct derivation to be $\mathbb{E}[T] = 1 + 2NZ(N + Z)$. An alternative derivation as in Ross (1988, p. 266) is now given. Define the Bernoulli

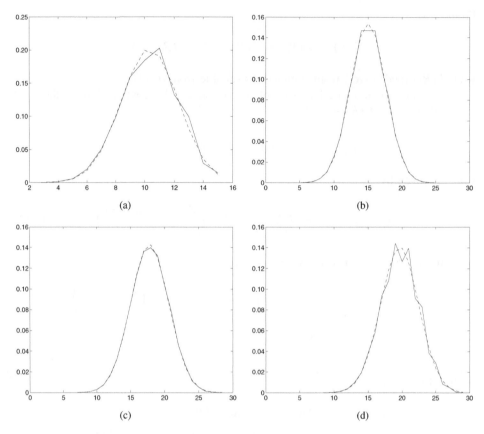

Figure 6.4 Probability of t runs versus t for $N = 14$ and (a) $Z = 7$, (b) $Z = 14$, (c) $Z = 21$ and (d) $Z = 28$ and overlaid with the normal approximation

r.v. S_i to be one if a run of zeros starts at the ith position in the sequence, and zero otherwise, so that $R_0 = \sum_{i=1}^{Z+N} S_i$ and $\mathbb{E}[R_0] = \sum_{i=1}^{Z+N} \mathbb{E}[S_i]$. Now, clearly, $\mathbb{E}[S_1] = Z/(Z+N)$ while, for $1 < i \leq Z + N$, $\mathbb{E}[S_i]$ is just the probability that a one occurs in position $i - 1$ and a zero occurs in position i in the sequence. Using (3.7), this is seen to be

$$\mathbb{E}[S_i] = \frac{N}{Z+N}\frac{Z}{Z+N-1}.$$

Thus,

$$\mathbb{E}[R_0] = \mathbb{E}[S_1] + (Z + N - 1)\mathbb{E}[S_2] = \frac{Z}{Z+N} + \frac{ZN}{Z+N}$$

and, from symmetry, $\mathbb{E}[R_1] = N/(Z+N) + ZN/(Z+N)$, so that

$$\mathbb{E}[T] = 1 + 2NZ(N+Z),$$

as was given in (6.13).

We now change our focus somewhat and consider a sequence of n i.i.d. trials, where each trial can take on one of k values, say $1, 2, \ldots, k$, and is a multinomial outcome with equal probabilities $1/k$. Interest centers on the probability $p_{m,n}(k)$ of observing at least one m-length sequence of ones. This can be written as

$$p_{m,n}(k) = 1 - \frac{t_{m,n}(k)}{k^n}, \qquad (6.15)$$

where $t_{m,n}(k)$ is the total number of sequences for which $1, 1, \ldots, 1$ (m times) does not occur. To demonstrate, take $k = m = 2$ and $n = 4$. All the possible realizations which do not include the sequence 1,1 are easily seen to be

$$2222, \ 1222, \ 2122, \ 2212, \ 2221, \ 1212, \ 2121, \ 1221,$$

so that $t_{2,4}(2) = 8$ and the probability is $p_{2,4}(2) = 1 - 8/2^4 = 1/2$. A closed-form expression for $t_{m,n}(k)$ was developed by Gani (1998), which we now outline.

First take $m = 2$ and let $t_n(k) := t_{2,n}(k)$ for convenience. Also let $a_n(k)$ be the number of the sequences among the $t_n(k)$ which end in a (necessarily single) 1. Then

$$t_{n+1}(k) = k t_n(k) - a_n(k),$$

which follows because there are $k t_n(k)$ possible 'candidate' sequences of length $n + 1$, but $a_n(k)$ of these end in 1,1, and so must be taken away. A similar analysis shows that

$$a_{n+1}(k) = t_n(k) - a_n(k),$$

so that

$$\begin{bmatrix} t_{n+1}(k) \\ a_{n+1}(k) \end{bmatrix} = \begin{bmatrix} k & -1 \\ 1 & -1 \end{bmatrix} \begin{bmatrix} t_n(k) \\ a_n(k) \end{bmatrix}, \qquad n = 1, 2, \ldots.$$

Then, as $t_1(k) = k$ and $a_1(k) = 1$,

$$\begin{bmatrix} t_n(k) \\ a_n(k) \end{bmatrix} = \begin{bmatrix} k & -1 \\ 1 & -1 \end{bmatrix}^{n-1} \begin{bmatrix} k \\ 1 \end{bmatrix}.$$

As

$$\begin{bmatrix} k \\ 1 \end{bmatrix} = \begin{bmatrix} k & -1 \\ 1 & -1 \end{bmatrix} \begin{bmatrix} 1 \\ 0 \end{bmatrix},$$

this can be written as

$$\begin{bmatrix} t_n(k) \\ a_n(k) \end{bmatrix} = \begin{bmatrix} k & -1 \\ 1 & -1 \end{bmatrix}^{n} \begin{bmatrix} 1 \\ 0 \end{bmatrix},$$

so that

$$t_n(k) = \begin{bmatrix} 1 & 0 \end{bmatrix} \begin{bmatrix} k & -1 \\ 1 & -1 \end{bmatrix}^n \begin{bmatrix} 1 \\ 0 \end{bmatrix}. \tag{6.16}$$

Because the eigenvalues of a 2×2 matrix are analytically tractable, (6.16) can also be expressed as (see Gani, 1998, p. 267)

$$t_n(k) = \frac{1}{2^{n+1}} \left[\left(1 + \frac{k+1}{u} \right) (k - 1 + u)^n + \left(1 - \frac{k+1}{u} \right) (k - 1 - u)^n \right], \tag{6.17}$$

where $u = \sqrt{(k-1)(k+3)}$.

For general m, $2 \le m \le n$, Gani extends the previous argument to show that

$$t_{m,n}(k) = \begin{bmatrix} 1 & 0 & \cdots & 0 \end{bmatrix} \mathbf{A}^n \begin{bmatrix} 1 \\ 0 \\ \vdots \\ 0 \end{bmatrix}, \quad \mathbf{A} = \begin{bmatrix} k & -1 & 0 & \cdots & 0 & 0 \\ 0 & 0 & 1 & 0 & \cdots & 0 \\ \vdots & & & \ddots & & \vdots \\ 0 & 0 & \cdots & 0 & 1 & 0 \\ 0 & 0 & \cdots & 0 & 0 & 1 \\ 1 & -1 & \cdots & \cdots & -1 & -1 \end{bmatrix}, \tag{6.18}$$

where $\mathbf{A}$ is $m \times m$.

To illustrate, take $n = 100$ and $k = 10$. Then, using (6.15) and (6.18), $p_{2,n}(k) = 0.5985$, $p_{3,n}(k) = 0.0848$ and $p_{4,n}(k) = 0.0087$.

It would also be useful to be able to compute $p_{m,n}(k)$ via simulation. First, this would allow us to confirm that (6.18) is at least possibly correct (simulation obviously cannot prove its validity); secondly, for large values of n and k, $t_{m,n}(k)$ and k^n could be too large for some computing platforms (for example, with $k = 20$ and $m = 3$, the exact calculation works in Matlab for $n = 236$, but fails for $n = 237$); and thirdly, other, more complicated schemes could be used, for which the exact solution is either difficult to derive, or intractable. A pertinent example of a (most likely) intractable case is when the equally likely assumption is relaxed.

To simulate, we would produce a large number, say s, n-length sequences, with each element in a sequence being a number in $1, 2, \ldots, k$, chosen randomly, with equal probability. That can be done in Matlab with the `unidrnd` function. Each sequence is then parsed to determine if it contains at least one m-length sequence of ones. This is straightforward to do with a FOR loop, though it will be very slow in a high-level language such as Matlab. Instead, it is faster to construct sums of m consecutive values in a sequence and check if any of them are equal to m. For $m = 2$, this would be accomplished by the following code.

```
n=100; k=10; sim=100000; b=zeros(sim,1);
for i=1:sim
  s = unidrnd(k,n,1); q = s(1:end-1) + s(2:end);
  b(i) = any(q==2);
end
```

Then $p_{m,n}(k)$ is approximated by mean(b). Building on the above code, a program for the general m-case, along with the exact computation, is shown in Program Listing 6.1.

```
function p=gani(n,k,m,sim)
if nargin<4, sim=0; end
if sim<=0 % do the exact way
  A=zeros(m,m); A(1,1)=k; A(1,2)=-1; A(m,1)=1;
  for i=2:m-1,A(i,i+1)=1; end,   for i=2:m, A(m,i)=-1; end
  a=zeros(m,1); a(1)=1; p = 1 - a'*(A^n)*a / k^n;
else % simulate to get an approximate answer
  b=zeros(sim,1);
  for i=1:sim
    s=unidrnd(k,n,1); q=s(1:end-m+1);
    for j=2:m, q=q+s(j:end-m+j); end
    b(i)=any(q==m);
  end
  p=mean(b);
end
```

Program Listing 6.1   Computes the probability of observing at least one m-length sequence of ones in n i.i.d. multinomial trials with k equally likely outcomes $1, 2, \ldots, k$

To demonstrate, we use the case with $n = 100$ and $k = 10$ from before. Based on 100 000 replications, a run of the program yielded $p_{2,n}(k) \approx 0.5978$, $p_{3,n}(k) \approx 0.08565$ and $p_{4,n}(k) \approx 0.00846$.

The next example is somewhat premature, as it involves several concepts which have not yet been introduced; we include it here because it illustrates a practical use of being able to calculate $p_{m,n}(k)$.

Example 6.5   In financial institutions, the *Value-at-Risk*, or VaR, is a popular measure of overall exposure to downside risk and, in one variety, involves forecasting a parametric density which describes (or purports to) the distribution of the asset's return on the subsequent business day. If, on the next day, the true return occurs too far in the left tail of the predictive density, then a 'hit' or 'violation' is said to have occurred. If the chosen tail probability, or *target probability* is, say, 5 %, then violations should occur, on average, with a frequency of 5 % (if the model used is correct). If 'significantly' more violations than the target occur over a period of time, the bank can be penalized by having to hold a higher capital reserve.

Let $\hat{F}_{t+1|t}$ denote the predictive c.d.f. at time t of the return at time $t + 1$, and r_{t+1} be the actual return at time $t + 1$. Then, under the (usually heroic) assumption that the model is correct, $U_{t+1} = \hat{F}_{t+1|t}(r_{t+1}) \sim \text{Unif}(0, 1)$, via the probability integral transform (discussed in Section 7.3). By dividing the interval $(0, 1)$ into 20 segments and setting $D_t = i$ if U_t falls into the ith interval, $t = 1, \ldots, n$, the D_t form an i.i.d. sequence of multinomial r.v.s with $k = 20$ equally likely values, and a violation occurs at time t when $D_t = 1$.

Interest centers on the probability that several violations occur in a row, which could be deemed as evidence that the bank's VaR model is not performing correctly.

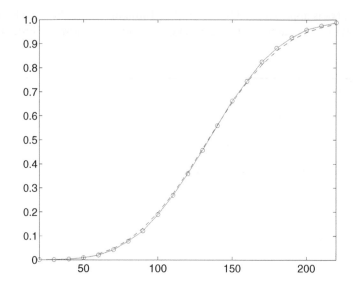

Figure 6.5 The probability $p_{5,n}$ that there is at least one occurrence of $k = 5$ birthdays in a row, based on n people, as a function of n. Solid line with circles shows simulated values; dashed line uses (6.15) and (6.18) as an approximation, based on noninteger k value, $k = 1/p(n)$, with $p(n)$ given in (6.19)

Typically, $n = 250$, which corresponds approximately to the number of trading days in a year. As we saw above, with $n = 250$ and $k = 20$, the exact method is no longer numerically possible to calculate in Matlab, so we use simulation. With 100 000 replications, $p_{2,250}(20) = 0.4494$, $p_{3,250}(20) = 0.0285$ and $p_{4,250}(20) = 0.00169$, showing that, while two violations in a row is quite likely, the occurrence of three in a row provides 'reasonable evidence' that the model is faulty, while four in a row should be cause for serious alarm. ∎

⊖ ***Example 6.6*** In a school class with n pupils, $n < 365$, a party is thrown on each day for which at least one of the n children has a birthday. What is the probability, $p_{m,n}$, that there will be a party m days in a row, $1 < m \leq n$? We make the simplifying assumptions (i) there are always 365 days in a year, (ii) birthdays are equally likely to occur on any of the 365 days, (iii) the n pupils have independent birthdays (i.e. no twins) and, most unrealistically, (iv) school is in session 365 days a year, i.e. no weekends or holidays. Even with these assumptions, this is not a trivial question.

The first attempt at an answer involves simulation. The program in Program Listing 6.2 does the job; it looks long, but only because there are many comments. This was used to obtain $p_{5,n}$ for a grid of n values, which is plotted as the solid line with circles in Figure 6.5.

Now consider the following approximation. Treat the 365 days as a sequence of i.i.d. Bernoulli r.v.s $D_1, \ldots, D_{365}$, with constant success probability $p(n)$ being the probability that at least one birthday occurs. (Note that, even though the pupils' birthdays

```
function prob = rowofbirthdays(m,n,C,sim)
if nargin<4, sim=20000; end
if nargin<3, C=365; end
boolv=zeros(sim,1);
for i=1:sim
  if mod(i,1000)==0, [n i], end
  bday = unidrnd(C,n,1); % these are the n birthdays
  bday = sort(bday);      % Just for debugging. Don't need to sort.
  tt=tabulate(bday); days=find(tt(:,2)>0); % remove redundancies
  %%%%%%%%%%%%%%%%%%%%%%%%%%%%%%%%%%%%%%%%%%%%%%%%%%%%%%%%%%
  %% This takes the circularity of the year into account.
  %% Just remove this block of code to ignore circularity
  %% Only relevant if there are birthdays on Jan1 and Dec31
  if (days(1)==1) & (days(end)==C)
    % now append to days-vector as many consecutive
    %  birthdays from Jan1 onwards as is necessary
    days=[days ; C+1]; % Jan1 is now the 366th day
    cnt=2; while cnt==days(cnt), days=[days;C+cnt]; cnt=cnt+1; end
  end
  %% Now we can parse days-vector without thinking about circularity
  %%%%%%%%%%%%%%%%%%%%%%%%%%%%%%%%%%%%%%%%%%%%%%%%%%%%%%%%%%
  vec=diff(days); % m birthdays in a row equiv to m-1 ones in a row
  % Now compare moving windows of vec to (1,1...,1),
  %  BUT, we can save lots of time by only searching where a one is!
  loc=find(vec==1); found=0; match=ones(m-1,1);
  for j=1:length(loc)
    pos=loc(j);
    if ( pos+(m-1) <= length(vec) ) % Are we too close to the end?
      tobecomp=vec(pos:pos+(m-1)-1);
      found=all(tobecomp==match);
    end
    % Once we find a match, no need to keep looking.
    if found, break, end
  end
  boolv(i)=found;
end
prob=mean(boolv);
```

Program Listing 6.2 Simulates the probability that there is at least one occurrence of m birthdays in a row, based on n people; assume that n people have independent birthdays, which are equally likely throughout the C-day year (choose C to be 365, but this can be a general 'balls in cells' setup)

are i.i.d., the D_i are not.) Similar to question 6.2.1 in Section 6.2.1 above,

$$p(n) = 1 - \left(\frac{364}{365}\right)^n. \tag{6.19}$$

We would like to use Gani's method, which would be applicable if the D_i were i.i.d., *and* if there were k equally likely possibilities. In this situation, there are two possibilities (at least one birthday occurs, or not), but these are certainly not equally likely. Instead, consider the following observation. For $n = 105$, $p(n)$ from (6.19) is, to four digits, 0.2503, so that $1/p(n) \approx 4$. If we take $k = 4$, and treat the sequence of days as multinomial with $k = 4$ equally likely possible outcomes, then Gani's method would yield the desired probability.

Thus, we would like to use Gani's formulae (6.15) and (6.18), taking k to be $1/p(n)$. However, this value of k is clearly not an integer. However, observe that, computationally speaking, (6.15) and (6.18) do not require k to be an integer. This does not imply that the formulae have meaning for noninteger k, but a continuity argument would suggest that they would. The computed values are shown in Figure 6.5 as the solid line, and are indeed extremely close to the 'true' values. They are also computed essentially instantaneously, which should be compared with the rather lengthy simulation time required to obtain 'exact' values. This speed aspect adds considerable value to the approximate method. Note that their graph starts around $n = 60$, because values below $n = 60$ lead to values of k which were too large for numeric computation. ∎

The general analysis of the distribution theory of runs is best conducted in a Markov chain setting. Further information on this and various applications of runs can be found in the book-length treatment by Balakrishnan and Koutras (2002) (see also Example 8.13).

6.4   Random variable decomposition

The previous examples showed how expressing a complicated r.v. as a sum of exchangeable Bernoulli r.v.s immensely simplified the expected value calculation. In this section, we examine such decompositions for the four basic sampling schemes.

6.4.1   Binomial, negative binomial and Poisson

If an urn contains N white and M black balls and n balls are consecutively and randomly withdrawn with replacement, then the r.v. giving the number of white balls drawn is binomial distributed. Each draw results in one of two possibilities, so that the trial outcomes are Bernoulli distributed and the probability $p = N/(N + M)$ remains constant for each. It follows that a binomial random variable with parameters n and p can be represented as a sum of n independent, Bernoulli distributed random variables, each with parameter p. That is, if $X \sim \text{Bin}(n, p)$, then $X = \sum_{i=1}^{n} X_i$, where $X_i \sim \text{Ber}(p)$, $i = 1, \ldots, n$, and $X_i \overset{\text{i.i.d.}}{\sim} \text{Ber}(p)$.

As $\mathbb{E}[X_i] = \Pr(X_i = 1) = p$ for all i, it follows directly from (6.1) that $\mathbb{E}[X] = np$. Similarly, from (6.2) and (5.25), $\mathbb{V}(X) = np(1 - p)$, because $\mathbb{V}(X_i) = p(1 - p)$.

⊙ ***Example 6.7***   As discussed in Section 5.3.1, let $\mathbf{X} = (X_1, \ldots, X_k) \sim \text{Multinom}(n, \mathbf{p})$, with $f_{\mathbf{X}}(\mathbf{x}) = n! \prod_{i=1}^{k} p_i^{x_i} / \prod_{i=1}^{k} x_i!$, and marginal distributions $X_i \sim \text{Bin}(n, p_i)$. It follows directly from the previous results that $E[X_i] = np_i$ and $\mathbb{V}(X_i) = np_i(1-p_i)$. For the covariances observe that, as $X_i + X_j \sim \text{Bin}(n, p_i + p_j)$,

$$\mathbb{V}(X_i + X_j) = n(p_i + p_j)(1 - p_i - p_j).$$

However, using (6.3), we have

$$2\,\text{Cov}(X_i, X_j) = \mathbb{V}(X_i + X_j) - \mathbb{V}(X_i) - \mathbb{V}(X_j),$$

so that

$$\text{Cov}(X_i, X_j) = \frac{1}{2}\left[n(p_i + p_j)(1 - p_i - p_j) - np_i(1-p_i) - np_j(1-p_j)\right],$$

i.e. $\text{Cov}(X_i, X_j) = -np_i p_j$. ∎

In a similar way to the previous discussion of the binomial, a bit of thought reveals that $X \sim \text{NBin}(r, p)$ can be decomposed into r i.i.d. geometric random variables, $X_1, \ldots, X_r$. If we only count the number of failures, then each X_i has mass function (4.16). In that case, it can be shown (see Example 4.22 and Problem 4.3) that $\mathbb{E}[X_i] = (1-p)/p$ and $\mathbb{V}(X_i) = (1-p)/p^2$ so that $\mathbb{E}[X] = r(1-p)/p$ and $\mathbb{V}(X) = r(1-p)/p^2$. Based on these results, it is easy to see that, if X now represents the total number of successes and failures, then $\mathbb{E}[X] = r/p$ and the variance remains unchanged.

This 'decomposing' of random variables is useful for determining the distribution of certain sums of random variables. In particular, it follows that, if $X_i \overset{\text{ind}}{\sim} \text{Bin}(n_i, p)$, then $X = \sum_i X_i \sim \text{Bin}(n, p)$, where $n = \sum_i n_i$. Likewise, if $X_i \overset{\text{ind}}{\sim} \text{NBin}(r_i, p)$, then $X = \sum_i X_i \sim \text{NBin}(r, p)$, where $r = \sum_i r_i$.

Example 6.8 In Section 4.2.6, r.v. Y represented the number of cereal boxes it was necessary to purchase in order to get at least one of each prize (or, in more general terms with respect to placing balls randomly into r cells, the number of balls required such that no cell is empty). Interest might also center more generally on r.v. Y_k, the number of balls needed so that k cells are not empty, $1 < k \le r$. Define G_i to be the number of balls needed in order to inhabit one of the remaining empty cells, $i = 0, 1, \ldots, k-1$. Clearly, $G_0 = 1$ and $G_i \overset{\text{ind}}{\sim} \text{Geo}(p_i)$ with density (4.18), $p_i = (r-i)/r$, expected value p_i^{-1} and $\mathbb{V}(G_i) = (1-p_i)/p_i^2$, $i = 1, \ldots, k-1$. Then, $Y_k = \sum_{i=0}^{k-1} G_i$,

$$\mathbb{E}[Y_k] = \sum_{i=0}^{k-1} \mathbb{E}[G_i] = 1 + \frac{r}{r-1} + \frac{r}{r-2} + \cdots + \frac{r}{r-k+1} = r\sum_{i=0}^{k-1}\frac{1}{r-i}$$

and, from (6.2),

$$\mathbb{V}(Y_k) = \sum_{i=0}^{k-1}\mathbb{V}(G_i) = \sum_{i=0}^{k-1}\frac{1 - \left(\frac{r-i}{r}\right)}{\left(\frac{r-i}{r}\right)^2} = r\sum_{i=1}^{k-1}\frac{i}{(r-i)^2}.$$

With $k = r$, $Y_k = Y$ and

$$\mathbb{E}[Y] = r \left(1 + \cdots + \frac{1}{r-1} + \frac{1}{r} \right) \xrightarrow{r \to \infty} r (\ln r + \gamma),$$

where

$$\gamma = \lim_{n \to \infty} \left(1 + \frac{1}{2} + \frac{1}{3} + \cdots + \frac{1}{n} - \log n \right) \approx 0.5772156649$$

is Euler's constant (see Examples A.38 and A.53). For the variance,

$$\mathbb{V}(Y) = r \sum_{i=1}^{r-1} \frac{i}{(r-i)^2} \xrightarrow{r \to \infty} \frac{r^2 \pi^2}{6}, \tag{6.20}$$

from Example A.40. The mass function of Y_k will be approximated in Volume II using a so-called saddlepoint approximation.   ∎

In the previous example, it is important to observe that r.v. Y_k does not specify which k cells are to be filled; just that some set of k cells happens to become nonempty first. This is quite different from the r.v. which models the required number of balls so that each of a *particular set* of k cells becomes nonempty. This is a special case of the negative multinomial distribution with equal cell probabilities $p_i = r^{-1}$ and such that sampling stops when a particular k cells (say the first k) each obtain at least one 'hit'.

Returning to the decomposition of r.v.s, less obvious is that, if $X_i \overset{\text{ind}}{\sim} \text{Poi}(\lambda_i)$, then $X = \sum_i X_i \sim \text{Poi}(\lambda)$, where $\lambda = \sum_i \lambda_i$. This is algebraically proven in Example 6.10 below, but can be intuitively thought of as follows. With the interpretation of a Poisson random variable being the limiting case of a binomial, let p and n_i be such that $n_i p = \lambda_i$ and $Y_i \overset{\text{ind}}{\sim} \text{Bin}(n_i, p)$, $\forall i$. Then, as $\sum_i Y_i \sim \text{Bin}(\sum_i n_i, p)$ can be approximated by a Poisson random variable with parameter $\lambda = p \sum_i n_i$, it follows (informally) that $X = \sum_i X_i \sim \text{Poi}(\lambda)$.

The computation of the moments in the binomial and negative binomial cases were easy because the with-replacement assumption allowed an i.i.d. decomposition. For the hypergeometric case considered next, such i.i.d. r.v.s are not available, but exchangeable ones are.

6.4.2   Hypergeometric

Let $X \sim \text{HGeo}(N, M, n)$ as in (4.14), i.e. from N white and M black balls, n are randomly drawn without replacement, and X denotes the number of white balls drawn. Of course, from the definition, we can try to simplify

$$\mathbb{E}[X^k] = \sum_{x=\max(0, n-M)}^{\min(n, N)} x^k \frac{\binom{N}{x} \binom{M}{n-x}}{\binom{N+M}{n}}$$

for $k = 1, 2$, which looks rather involved. Instead, to calculate the mean, similar to Feller (1957, p. 218), define Y_i as

$$Y_i = \begin{cases} 1, & \text{if the } i\text{th ball drawn is white,} \\ 0, & \text{otherwise,} \end{cases}$$

$i = 1, \ldots, n$. Observe that the Y_i are exchangeable and that $X = \sum_{i=1}^{n} Y_i$. As the ith ball drawn is equally likely to be any of the $N + M$, it follows that $\mathbb{E}[Y_i] = N/(N + M)$ and

$$\mathbb{E}[X] = \mathbb{E}\left[\sum_{i=1}^{n} Y_i\right] = \sum_{i=1}^{n} \mathbb{E}[Y_i] = \frac{nN}{N + M}. \tag{6.21}$$

For the variance, use (6.2) and the exchangeability of the Y_i to get

$$\mathbb{V}(X) = \sum_{i=1}^{n} \mathbb{V}(Y_i) + \sum\sum_{i \neq j} \text{Cov}(Y_i, Y_j) = n\mathbb{V}(Y_1) + 2\binom{n}{2}\text{Cov}(Y_1, Y_2)$$

$$= n\frac{N}{N + M}\left(1 - \frac{N}{N + M}\right) + 2\binom{n}{2}[\mathbb{E}[Y_1 Y_2] - \mathbb{E}[Y_1]\mathbb{E}[Y_2]],$$

where the variance of Y_1 follows because it is a Bernoulli event and thus has variance $p(1 - p)$, where $p = \Pr(Y_1 = 1) = \mathbb{E}[Y_1]$. Similarly, $\mathbb{E}[Y_1 Y_2]$ is just

$$\Pr(Y_1 = 1, Y_2 = 1) = \frac{N}{N + M}\frac{N - 1}{N + M - 1},$$

so that

$$\mathbb{V}(X) = n\frac{N}{N + M}\left(1 - \frac{N}{N + M}\right) + 2\binom{n}{2}\left[\frac{N}{N + M}\frac{N - 1}{N + M - 1} - \left(\frac{N}{N + M}\right)^2\right]$$

$$= n\frac{N}{N + M}\frac{M}{N + M} + n(n - 1)\left[\frac{N}{N + M}\frac{N - 1}{N + M - 1} - \left(\frac{N}{N + M}\right)^2\right]$$

$$= \frac{nN}{N + M}\left[\frac{M}{N + M} + (n - 1)\left(\frac{N - 1}{N + M - 1} - \frac{N}{N + M}\right)\right]$$

$$= \frac{nN}{N + M}\left(\frac{M}{N + M}\frac{N + M - n}{N + M - 1}\right). \tag{6.22}$$

Notice that, for $n = 1$, X is Bernoulli with $p = N/(N + M)$ and $\mathbb{V}(X)$ reduces to $p(1 - p)$ as it should. For small n (relative to $N + M$), $\mathbb{V}(X) \approx np(1 - p)$, or that of a binomially distributed random variable.

Example 6.9   It is instructive to compare the mean and variance of the binomial and hypergeometric sampling schemes. For an urn containing w white and b black balls,

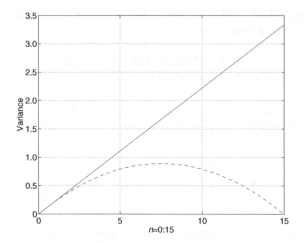

Figure 6.6 $\mathbb{V}(X)$ (solid) and $\mathbb{V}(Y)$ (dashed)

let the random variables X and Y denote the number of white balls drawn when $n \leq w + b$ balls are drawn with and without replacement, respectively. Then, with $p = w/(w+b)$, $X \sim \text{Bin}(n, p)$ and $Y \sim \text{HGeo}(w, b, n)$,

$$\mathbb{E}[X] = np = \frac{nw}{w+b} = \mathbb{E}[Y],$$

$$\mathbb{V}(X) = np(1-p) = \frac{nw}{w+b}\left(1 - \frac{w}{w+b}\right) = \frac{nwb}{(w+b)^2}$$

and

$$\mathbb{V}(Y) = \frac{nw}{w+b}\frac{b}{w+b}\frac{w+b-n}{w+b-1} \leq \mathbb{V}(X),$$

with strict equality holding only for $n = 1$. For $w = 10$ and $b = 5$, Figure 6.6 plots both variances. Clearly, $\mathbb{V}(X)$ is maximized at $n = 15$ and differentiating shows that, for $n = \{7, 8\}$, $\mathbb{V}(Y) = 8/9$ is a maximum.   ∎

6.4.3   Inverse hypergeometric*

An urn contains w white and b black balls. Let the random variables X_1 and Y_1 be the number of balls drawn until (and including) the first white ball appears, with and without replacement, respectively. Interest mainly centers on $Y_1 \sim \text{IHGeo}(1, w, b)$, while X_1 serves to provide a comparison with a similar sampling scheme. In particular

$$\Pr(X_1 = x) = p(1-p)^{x-1}\mathbb{I}_{\{1,2,\dots\}}(x),$$

i.e. X_1 follows the geometric distribution (4.18) with expected value p^{-1}, where $p = w/(w+b)$.

For Y_1, we demonstrate four ways to show that

$$\mathbb{E}[Y_1] = \frac{w+b+1}{w+1}.$$

Note that this is the general solution to Problem 4.10 and also that the expected number of balls *not including* the white ball is obviously just $\mathbb{E}[Y_1] - 1 = b/(w+1)$.

(1) *From the definition* By writing out $\Pr(Y_1 = y)$ for $y = 1, 2$ and 3, one sees that, in general,

$$\Pr(Y_1 = y) = \frac{b}{w+b} \frac{b-1}{w+b-1} \cdots \frac{b-(y-2)}{w+b-(y-2)} \frac{w}{w+b-(y-1)}$$

$$= \frac{wb_{[y-1]}}{(w+b)_{[y]}},$$

so that

$$\mathbb{E}[Y_1] = \sum_{y=1}^{b+1} y f_Y(y) = w \sum_{y=1}^{b+1} y \frac{b_{[y-1]}}{(w+b)_{[y]}}.$$

Thus,

$$\mathbb{E}[Y_1] = w \sum_{y=1}^{b+1} y \frac{b_{[y-1]}}{(w+b)_{[y]}} = \sum_{y=1}^{b+1} y \frac{\binom{w}{1}\binom{b}{y-1}}{\binom{w+b}{y}} \frac{(y-1)!}{y!}$$

$$= \sum_{y=1}^{b+1} \frac{\binom{w}{1}\binom{b}{y-1}}{\binom{w+b}{y}} = \frac{w+b+1}{w+1},$$

where the last combinatoric equality can (presumably) be shown by induction.

(2) *Decomposition into exchangeable r.v.s* (Ross, 1988, p. 267) Label the black balls as $b_1, \ldots, b_b$ and define the random variables

$$Z_i = \begin{cases} 1, & \text{if } b_i \text{ is withdrawn before any of the white balls,} \\ 0, & \text{otherwise,} \end{cases}$$

$i = 1, \ldots b$, so that $Y_1 = 1 + \sum_{i=1}^{b} Z_i$ and $\mathbb{E}[Y_1] = 1 + \sum_{i=1}^{b} \Pr(Z_i = 1)$. Because each of $w + 1$ balls (the w white and b_i) have an equally likely chance of being the first *of this set* to be withdrawn, $\Pr(Z_i = 1) = 1/(w+1)$ and

$$\mathbb{E}[Y_1] = 1 + \frac{b}{w+1} = \frac{w+b+1}{w+1}.$$

Note that the Z_i are not independent but are exchangeable.

(3) *Conditioning and recursion* (Ross, 1988, p. 288)[1] Let $Y = Y_1 - 1$ be the number of black balls which are chosen before the first white ball, and Z be such that

---

[1] This solution involves a rudimentary application of conditional expectation which has not been discussed yet, but the technique should nevertheless be easy to follow, or can be returned to after reading Section 8.2.3.

$Z = 1$ if the first ball chosen is white, and zero otherwise. To make explicit the dependence on w and b, define $M_{w,b} = \mathbb{E}[Y]$. Clearly, $\mathbb{E}[Y \mid Z = 1] = 0$. If $Z = 0$, then $\mathbb{E}[Y \mid Z = 0] = 1 + M_{w,b-1}$ because the sampling scheme is the same, but with $b - 1$ black balls. Thus, conditioning on the first draw,

$$M_{w,b} = \mathbb{E}[Y] = \mathbb{E}[Y \mid Z = 1]\Pr(Z = 1) + \mathbb{E}[Y \mid Z = 0]\Pr(Z = 0)$$

$$= \frac{b}{w + b}\left(1 + M_{w,b-1}\right)$$

and, using the boundary condition $M_{w,0} = 0$, this can be iterated to show that $M_{w,b} = b/(w + 1)$ so that

$$\mathbb{E}[Y_1] = 1 + M_{w,b} = \frac{w + b + 1}{w + 1}.$$

(4) *Simple, brilliant intuition* (Mosteller, 1965, p. 61) Imagine the $w + b$ balls are arranged in a row such that the w white balls partition the b black balls into $w + 1$ segments. On average, we expect the length of each segment to consist of $b/(w + 1)$ black balls so that, to get to the first white ball, we would need to pick $b/(w + 1) + 1$ balls.

Finally, comparing the expected values under the two sampling schemes,

$$\mathbb{E}[X_1] = \frac{w + b}{w} > \frac{w + b + 1}{w + 1} = \mathbb{E}[Y_1], \qquad (6.23)$$

and note that, as w and b increase, $\mathbb{E}[Y_1] \to \mathbb{E}[X_1]$.

Now consider $\mathbb{V}(Y_1)$ and $\mathbb{V}(X_1)$. Using the notation in method (2) above,

$$\mathbb{V}(Y_1) = \mathbb{V}\left(\sum_{i=1}^{b} Z_i\right) = \sum_{i=1}^{b}\mathbb{V}(Z_i) + \sum_{i \neq j}\text{Cov}\left(Z_i Z_j\right)$$

where $\text{Cov}\left(Z_i Z_j\right) = \mathbb{E}\left[Z_i Z_j\right] - \mathbb{E}[Z_i]\mathbb{E}\left[Z_j\right]$ and, for $i \neq j$,

$$\mathbb{E}\left[Z_i Z_j\right] = \Pr\left(\text{both } b_i \text{ and } b_j \text{ are drawn before any whites}\right)$$
$$= 2(w + 2)^{-1}(w + 1)^{-1}.$$

As the Z_i are Bernoulli with probability $(w + 1)^{-1}$,

$$\mathbb{V}(Y_1) = b \cdot \frac{1}{w + 1}\left(1 - \frac{1}{w + 1}\right) + b(b - 1)\left[\frac{2}{(w + 2)(w + 1)} - \left(\frac{1}{w + 1}\right)^2\right]$$

$$= \frac{bw(w + b + 1)}{(w + 2)(w + 1)^2}.$$

As the variance of a geometric random variable (whether including the 'success' or not) is given by $(1 - p) p^{-2} = b (w + b) w^{-2}$,

$$V(X_1) = \frac{b(w+b)}{w^2} > \frac{bw(w+b+1)}{(w+2)(w+1)^2} = V(Y_1).$$ (6.24)

As n and w increase, $Var(Y_1) \to V(X_1)$.

Now let X_k and Y_k denote the number of balls drawn until the kth white ball appears, with and without replacement, respectively. The sampling scheme indicates that X_k follows the negative binomial distribution (4.20) (defined to *include* the k white balls). As in Section 6.4.1, observing that X_k is the sum of k independent and identically distributed geometric random variables with density (4.18) and $p = w/(w + b)$, it follows that

$$E[X_k] = \frac{k}{p} = k\frac{w+b}{w}.$$

For computing the expected value of Y_k, we illustrate two methods.

(1) Generalizing the fourth method above, it follows directly that

$$E[Y_k] = k\left(\frac{w+b+1}{w+1}\right),$$ (6.25)

so that, as in the $k = 1$ case, $E[X_k] > E[Y_k]$, and they converge as w and b increase.

(2) (Rohatgi, 1984, p. 369(27)) First note that, in order to draw k white balls without replacement, the probability of drawing exactly y balls means that the yth draw is a white ball and in the previous $y - 1$ draws, there must have been exactly $k - 1$ white balls. Thus,

$$Pr(Y_k = y) = \frac{\binom{w}{k-1}\binom{b}{y-k}}{\binom{w+b}{y-1}} \cdot \frac{w - (k-1)}{w + b - (y - 1)}, \quad y = k, k + 1, \ldots, b + k,$$

or, rewriting,

$$Pr(Y_k = y) = \binom{y-1}{k-1}\frac{\binom{w+b-y}{w-k}}{\binom{w+b}{w}}, \quad y = k, k + 1, \ldots, b + k$$

as in (4.23). As this is a valid density, it follows that

$$\binom{w+b}{w} = \sum_{y=k}^{b+k} Pr(Y_k = y) = \sum_{y=k}^{b+k}\binom{y-1}{k-1}\binom{w+b-y}{w-k}.$$ (6.26)

With

$$y\binom{y-1}{k-1} = \frac{y!}{(k-1)!\,(y-k)!} = \frac{ky!}{k!\,(y-k)!} = k\binom{y}{k}$$

and $x = y + 1$, $j = k + 1$, this gives

$$\mathbb{E}\,[Y_k] = \sum_{y=k}^{b+k} y\binom{y-1}{k-1}\frac{\binom{w+b-y}{w-k}}{\binom{w+b}{w}} = \frac{k}{\binom{w+b}{w}} \sum_{y=k}^{b+k}\binom{y}{k}\binom{w+b-y}{w-k}$$

$$= \frac{k}{\binom{w+b}{w}} \sum_{x=j}^{b+j}\binom{x-1}{j-1}\binom{(w+b+1)-x}{(w+1)-j} = \frac{k}{\binom{w+b}{w}}\binom{w+b+1}{w+1}$$

$$= k\frac{w+b+1}{w+1} \tag{6.27}$$

from (6.26).

Now onto the variance. By decomposing the negative binomial variable X_k into k i.i.d. geometric random variables, it follows that

$$\mathbb{V}\,(X_k) = k\,(1-p)\,p^{-2} = k\frac{b\,(w+b)}{w^2}. \tag{6.28}$$

For Y_k, similar to the above development,

$$\mathbb{E}\,[Y_k\,(Y_k+1)]$$

$$= \sum_{y=k}^{b+k} y\,(y+1)\binom{y-1}{k-1}\frac{\binom{w+b-y}{w-k}}{\binom{w+b}{w}} = \frac{k\,(k+1)}{\binom{w+b}{w}} \sum_{y=k}^{b+k}\binom{y+1}{k+1}\binom{w+b-y}{w-k}$$

$$= \frac{k\,(k+1)}{\binom{w+b}{w}} \sum_{x=j}^{b+j}\binom{x-1}{j-1}\binom{(w+b+2)-x}{(w+2)-j} = \frac{k\,(k+1)}{\binom{w+b}{w}}\binom{w+b+2}{w+2}$$

$$= k\,(k+1)\,\frac{(w+b+1)\,(w+b+2)}{(w+1)\,(w+2)},$$

because

$$y\,(y+1)\binom{y-1}{k-1} = \frac{(y+1)!}{(k-1)!\,(y-k)!} = \frac{k\,(k+1)\,(y+1)!}{(k+1)!\,(y-k)!} = k\,(k+1)\binom{y+1}{k+1}$$

and from equality (6.26) with $x = y + 2$ and $j = k + 2$. Then,

$$\mathbb{V}\,(Y_k) = \mathbb{E}\,[Y_k\,(Y_k+1)] - \mathbb{E}\,[Y_k] - (\mathbb{E}\,[Y_k])^2$$

$$= k\,(k+1)\,\frac{(w+b+1)\,(w+b+2)}{(w+1)\,(w+2)} - k\frac{w+b+1}{w+1} - \left(k\frac{w+b+1}{w+1}\right)^2$$

$$= kb\frac{(w+b+1)\,(w-k+1)}{(w+2)\,(w+1)^2}. \tag{6.29}$$

Comparing (6.28) and (6.29), we see that, as in the $k = 1$ case, $\mathbb{V}(X_k) > \mathbb{V}(Y_k)$. Note that only for relatively small k will they converge as w and b increase.

6.5 General linear combination of two random variables

Notice that, for the binomial and negative binomial cases above, if the X_i are still independent but p_i differs among the X_i, the result no longer holds. In this case, the mass function of $\sum_i X_i$ is more complicated. For example, if $X_i \sim \text{Bin}(n_i, p_i)$, $i = 1, 2$, then the mass function of $X = X_1 + X_2$, or the *convolution* of X_1 and X_2, can be written as

$$
\begin{aligned}
\Pr(X_1 + X_2 = x) &= \sum_{i=0}^{x} \Pr(X_1 = i) \Pr(X_2 = x - i) \\
&= \sum_{i=0}^{x} \Pr(X_1 = x - i) \Pr(X_2 = i).
\end{aligned}
$$

(6.30)

This makes sense because X_1 and X_2 are independent and, in order for X_1 and X_2 to sum to x, it must be the case that one of the events $\{X_1 = 0, X_2 = x\}$, $\{X_1 = 1, X_2 = x - 1\}$, ..., $\{X_1 = x, X_2 = 0\}$ must have occurred. These events partition the event $\{X = x\}$.

A similar argument verifies that the c.d.f. of $X = X_1 + X_2$ is given by

$$
\Pr(X \leq x) = \sum_{i=0}^{x} \Pr(X_1 = i) \Pr(X_2 \leq x - i) = \sum_{i=0}^{x} \Pr(X_1 \leq x - i) \Pr(X_2 = i).
$$

(6.31)

This result extends to any two discrete independent random variables, although if both do not have bounded support, the sums in (6.30) and (6.31) will be infinite. It should also be clear that (6.30) can be generalized to the sum of three, four or more (discrete and independent) random variables, but will then involve double, triple, etc., sums and become computationally inefficient. Volume II will discuss other methods which can more easily handle these generalizations.

◎ ***Example 6.10*** Let $X_i \stackrel{\text{ind}}{\sim} \text{Poi}(\lambda_i)$ and consider the distribution of $Y = X_1 + X_2$. With $\lambda = \lambda_1 + \lambda_2$,

$$
\begin{aligned}
\Pr(Y = y) &= \sum_{i=-\infty}^{\infty} \Pr(X_1 = i) \Pr(X_2 = y - i) \\
&= \sum_{i=0}^{y} \frac{e^{-\lambda_1} \lambda_1^i}{i!} \frac{e^{-\lambda_2} \lambda_2^{y-i}}{(y-i)!} = e^{-(\lambda_1 + \lambda_2)} \sum_{i=0}^{y} \frac{\lambda_1^i}{i!} \frac{\lambda_2^{y-i}}{(y-i)!}
\end{aligned}
$$

$$= \frac{e^{-(\lambda_1+\lambda_2)}}{y!} \sum_{i=0}^{y} \frac{y!}{i!\,(y-i)!} \lambda_1^i \lambda_2^{y-i} = \frac{e^{-(\lambda_1+\lambda_2)}}{y!} (\lambda_1 + \lambda_2)^y$$

$$= \frac{e^{-\lambda} \lambda^y}{y!},$$

where the second to last equality follows from the binomial theorem. It follows that $\sum_{i=1}^{n} X_i \sim \text{Poi}(\lambda)$, where $\lambda = \sum_{i=1}^{n} \lambda_i$. ∎

Next, consider the more general linear combination $C = a_0 + a_1 X_1 + a_2 X_2$ for constant values $a_0, a_1, a_2 \in \mathbb{R}$. In order to avoid trivial cases, we assume at least one of the variables a_1 and a_2 to be different from 0. The same reasoning as above shows that the p.m.f. $\Pr(C = c)$ is given by

$$\underbrace{\sum_{i=-\infty}^{\infty} \Pr(X_1 = i) \Pr\left(X_2 = \frac{c - ia_1 - a_0}{a_2}\right)}_{\text{if } a_2 \neq 0}$$

$$= \underbrace{\sum_{i=-\infty}^{\infty} \Pr\left(X_1 = \frac{c - ia_2 - a_0}{a_1}\right) \Pr(X_2 = i)}_{\text{if } a_1 \neq 0},$$

while the c.d.f. is

$$\Pr(C \leq c) = \begin{cases} \sum_{i=-\infty}^{\infty} \Pr(X_1 = i) \Pr\left(X_2 \leq \frac{c-ia_1-a_0}{a_2}\right), & \text{if } a_2 > 0, \\ \sum_{i=-\infty}^{\infty} \Pr(X_1 = i) \Pr\left(X_2 \geq \frac{c-ia_1-a_0}{a_2}\right), & \text{if } a_2 < 0 \end{cases}$$

or, equivalently,

$$\Pr(C \leq c) = \begin{cases} \sum_{i=-\infty}^{\infty} \Pr\left(X_1 \leq \frac{c-ia_2-a_0}{a_1}\right) \Pr(X_2 = i), & \text{if } a_1 > 0, \\ \sum_{i=-\infty}^{\infty} \Pr\left(X_1 \geq \frac{c-ia_2-a_0}{a_1}\right) \Pr(X_2 = i), & \text{if } a_1 < 0. \end{cases}$$

If $a_1 = 1$, $a_2 = -1$ and $a_0 = 0$, we obtain the important special case of $D = X_1 - X_2$ with

$$\boxed{\begin{aligned} \Pr(D = d) &= \sum_{i=-\infty}^{\infty} \Pr(X_1 = i) \Pr(X_2 = i - d) \\ &= \sum_{i=-\infty}^{\infty} \Pr(X_1 = d + i) \Pr(X_2 = i) \end{aligned}}$$
(6.32)

and

$$\Pr(D \le d) = \sum_{i=-\infty}^{\infty} \Pr(X_1 = i) \Pr(X_2 \ge i - d) = \sum_{i=-\infty}^{\infty} \Pr(X_1 \le d + i) \Pr(X_2 = i).$$

$$(6.33)$$

◎ *Example 6.11*   With $X_i \overset{\text{ind}}{\sim} \text{Poi}(\lambda_i)$, $i = 1, 2$, and $D = X_1 - X_2$, the first equation in (6.32) gives

$$\Pr(D = d) = \sum_{i=-\infty}^{\infty} \Pr(X_1 = i) \Pr(X_2 = i - d) = e^{-(\lambda_1 + \lambda_2)} \sum_{i=d}^{\infty} \frac{\lambda_1^i}{i!} \frac{\lambda_2^{i-d}}{(i - d)!},$$

a somewhat disappointing result, when compared to the pleasant expression obtained when summing Poisson r.v.s. By multiplying by $1 = (\lambda_1/\lambda_2)^{d/2} (\lambda_1/\lambda_2)^{-d/2}$ and using $j = i - d$, we arrive at the still-not-so-attractive expression

$$\Pr(D = d) = e^{-(\lambda_1 + \lambda_2)} \left(\frac{\lambda_1}{\lambda_2} \right)^{d/2} \sum_{j=0}^{\infty} \frac{(\lambda_1 \lambda_2)^{j+d/2}}{j! \, \Gamma(d + j + 1)}.$$

However, this can now be expressed quite concisely in terms of the *modified Bessel function of the first kind of order* v,[2] given by

$$I_v(x) = \left(\frac{x}{2} \right)^v \sum_{j=0}^{\infty} \frac{(x^2/4)^j}{j! \, \Gamma(v + j + 1)}$$

$$(6.34)$$

or, if v is an integer, then

$$I_n(x) = \frac{1}{\pi} \int_0^{\pi} \exp\{x \cos\theta\} \cos(n\theta) \, d\theta, \qquad n \in \mathbb{Z}.$$

Then, as

$$I_d\left(2\sqrt{\lambda_1 \lambda_2}\right) = \left(\frac{2\sqrt{\lambda_1 \lambda_2}}{2} \right)^d \sum_{j=0}^{\infty} \frac{\left(\left(2\sqrt{\lambda_1 \lambda_2}\right)^2/4 \right)^j}{j! \, \Gamma(d + j + 1)} = \sum_{j=0}^{\infty} \frac{(\lambda_1 \lambda_2)^{j+d/2}}{j! \, \Gamma(d + j + 1)},$$

---

[2] The Bessel functions, after Friedrich Wilhelm Bessel (1784–1846), belong to the class of so-called *special functions*, and are of great importance in applied mathematics, physics, engineering and, as we see here, also arise in probabilistic models. See Davis (1989, Section 4.6), Jones (2001, Section 13.G) and Lebedev (1972, Chapter 5) for a nice account of Bessel functions, and Andrews, Askey and Roy (1999) for a large and quite advanced treatment of numerous special functions. Of course, Abramowitz and Stegun (1972) contains all the relevant formulae, but without derivation.

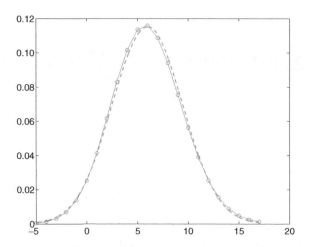

Figure 6.7   Mass function of $D = X_1 - X_2$ (solid), where $X_i \overset{\text{ind}}{\sim} \text{Poi}(\lambda_i)$, with $\lambda_1 = 9$ and $\lambda_2 = 3$; overlaid is the normal p.d.f. with matching mean and variance (dashed)

we obtain the better looking

$$\Pr(D = d) = e^{-(\lambda_1 + \lambda_2)} \left(\frac{\lambda_1}{\lambda_2} \right)^{d/2} I_d \left(2\sqrt{\lambda_1 \lambda_2} \right). \tag{6.35}$$

Many computer packages offer fast and accurate routines to evaluate the Bessel functions, so that the expression for $\Pr(D = d)$ is indeed 'essentially closed-form'. In Matlab, (6.34) is evaluated with `besseli(v,x)`.

To illustrate, imagine that, while reading and evaluating a new probability book, you keep a score: plus one point for each stroke of didactic genius, minus one point for each typographical error. Assume these are independent events which occur as Poisson r.v.s with rates nine and three, respectively. The mass function is shown in Figure 6.7. Overlaid is the normal p.d.f. with mean $\mathbb{E}[X_1 - X_2] = \lambda_1 - \lambda_2 = 6$ and variance $\mathbb{V}(X_1 - X_2) = \lambda_1 + \lambda_2 = 12$, which is seen to approximate the p.m.f. of D very well. It could be used quickly and accurately to approximate, say, the probability of the book receiving a negative score. ∎

If $Q = X_1/X_2$ and $\Pr(X_2 = 0) = 0$, then the p.m.f. can be expressed as

$$\Pr(X_1/X_2 = q) = \Pr(X_1 = qX_2) = \Pr(X_1 - qX_2 = 0)$$

which is a special case of C with $a_1 = 1$, $a_2 = -q$, $a_0 = 0$ and $c = 0$. The mass function is thus

$$\Pr(Q = q) = \sum_{i=-\infty}^{\infty} \Pr(X_1 = i) \Pr\left(X_2 = \frac{i}{q} \right) = \sum_{i=-\infty}^{\infty} \Pr(X_1 = iq) \Pr(X_2 = i).$$

However, we can derive the c.d.f. of Q directly from the above expressions for $\Pr(C \leq c)$ only if the support of X_2 is positive. In general,

$$
\begin{aligned}
\Pr(Q \leq q) &= \Pr(X_1/X_2 \leq q) = \Pr(X_1/X_2 \leq q, X_2 < 0) + \Pr(X_1/X_2 \leq q, X_2 > 0) \\
&= \Pr(X_1 \geq qX_2, X_2 < 0) + \Pr(X_1 \leq qX_2, X_2 > 0) \\
&= \sum_{i=-\infty}^{-1} \Pr(X_1 \geq iq)\Pr(X_2 = i) + \sum_{i=1}^{\infty} \Pr(X_1 \leq iq)\Pr(X_2 = i). \quad (6.36)
\end{aligned}
$$

Similar arguments for $M = X_1 \cdot X_2$ yield

$$
\Pr(M = m) = \begin{cases} \sum_{i=-m,i\neq 0}^{m} \Pr\left(X_1 = \frac{m}{i}\right)\Pr(X_2 = i), & \text{if } m \neq 0, \\ \Pr(X_1 = 0) + \Pr(X_2 = 0) - \Pr(X_1 = 0)\Pr(X_2 = 0), & \text{if } m = 0. \end{cases}
$$

⊖ **Example 6.12** (Example 4.15 cont.) Our online bank still assumes that the number of callers in a particular one-minute interval, random variable $C = C(\lambda)$, is approximately Poisson with parameter λ, where λ can be taken to be the total number of customers times the probability that a customer will call. However, assume that λ changes from day to day, influenced by the current amount and type of economic and business related news. An estimate of λ is provided to the bank in the early morning by a specialized team of highly paid statisticians.

The bank's goal is still to ensure that a random caller immediately gets through to an operator 95 % of the time. One obvious way to do this is to obtain an upper limit on λ from the statisticians, thus giving the 'worst-case scenario' so that the bank can just employ the corresponding necessary number of telephone operators every day. To save money, management has a better plan. They maintain a large pool of trained college students who work the phones part time. A mass e-mail is sent in the morning to a random batch of n of the students requesting that they come to work; however, a particular student can only show up with probability p. Assume that each student's decision to work is independent of the others. How large should n be for a given estimate of λ?

Let S be the actual number of students who show up; from the independence assumption, $S \sim \text{Bin}(n, p)$. If $p = 1$, $S = n$ and we take n to be the smallest integer satisfying $\Pr(C \leq n) \geq 0.95$ as in Example 4.15. If $0 < p < 1$, we require the smallest integer n such that $\Pr(C \leq S) \geq 0.95$ or $\Pr(S < C) = \Pr(D < 0) \leq 0.05$, where $D := S - C$. Assuming that C and S are independent, the first expression in (6.33) gives the c.d.f. of D as

$$
\begin{aligned}
F_D(0) &= \sum_{i=-\infty}^{\infty} \Pr(S = i)\Pr(C \geq i) \\
&= \sum_{i=0}^{\infty} f_S(i)[1 - F_C(i-1)] = 1 - \sum_{i=1}^{\infty} f_S(i) F_C(i-1)
\end{aligned}
$$

while the second expression gives

$$F_D(0) = \sum_{i=-\infty}^{\infty} \Pr(S \le i) \Pr(C = i)$$

$$= \sum_{i=0}^{\infty} F_S(i) f_C(i) = \sum_{i=0}^{\infty} \left[\sum_{j=0}^{i} \binom{n}{j} p^j (1-p)^{n-j} \right] \frac{e^{-\lambda} \lambda^i}{i!}.$$

The c.d.f. $F_D(0)$ cannot be expressed in closed form and must be numerically resolved for given values of p, λ and n. A reasonable initial guess is to first compute the smallest cut-off c^* such that $\Pr(C \le c^*) \ge 0.95$ and then choose n such that $\mathbb{E}[S] = c^*$, or $n = c^*/p$.

For $\lambda = 500$ and $p = 0.75$, the initial guess of n is $537/0.75 = 716$. Trial and error suggests that the sum in $F_D(0)$ can be truncated at about 600 for reasonable accuracy with these parameters; to be safe, we use an upper limit of 1000 and then find that $F_D(0) = 0.0499$ for $n = 723$, with $F_D(0) = 0.0745$ for $n = 716$.

Another way of obtaining an approximate answer is via simulation. A large number, say B, of independent draws from both S and C are simulated and the proportion $P = B^{-1} \sum_{i=1}^{B} \mathbb{I}(S_i < C_i)$ is computed. As $\sum_{i=1}^{B} \mathbb{I}(S_i < C_i) \sim \text{Bin}(B, \psi)$, where $\psi = \Pr(S < C)$ depends on n, B can be specified (approximately) such that $\mathbb{V}(P)$ does not exceed a certain value. For $B = 5000$ we obtain $F_D(0) \approx 0.0476$ for $n = 723$ and $F_D(0) \approx 0.069$ for $n = 716$. Clearly, B must be chosen to be larger in order to get more reliable estimates.

For this bank example, it seems reasonable that S and C are independent. However, in general, one might postulate that S and C are correlated or, more specifically, p and λ. In this case, expressions for F_D are far more complicated – if they exist at all, and simulation might be the only reasonable alternative. ∎

6.6 Problems

Smooth seas do not make skillful sailors.   (African proverb)

Tomorrow's battle is won during today's practice.   (Samurai maxim)

The fight is lost in the gym, not in the ring.   (Boxing maxim)

The time to repair the roof is when the sun is shining.   (John F. Kennedy)

6.1. ★ Using (6.30), derive the distribution of $Y = X_1 + X_2$, where

(a) $X_i \stackrel{\text{ind}}{\sim} \text{Bin}(n_i, p)$,

(b) $X_i \stackrel{\text{ind}}{\sim} \text{NBin}(r_i, p)$.

6.2. How many times do you expect to have to roll a six-sided die until:

(a) the numbers 1, 2 and 3 appear at least once;

(b) any three numbers appear at least once?

6.3. ★ Recall the birthday application in Section 6.2.1. Imagine inspecting each of the $\binom{r}{2}$ pairs of people and noting whether both have the same birthday.
(Ross, 1988, pp. 131 and 144(14))

(a) This is clearly not a binomial experiment, because trials are not independent; but if they were, what would the parameters of the binomial model, n and p, be?

(b) Consider the dependence between trials. In particular, denote $E_{i,j}$ as the event that persons i and j have the same birthday, and compute:
  (i) $\Pr\left(E_{1,2} \mid E_{3,4}\right)$,
  (ii) $\Pr\left(E_{1,3} \mid E_{1,2}\right)$ and
  (iii) $\Pr\left(E_{2,3} \mid E_{1,2} \cap E_{1,3}\right)$.

(c) Assuming that the dependence between the $\binom{r}{2}$ trials is weak enough, we could approximate the number of 'successes' (pairs of people with the same birthday) as a binomial distribution or, much simpler, as a Poisson because p is very small. Using the Poisson, find an approximate expression for the smallest r such that $p_r \geq p^*$.

(d) Now let $p_{r,k}$ be the probability that, among r people, at least k, $0 < k < r$ have the same birthday. Using the Poisson approximation, show that the smallest r such that $p_{r,3} \geq 0.5$ is 84.

6.4. ★ Recall our six vacationing couples from Example 3.8.

(a) Letting X be the random variable denoting the number of man–woman roommate pairs, compute $\mathbb{E}[X]$ and $\mathbb{V}(X)$.
(similar to Ross, 1988, p. 331(44))

(b) Now let X denote the number of husband–wife roommate pairs. Compute $\mathbb{E}[X]$ and $\mathbb{V}(X)$.

6.5. ★ Show (6.9) by expressing the binomial coefficient $\binom{Z+1}{r}$ as a ratio of a descending factorial and $r!$, multiplying and dividing by $\binom{Z+1-g}{r-g}$, simplifying, and then using (1.55).

6.6. ★ ★ Prove (6.14).

6.7. ★ ★ Consider the distribution of $X = \sum_{i=1}^{3} X_i$ where $X_i \sim \text{NBin}(r_i, p_i)$, $i = 1, 2, 3$. This can be thought of as a waiting-time distribution which consists of three stages, as shown in Figure 6.8. The total 'time' or number of trials to get through all three nodes is $X + \sum_{i=1}^{3} r_i$.

(a) Give an expression for the p.m.f. and c.d.f. of X.

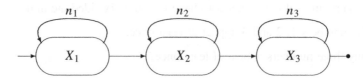

Figure 6.8   Three independent nodes, each following a negative binomial distribution

(b) For $r_i = 4i$ and $p_i = i/4$, calculate the mean and variance of X.

(c) Program and plot the exact density of X for $r_i = 4i$ and $p_i = i/4$ and, on the same graph, overlay a plot of the normal density with the same mean and variance as X. Is the normal a good approximation?

6.8. ★ Every morning, Martin and I individually predict a (same-sized) range of the daily closing price of a particular stock. If the true closing stock price is contained in my range, I get one 'point', and likewise for Martin. I am correct (i.e. the true price falls in my range) with probability p_1, and Martin is correct with probability p_2. (Notice that we can both be correct, both be wrong, etc.) Assume that our successes are not related day to day, nor to one another. Let S_1 be the sum of my points and S_2 be the sum of Martin's points. We compete for one month (assume 20 trading days per month). How are S_1 and S_2 distributed?

(a) Write an equation for the probability that I get more points than Martin, i.e. $\Pr(S_1 > S_2)$.

(b) Write an expression for the probability that I get at least twice as many points as Martin.

(c) Calculate the expected value and variance of the event $\{S_1 > S_2\}$.

(d) Define S_{1i} and S_{2i} to be the sums of points we each get every week, $i = 1, \ldots, 4$, i.e. $S_1 = \sum_{i=1}^{4} S_{1i}$, and likewise for S_2. Suppose now that only the outcome of the event $\{S_{1i} > S_{2i}\}$ is reported every week, instead of the actual sums. More formally, define

$$B_i := \begin{cases} 0 & \text{if } S_{1i} \leq S_{2i} \\ 1 & \text{if } S_{1i} > S_{2i} \end{cases}$$

and $B := \sum_{i=1}^{4} B_i$. How is B distributed, and what is the probability that I 'beat' Martin, i.e. what is $\Pr(B > 2)$? What is $\mathbb{V}(B)$?

(e) If $p_1 > p_2$, then, intuitively, why is $\Pr(S_1 > S_2) > \Pr(B > 2)$?

6.9. ★ ★ A professor, about to administer an exam to her n students, discovers that the students have seated themselves randomly throughout the room, instead of alphabetically, as she would like. (We know that the probability that exactly m seats are correctly filled is given by the solution to the problem of coincidences in (2.18).)

To rectify the situation, the professor proceeds as follows: Starting with the first seat, if a wrong person is there, that person goes to the front of the room (and waits until his or her seat is 'called') and the correct person is placed in the seat. Let N_{kn} be the number of students at the front of the class after k chairs have been called, $1 \leq k < n$. We wish to calculate $\Pr(N_{kn} = j)$ and $\mathbb{E}[N_{kn}]$.[3]

Use the convention that person A belongs in chair one, person B in chair two, etc., and let N_{kn} be the number of people standing at 'time' k (i.e. after k names have been called).

(a) Using basic principles, work out the mass function for N_{kn} for $k = 1, 2, 3$, and, based on the definition of expected value, show that

$$\mathbb{E}[N_{1n}] = \frac{n-1}{n}, \ n \geq 2, \quad \mathbb{E}[N_{2n}] = 2\frac{n-2}{n}, \ n \geq 4,$$

$$\mathbb{E}[N_{3n}] = 3\frac{n-3}{n}, \ n \geq 6.$$

(b) Propose a generalization of the above result for general k, determine how N_{kn} is distributed, and that

$$\mathbb{E}[N_{kn}] = k\frac{n-k}{n}.$$

(c) Based on the last result, think about how the mass function of N_{kn} could have been established *without the above work*.

(d) This problem, like most we encounter, is also amenable to simulation. It is also instructive, because a natural data structure to maintain the list of students at the front of the class is a set, which is supported by Matlab. In particular, Matlab has several useful built-in set functions (mostly with self-explaining names) such as `union`, `setdiff` and `ismember`. Write a program, `V=classroom(n,sim,kmax)`, which can determine the mass function, and expected value, of N_{kn}, via simulation.

---

[3] The question of determining $\mathbb{E}[N_{kn}]$ appears in Ghahramani (2000, p. 381), along with the answer of $k(n-k)/n$ (but not the derivation) and he cites Michael Khoury, US Mathematics Olympiad Member, for posing the problem for a mathematics competition.

PART III

CONTINUOUS RANDOM VARIABLES

7

Continuous univariate random variables

[I] have been struck by the manner in which the history of science gets created and attributions accepted. Instances of this may be found in such widespread notions as that Leibniz was the first to use differentials, that Gauss was the discoverer of the normal curve of errors, that Lagrange invented Lagrange's formula, that Bessel originated Bessel's method of interpolation, that Bravais introduced the coefficient of correlation, or that James Bernoulli demonstrated James Bernoulli's Theorem. (Karl Pearson)[1]

Reproduced by permission of Oxford Journals

In many branches of statistics and other fields, continuous distributions offer reasonable and tractable approximations to the underlying process of interest even though virtually all phenomena are, at some level, ultimately discrete. This chapter introduces the most widely used continuous univariate distributions and discusses the method of transformation. Further continuous univariate distributions of a more advanced nature will be considered in Volume II. A comprehensive reference work discussing all the major univariate distributions in statistics as well as numerous more specialized ones is the two volume set *Continuous Univariate Distributions*, 2nd edn, by Johnson, Kotz and Balakrishnan (1994, 1995). They provide a wealth of information, including more recent contributions to the literature.

7.1 Most prominent distributions

Below we discuss the most common and most important continuous univariate r.v.s; they will be encountered numerous times later and form the basis of a large part of statistical inference. In general, when encountering a new distribution, it is valuable to know (or find out) particular characteristics of importance including the following.

---

[1] From Pearson, K., 'James Bernoulli's Theorem', *Biometrika*, Vol. **17**, No. 3/4, Dec., 1925.

Fundamental Probability: A Computational Approach   M.S. Paolella
© 2006 John Wiley & Sons, Ltd

(1) *Its theoretical importance.* The normal distribution, for example, arises in a plethora of applications because of its association with the so-called central limit theorem (discussed later). Some of the more fundamental theoretical aspects of the distributions outlined here will be presented in this and later chapters, while detailed treatments can be found in more advanced texts such as Johnson, Kotz and Balakrishnan (1994, 1995) and others listed below.

(2) *Its use in applications.* Many distributions are often associated with a specific application, such as Weibull for lifetime studies, Pareto for income analysis and Student's *t* for the common '*t*-test'. However, there often exists a large number of less common uses for certain distributions. The Weibull, Pareto and *t*, for example, are all useful in modeling the probability of 'extreme events', such as those associated with the returns on financial assets such as stocks and currency exchange rates.

(3) *How its functional form came about.* For example, it might

- be a 'base' distribution arising from mathematical simplicity, e.g. the uniform or exponential;

- arise or be strongly associated with a particular application, e.g. the Cauchy in a geometrical context, the Pareto for income distribution, the *F* in the analysis of variance;

- be a 'natural' or obvious generalization of a base distribution, such as the Laplace or Weibull from the exponential distribution;

- be a function of other, simpler r.v.s, such as the gamma as a sum of exponentials, or Student's *t* as a ratio of a normal and weighted chi-square;

- be a *limiting* distribution, such as the Poisson, normal and Gumbel.

(4) *The extent to which certain characteristics can be easily ascertained and/or computed*, e.g. functions such as the expected value and higher moments, quantiles and the c.d.f.; and properties such as whether the distribution is unimodal, closed under addition, member of the exponential family, etc.

(5) *Recognizing what role the associated parameters play.* Continuous distributions often have a *location* parameter which shifts the density and a *scale* parameter which stretches or shrinks the density. Further ones are referred to generically as *shape* parameters but, for any particular distribution, often have standard names, e.g. the degrees of freedom for Student's *t*.

(6) *The behavior of the c.d.f. far into the* tails *of the distribution.* This is discussed below in association with the Pareto density.

Regarding the location and scale parameters just mentioned, if X is a continuous random variable with p.d.f. $f_X(x)$, then the *linearly transformed* random variable $Y = \sigma X + \mu$, $\sigma > 0$, has density

$$f_Y(y) = \frac{1}{\sigma} f_X\left(\frac{y - \mu}{\sigma}\right) \qquad (7.1)$$

(see Section 7.3 below). The distributions of X and Y are said to be members of the same *location–scale family*, with *location parameter* μ and *scale parameter* σ.

The *kernel* of a p.d.f. is that part of it which involves only the variables associated with the r.v.s of interest, e.g. x in the above descriptions, and not the remaining quantities, which are just *constants of integration*. For example, the kernel of the standard normal density (4.8) is $\exp(-z^2/2)$. We could, then, just write $f_Z(z) \propto \exp(-z^2/2)$. Similarly, for the Student's t density discussed below, the kernel is $(1 + x^2/\nu)^{-(\nu+1)/2}$.

The previous comments should be kept in mind while studying the following set of distributions. Table 7.1 just lists their name and number used in the subsequent list. Tables C.4 to C.7 summarize their major properties and also use the same numbering.

Table 7.1   List of continuous distributions

(1)	uniform	(7)	Cauchy	(13)	F
(2)	exponential	(8)	Pareto I	(14)	log–normal
(3)	gamma	(9)	Pareto II	(15)	logistic
(4)	beta	(10)	normal	(16)	Gumbel
(5)	Laplace	(11)	chi-square	(17)	inverse Gaussian
(6)	Weibull	(12)	Student's t	(18)	hyperbolic

(1) Uniform, Unif (a, b): the family of continuous uniform distributions is indexed by two parameters $a, b \in \mathbb{R}$ with $a < b$ and has density

$$f_{\text{Unif}}(x; a, b) = \frac{1}{b - a} \mathbb{I}_{(a,b)}(x) \text{ and}$$
$$F_{\text{Unif}}(x; a, b) = \frac{x - a}{b - a} \mathbb{I}_{(a,b)}(x) + \mathbb{I}_{[b,\infty)}(x)$$

The uniform is arguably the simplest distribution and is used for modeling situations in which events of equal length in (a, b) are equally likely to occur. Most prominent is the Unif $(0, 1)$ because it plays a crucial role in simulating random variables of various types, as was first discussed in Example 4.5 of Section 4.1.2.

(2) Exponential, Exp(λ): $\lambda \in \mathbb{R}_{>0}$ with density

$$f_{\text{Exp}}(x; \lambda) = \lambda e^{-\lambda x} \mathbb{I}_{(0,\infty)}(x)$$

and distribution function

$$F_{\text{Exp}}(x; \lambda) = \int_0^x \lambda e^{-\lambda t} dt = \left(1 - e^{-\lambda x}\right) \mathbb{I}_{(0,\infty)}(x).$$

Referring to (7.1), observe that $1/\lambda$ is a scale parameter.

Let $Z \sim \text{Exp}(1)$. Then, integration by parts with $u = z^r$ and $dv = e^{-z} dz$ gives

$$\mathbb{E}\left[Z^r\right] = \int_0^\infty z^r e^{-z} dz = -z^r e^{-z}\Big|_0^\infty - \int_0^\infty -e^{-z} r z^{r-1} dz = r \int_0^\infty z^{r-1} e^{-z} dz,$$
(7.2)

where $0 = \lim_{z \to \infty} z^r e^{-z}$ follows from (A.36) or repeated application of l'Hôpital's rule,

$$\lim_{z \to \infty} \frac{z^r}{e^z} = \lim_{z \to \infty} \frac{r z^{r-1}}{e^z} = \cdots = \lim_{z \to \infty} \frac{r(r-1)\cdots 2 \cdots 1}{e^z} = 0.$$
(7.3)

Recursive repetition of the integration by parts in (7.2) gives

$$\mathbb{E}\left[Z^r\right] = r(r-1)\cdots(1) \int_0^\infty e^{-z} dz = r!, \qquad r \in \mathbb{N}.$$
(7.4)

Less work is involved and a more general result is obtained by using the gamma function, $\Gamma(\cdot)$, which is discussed in Section 1.5. In particular, for $r \in \mathbb{R}_{>0}$,

$$\mathbb{E}\left[Z^r\right] = \int_0^\infty z^{(r+1)-1} e^{-z} dz = \Gamma(r+1)$$

which, recalling that $\Gamma(r+1) = r!$ for integer $r > 0$, verifies (7.4). If $X = Z/\lambda$, then, from (7.1), $X \sim \text{Exp}(\lambda)$ and $\mathbb{E}[X^r] = r!/\lambda^r$. As important special cases,

$$\mathbb{E}[X] = \frac{1}{\lambda} \quad \text{and} \quad \mathbb{V}(X) = \mathbb{E}\left[X^2\right] - (\mathbb{E}[X])^2 = \frac{2}{\lambda^2} - \frac{1}{\lambda^2} = \frac{1}{\lambda^2}.$$
(7.5)

As first mentioned in Example 4.6, the exponential distribution plays an important role in modeling the lifetimes of various entities (physical components, living creatures) or, more generally, the time until some prescribed event occurs (see Section 4.5). The exponential is typically a special case of more sophisticated distributions which are used in this context (e.g. the gamma and Weibull, introduced below). The volume edited by Balakrishnan and Basu (1995) is dedicated to the theory and applications of the exponential distribution.

(3) Gamma, Gam (α, β), $\alpha, \beta \in \mathbb{R}_{>0}$: the location-zero, scale-one density is given by

$$f_{\text{Gam}}(x; \alpha) = \frac{1}{\Gamma(\alpha)} x^{\alpha-1} \exp(-x) \mathbb{I}_{(0,\infty)}(x),$$

and with a scale change,

$$f_{\text{Gam}}(x; \alpha, \beta) = \frac{\beta^\alpha}{\Gamma(\alpha)} x^{\alpha-1} \exp(-\beta x) \mathbb{I}_{(0,\infty)}(x),$$
(7.6)

which is also expressible as

$$\underbrace{\frac{\beta^\alpha}{\Gamma(\alpha)}}_{a(\alpha,\beta)} \underbrace{\mathbb{I}_{(0,\infty)}(x)}_{b(x)} \exp[\underbrace{(\alpha-1)}_{c_1(\alpha,\beta)} \underbrace{\ln x}_{d_1(x)}] \exp(\underbrace{-\beta}_{c_2(\alpha,\beta)} \underbrace{x}_{d_2(x)}), \tag{7.7}$$

showing that it is a member of the exponential family of distributions (see Section 4.1.2). Observe that β is a scale parameter[2] and, with $\alpha = 1$, reduces to the exponential distribution. Densities for $\beta = 1$ and three values of α are shown in Figure 7.1.

Let $Z \sim \text{Gam}(\alpha, 1)$. The c.d.f. is given by the incomplete gamma ratio

$$F_Z(z; \alpha, 1) = \Pr(Z \le z) = \overline{\Gamma}_z(\alpha) = \frac{\Gamma_z(\alpha)}{\Gamma(\alpha)}, \tag{7.8}$$

where

$$\Gamma_x(a) = \int_0^x t^{a-1} e^{-t} dt$$

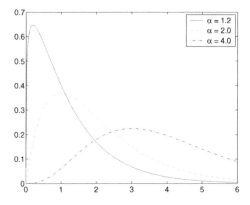

Figure 7.1 Gamma p.d.f. with shape α and scale one

---

[2] Actually, from (7.1), $1/\beta$ is the scale parameter. As such, it might make more sense to write the density as

$$f_{\text{Gam}}(x; \alpha, \sigma) = \frac{1}{\sigma^\alpha \Gamma(\alpha)} x^{\alpha-1} \exp(-x/\sigma) \mathbb{I}_{(0,\infty)}(x),$$

where $\sigma = 1/\beta$ is a genuine scale parameter. Both notational conventions are popular in the literature. It is useful to know that Matlab uses the latter, i.e. in its functions for random variable generation (gamrnd), p.d.f. (gampdf), c.d.f. (gamcdf) and inverse c.d.f. (gaminv), the 'second distribution parameter' is a genuine scale parameter, whereas S-plus® uses notation (7.6) in its functions rgamma, dgamma, pgamma and qgamma, and refers to the inverse scale parameter as the 'rate'. Similar comments obviously hold when working with the exponential distribution.

is the incomplete gamma function; see (1.43) and the subsequent discussion on its computation.

The c.d.f. of the scaled gamma, $X = Z/\beta \sim \text{Gam}(\alpha, \beta)$ is

$$F_X(x; \alpha, \beta) = \Pr(X \le x) = \Pr(Z/\beta \le x) = \Pr(Z \le x\beta) = \frac{\Gamma_{x\beta}(\alpha)}{\Gamma(\alpha)}.$$

For the moments, with $u = \beta x$,

$$\int_0^\infty x^{\alpha-1+k} e^{-\beta x} dx = \beta^{-1-(\alpha-1+k)} \int_0^\infty u^{\alpha-1+k} e^{-u} du = \beta^{-(\alpha+k)} \Gamma(k+\alpha),$$

so that, from (4.36),

$$\mathbb{E}[X^k] = \frac{\Gamma(k+\alpha)}{\beta^k \Gamma(\alpha)}, \quad k > -\alpha. \tag{7.9}$$

In particular, for $k = 0$, we see that the density integrates to one; for $k = 1$, it follows that $\mathbb{E}[X] = \alpha/\beta$; with $k = 2$ and using (4.40), we find

$$\mathbb{V}(X) = \alpha(1+\alpha)/\beta^2 - (\alpha/\beta)^2 = \alpha/\beta^2.$$

While this calculation might be instructive, note that it is easier to work with $Z \sim \text{Gam}(\alpha, 1)$, with mean and variance α. Then, with $X = Z/\beta \sim \text{Gam}(\alpha, \beta)$, $\mathbb{E}[X] = \mathbb{E}[Z]/\beta = \alpha/\beta$, and $\mathbb{V}(X) = \mathbb{V}(Z)/\beta^2$. The third and fourth moments will be derived later (using the so-called moment generating function), which gives the skewness and kurtosis coefficients

$$\frac{\mu_3}{\mu_2^{3/2}} = \frac{2\alpha/\beta^3}{(\alpha/\beta^2)^{3/2}} = \frac{2}{\sqrt{\alpha}} \quad \text{and} \quad \frac{\mu_4}{\mu_2^2} = \frac{3\alpha(2+\alpha)/\beta^4}{(\alpha/\beta^2)^2} = \frac{3(2+\alpha)}{\alpha}.$$

Let $X \sim \text{Gam}(\alpha, \beta)$ with $\alpha \in \mathbb{N}$. Then, with $u = t^{\alpha-1}$ and $dv = e^{-\beta t} dt$, successive integration by parts yields

$$\begin{aligned}
\Pr(X > x) &= \frac{\beta^\alpha}{(\alpha-1)!} \int_x^\infty t^{\alpha-1} e^{-\beta t} dt \\
&= \frac{\beta^{\alpha-1} x^{\alpha-1} e^{-\beta x}}{(\alpha-1)!} + \frac{\beta^{\alpha-1}}{(\alpha-2)!} \int_x^\infty t^{\alpha-2} e^{-\beta t} dt \\
&= \frac{\beta^{\alpha-1} x^{\alpha-1} e^{-\beta x}}{(\alpha-1)!} + \frac{\beta^{\alpha-2} x^{\alpha-2} e^{-\beta x}}{(\alpha-2)!} + \cdots + \beta x e^{-\beta x} + e^{-\beta x} \\
&= \sum_{i=0}^{\alpha-1} \frac{e^{-\beta x}(\beta x)^i}{i!} = \Pr(Y \le \alpha-1) = F_Y(\alpha-1), \tag{7.10}
\end{aligned}$$

Figure 7.2 Poisson process

where $Y \sim \text{Poi}(\beta x)$. Rearranging and replacing x with t and β with λ, this can also be written

$$F_X(t; \alpha, \lambda) = \Pr(Y \geq \alpha; \lambda t). \tag{7.11}$$

This relation between the c.d.f.s of gamma and Poisson distributed random variables can be interpreted as follows. Assume that Y, the random variable describing the number of events occurring through time, can be modeled as a Poisson process with intensity λ, so that $\Pr(Y \geq \alpha; \lambda t)$ is the probability of observing at least α events by time t. Now observe that the event 'at least α events occur by time t' is equivalent to the event 'the time at which the αth event occurs is less than t'. This is best seen by using Figure 7.2 which shows the time axis t along which the occurrence times of the first four events are shown, labeled $T_1, \ldots, T_4$. For example, for any value t, event $T_4 \leq t$ is equivalent to event 'at least four events occur by time t'. Then, via (7.11) and to the extent that a process of interest unfolds according to a Poisson process, we can envisage modeling the time until particular events of interest occur (or waiting times) as a gamma random variable.

(4) Beta, Beta (p, q), $p, q \in \mathbb{R}_{>0}$:

$$f_{\text{Beta}}(x; p, q) = \frac{1}{B(p, q)} x^{p-1} (1 - x)^{q-1} \mathbb{I}_{[0,1]}(x),$$

which can be written as a member of the exponential family:

$$\underbrace{\frac{1}{B(p, q)}}_{a(p,q)} \underbrace{\mathbb{I}_{[0,1]}(x)}_{b(x)} \exp[\underbrace{(p-1)}_{c_1(p,q)} \underbrace{\ln x}_{d_1(x)}] \exp[\underbrace{(q-1)}_{c_2(p,q)} \underbrace{\ln(1-x)}_{d_1(x)}].$$

The c.d.f. is

$$F_{\text{Beta}}(x; p, q) = \frac{B_x(p, q)}{B(p, q)} \mathbb{I}_{[0,1]}(x) + \mathbb{I}_{(1,\infty)}(x),$$

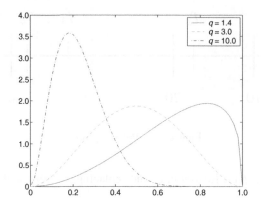

Figure 7.3 Beta p.d.f. with $p = 3$

where

$$B_x (p, q) = \mathbb{I}_{[0,1]} (x) \int_0^x t^{p-1} (1 - t)^{q-1} \, dt$$

is the incomplete beta function. The normalized function $B_x (p, q) / B (p, q)$ is the incomplete beta ratio, which we denote by $\overline{B}_x (p, q)$.
If $p = q = 1$, the beta distribution reduces to that of a Unif $(0, 1)$ random variable. Figure 7.3 shows the p.d.f. for $p = 3$ and three values of q.
The kth raw moment of $X \sim$ Beta (p, q) is simply

$$\mathbb{E}[X^k] = \frac{1}{B (p, q)} \int_0^1 x^{p+k-1} (1 - x)^{q-1} \, dx = \frac{B (p + k, q)}{B (p, q)}$$
$$= \frac{\Gamma (p + q)}{\Gamma (p)} \frac{\Gamma (p + k)}{\Gamma (p + k + q)},$$

for $k > -p$. Using the recursive nature of the gamma function, we see that, for $k = 1$,

$$\mathbb{E} [X] = \frac{p}{p + q} \tag{7.12}$$

and $\mathbb{E}\left[X^2\right] = p (p + 1) / (p + q) (p + q + 1)$ so that, from (4.40),

$$\mathbb{V} (X) = \frac{p (p + 1)}{(p + q) (p + q + 1)} - \frac{p^2}{(p + q)^2} = \frac{pq}{(p + q)^2 (p + q + 1)}. \tag{7.13}$$

Recalling the discussion in Section 4.4, the skewness and kurtosis of the beta are given by

$$\frac{\mu_3}{\mu_2^{3/2}} = \frac{2(q - p)(p + q + 1)^{1/2}}{(p + q + 2)(pq)^{1/2}},$$

$$\frac{\mu_4}{\mu_2^2} = \frac{3(p+q+1)[2(p+q)^2 + pq(p+q-6)]}{pq(p+q+2)(p+q+3)}. \tag{7.14}$$

⊖ **Example 7.1** Let $X \sim \text{Beta}(p, q)$. Then, for $k, m > 0$,

$$\mathbb{E}\left[\frac{1}{(x+k)^m}\right] = \frac{1}{B(p,q)} \int_0^1 \frac{x^{p-1}(1-x)^{q-1}}{(x+k)^m} dx,$$

which appears difficult to solve, unless $m = p + q$, in which case (1.64) implies that

$$\mathbb{E}\left[\frac{1}{(x+k)^{p+q}}\right] = \frac{1}{(1+k)^p k^q}. \tag{7.15}$$

As $(x+k)^m$ is positive and decreasing on $(0, 1)$, the Bonnet mean-value theorem (A.59) can be applied, with $g(x)$ as the beta density and $f(x) = (x+k)^{-m}$. It shows that there exists $c \in (0, 1)$ such that

$$\mathbb{E}\left[\frac{1}{(x+k)^m}\right] = \frac{1}{k^m} \frac{1}{B(p,q)} \int_0^c x^{p-1}(1-x)^{q-1} dx = \frac{\overline{B}_c(p,q)}{k^m},$$

and this r.h.s. is obviously bounded by $1/k^m$. Thus, we have an upper bound to the expectation. As $k \to 0^+$, the upper bound grows without bound, but for $k = 0$, $\mathbb{E}[x^{-m}]$ exists for $p > m$, i.e. the bound becomes useless as $k \to 0^+$. However, as k grows, the bound becomes increasingly sharp, which can be seen from (7.15), i.e. as k grows, $\mathbb{E}[(x+k)^{p+q}] \to k^{-(p+q)}$. ∎

Remark: A result which at this point is not at all obvious is that, if $X \sim \text{Beta}(a, b)$, then, for integer $b \geq 1$,

$$\mathbb{E}[\ln X] = \int_0^1 (\ln u) u^{a-1}(1-u)^{b-1} du$$

$$= -\frac{1}{b\binom{a+b-1}{b}} \sum_{i=0}^{b-1} \frac{1}{a+b-1-i},$$

as will be shown in Volume II when working with order statistics. ∎

We will occasionally write the beta as a four-parameter family, $\text{Beta}(p, q, a, b)$ with additional parameters $a, b \in \mathbb{R}$, $a < b$, specifying the end points of the distribution (as in the uniform). That is,[3]

$$f_{\text{Beta}}(x; p, q, a, b) = \frac{(x-a)^{p-1}(b-x)^{q-1}}{B(p,q)(b-a)^{p+q-1}} \mathbb{I}_{(a,b)}(x),$$

---

[3] There is another generalization of the beta which also has two additional parameters: the doubly noncentral beta, discussed in Volume II, for which we also write $\text{Beta}(n_1, n_2, \theta_1, \theta_2)$. The context should make it quite clear which of the two is meant.

and, for $a < x < b$,

$$F_{\text{Beta}}(x; p, q, a, b) = F_{\text{Beta}}\left(\frac{x - a}{b - a}; p, q, 0, 1\right).$$

This distribution is quite flexible and can be used for modeling continuous outcomes with finite support. It can also serve as an approximation to the distribution of certain statistics when their exact densities are unavailable.[4] The Beta (p, q, a, b) does not belong to the exponential family (and, hence, neither does the Unif (a, b) distribution). If $Y \sim \text{Beta}(p, q, a, b)$, then $X = (Y - a) / (b - a) \sim \text{Beta}(p, q, 0, 1)$ so that

$$\mathbb{E}[Y^k] = \mathbb{E}\left[((b - a)X + a)^k\right] = \sum_{i=0}^{k} \binom{k}{i} a^{k-i} (b - a)^i \mathbb{E}[X^i].$$

As Y is just a location–scale transformation of X, the skewness and kurtosis of Y are the same as those for X.

(5) Laplace, after Pierre Simon Laplace (1749–1827) or double exponential, Lap (μ, σ) or DExp (μ, σ), $\mu \in \mathbb{R}$ and $\sigma \in \mathbb{R}_{>0}$:

$$f_{\text{Lap}}(x; 0, 1) = \frac{\exp(-|x|)}{2}$$

and

$$F_{\text{Lap}}(x; 0, 1) = \frac{1}{2}\begin{cases} e^x, & \text{if } x \le 0, \\ 2 - e^{-x}, & \text{if } x > 0. \end{cases} \tag{7.16}$$

Symmetry implies that the odd moments of a Laplace random variable L are zero, and that the even moments are the same as those of the exponential, as given in (7.4), namely $\mathbb{E}[L^r] = r!$, $r = 0, 2, 4, \ldots$.

Because μ and σ are, respectively, location and scale parameters, the density of Lap (μ, σ) is obtained using (7.1), i.e.

$$f_{\text{Lap}}(y; \mu, \sigma) = \frac{1}{2\sigma} \exp\left(-\frac{|y - \mu|}{\sigma}\right) \tag{7.17}$$

and, with $X \sim \text{Lap}(0, 1)$ and $Y = \sigma X + \mu$, $F_Y(y; \mu, \sigma)$ is given by

$$\Pr(Y \le y) = \Pr(\sigma X + \mu \le y) = \Pr\left(X \le \frac{y - \mu}{\sigma}\right) = F_X\left(\frac{y - \mu}{\sigma}\right), \tag{7.18}$$

where F_X is given in (7.16). The density $f_{\text{Lap}}(x; \mu, \sigma)$ is not a member of the exponential family, but it is easy to see that $f_{\text{Lap}}(x; 0, \sigma)$ does belong.

---

[4] See, for example, Henshaw (1966) and Axelrod and Glosup (1994).

The Laplace is of both theoretical and practical interest. In particular, it arises in a statistical context with regard to the optimality of the sample median as an estimator of the location parameter of an i.i.d. sample. Also, the difference of two i.i.d. exponential r.v.s will be shown in Example 9.5 to follow a Laplace distribution. More practically, the Laplace is sometimes found to offer a much better description of data than the normal, such as for returns on financial instruments.[5] A wealth of further information on the Laplace distribution can be found in Johnson, Kotz and Balakrishnan (1995) and Kotz, Podgorski and Kozubowski (2001).

(6) Weibull, after Waloddi Weibull (1887–1979), $\text{Weib}(\beta, x_0, \sigma)$, $\beta, \sigma \in \mathbb{R}_{>0}$ and $x_0 \in \mathbb{R}$: the density of the location-zero, scale-one distribution is given by

$$f_{\text{Weib}}(x; \beta, 0, 1) = \beta x^{\beta-1} \exp\left(-x^\beta\right) \mathbb{I}_{(0,\infty)}(x),$$

and, from (7.1), the density of $\text{Weib}(\beta, x_0, \sigma)$ is given by

$$f_{\text{Weib}}(x; \beta, x_0, \sigma) = \frac{\beta}{\sigma} \left(\frac{x - x_0}{\sigma}\right)^{\beta-1} \exp\left[-\left(\frac{x - x_0}{\sigma}\right)^\beta\right] \mathbb{I}_{(x_0,\infty)}(x), \quad (7.19)$$

where x_0 and σ are location and scale parameters, respectively. It is often used in lifetime studies but plays an important role in other contexts as well. With $\beta = 1$ and $x_0 = 0$, $\text{Weib}(\beta, x_0, \sigma)$ reduces to the exponential distribution. Figures 7.4 and 7.5 plot the location-zero, scale-one Weibull density for several values of β. Figure 7.5 concentrates on the tails of the distribution, and shows the fat-tailed nature of the Weibull when $\beta < 1$; see the discussion of the Pareto distribution below for the definition of fat tails.

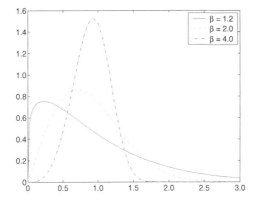

Figure 7.4 Weibull p.d.f. with $x_0 = 0$ and $\sigma = 1$

---

[5] See, for example, Granger and Ding (1995), Mittnik, Paolella and Rachev (1998), Kotz, Podgorski and Kozubowski (2001) and the references therein.

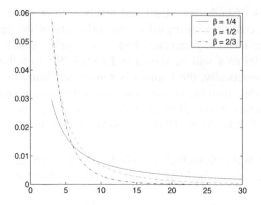

Figure 7.5   Weibull p.d.f. with $x_0 = 0$ and $\sigma = 1$

Its c.d.f. is given by

$$F_{\text{Weib}}(x; \beta, 0, 1) = 1 - \exp\left(-x^\beta\right) \mathbb{I}_{(0,\infty)}(x)$$

and

$$F_{\text{Weib}}(x; \beta, x_0, \sigma) = 1 - \exp\left[-\left(\frac{x - x_0}{\sigma}\right)^\beta\right] \mathbb{I}_{(x_0,\infty)}(x) \qquad (7.20)$$

and, for $\text{Weib}(b, 0, s)$, substituting $u = (x/s)^b$ and simplifying gives

$$\mathbb{E}\left[X^p\right] = s^p \int_0^\infty u^{p/b} \exp\left(-u\right) du = s^p \Gamma\left(1 + \frac{p}{b}\right), \qquad (7.21)$$

which exists for $p > -b$.

(7) Cauchy,[6] $\text{Cau}(\mu, \sigma)$, $\mu \in \mathbb{R}$ and $\sigma \in \mathbb{R}_{>0}$: the density is

$$f_{\text{Cau}}(c; 0, 1) = \frac{1}{\pi} \cdot \frac{1}{1 + c^2},$$

and the density of $\text{Cau}(\mu, \sigma)$ can be calculated via (7.1) using the fact that μ and σ are location and scale parameters, respectively. The integral of f_{Cau} is well known (see Example A.6), so that the c.d.f. is

$$F_{\text{Cau}}(c; 0, 1) = \frac{1}{\pi} \int_{-\infty}^{c} \frac{1}{1 + x^2} dx \qquad (7.22)$$

$$= \frac{1}{2} + \frac{1}{\pi} \arctan(c), \qquad (7.23)$$

---

[6] After the prolific Augustin Louis Cauchy (1789–1857), author of 789 mathematical papers. As with most naming conventions though, the distribution and its properties were known and used by others, such as Poisson.

which will also be demonstrated geometrically in Example 8.18. As a check, recall that $\lim_{x\to\infty} \arctan(x) = \pi/2$, so indeed, $\lim_{x\to\infty} F_X(x) = 1$.
The c.d.f. can also be expressed in terms of the incomplete beta ratio $\bar{B}_w(\cdot, \cdot)$ (see Section 1.5.2). First consider $c < 0$, and substitute $y = (1 + x^2)^{-1}$ into (7.22). Note that x has a unique solution when $c < 0$, i.e.

$$y = \frac{1}{1+x^2}, \quad x = -\frac{1}{y}\sqrt{y(1-y)}, \quad dx = \frac{1}{2y\sqrt{y(1-y)}}dy,$$

so that, with $w = 1/(1 + c^2)$,

$$F_{\text{Cau}}(c) = \frac{1}{\pi}\int_{-\infty}^{c}\frac{1}{1+x^2}dx = \frac{1}{2\pi}\int_0^w y^{-\frac{1}{2}}(1-y)^{-\frac{1}{2}}dy = \frac{1}{2\pi}B_w\left(\frac{1}{2},\frac{1}{2}\right);$$

but

$$B\left(\frac{1}{2},\frac{1}{2}\right) = \frac{\Gamma(1/2)\Gamma(1/2)}{\Gamma(1)} = \pi,$$

so that

$$F_{\text{Cau}}(c) = \frac{1}{2}\bar{B}_w\left(\frac{1}{2},\frac{1}{2}\right), \quad w = \frac{1}{1+c^2}, \quad c < 0. \tag{7.24}$$

For $c > 0$, the symmetry of the Cauchy density about zero implies that

$$F_{\text{Cau}}(c) = 1 - F_{\text{Cau}}(-c).$$

The Cauchy is an important theoretical distribution which will appear in several contexts in later chapters. It is also a special case of the Student's t distribution (see below) and the so-called stable Paretian distribution (developed in Volume II). Example A.27 shows that the integral corresponding to its expected value diverges, so that moments of order one and higher do not exist.

(8) Type I Pareto (or Pareto distribution of the first kind), after Vilfredo Pareto (1848–1923), $\text{Par I}(\alpha, x_0)$, $\alpha, x_0 \in \mathbb{R}_{>0}$: the c.d.f. is

$$F_{\text{Par I}}(x; \alpha, x_0) = \left[1 - \left(\frac{x_0}{x}\right)^\alpha\right]\mathbb{I}_{(x_0,\infty)}(x) \qquad \text{(Type I)}$$

which is easily seen to satisfy the four required properties of a c.d.f. given in Section 4.1.1, i.e. (i) $\forall x \in \mathbb{R}$, $0 \le F(x) \le 1$, (ii) F is nondecreasing, (iii) F is right continuous, and (iv) $\lim_{x\to-\infty} F(x) = 0$ and $\lim_{x\to\infty} F(x) = 1$. Differentiating the c.d.f. gives

$$f_{\text{Par I}}(x; \alpha, x_0) = \alpha x_0^\alpha x^{-(\alpha+1)}\mathbb{I}_{(x_0,\infty)}(x).$$

Location and scale parameters could also be introduced in the usual way.

Besides its classical use for modeling income and wealth, the Pareto arises in several contexts within hydrology and astronomy (see, for example, Whittle, 1992, p. 185(4), Johnson, Kotz and Balakrishnan, 1994, Chapter 20, and the references therein). The type I Pareto will also play an important role in the study of *extreme value theory*.

The moments of $X \sim \mathrm{Par\,I}\,(\alpha, x_0)$ are given by

$$\mathbb{E}\left[X^m\right] = \alpha x_0^\alpha \int_{x_0}^\infty x^{-\alpha-1+m}\,\mathrm{d}x = \frac{\alpha x_0^\alpha}{m-\alpha}\,x^{m-\alpha}\Big|_{x_0}^\infty = \frac{\alpha}{\alpha-m}x_0^m, \quad m < \alpha,$$

(7.25)

and do not exist for $m \geq \alpha$. In particular,

$$\mathbb{E}[X] = \frac{\alpha}{\alpha-1}x_0, \quad \mathbb{V}(X) = \frac{\alpha}{(\alpha-1)^2(\alpha-2)}x_0^2.$$

(7.26)

In various applications, the *survivor function* of r.v. X, defined by

$$\overline{F}_X(x) = 1 - F_X(x) = \Pr(X > x),$$

(7.27)

is of particular interest. For $X \sim \mathrm{Par\,I}\,(\alpha, x_0)$, $\overline{F}(x) = Cx^{-\alpha}$, where $C = x_0^\alpha$. It turns out that the survivor function for a number of important distributions is asymptotically of the form $Cx^{-\alpha}$ as x increases, where C denotes some constant. If this is the case, we say that the right *tail* of the density is *Pareto-like* or that the distribution has *power tails* or *fat tails*. Somewhat informally, if $\overline{F}_X(x) \approx Cx^{-\alpha}$, then, as $\mathrm{d}\left(1 - Cx^{-a}\right)/\mathrm{d}x \propto x^{-(a+1)}$, it follows from (7.25) that the maximally existing moment of X is bounded above by α.

⊖ *Example 7.2* Let the p.d.f. of r.v. R be given by

$$f_R(r) = \frac{ar^{a-1}}{(r+1)^{a+1}}\mathbb{I}_{(0,\infty)}(r).$$

(7.28)

Figure 7.6 plots the density for two different values of α. To see that this is a valid density, let $u = r/(r+1)$ (and, thus, $r = u/(1-u)$, $r+1 = (1-u)^{-1}$ and $\mathrm{d}r = (1-u)^{-2}\,\mathrm{d}u$) so that

$$\int_0^\infty f_R(r)\,\mathrm{d}r = a\int_0^\infty r^{a-1}(r+1)^{-(a+1)}\,\mathrm{d}r$$

$$= a\int_0^1 \left(\frac{u}{1-u}\right)^{a-1}\left(\frac{1}{1-u}\right)^{-(a+1)}(1-u)^{-2}\,\mathrm{d}u$$

$$= a\int_0^1 u^{a-1}\,\mathrm{d}u = a\left.\frac{u^a}{a}\right|_0^1 = 1.$$

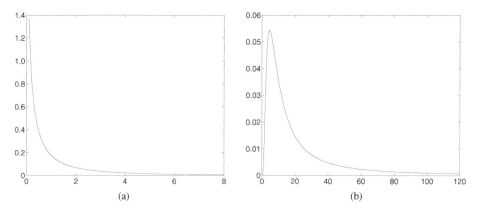

Figure 7.6 Density (7.28) for (a) $\alpha = 0.5$ and (b) $\alpha = 10$

More generally, the same substitution yields

$$\mathbb{E}\left[R^m\right] = a \int_0^\infty r^{a+m-1} (r+1)^{-(a+1)} \, dr = a \int_0^1 u^{a+m-1} (1-u)^{-m} \, du$$

$$= aB(a+m, 1-m) = a \frac{\Gamma(a+m)\Gamma(1-m)}{\Gamma(a+1)},$$

which exists for $1 - m > 0$ or $m < 1$, i.e. the mean and higher moments do not exist. For the c.d.f., note that

$$\frac{d}{dt}\left(\frac{t}{t+1}\right)^a = at^{a-1}(t+1)^{-(a+1)}, \quad \text{i.e.} \quad F_R(t) = \left(\frac{t}{t+1}\right)^a \mathbb{I}_{(0,\infty)}(t)$$

and, for relatively large t, i.e. far into the right tail,

$$F_R(t) = \exp\left[\ln F_R(t)\right] = \exp\left[-a\ln\left(1+t^{-1}\right)\right] \approx \exp\left(-a/t\right) \approx 1 - a/t,$$

i.e. $\overline{F}_R(t) = \Pr(R > t) \approx a/t \propto t^{-1}$. From the previous discussion on Pareto-like tails, the form of $\overline{F}_R$ suggests that moments of order less than one exist, which agrees with the direct calculation of $\mathbb{E}[R^m]$ above. ∎

For many distributions, all positive moments exist, such as for the gamma and normal. These are said to have *exponential tails*. For instance, if $X \sim \text{Exp}(\lambda)$, then $\overline{F}_X(x) = e^{-\lambda x}$. Figure 7.7 shows the survivor function for (a) a standard normal r.v. and (b) the Pareto survivor function with $\alpha = 1$ (and $x_0 = 1$). The shape of each graph is the same for any interval far enough into the tail. While the normal c.d.f. (or that of any distribution with exponential right tail) dies off rapidly, the c.d.f. with power tails tapers off slowly. For the latter, there is enough mass in the tails of the distribution so that the probability of extreme events never becomes negligible. This is why the expected value of X raised to a sufficiently large power will fail to exist. Thus, in Figure 7.6, while the p.d.f. clearly converges to zero as x increases, the rate at which it converges is too slow for the mean (and higher moments) to exist.

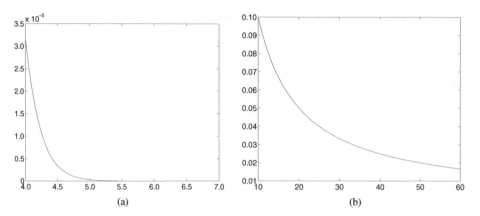

Figure 7.7 Comparing (a) an exponential tail and (b) a power tail

(9) Type II Pareto, Par II (b), $b \in \mathbb{R}_{>0}$: the c.d.f. is

$$F_{\text{Par II}}(x; b) = \left[1 - \left(\frac{1}{1+x} \right)^b \right] \mathbb{I}_{(0,\infty)}(x) \qquad \text{(Type II)}.$$

As for the type I Pareto, this satisfies the four required properties of a c.d.f. Differentiating,

$$f_{\text{Par II}}(x; b) = b(1+x)^{-(b+1)} \mathbb{I}_{(0,\infty)}(x).$$

A scale parameter $c \in \mathbb{R}_{>0}$ can be introduced in the usual way, giving

$$f_{\text{Par II}}(x; b, c) = \frac{b}{c} \left(\frac{c}{c+x} \right)^{b+1} \mathbb{I}_{(0,\infty)}, \quad b, c \in \mathbb{R}_{>0}, \qquad (7.29)$$

written $X \sim \text{Par II}(b, c)$. It is not apparent why the functional form (7.29) should be of interest; however, it arises naturally as a so-called mixture model (detailed in Volume II), where the mixing components are gamma distributed r.v.s, which themselves appear ubiquitously in applications.

As $\lim_{x \to \infty} F_X(x) = 1$, the fact that $\int f_X = 1$ follows without calculation; however, we require it for calculating the moments of X. With $u = (1+x)^{-1}$, $x = (1-u)/u$ and $dx = -u^{-2}du$, we indeed have

$$\frac{1}{b} \int_0^\infty f_X(x; b, 1) \, dx = \int_0^\infty \frac{dx}{(1+x)^{b+1}} = -\int_1^0 u^{b-1} du = \int_0^1 u^{b-1} du = \frac{1}{b}. \qquad (7.30)$$

For the expectation, assume that $b > 1$, so that (i) via integration by parts with $u = x$ ($du = dx$) and $dv = (1+x)^{-(b+1)} dx$ [$v = -(1+x)^{-b}/b$], (ii) from (7.30) with parameter $b - 1$ instead of b, and (iii) applying l'Hôpital's rule to the limit (with $b > 1$),

$$\int_0^\infty \frac{x \, dx}{(1+x)^{b+1}} = -\frac{x}{b}(1+x)^{-b}\Big|_0^\infty + \frac{1}{b}\int_0^\infty (1+x)^{-b} \, dx$$

$$= -\frac{1}{b}\lim_{x\to\infty}\frac{x}{(1+x)^b} + \frac{1}{b}\frac{1}{b-1} = \frac{1}{b}\frac{1}{b-1}, \quad b > 1, \quad (7.31)$$

so that $\mathbb{E}[X] = 1/(b-1)$. If $b > 2$, then a similar calculation with $u = x^2$ and $dv = (1+x)^{-(b+1)} \, dx$ and using (7.31) with b in place of $b+1$ gives

$$\int_0^\infty \frac{x^2 \, dx}{(1+x)^{b+1}} = -\frac{x^2}{b}(1+x)^{-b}\Big|_0^\infty + \frac{2}{b}\int_0^\infty x \, (1+x)^{-b} \, dx$$

$$= -\frac{1}{b}\lim_{x\to\infty}\frac{x^2}{(1+x)^b} + \frac{2}{b}\left(\frac{1}{b-1}\frac{1}{b-2}\right)$$

$$= \frac{2}{b(b-1)(b-2)}, \quad b > 2,$$

from which the variance can be computed; in particular,

$$\boxed{\mathbb{E}[X] = \frac{1}{b-1}, \quad \mathbb{V}(X) = \frac{b}{(b-1)^2(b-2)}.} \quad (7.32)$$

(Compare with the type I Pareto moments (7.26) for $x_0 = 1$.) Repeated use of the previous calculations yields the general pattern $\mathbb{E}[X^k] = k!/(b-1)_{[k]}$ for $k < b$.

(10) Normal or Gaussian, after Johann Carl Friedrich Gauss (1777–1855), $N(\mu, \sigma^2)$, $\mu \in \mathbb{R}$ and $\sigma \in \mathbb{R}_{>0}$: with density

$$\boxed{f_N(x; \mu, \sigma) = \frac{1}{\sqrt{2\pi}\sigma}\exp\left[-\frac{1}{2}\left(\frac{x-\mu}{\sigma}\right)^2\right].} \quad (7.33)$$

The fact that the density integrates to one is shown in Example A.79.
Let $Z \sim N(0, 1)$, i.e. standard normal. Because its c.d.f. arises so often, a notation for it has become practically universal, this being $\Phi(z)$. From the Taylor series expression of e,

$$\exp\left(-z^2/2\right) = \sum_{i=0}^\infty (-1)^i \frac{z^{2i}}{2^i \, i!},$$

so that, integrating termwise,

$$\Phi(z) = F_Z(z) = F_Z(0) + \frac{1}{\sqrt{2\pi}}\int_0^z e^{-t^2/2} dt$$

$$= \frac{1}{2} + \frac{1}{\sqrt{2\pi}}\sum_{i=0}^\infty (-1)^i \frac{z^{2i+1}}{(2i+1)\, 2^i \, i!}, \quad (7.34)$$

which can be used to numerically evaluate $F_Z(z)$ by truncating the infinite sum. For example, with $z = -1.64485362695147$, $F_Z(z) = 0.05$ to 14 significant digits. Using (7.34) with six terms in the sum yields 0.0504; with 11 terms, 0.049999982.

The ability to compute the inverse c.d.f., $\Phi^{-1}(p)$, $0 < p < 1$, is also of value. Efficient methods exist for this calculation (see Kennedy and Gentle, 1980; Press, Teukolsky, Vetterling and Flannery, 1989, and the references therein), including an easily computed and highly accurate approximation recently given by Moro (1995). A statement of Moro's procedure is most usefully given directly as a Matlab program; this is shown in Program Listing 7.1. The discrepancy of Moro's method and the method used in Matlab (which is accurate to machine precision) is shown in Figure 7.8.

```
function x=moro(pvec)
a = [ 2.50662823884, -18.61500062529,  41.39119773534, -25.44106049637];
b = [-8.47351093090,  23.08336743743, -21.06224101826,   3.13082909833];
c = [ 0.3374754822726147, 0.9761690190917186, 0.1607979714918209, ...
      0.0276438810333863, 0.0038405729373609, 0.0003951896511919, ...
      0.0000321767881768, 0.0000002888167364, 0.0000003960315187];
pl=length(pvec); x=zeros(pl,1);
for i=1:pl
  p=pvec(i); y=p-0.5;
  if abs(y)<0.42
    r=y^2; num = y * sum(a .* r.^(0:3)); den = 1 + sum(b .* r.^(1:4));
    x(i)=num/den;
  else
    if y<0, r=p; else r=1-p; end
    s = log(-log(r)); sj=s.^(0:8); t = sum(c .* sj);
    if p>0.5, x(i)=t; else x(i)=-t; end
  end
end
```

Program Listing 7.1   Moro's (1995) algorithm for computing $\Phi^{-1}(p)$, $0 < p < 1$

Of great importance is the fact that, if $X \sim N(\mu, \sigma^2)$, then $Z = (X - \mu)/\sigma \sim N(0, 1)$; random variables X and Z are members of the normal location–scale family, with μ being the location parameter, and σ the scale parameter. This follows from (7.1) and the density (7.33). According to our notation used so far, one would expect to write $N(\mu, \sigma)$ as the abbreviation for the normal distribution with location μ and scale σ; however, it is standard convention to write $N(\mu, \sigma^2)$ instead, and we do so as well.

The first four moments are

$$\mu_1 = \mathbb{E}[X] = \mu, \quad \mu_2 = \mathbb{V}(X) = \sigma^2, \quad \mu_3 = 0, \quad \mu_4 = 3\sigma^4,$$

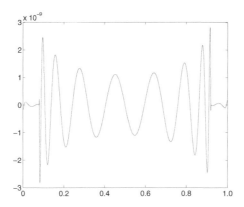

Figure 7.8   Discrepancy of Moro's method and the method used in Matlab for computing $\Phi^{-1}(p)$, $0 < p < 1$

as will be shown in the next example. First, however, we use the mean and variance results to establish a very useful property of normal r.v.s:

If $X \sim \mathrm{N}\left(\mu, \sigma^2\right)$, then, for constant $a \in \mathbb{R}$, $aX \sim \mathrm{N}\left(a\mu, a^2\sigma^2\right)$.   (7.35)

To see why this is true, first observe that casual inspection of the transformation result (7.1) shows that random variable aX follows *some* normal distribution. Its mean is $\mathbb{E}[aX] = a\mathbb{E}[X] = a\mu$ from the linearity of expectation property (4.37), and its variance is $\mathbb{V}(aX) = a^2\mathbb{V}(X) = a^2\sigma^2$ from (4.41). The result now follows because the normal distribution is *characterized* (i.e. uniquely determined) by its two parameters, which themselves are uniquely determined by (i.e. are related by a one-to-one transformation with) the mean and variance.

⊖ ***Example 7.3***   The even moments of a standard normal, i.e. $Z \sim \mathrm{N}\,(0, 1)$, are

$$
\mathbb{E}\left[Z^{2r}\right] = \int_{-\infty}^{\infty} z^{2r} f_Z\,(z)\,\mathrm{d}z
$$

$$
= \frac{2}{\sqrt{2\pi}} \int_{0}^{\infty} z^{2r} \mathrm{e}^{-\frac{1}{2}z^2}\,\mathrm{d}z = \frac{2^{r+1-1/2}}{\sqrt{2\pi}} \int_{0}^{\infty} u^{r-1/2} \mathrm{e}^{-u}\,\mathrm{d}u
$$

$$
= \frac{2^r \Gamma\left(r + \frac{1}{2}\right)}{\sqrt{\pi}}
$$

for $r \in \mathbb{N}$, where $u = z^2/2$, $z = (2u)^{1/2}$ (because z is positive) and $\mathrm{d}z = (2u)^{-1/2}\,\mathrm{d}u$. That is, for $s = 2r$ and recalling that $\Gamma\,(a+1) = a\Gamma\,(a)$ and $\Gamma\,(1/2) = \sqrt{\pi}$,

$$
\mathbb{E}\left[Z^s\right] = \frac{1}{\sqrt{\pi}} 2^{s/2}\Gamma\left[\frac{1}{2}\,(1+s)\right] = (s-1)\,(s-3)\,(s-5)\cdots \cdot 3 \cdot 1.   (7.36)
$$

This can also be written

$$\mathbb{E}\left[Z^s\right] = \mathbb{E}\left[Z^{2r}\right] = \frac{(2r)!}{2^r r!}, \tag{7.37}$$

which follows because

$$M(r) := \frac{(2r)!}{2^r r!} = \left[\frac{(2r-1)2r}{2r}\right] \frac{[2(r-1)]!}{2^{r-1}(r-1)!} = (2r-1)M(r-1),$$

e.g.

$$\mathbb{E}[Z^6] = \mathbb{E}[Z^{2\cdot 3}] = M(3) = 5 \cdot M(2) = 5 \cdot 3 \cdot M(1) = 5 \cdot 3 \cdot 1.$$

With $X = \sigma Z + \mu \sim N\left(\mu, \sigma^2\right)$,

$$\mathbb{E}\left[(X - \mu)^{2r}\right] = \sigma^{2r}\mathbb{E}\left[Z^{2r}\right] = \left(2\sigma^2\right)^r \pi^{-1/2}\Gamma\left(r + \frac{1}{2}\right). \tag{7.38}$$

(The reader should directly check that (7.38) reduces to $3\sigma^4$ for $r = 2$.) An expression for the even raw moments of X can be obtained via (7.37) and the binomial formula applied to $(\sigma Z + \mu)^{2r}$.
For odd moments, similar calculations give[7]

$$\int_{-\infty}^{\infty} z^{2r+1} f_Z(z) \, dz = \frac{1}{\sqrt{2\pi}} \int_{-\infty}^{0} z^{2r+1} e^{-\frac{1}{2}z^2} dz + \frac{1}{\sqrt{2\pi}} \int_{0}^{\infty} z^{2r+1} e^{-\frac{1}{2}z^2} dz$$

$$= -\frac{2^r \Gamma(r+1)}{\sqrt{2\pi}} + \frac{2^r \Gamma(r+1)}{\sqrt{2\pi}} = 0.$$

Thus, for example, the skewness and kurtosis of X are zero and three, respectively, recalling that those measures are location and scale invariant.
To calculate $\mathbb{E}|Z| := \mathbb{E}[|Z|]$, use (4.36) and the same u substitution as above to give

$$\mathbb{E}|Z| = \frac{1}{\sqrt{2\pi}} \int_{-\infty}^{\infty} |z| f_Z(z) \, dz = \frac{2}{\sqrt{2\pi}} \int_{0}^{\infty} z e^{-\frac{1}{2}z^2} dz$$

$$= \frac{2}{\sqrt{2\pi}} \int_{0}^{\infty} (2u)^{1/2} e^{-u} (2u)^{-1/2} \, du = \sqrt{\frac{2}{\pi}}, \tag{7.39}$$

where $\int_{0}^{\infty} e^{-u} du = 1$. ∎

---

[7] It should be clear from the symmetry of f_Z that $\int_{-k}^{k} z^{2r+1} f_Z(z) \, dz = 0$ for any $k > 0$ (see Section A.2.3.3). Recall also from Section A.2.3.3 that, for a general density f_X, in order to claim that $\int_{-\infty}^{\infty} x^{2r+1} f_X(x) \, dx = 0$, both $\lim_{k \to \infty} \int_{0}^{k} x^{2r+1} f_X(x) \, dx$ and $-\lim_{k \to \infty} \int_{-k}^{0} x^{2r+1} f_X(x) \, dx$ must converge to the same finite value.

While the normal c.d.f. is not expressible in closed form, the survivor function for $Z \sim N(0, 1)$ does have an upper bound in the right tail. For $t > 0$ and with $u = z^2/2$,

$$\Pr(Z \geq t) = \frac{1}{\sqrt{2\pi}} \int_t^\infty \exp\left(-\frac{1}{2}z^2\right) dz$$

$$\leq \frac{1}{\sqrt{2\pi}} \frac{1}{t} \int_t^\infty z \exp\left(-\frac{1}{2}z^2\right) dz \qquad \left(\text{because } \frac{z}{t} > 1\right)$$

$$= \frac{1}{\sqrt{2\pi}} \frac{1}{t} \int_{t^2/2}^\infty \exp(-u) \, du = \frac{1}{\sqrt{2\pi}} \frac{1}{t} \exp\left(-t^2/2\right).$$

The term $\exp\{-t^2/2\}$ clearly goes to zero far faster than t^{-1}, so that Z has exponential tails.

To derive a lower bound, note that $(1 - 3z^{-4}) \exp(-z^2/2) < \exp(-z^2/2)$ implies, for any t, that (with $A(t) = \exp(-t^2/2)$),

$$A(t)\frac{t^2 - 1}{t^3} = \int_t^\infty \left(1 - 3z^{-4}\right) A(z) \, dz < \int_t^\infty A(z) \, dz = \overline{\Phi}(t) \sqrt{2\pi},$$

showing that[8]

$$\frac{1}{\sqrt{2\pi}} \exp\left(-\frac{t^2}{2}\right) \left(\frac{1}{t} - \frac{1}{t^3}\right) < \overline{\Phi}(t) < \frac{1}{\sqrt{2\pi}} \frac{1}{t} \exp\left(-\frac{t^2}{2}\right), \qquad t > 0. \quad (7.40)$$

It also follows that the r.h.s. is the limit of $\overline{\Phi}(t)$ as $t \to \infty$. A slightly more accurate approximation of $\overline{\Phi}(t)$ for $t > 5.5$ is given in (7.79) in Problem 7.24. The normal distribution, sometimes referred to as the 'bell curve', enjoys certain properties which render it the most reasonable description for modeling a large variety of stochastic phenomena. In particular, under certain general conditions, the cumulative effect of many independent random variables tends to a normal distribution, which is formally stated as the *central limit theorem*, named (presumably) for its central role in probability and statistical theory. It will be detailed in Volume II, and has its origins in the memoir read by Laplace in 1810 to the Academy in Paris.

Remark: It is often noted that the normal distribution was also derived by De Moivre and Laplace (see, for example, Bennett, 1998, for details), and, as cultural allegiance would have it, is often referred to in French literature as the 'law of De Moivre–Laplace' or the 'second law of Laplace' (as opposed to the first law of Laplace, which is what we call the Laplace or double exponential distribution).[9] ∎

---

[8] See Feller (1968, pp. 175, 193) and Lebedev (1972, Section 2.2) for more detail and further expressions.

[9] These naming conventions could undoubtedly be reconciled in a quiescent and civil manner by just universally adopting the name 'Adrian distribution', after the American Robert Adrian (who actually discovered it one year before Gauss).

The following example illustrates one of its many uses in a biostatistical context and gives another inequality involving the c.d.f. Φ.

⊖ **Example 7.4** (Yao and Iyer, 1999) Interest centers on measuring the extent to which a generically manufactured drug, T (for test drug) is interchangeable with a brand name drug, R (for reference drug) for individual patients. Let $X_R \sim N\left(\mu_R, \sigma_R^2\right)$ describe the amount of the relevant chemical absorbed into the patient's bloodstream when using drug R. This is referred to as the *bioavailability* of drug R for the particular patient. Similarly, let $X_T \sim N\left(\mu_T, \sigma_T^2\right)$ be the bioavailability of drug T, for the same patient. The *therapeutic window* describes the range of drug R's concentration in the bloodstream which is deemed beneficial to the patient, and is taken to be $\mu_R \pm z\sigma_R$, where $z > 0$ is an unknown, patient-specific value. Then, value

$$p_R := \Pr\left(\mu_R - z\sigma_R < X_R < \mu_R + z\sigma_R\right)$$

provides a measure of the benefit of drug R. As $(X_R - \mu_R)/\sigma_R \sim N(0, 1)$,

$$p_R = \Pr\left(-z < \frac{X_R - \mu_R}{\sigma_R} < z\right) = \Phi(z) - \Phi(-z).$$

Similarly, for drug T,

$$p_T = \Pr\left(\mu_R - z\sigma_R < X_T < \mu_R + z\sigma_R\right)$$

$$= \Pr\left(\frac{\mu_R - z\sigma_R - \mu_T}{\sigma_T} < \frac{X_T - \mu_T}{\sigma_T} < \frac{\mu_R + z\sigma_R - \mu_T}{\sigma_T}\right)$$

$$= \Phi\left(\frac{\mu_R + z\sigma_R - \mu_T}{\sigma_T}\right) - \Phi\left(\frac{\mu_R - z\sigma_R - \mu_T}{\sigma_T}\right).$$

Drug T is deemed *bioequivalent* if the ratio p_T/p_R is close enough to (or greater than) unity.

However, because z is patient-specific, the value

$$B = \inf_{z \in \mathbb{R}_{>0}} \frac{p_T}{p_R}$$

is of more use. Based on this, the generic drug T can be accepted for use if B is greater than some prespecified value γ (dictated by the relevant regulatory agency). Yao and Iyer (1999) showed that

$$B = \min\left[1, \frac{\sigma_R\sqrt{2\pi}}{\sigma_T}\phi\left(\frac{\mu_T - \mu_R}{\sigma_T}\right)\right],$$

where $\phi(x)$ is the standard normal p.d.f. evaluated at x. This follows from their general result

$$\frac{\Phi\left(\frac{z-\mu}{\sigma}\right) - \Phi\left(\frac{-z-\mu}{\sigma}\right)}{\Phi(z) - \Phi(-z)} > \min\left[1, \frac{\sqrt{2\pi}}{\sigma}\phi\left(\frac{\mu}{\sigma}\right)\right]$$

for $(\mu, \sigma) \neq (0, 1)$ and $z \in \mathbb{R}_{>0}$. ∎

The following three distributions, χ^2, t and F, are of utmost importance for statistical inference involving the wide class of *normal linear models* (which includes the familiar two-sample t test, regression analysis, ANOVA, random-effects models, as well as many econometrics models such as for modeling panel and time series data).

(11) Chi-square with k *degrees of freedom*, $\chi^2(k)$ or χ_k^2, $k \in \mathbb{R}_{>0}$: the density is given by

$$f(x; k) = \frac{1}{2^{k/2}\Gamma(k/2)} x^{k/2-1} e^{-x/2} \mathbb{I}_{(0,\infty)}(x) \qquad (7.41)$$

and is a special case of the gamma distribution with $\alpha = k/2$ and $\beta = 1/2$. With $k = 2$, (7.41) reduces to the p.m.f. of an Exp(1/2) random variable, i.e. an exponential with scale 2. In most statistical applications, $k \in \mathbb{N}$.

⊖ **Example 7.5** The moments $\mathbb{E}[X^s]$ for $X \sim \chi_k^2$ follow from those of the gamma distribution or, with $u = x/2$,

$$\int_0^\infty x^{k/2+s-1} e^{-x/2} dx = 2^{k/2+s} \int_0^\infty u^{k/2+s-1} e^{-u} du = 2^{k/2+s} \Gamma(k/2 + s)$$

for $k/2 + s > 0$, so that

$$\mathbb{E}[X^s] = 2^s \frac{\Gamma(k/2 + s)}{\Gamma(k/2)}, \qquad s > -\frac{k}{2}. \qquad (7.42)$$

In particular, $\mathbb{E}[X] = k$ and $\mathbb{E}[X^2] = (k+2)k$, which leads to $\mathbb{V}(X) = 2k$. More generally, for $s \in \mathbb{N}$, (7.42) implies

$$\mathbb{E}[X^s] = 2^{s-1} \frac{(k+2s-2)\Gamma\left(\frac{k}{2}+s-1\right)}{\Gamma\left(\frac{k}{2}\right)} = [k + 2(s-1)]\mathbb{E}[X^{s-1}],$$

a recursion starting with $\mathbb{E}[X^0] = 1$. Thus, $\mathbb{E}[X^1] = k$, $\mathbb{E}[X^2] = k(k+2)$, and, in general,

$$\mathbb{E}[X^s] = k(k+2)(k+4)\cdots[k+2(s-1)], \qquad s \in \mathbb{N}. \qquad (7.43)$$

Also,

$$\mathbb{E}[X^{-1}] = 1/(k-2), \qquad k > 2, \qquad (7.44)$$

which will be of occasional use. ■

(12) Student's t with n *degrees of freedom*, abbreviated $t(n)$ or t_n, $n \in \mathbb{R}_{>0}$:[10]

$$f_t(x; n) = \frac{\Gamma\left(\frac{n+1}{2}\right) n^{\frac{n}{2}}}{\sqrt{\pi}\,\Gamma\left(\frac{n}{2}\right)} \left(n + x^2\right)^{-\frac{n+1}{2}} = K_n \left(1 + x^2/n\right)^{-\frac{n+1}{2}}, \qquad (7.45)$$

---

[10] After William Sealy Gosset (1876–1937), who wrote under the name 'Student' while he was a chemist with Arthur Guinness Son and Company. See also Lehmann (1999) for details on Gosset's contributions to statistics and Boland (1984) for a short biography of Gosset.

where the constant of integration in the latter expression is given by

$$K_n = \frac{n^{-\frac{1}{2}}}{B\left(\frac{n}{2}, \frac{1}{2}\right)}.$$

In most statistical applications, $n \in \mathbb{N}$. If $n = 1$, then the Student's t distribution reduces to the Cauchy distribution while, as $n \to \infty$, it approaches the normal. To see this, recall that $e^{-k} = \lim_{n\to\infty} (1 + k/n)^{-n}$ for $k \in \mathbb{R}$, which, applied to the kernel of the density, gives

$$\lim_{n\to\infty} \left(1 + x^2/n\right)^{-\frac{n+1}{2}} = \lim_{n\to\infty} \left(1 + x^2/n\right)^{-\frac{n}{2}} = \left[\lim_{n\to\infty} \left(1 + x^2/n\right)^{-n}\right]^{1/2} = e^{-\frac{1}{2}x^2}.$$

This is enough to establish the result, because the integration constant (all terms not involving x) can be ignored (remember that the p.d.f. must integrate to one). Just to check, applying Stirling's approximation $\Gamma(n) \approx \sqrt{2\pi} n^{n-1/2} \exp(-n)$ to the integration constant K_n yields

$$K_n \approx n^{-\frac{1}{2}} \frac{\sqrt{2\pi} \left(\frac{n+1}{2}\right)^{\frac{n+1}{2}-1/2} \exp\left(-\frac{n+1}{2}\right)}{\sqrt{\pi}\sqrt{2\pi} \left(\frac{n}{2}\right)^{\frac{n}{2}-1/2} \exp\left(-\frac{n}{2}\right)} = \exp\left(-\frac{1}{2}\right) \frac{1}{\sqrt{2\pi}} \left(\frac{n+1}{n}\right)^{n/2}.$$

The result follows because $\lim_{n\to\infty} \left(1 + n^{-1}\right)^{n/2} = e^{1/2}$.

Let $T \sim t_n$. The mean of T is zero for $n > 1$, but does not otherwise exist. This, and the calculation of higher moments is shown in Problem 7.1. Example 9.7 will also demonstrate that it suffices to know the moments of standard normal and chi-square r.v.s. In particular, we will see later that

$$\mathbb{V}(T) = \frac{n}{n-2}, \qquad n > 2. \tag{7.46}$$

It is easy to see from (7.45) that as $|x| \to \infty$, $f_t(x; n) \propto |x|^{-(n+1)}$, which is similar to the type I Pareto, showing that the maximally existing moment of the Student's t is bounded above by n.

For $t < 0$, the c.d.f. is given by

$$F_T(t) = \frac{1}{2}\bar{B}_L\left(\frac{n}{2}, \frac{1}{2}\right), \qquad L = \frac{n}{n+t^2}, \qquad t < 0, \tag{7.47}$$

which generalizes the Cauchy c.d.f. expression (7.24). Problem 7.1 details the derivation of this expression. For $t > 0$, the symmetry of the t density about zero implies that $F_T(t) = 1 - F_T(-t)$.

(13) Fisher, or F, after Sir Ronald Aylmer Fisher (1890–1962),[11] with n_1 numerator and n_2 denominator *degrees of freedom*, abbreviated F (n_1, n_2) or F_{n_1, n_2}, $n_1, n_2 \in$

---

[11] In addition to the biographical information available in the internet, Fisher plays a prominent role in the book by Salsburg (2001), where some quite interesting reading about him can be found.

$\mathbb{R}_{>0}$:

$$f_F(x; n_1, n_2) = \frac{n}{B\left(\frac{n_1}{2}, \frac{n_2}{2}\right)} \frac{(nx)^{n_1/2-1}}{(1+nx)^{(n_1+n_2)/2}}, \quad n = \frac{n_1}{n_2}, \tag{7.48}$$

and

$$F_F(x; n_1, n_2) = \overline{B}_y\left(\frac{n_1}{2}, \frac{n_2}{2}\right), \quad y = \frac{n_1 x}{n_2 + n_1 x} = \frac{nx}{1 + nx}. \tag{7.49}$$

In most statistical applications, $n_1, n_2 \in \mathbb{N}$. Similar to the t, the moments of $X \sim F$ are straightforward to calculate using the representation given in Example 9.8. The functional forms f_t and f_F admittedly do not suggest themselves as 'naturally' as do some of the previously discussed ones. These densities actually arise as particular *functions* of several independent normal random variables, the derivations of which are postponed until Chapter 8.

7.2 Other popular distributions

This section provides some detail on several more distributions which are very prominent in specific applications, but can be skipped on a first reading.

(14) Log–normal, $\text{LN}(\theta, \sigma, \zeta)$, $\theta, \zeta \in \mathbb{R}$, $\sigma \in \mathbb{R}_{>0}$:

$$f_{\text{LN}}(x; \theta, \sigma, \zeta) = \left[\sigma\sqrt{2\pi}(x-\theta)\right]^{-1} \exp\left\{-\frac{1}{2\sigma^2}[\ln(x-\theta) - \zeta]^2\right\} \mathbb{I}_{(\theta,\infty)}(x). \tag{7.50}$$

The name indicates a relation with the normal distribution, although it is a bit misleading: if $Z \sim \text{N}(0, 1)$, then the distribution of $\exp(Z)$ is log–normal. Example 7.9 below shows how to derive the density and the moments. Figure 7.9 plots the log–normal density for $\theta = 0$, $\zeta = -1/2$ and three values of σ.

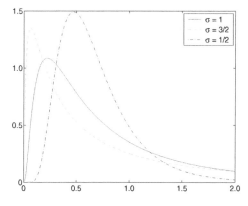

Figure 7.9 Log–normal p.d.f. (7.50) with $\theta = 0$ and $\zeta = -1/2$

Recall (7.1), which instructs how to introduce location and scale parameters into a distribution. Let $X \sim \mathrm{LN}\left(0, \sigma^2, \zeta\right)$; the p.d.f. of the scale-transformed random variable $Y = sX$ is then given by

$$
f_Y(y) = s^{-1} f_X\left(\frac{y}{s}\right) = \left[\sigma\sqrt{2\pi}\, y\right]^{-1} \exp\left[-\frac{1}{2\sigma^2}\left(\ln y - \zeta^*\right)^2\right],
$$

where $\zeta^* = \ln s + \zeta = \ln\left[s \exp(\zeta)\right]$. This shows that scaling X by s is redundant, and that the scaling is embodied in parameter ζ.

The log–normal is associated with numerous applications in a wide variety of fields; see, for example, Crow and Shimizu (1988) and Johnson, Kotz and Balakrishnan (1994).

(15) Logistic, Log $(x; \alpha, \beta)$, with $\alpha \in \mathbb{R}$ as a location and $\beta \in \mathbb{R}_{>0}$ as a scale parameter:

$$
f_{\mathrm{Log}}(x; \alpha, \beta) = \beta^{-1} \exp\left(-\frac{x - \alpha}{\beta}\right)\left[1 + \exp\left(-\frac{x - \alpha}{\beta}\right)\right]^{-2} \tag{7.51}
$$

and c.d.f.

$$
F_{\mathrm{Log}}(x) = \left[1 + \exp\left(-\frac{x - \alpha}{\beta}\right)\right]^{-1}.
$$

Figure 7.10 plots (7.51) as well as the p.d.f. of the location-zero, scale-one density

$$
f_{\mathrm{GLog}}(x; p, q, 0, 1) = \frac{1}{B(p, q)} \frac{e^{-qx}}{(1 + e^{-x})^{p+q}}, \quad p, q \in \mathbb{R}_{>0} \tag{7.52}
$$

which generalizes (7.51) (take $p = q = 1$). There are numerous applications of the logistic distribution and its generalization (see Balakrishnan (1992) and Johnson, Kotz and Balakrishnan (1994) for a full account). Moments of the logistic and the generalization (7.52) will be derived in Volume II.

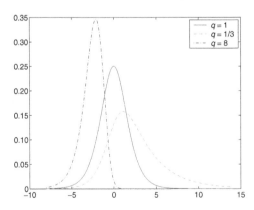

Figure 7.10   Location-zero, scale-one logistic density (7.51) (solid) and generalization (7.52) with $p = 1$

(16) Gumbel or extreme value distribution, Gum (λ): this arises when studying the behavior of the largest values of a sample. The c.d.f. is given by

$$F_{\text{Gum}}(x; a, b) = \exp\left[-\exp\left(-\frac{x-a}{b}\right)\right];$$

the location-zero, scale-one p.d.f. is

$$f_{\text{Gum}}(x; 0, 1) = \frac{d}{dx} F_{\text{Gum}}(x; 0, 1) = \exp\left(-x - e^{-x}\right),$$

and, as usual, with location and scale parameters a and b,

$$f_{\text{Gum}}(x; a, b) = \frac{1}{b} \exp\left(-z - e^{-z}\right), \quad z = \frac{x-a}{b}. \tag{7.53}$$

In some situations, the distribution of the extremes is more important than the distribution of, say, the average behavior of some process. This will be discussed in more detail in Volume II.

(17) Inverse Gaussian, IG (μ, λ), $\mu, \lambda \in \mathbb{R}_{>0}$: there are several useful parameterizations for this density, and we denote different ones by adding a subscript to the inverse Gaussian abbreviation IG. One popular parameterization is

$$f_{\text{IG}_2}(x; \mu, \lambda) = \left(\frac{\lambda}{2\pi x^3}\right)^{1/2} \exp\left[-\frac{\lambda}{2\mu^2 x}(x-\mu)^2\right] \mathbb{I}_{(0,\infty)}(x), \tag{7.54}$$

from which the c.d.f.

$$F_{\text{IG}_2}(x; \mu, \lambda) = \Phi\left[\sqrt{\frac{\lambda}{x}}\left(\frac{x}{\mu} - 1\right)\right] + \exp\left(\frac{2\lambda}{\mu}\right)\Phi\left[-\sqrt{\frac{\lambda}{x}}\left(\frac{x}{\mu} + 1\right)\right],$$

can be ascertained, where $\Phi(\cdot)$ denotes the c.d.f. of the standard normal distribution. The mean and variance are given by μ and μ^3/λ, respectively. It is easy to show that (7.54) belongs to the exponential family. Figure 7.11 plots (7.54) for several parameter values.

Recalling (7.1), a location parameter μ of a density appears in the p.d.f. as being subtracted from the ordinate x. Thus, parameter μ in (7.54) is not a location parameter because of its appearance in the fraction in the exponent term. Also, from Figure 7.11(b), it is clear that changes in μ do not result in a shift of the density. Similarly, one can see from (7.54) that λ is not a scale parameter, which is also clear from Figure 7.11(a), which shows that λ affects the overall shape of the p.d.f., most notably the skewness and tail thickness. Thus, it would appear that a scale parameter could be introduced, but this would be redundant. Letting $X \sim \text{IG}_2(\mu, \lambda)$ and $Y = aX$, the scale transformation in (7.1) reveals that

$$f_Y(y) = \frac{1}{a} f_X\left(\frac{y}{a}\right) = \left(\frac{a\lambda}{2\pi y^3}\right)^{1/2} \exp\left[-\frac{\lambda}{2\mu^2 ya}(y - \mu a)^2\right] \mathbb{I}_{(0,\infty)}(y),$$

i.e. that $Y \sim \text{IG}_2(a\mu, a\lambda)$.

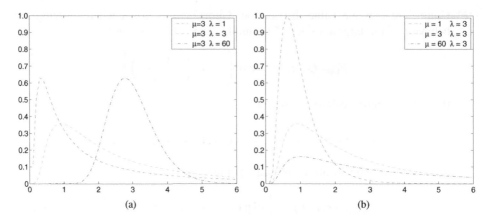

Figure 7.11   Inverse Gaussian p.d.f. (7.54)

Another useful notation for the inverse Gaussian is

$$f_{IG_1}(x; \chi, \psi) = \left(\frac{\chi}{2\pi}\right)^{1/2} e^{\sqrt{\chi\psi}} x^{-3/2} \exp\left[-\frac{1}{2}(\chi x^{-1} + \psi x)\right] \mathbb{I}_{(0,\infty)}(x), \quad (7.55)$$

for $\chi > 0$ and $\psi > 0$. This form allows it to be recognized more clearly as a special case of the *generalized inverse Gaussian*, or GIG, distribution, as will be detailed in Volume II. The moments of an inverse Gaussian random variable are also given there.

In addition to the competent overview provided by Johnson, Kotz and Balakrishnan (1994), a wealth of information on this distribution and the myriad of its applications can be found in the texts by Jørgensen (1982), Chhikara and Folks (1989), Seshadri (1994, 1999) and Forsberg (2002).

◎ ***Example 7.6***   Observe that $\lim_{\mu\to\infty} (x - \mu)^2 / \mu^2 = 1$. Thus, the limiting density in (7.54) as $\mu \to \infty$ becomes

$$f(x; \lambda) = \left(\frac{\lambda}{2\pi x^3}\right)^{1/2} \exp\left(-\frac{\lambda}{2x}\right) \mathbb{I}_{(0,\infty)}(x). \quad (7.56)$$

This is referred to as the *Lévy distribution*, after Paul Lévy. Taking $\psi = 0$ in (7.55) also yields this. The Lévy distribution is also a special case of the stable Paretian family of densities which will be discussed later (see also Example 7.15 below).   ∎

(18) Hyperbolic distribution, introduced by Barndorff-Nielsen (1977, 1978) for use in empirical studies in geology. As with the inverse Gaussian, there are several common parameterizations of the hyperbolic p.d.f. One of them, as a location-zero, scale-one p.d.f., is given by

$$f_{Hyp_2}(z; p, q, 0, 1) = C_1 \exp\left[C_2\left(p\sqrt{1 + z^2} - qz\right)\right], \quad (7.57)$$

where

$$C_1 = \frac{\sqrt{p^2 - q^2}}{2pK_1\left(p^{-2} - 1\right)}, \qquad C_2 = \frac{p^2 - 1}{\sqrt{p^2 - q^2}\, p^2}$$

for $0 < p \le 1$ and $|q| < p$, and K_1 is a Bessel function (see below). Another parameterization is

$$f_{\mathrm{Hyp}_1}(z; \phi, \gamma, \delta, \mu) = K_{\phi,\gamma,\delta} \exp\{H\},$$
$$H = -\frac{1}{2}(\phi + \gamma)\sqrt{\delta^2 + (z - \mu)^2} + \frac{1}{2}(\phi - \gamma)(z - \mu)$$

(7.58)

for $\phi, \gamma, \delta \in \mathbb{R}_{>0}$, $\mu \in \mathbb{R}$. (Another, related parameterization will be presented in Volume II.) Observe that (7.58) is asymmetric for $\phi \ne \gamma$, and symmetric otherwise, and that H is a hyperbola, hence the name. The constant of integration is $K_{\phi,\gamma,\delta}$, given by

$$K_{\phi,\gamma,\delta}^{-1} = \delta\left(\phi^{-1} + \gamma^{-1}\right)\sqrt{\phi\gamma}\, K_1\left(\delta\sqrt{\phi\gamma}\right),$$

where K_1 is the *modified Bessel function of the third kind with index* $v = 1$.

Remark: We gather here some useful facts on modified Bessel function of the third kind, $K_v(z)$, with real argument $z \in \mathbb{R}_{>0}$. It is also referred to as the Basset function, Bessel's function of the second kind of imaginary argument, MacDonald's function, and the modified Hankel function. When $v = 1$,

$$K_1(z) = z \int_0^\infty \frac{\cos t}{\left(t^2 + z^2\right)^{3/2}}\, dt.$$

In the general case,

$$K_v(z) = \frac{\Gamma(v + 1/2)(2z)^v}{\sqrt{\pi}} \int_0^\infty \frac{\cos t}{\left(t^2 + z^2\right)^{v+1/2}}\, dt, \qquad v > -\frac{1}{2}, \qquad (7.59)$$

$$K_v(z) = \frac{\sqrt{\pi}}{\Gamma(v + 1/2)} \left(\frac{z}{2}\right)^v \int_1^\infty e^{-zt}\left(t^2 - 1\right)^{v-1/2}\, dt, \qquad v > -\frac{1}{2}, \qquad (7.60)$$

and

$$K_v(z) = \frac{1}{2}\left(\frac{z}{2}\right)^v \int_0^\infty \exp\left(-t - \frac{z^2}{4t}\right) t^{-(v+1)}\, dt$$
$$= \frac{1}{2}\int_0^\infty u^{v-1} \exp\left[-\frac{z}{2}\left(\frac{1}{u} + u\right)\right]\, du, \qquad (7.61)$$

having used $u = z/(2t)$. See, for example, Lebedev (1972, Section 5.10, in particular pp. 114 and 119).

It can be shown that $K_v(z) = K_{-v}(z)$, and that

$$\lim_{z\downarrow 0} K_v(z) = \Gamma(v)2^{v-1}z^{-v}, \quad v > 0, \tag{7.62}$$

(Lebedev, 1972, p. 111), which for $v = 1$ reduces to

$$\lim_{z\to 0} K_1(z) = z^{-1}. \tag{7.63}$$

In Matlab, $K_v(z)$ is computed with the built-in function besselk(v,z). Robert (1990) discusses several applications of Bessel functions in probability and statistics.   ∎

Using (7.63), for $\phi = \gamma = 1$, $\mu = 0$ and $\delta \to 0$,

$$f(z; 1, 1, \delta, 0) = \frac{1}{2\delta K_1(\delta)} \exp\left(-\sqrt{\delta^2 + z^2}\right) \to \frac{1}{2}e^{-|z|},$$

which is the Laplace distribution.

Now let $\mu = 0$, set $\phi = \gamma = \delta$, and let $\delta \to \infty$. Then, using the Taylor series approximation

$$\sqrt{\delta^2 + z^2} \approx \delta + \frac{1}{2}\frac{z^2}{\delta},$$

which becomes increasingly accurate as δ grows, (7.58) simplifies to

$$\frac{1}{2\delta K_1(\delta^2)} \exp\left(-\delta\sqrt{\delta^2 + z^2}\right) = \frac{\exp(-\delta^2)}{2\delta K_1(\delta^2)} \exp\left(-\frac{1}{2}z^2\right),$$

which has the same kernel as the standard normal density. This suggests that

$$\lim_{\delta\to\infty} \frac{\exp(-\delta^2)}{2\delta K_1(\delta^2)} = \frac{1}{\sqrt{2\pi}} \quad \text{or} \quad K_1(z) \stackrel{z\to\infty}{\to} B(z) := \sqrt{2\pi}\,\frac{\exp(-z)}{2\sqrt{z}}$$

(in the sense that the ratio of $K_1(z)$ to $B(z)$ converges to unity as z grows), which appears true via numeric calculation. More generally, by letting

$$\delta \to \infty, \quad \delta(\phi\gamma)^{-1/2} \to \sigma^2 > 0 \quad \text{and} \quad (\phi - \gamma) \to 0,$$

density (7.58) approaches the $N(\mu, \sigma^2)$ p.d.f.

The moments of the hyperbolic distribution are given in Volume II, where they are derived in the context of a more general distribution which nests the hyperbolic. Parameter δ in (7.58) is 'almost' a scale parameter in the sense that, for $Z \sim f(\phi, \gamma, \delta, \mu)$, use of (7.1) shows that $Z/\delta \sim f(\delta\phi, \delta\gamma, 1, \mu)$. To eliminate the dependence of ϕ and γ on δ, we use the following parameterization, as suggested in Küchler, Neumann, Sørensen and Streller (1999) and similar to that in

Barndorff-Nielsen, Blæsild, Jensen and Sørensen (1985):

$$p = \left(1 + \delta\sqrt{\phi\gamma}\right)^{-\frac{1}{2}}, \qquad q = \frac{\phi - \gamma}{\phi + \gamma}p,$$

where $0 < p \le 1$ and $|q| < p$. Solving (use Maple) gives

$$\phi = \frac{1 - p^2}{\delta p^2}\sqrt{\frac{p + q}{p - q}}, \qquad \gamma = \frac{1 - p^2}{\delta p^2}\sqrt{\frac{p - q}{p + q}}.$$

Substituting and simplifying shows that

$$\phi + \gamma = -2\frac{p^2 - 1}{\sqrt{p^2 - q^2}\,p\delta}, \qquad \phi - \gamma = -2\frac{p^2 - 1}{\sqrt{p^2 - q^2}}\frac{q}{\delta p^2},$$

$$\phi^{-1} + \gamma^{-1} = -2\delta\frac{p^3}{\sqrt{p^2 - q^2}\left(p^2 - 1\right)}, \qquad \sqrt{\phi\gamma} = \frac{\left(1 - p^2\right)}{\delta p^2},$$

which leads to

$$f(z; p, q, \delta, \mu) = \frac{\sqrt{p^2 - q^2}}{(2\delta p)\,K_1\left(p^{-2} - 1\right)} \tag{7.64}$$

$$\times \exp\left[\left(\frac{p^2 - 1}{\sqrt{p^2 - q^2}\,p^2\delta}\right)\left(p\sqrt{\delta^2 + (z - \mu)^2} - q\,(z - \mu)\right)\right],$$

for $\mu \in \mathbb{R}$, $\delta \in \mathbb{R}_{>0}$, $0 < p \le 1$ and $|q| < p$, which is the location–scale version of (7.57). Observe that the parameter range of p and q forms a triangle; it is referred to as the *shape triangle* (Barndorff-Nielsen, Blæsild, Jensen and Sørensen, 1985).

For $q = 0$, the density is symmetric; for $q < 0$ $(q > 0)$ it is left (right) skewed. As $p \to 0$, $f(z)$ approaches a normal density, and with $p = 1$, $q = 0$, it coincides with the (location–scale) Laplace. Hence, this parameterization has the further advantage that p and q can be interpreted as measures of kurtosis and skewness, respectively. Observe also that a 'skewed Laplace' distribution is possible with $p = 1$ and $q \ne 0$.

Figure 7.12 illustrates the hyperbolic p.d.f. (7.64) for several parameter constellations.

7.3 Univariate transformations

If X is a continuous random variable with p.d.f. f_X and g is a continuous differentiable function with domain contained in the range of X and $dg/dx \ne 0 \ \forall x \in \mathcal{S}_X$, then f_Y,

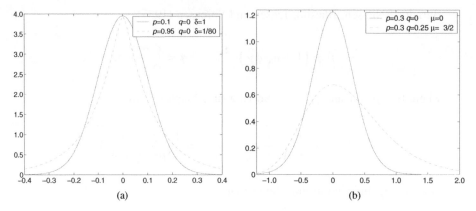

Figure 7.12 Hyperbolic p.d.f. (7.64) with (a) $\mu = 0$ and (b) $\delta = 1$

the p.d.f. of $Y = g(X)$, can be calculated by

$$f_Y(y) = f_X(x) \left| \frac{dx}{dy} \right|, \qquad (7.65)$$

where $x = g^{-1}(y)$ is the inverse function of Y. This can be intuitively understood by observing that

$$f_X(x) \, \Delta \, x \approx \Pr[X \in (x, x + \Delta \, x)] \approx \Pr[Y \in (y, y + \Delta \, y)] \approx f_Y(y) \, \Delta \, y$$

for small $\Delta \, x$ and $\Delta \, y$, where $\Delta \, y = g(x + \Delta \, x) - g(x)$ depends on g, x and $\Delta \, x$.

More formally, first note that, if $dg/dx > 0$, then $\Pr[g(X) \leq y] = \Pr[X \leq g^{-1}(y)]$ and, if $dg/dx < 0$, then $\Pr[g(X) \leq y] = \Pr[X \geq g^{-1}(y)]$ (plot $g(x)$ versus x to quickly see this). Now, if $dg/dx > 0$, then differentiating

$$F_Y(y) = \Pr[g(X) \leq y] = \Pr[X \leq g^{-1}(y)] = \int_{-\infty}^{g^{-1}(y)} f_X(x) \, dx$$

with respect to y gives[12]

$$f_Y(y) = f_X[g^{-1}(y)] \frac{dg^{-1}(y)}{dy} = f_X(x) \frac{dx}{dy},$$

recalling that $x = g^{-1}(y)$. Similarly, if $dg/dx < 0$, then differentiating

$$F_Y(y) = \Pr[g(X) \leq y] = \Pr[X \geq g^{-1}(y)] = \int_{g^{-1}(y)}^{\infty} f_X(x) \, dx$$

---

[12] Recall that, if $I = \int_{\ell(y)}^{h(y)} f(x) \, dx$, then

$$\frac{\partial I}{\partial y} = f[h(y)] \frac{dh}{dy} - f[\ell(y)] \frac{d\ell}{dy}.$$

This is a special case of Leibniz' rule; see Section A.3.4 for derivation and details.

gives

$$f_Y(y) = -f_X\left(g^{-1}(y)\right)\frac{dg^{-1}(y)}{dy} = -f_X(x)\frac{dx}{dy}.$$

Transformation (7.65) is perhaps best remembered by the statement

$$\boxed{f_X(x)\,dx \approx f_Y(y)\,dy}$$

and that $f_Y(y) > 0$.

Our description so far has involved one-to-one transformations from the entire support of X, and are sometimes referred to as such. The following section provides several examples of the technique, while the subsequent section generalizes this to the many-to-one case.

7.3.1 Examples of one-to-one transformations

◎ **Example 7.7** Let $U \sim \text{Unif}(0, 1)$ and define $Y = -\ln U$. Then, with $u = \exp(-y)$, (7.65) implies

$$f_Y(y) = f_U(u)\left|\frac{du}{dy}\right| = \mathbb{I}_{(0,1)}\left(\exp(-y)\right)\left(e^{-y}\right) = e^{-y}\mathbb{I}_{(0,\infty)}(y),$$

so that $Y \sim \text{Exp}(1)$. ■

⊛ **Example 7.8** (location–scale family) Let X have density f_X. Then for $\mu \in \mathbb{R}$ and $\sigma \in \mathbb{R}_{>0}$, it follows directly from (7.65) with $x = (y - \mu)/\sigma$ that

$$\boxed{Y = \sigma X + \mu, \qquad f_Y(y) = f_X(x)\left|\frac{dx}{dy}\right| = \frac{1}{\sigma}f_X\left(\frac{y-\mu}{\sigma}\right),}$$

a simple result of utmost importance. ■

⊖ **Example 7.9** If Z is standard normal, then the density of $X = \exp(Z\sigma + \zeta) + \theta$ is, from (7.65), given by

$$f_X(x) = f_Z(z)\frac{dz}{dx} = \frac{1}{\sqrt{2\pi}}\frac{1}{(x-\theta)\sigma}\exp\left\{-\frac{1}{2\sigma^2}[\ln(x-\theta) - \zeta]^2\right\}\mathbb{I}_{(\theta,\infty)}(x).$$

Using (4.36), the rth moment of $X - \theta$ is given by

$$\mathbb{E}\left[(X - \theta)^r\right] = \mathbb{E}\left[(\exp(Z\sigma + \zeta))^r\right] = \mathbb{E}\left[\exp(rZ\sigma + r\zeta)\right]$$

$$= \exp(r\zeta)\int_{-\infty}^{\infty}\exp(r\sigma Z)f_Z(z)\,dz$$

$$= \exp(r\zeta)\exp\left(\frac{1}{2}r^2\sigma^2\right) = \exp\left(r\zeta + \frac{1}{2}r^2\sigma^2\right),$$

where $\mathbb{E}\left[e^{tZ}\right]$ is derived in Problem 7.17. Thus, with $w = \exp(\sigma^2)$,

$$\mathbb{E}\left[X - \theta\right] = \exp\left(\zeta + \frac{1}{2}\sigma^2\right) = e^\zeta w^{1/2}$$

and, from (4.40),

$$\mathbb{V}\left(X - \theta\right) = e^{2\zeta} w^2 - e^{2\zeta} w = e^{2\zeta} w\left(w - 1\right).$$

Higher moments can be found in Johnson, Kotz and Balakrishnan (1994, p. 212). ∎

⊖ **Example 7.10** A stick is broken at a point which is equally likely to be anywhere along the length of the stick. Let the r.v. R denote the ratio of the shorter to the longer of the resulting segments. Assume without loss of generality that the stick has length one. Let the r.v. X be the break-point, i.e. the length of the left segment. If $X < 1/2$, then the ratio is $X/(1 - X)$, while if $X > 1/2$, then the ratio is $(1 - X)/X$. The expected ratio is, using (4.36), given by

$$\mathbb{E}\left[R\right] = \int_0^{1/2} \frac{x}{1 - x} dx + \int_{1/2}^1 \frac{1 - x}{x} dx = 2 \int_{1/2}^1 \frac{1 - x}{x} dx = 2 \ln 2 - 1 \approx 0.3863.$$

It might be more reasonable to assume that the stick tends to get broken near the middle. To incorporate this, assume that $X \sim$ Beta(a, a) for $a \geq 1$. As a increases, the break-point converges to the center of the stick and $\mathbb{E}\left[R\right] \to 1$. As a function of a, $\mathbb{E}\left[R\right]$ is given by

$$\frac{1}{B\left(a, a\right)} \int_0^{1/2} x^{a-1} \left(1 - x\right)^{a-1} \frac{x}{1 - x} dx + \frac{1}{B\left(a, a\right)} \int_{1/2}^1 x^{a-1} \left(1 - x\right)^{a-1} \frac{1 - x}{x} dx$$

$$= \frac{2}{B\left(a, a\right)} \int_0^{1/2} x^{a+1-1} \left(1 - x\right)^{a-1-1} dx$$

$$= 2 \frac{B_{1/2}\left(a + 1, a - 1\right)}{B\left(a, a\right)}.$$

In particular, for $a = 3$,

$$2 \frac{\Gamma\left(6\right)}{\Gamma\left(3\right) \Gamma\left(3\right)} \int_0^{1/2} x^3 \left(1 - x\right) dx = \frac{9}{16} = 0.5625.$$

The mth raw moment is similarly obtained, giving

$$\mathbb{E}\left[R^m\right] = 2 \frac{B_{1/2}\left(a + m, a - m\right)}{B\left(a, a\right)}, \quad a > m.$$

Thus, the variance exists for $a > 2$ and can be calculated from the relation $\mathbb{V}\left(R\right) = \mathbb{E}\left[R^2\right] - \mathbb{E}\left[R\right]^2$. Figure 7.13 plots the expected value along the stick with plus and minus twice the variance for a between 2.01 and 100. ∎

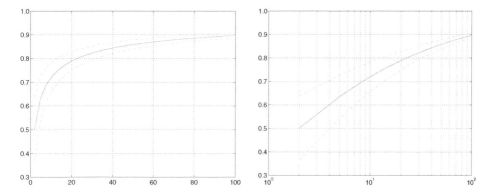

Figure 7.13 (a) $\mathbb{E}[R]$ (solid) and $\mathbb{E}[R] \pm 2\mathbb{V}(R)$ (dashed) versus a, $a = 2.01, 3.01, \dots, 100.01$; (b) uses a logarithmic (base 10) scale

7.3.2 Many-to-one transformations

It is sometimes necessary to split the range of integration at points where g is not differentiable. This is sometimes referred to as a *many-to-one transformation*. Instead of stating a general theorem, the following examples will illustrate the idea, all using the most common situation, in which the support needs to be split into two pieces. Further detail and generalizations are provided by Rohatgi (1976, pp. 73–4) and Glen, Leemis and Drew (1997).

Example 7.11 (Example 7.10 cont.) In the broken stick example above, the formula for R depends on X, i.e. $R = X/(1 - X)$ if $X < 1/2$ and $R = (1 - X)/X$ if $X > 1/2$. To derive the density of R, first split the support of X as $(0, 1/2]$ and $(1/2, 1)$ so that

$$f_R(r) = f_X(x) \left| \frac{dx}{dr} \right| \mathbb{I}_{(0,1/2]}(x) + f_X(x) \left| \frac{dx}{dr} \right| \mathbb{I}_{(1/2,1)}(x).$$

For $x < 1/2$, $r = x/(1 - x)$, $x = r/(1 + r)$ and $dx = (1 + r)^{-2} dr$. Likewise, for $x > 1/2$, $r = (1 - x)/x$, $x = 1/(1 + r)$ and $dx = -(1 + r)^{-2} dr$. In both cases, $0 < r < 1$. Then $B(a, a) f_R(r)$ is given by

$$\left(\frac{r}{1+r} \right)^{a-1} \left(\frac{1}{1+r} \right)^{a-1} \left(\frac{1}{1+r} \right)^2 + \left(\frac{1}{1+r} \right)^{a-1} \left(\frac{r}{1+r} \right)^{a-1} \left(\frac{1}{1+r} \right)^2$$

or

$$f_R(r) = \frac{2}{B(a, a)} r^{a-1} (1 + r)^{-2a} \mathbb{I}_{(0,1)}(r).$$

Problem 7.4 verifies that $\int r f_R(r) \, dr$ yields the same expression for $\mathbb{E}[R]$ as was found using the direct method in Example 7.10. ∎

⊛ ***Example 7.12***   Let $Y = X^2$ and $X \sim N(0, 1)$. With $y = x^2$, split up the x region as $x = \pm\sqrt{y}$ so that, from (7.65),

$$f_Y(y) = f_X(x) \left| \frac{dx}{dy} \right|$$

$$= \frac{1}{\sqrt{2\pi}} \exp\left[-\frac{1}{2} \left(\sqrt{y}\right)^2 \right] \frac{y^{-1/2}}{2} \mathbb{I}_{(0,\infty)} \left(\sqrt{y}\right)$$

$$+ \frac{1}{\sqrt{2\pi}} \exp\left[-\frac{1}{2} \left(-\sqrt{y}\right)^2 \right] \frac{y^{-1/2}}{2} \mathbb{I}_{(-\infty,0)} \left(-\sqrt{y}\right)$$

$$= \frac{1}{2\sqrt{2\pi}} y^{-1/2} e^{-y/2} \mathbb{I}_{(0,\infty)}(y) + \frac{1}{2\sqrt{2\pi}} y^{-1/2} e^{-y/2} \mathbb{I}_{(0,\infty)}(y)$$

$$= \frac{1}{\sqrt{2\pi}} y^{-1/2} e^{-y/2} \mathbb{I}_{(0,\infty)}(y),$$

which, comparing with (7.41), shows that $Y \sim \chi_1^2$. ∎

⊖ ***Example 7.13***   Let $Y = X^2$ and $X \sim \text{Lap}(0, \sigma)$. In a similar way to Example 7.12, from (7.65),

$$f_Y(y) = \frac{1}{2\sigma} \exp\left(-\left| -\sqrt{y} \right| \sigma^{-1} \right) \left| -\frac{1}{2} y^{-1/2} \right| \mathbb{I}_{(-\infty,0)} \left(-\sqrt{y}\right)$$

$$+ \frac{1}{2\sigma} \exp\left(-\left| \sqrt{y} \right| \sigma^{-1} \right) \left| \frac{1}{2} y^{-1/2} \right| \mathbb{I}_{(0,\infty)} \left(\sqrt{y}\right)$$

$$= \left(2\sigma \sqrt{y}\right)^{-1} \exp\left(-\frac{\sqrt{y}}{\sigma} \right) \mathbb{I}_{(0,\infty)}(y).$$

Similarly, for $Z = |X|$,

$$f_Z(z) = \frac{1}{2\sigma} \exp\left(-\left| -z \right| \sigma^{-1} \right) |-1| \mathbb{I}_{(-\infty,0)}(-z) + \frac{1}{2\sigma} \exp\left(-\left| z \right| \sigma^{-1} \right) |1| \mathbb{I}_{(0,\infty)}(z)$$

$$= \sigma^{-1} \exp\left(-\frac{z}{\sigma} \right) \mathbb{I}_{(0,\infty)}(z)$$

or $Z \sim \text{Exp}(1/\sigma)$. Thus $V = |X|/\sigma = Z/\sigma \sim \text{Exp}(1)$. This should not be surprising, given the nature of the Laplace (double exponential) distribution. ∎

⊖ ***Example 7.14***   Let $X \sim \text{Cauchy}(0, 1)$ and consider the random variable $Y = X^2$. The c.d.f. of Y can be computed as

$$F_Y(y) = \Pr(Y < y) = \Pr(X^2 < y) = \Pr\left(-\sqrt{y} < X < \sqrt{y}\right)$$

$$= F_X\left(\sqrt{y}\right) - F_X\left(-\sqrt{y}\right)$$

while the density of Y is given by $\partial F_Y(y)/\partial y$, or

$$f_Y(y) = \frac{1}{2\sqrt{y}} \left[f_X\left(\sqrt{y}\right) + f_X\left(-\sqrt{y}\right) \right] = \pi^{-1} y^{-\frac{1}{2}} (1+y)^{-1} \mathbb{I}_{(0,\infty)}(y).$$

To verify that this is a proper density, let

$$u = (1+y)^{-1}, \ y = (1-u)\,u^{-1} \text{ and } dy/du = -u^{-2}$$

so that

$$\int_0^\infty f_Y(y)\,dy = \pi^{-1}\int_1^0 \frac{-u\sqrt{u}}{\sqrt{1-u}}\frac{1}{u^2}du = \pi^{-1}\int_0^1 u^{-\frac{1}{2}}(1-u)^{-\frac{1}{2}}\,du$$

$$= \pi^{-1} B\left(\frac{1}{2},\frac{1}{2}\right) = \pi^{-1}\frac{\Gamma\left(\frac{1}{2}\right)\Gamma\left(\frac{1}{2}\right)}{\Gamma(1)} = 1.$$

Note that $Y \sim F(y; 1, 1)$, i.e. an F random variable with one numerator and one denominator degree of freedom. A similar calculation shows that

$$\mathbb{E}\left[y^h\right] = \pi^{-1}\Gamma\left(\frac{1}{2}-h\right)\Gamma\left(\frac{1}{2}+h\right), \qquad |h| < 1/2. \qquad \blacksquare$$

⊙ **Example 7.15** Let $Z \sim N(0, 1)$. To compute the density of $Y = Z^{-2}$, let $z = +y^{-1/2}$ and $dz = -(1/2)\,y^{-3/2}$, so that the symmetry of the normal distribution and (7.65) imply

$$f_Y(y) = 2 f_Z(z)\left|\frac{dz}{dy}\right| = 2\frac{1}{\sqrt{2\pi}}\exp\left(-\frac{1}{2}y^{-1}\right)\frac{1}{2}y^{-3/2}\mathbb{I}_{(0,\infty)}(y)$$

$$= (2\pi)^{-1/2}\,y^{-3/2}e^{-1/(2y)}\mathbb{I}_{(0,\infty)}(y).$$

This is again the Lévy distribution, which was shown to be a limiting case of the inverse Gaussian distribution in Example 7.6. Also, for $y > 0$,

$$\Pr(Y \le y) = \Pr\left(Z^{-2} \le y\right) = \Pr\left(Z^2 > y^{-1}\right)$$

$$= 1 - \Pr\left(-y^{-1/2} \le Z \le y^{-1/2}\right) = 2\Phi\left(-y^{-1/2}\right). \qquad \blacksquare$$

7.4 The probability integral transform

An important result for continuous random variable X is the *probability integral transform* of X. This is defined by $Y = F_X(X)$, where $F_X(t) = \Pr(X \le t)$. Here, $F_X(X)$ is not to be interpreted as $\Pr(X \le X) = 1$, but rather as a random variable Y defined as the transformation of X, the transformation being the function F_X.

Assume that F_X is strictly increasing. Then $F_X(x)$ is a one-to-one function for $x \in (0, 1)$ so that

$$F_Y(y) = \Pr(Y \le y) = \Pr\left[F_X(X) \le y\right] = \Pr\left[F_X^{-1}(F_X(X)) \le F_X^{-1}(y)\right]$$

$$= \Pr\left[X \le F_X^{-1}(y)\right] = F_X\left[F_X^{-1}(y)\right] = y, \qquad (7.66)$$

showing that $Y \sim \text{Unif}(0, 1)$.

Remark: For values of y outside of $(0, 1)$, note that, if $y \geq 1$, then

$$1 = \Pr\left[F_X(X) \leq y\right] = F_Y(y)$$

and, if $y \leq 0$, then

$$0 = \Pr\left[F_X(X) \leq y\right],$$

as F_X is a c.d.f.

The result also holds if F_X is continuous but only nondecreasing. This still rules out discrete jumps in the c.d.f., but allows the c.d.f. to be constant over some interval subset of $[0, 1]$. In that case, however, F_X is no longer one-to-one and F_X^{-1} is no longer unique. This can be remedied by defining $F_X^{-1}(y)$ such that a unique value is returned. The usual choice is to take $F_X^{-1}(y)$ as $\inf\left[x : F_X(x) \geq y\right]$. ∎

7.4.1 Simulation

The probability integral transform is of particular value for simulating random variables. By applying F^{-1} to both sides of $Y = F_X(X) \sim \text{Unif}(0, 1)$, we see that, if $Y \sim \text{Unif}(0, 1)$, then $F^{-1}(Y)$ is a realization of a random variable with c.d.f. F. For example, from the exponential c.d.f. $y = F(x) = F_{\text{Exp}}(x; 1) = 1 - e^{-x}$, we have $F^{-1}(y) = -\ln(1 - y)$. Thus, taking $Y \sim \text{Unif}(0, 1)$, it follows from the probability integral transform that $-\ln(1 - Y) \sim \text{Exp}(1)$. As $1 - Y$ is also uniformly distributed, we can use $-\ln(Y)$ instead. Recall also Example 7.7, which showed that, if $U \sim \text{Unif}(0, 1)$, then $Z = -\ln U \sim \text{Exp}(1)$.

As discussed in Section 2.2.2, we *assume* the availability of a viable uniform random number generator, so that a stream of i.i.d. uniform r.v.s, $U_1, \ldots, U_n$, can be generated quickly, from which we can compute $Z_1 = -\ln U_1, Z_2 = -\ln U_2, \ldots, Z_n = -\ln U_n$, which are then i.i.d. exponential r.v.s with rate one. From Example 7.8, taking $X_i = Z_i/\lambda$ for some value $\lambda > 0$ results in a set of i.i.d. r.v.s $X_i \overset{\text{i.i.d.}}{\sim} \text{Exp}(\lambda)$.

In principle, this method can be applied to generate any random variable with a well-defined c.d.f. F and computable inverse c.d.f. F^{-1}. For example, even though closed-form expressions do not exist, algorithms have been developed to compute F^{-1} corresponding to standard normal and Student's t distributions to machine precision (in Matlab, these are computed by calling functions `norminv` and `tinv`, respectively), or at least to a reasonably high degree of accuracy, such as Moro's (1995) method for the normal c.d.f., as illustrated above.[13]

Thus, such r.v.s are easily generated; for example, calling `t=tinv(rand(1000, 1),4);` in Matlab yields a vector of 1000 i.i.d. Student's t r.v.s with four degrees of freedom. While this method is theoretically always applicable, it relies on having at ones disposal fast and accurate numeric methods for the inverse c.d.f., which may not always be available. It turns out that, for almost all random variables of interest, there

---

[13] In Matlab, type `help stats` and then look under the headings 'Critical Values of Distribution Functions' and 'Random Number Generators' to see which distributions are supported in this regard by Matlab.

are other ways of efficiently generating an i.i.d. sample (see, for example, Thompson (2000) and Ross (2002)) for excellent textbook presentations. In Matlab, for example, there are the functions `randn` and `trnd` for generating i.i.d. normal and Student's *t* r.v.s, respectively, which use more efficient methods than c.d.f. inversion to produce the sample.

7.4.2 Kernel density estimation

Along with the ability to generate r.v.s, it is desirable to see a graphical depiction of them. This can be accomplished with a standard histogram, but there is a well-developed methodology of how to optimally 'connect the bars' of a histogram to produce a smooth curve. This results in an approximation, or estimate, of the underlying continuous p.d.f., and is scaled such that the area under the curve is unity, as a true p.d.f. requires.

This is the method of *kernel density estimation*. It necessitates specification of a *smoothing constant*, which parallels the bin-width choice for a histogram. Often, in practice, one uses a *data driven* value for the smoothing constant, i.e. a (usually simple) function of the data which will produce reasonable results for data sets whose underlying theoretical distribution function satisfies certain properties.

For now, we just mention that there are several good books on kernel density estimation; highly recommended places to start include Silverman (1986), Wand and Jones (1995) and Bowman and Azzalini (1997). Program Listing 7.2 shows a program to compute the kernel density estimate of a given set of data which incorporates common choices for the data driven smoothing constant.

```
function [pdf,grd] = kerngau(x,h)
nobs=length(x);
switch h
  case -1, hh=1.06 * std(x) * nobs^(-0.2);
  case -2, hh=0.79 * iqr(x) * nobs^(-0.2);
  case -3, hh=0.90 * min([std(x) iqr(x)/1.34]) * nobs^(-0.2);
  otherwise, hh=h;
end
disp(['smoothing parameter = ',num2str(hh)])
x=sort(x); spdf=zeros(length(x),1);
for k=1:nobs
  xx=(x-x(k))/hh;  pdf=(1/sqrt(2*pi))*exp(-(xx.^2)/2);  spdf=spdf+pdf;
end
pdf=spdf/(nobs*hh); grd=x;
```

Program Listing 7.2 Kernel density estimate for univariate data. Parameter h controls the degree of smoothing. Three data driven choices for it are available in the program; pass h=-1 for approximate normal data, h=-2 for fat-tailed data and h=-3 for a compromise between the two. Function std is part of Matlab, and computes the sample standard deviation; function iqr is part of Matlab's statistics toolbox, and computes the sample interquartile range, which is the difference between the 25th and 75th percentiles of the data

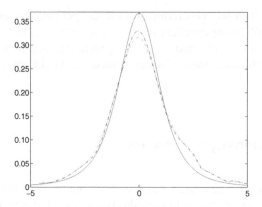

Figure 7.14 True p.d.f. of the Student's t density with three degrees of freedom (solid), and the kernel density estimates based on a simulated sample with 300 observations using a data driven smoothing parameter which is optimal for approximately normal data (dashed) and for fat-tailed data (dash dot)

⊖ *Example 7.16* Figure 7.14 plots the t_3 p.d.f., i.e. Student's t p.d.f. for three degrees of freedom, and two kernel density estimates based on (the same) random sample of 300 simulated i.i.d. t_3 data points. These were drawn using Matlab's function `trnd`.

The dashed line corresponds to the kernel density with smoothing parameter suitable for Gaussian data (pass parameter `h` in program Listing 7.2 as -1) and the dash–dot line corresponds to the fat-tailed choice (pass `h` as -2). The latter is indeed somewhat better in the center of the density, though otherwise, there is little difference between the two. ∎

7.5 Problems

Never seem more learned than the people you are with. Wear your learning like a pocket watch and keep it hidden. Do not pull it out to count the hours, but give the time when you are asked. (Lord Chesterfield)

If you have an important point to make, don't try to be subtle or clever. Use a pile driver. Hit the point once. Then come back and hit it again. Then hit it a third time – a tremendous whack. (Sir Winston Churchill)

7.1. ★ ★ Let $T \sim t_n$.

(a) Show that $\mathbb{E}[T] = 0$ if $n > 1$ and does not otherwise exist.

(b) Calculate $\mathbb{E}\left[\,|T|^k\,\right]$.

(c) Determine which moments exist.

(d) Show (7.46).

(e) Show (7.47).

7.2. Derive a closed-form expression for the upper $(1 - \alpha)$th quantile of the $F(2, n)$ distribution.

7.3. Show directly that the Gam $(2, 1)$ density $f_X(x) = xe^{-x}\mathbb{I}_{(0,\infty)}(x)$ is valid.

7.4. ★ In Example 7.11, verify that the density of R integrates to one and use it to compute $\mathbb{E}[R]$.

7.5. ★ If $C \sim \chi_\nu^2$, calculate the density of $X \sim \sqrt{C/\nu}$. Recalling Example 7.12, note that, for $\nu = 1$, X is the absolute value of a standard normal r.v., which is also referred to as the *folded normal*. Compute the mean and variance of X in this case.

7.6. ★ Let $X \sim N(\mu, \sigma^2)$. Derive the density of $Y = X^2$ and show that

$$\mathbb{E}[Y] = \sigma^2 + \mu^2 \quad \text{and} \quad \mathbb{V}(Y) = 2\sigma^2 (\sigma^2 + 2\mu^2). \tag{7.67}$$

Moments $\mathbb{E}\left[|X/\sigma|^r\right]$ for $r \in \mathbb{R}_{>0}$ will be derived in Volume II in the context of the noncentral chi-square distribution.

7.7. ★ Let $X \sim F(n_1, n_2)$. For $a > 0$ and $b > 0$, derive the density of $Y = aX^b$, give an expression in terms of the incomplete beta function for the c.d.f. of Y and calculate $\mathbb{E}[Y^r]$ as a function of the raw moments of X.

7.8. Occasionally, there is a need to evaluate $\mathbb{E}\left[X^{-1}\right]$. Piegorsch and Casella (1985)[14] have shown that, if X has a continuous density $f_X(x)$ with support $S = (0, \infty)$, then a sufficient condition for the existence of $\mathbb{E}\left[X^{-1}\right]$ is

$$\lim_{x \to 0+} \frac{f_X(x)}{x^r} < \infty \tag{7.68}$$

for some $r > 0$. Under the same conditions, Hannan (1985) showed that

$$\lim_{x \to 0+} |\log x|^r f_X(x) < \infty$$

for some $r > 1$ is also sufficient for the existence of $\mathbb{E}\left[X^{-1}\right]$.

Let $Z \sim \text{Gam}(\alpha, \beta)$. For what α does $\mathbb{E}\left[Z^{-1}\right]$ exist?

7.9. ★ ★ Let α, β be positive real numbers with $\alpha > 2$ and $Z \sim \text{Gam}(\alpha, \beta)$, with

$$f_Z(z; \alpha, \beta) = \frac{\beta^\alpha}{\Gamma(\alpha)} z^{\alpha-1} \exp(-\beta z) \mathbb{I}_{(0,\infty)}(z).$$

Random variable $Y = 1/Z$ is said to follow an *inverse gamma distribution*, denoted by $Y \sim \text{IGam}(\alpha, \beta)$.

---

[14] See also Piegorsch and Casella (2002) for an update which extends their original results.

(a) Compute $\mathbb{E}[Z]$ and $\mathbb{E}[Z^{-1}]$.

(b) Derive $f_Y(y; \alpha, \beta)$, the p.d.f. of Y, and interpret parameter β.

(c) Derive a useful expression for the c.d.f. of Y.

(d) Derive a useful expression for μ'_r, the rth raw moment of Y. Simplify for $\mathbb{E}[Y]$ and $\mathbb{V}(Y)$.

(e) When are $\mathbb{E}[Z]$ and $\mathbb{E}[Z^{-1}]$ reciprocals of one another, i.e. what conditions on α and β are necessary so that $\mathbb{E}[1/Z] = 1/\mathbb{E}[Z]$?

7.10. ★ Show that

$$I = \int_0^2 x e^{-(x-1)^2} dx \approx 1.49$$

without numerically integrating I, but using only a value of Φ.

7.11. Evaluate $I = \int_{-\infty}^{\infty} x^2 e^{-x^2} dx$.

7.12. ★ Let X be a positive continuous r.v. with p.d.f. f_X and c.d.f. F_X.

(a) Show that a necessary condition for $\mathbb{E}[X]$ to exist is $\lim_{x \to \infty} x[1 - F_X(x)] = 0$. Use this to show that the expected value of a Cauchy random variable does not exist.

(b) Prove via integration by parts that $\mathbb{E}[X] = \int_0^{\infty} [1 - F_X(x)] dx$ if $\mathbb{E}[X] < \infty$.

7.13. ★ Let X be a continuous random variable with p.d.f. f_X, c.d.f. F_X and finite nth raw moment, $n \geq 1$.

(a) Assume that X is positive, i.e. $\Pr(X > 0) = 1$. Prove that

$$\mathbb{E}[X] = \int_0^{\infty} [1 - F_X(x)] dx \qquad (7.69)$$

by expressing $1 - F_X$ as an integral, and reversing the integrals.

(b) Generalize (7.69) to

$$\mathbb{E}[X^n] = \int_0^{\infty} n x^{n-1} [1 - F_X(x)] dx. \qquad (7.70)$$

(c) Relaxing the positivity constraint, generalize (7.69) to

$$\mathbb{E}[X] = \int_0^{\infty} [1 - F_X(x)] dx - \int_{-\infty}^0 F_X(x) dx \,. \qquad (7.71)$$

Even more generally, it can be shown that

$$\mathbb{E}\left[(X - \mu)^n\right] = \int_{\mu}^{\infty} n\, (x - \mu)^{n-1}\,[1 - F_X\,(x)]\,dx$$

$$- \int_{-\infty}^{\mu} n\, (x - \mu)^{n-1}\, F_X\,(x)\,dx, \qquad (7.72)$$

which the reader can also verify.

7.14. Let $X \sim \text{Beta}\,(a, b)$.

(a) For $a > 2$ and $b > 2$, show that the mode of X is given by $x = \frac{a-1}{a+b-2}$.

(b) Write an equation in x for the median of X when $a = 3$ and $b = 2$ and verify that $x \approx 0.614272$.

(c) Find a and b such that X is symmetric with a standard deviation of 0.2.

(d) ★ If I believe that $X \sim \text{Beta}\,(a, b)$ with mode at 5/8 and a standard deviation of approximately 0.1, what integer values of parameters a and b could I use?

7.15. Let $X \sim \text{N}\,(0, 1)$. Calculate $\mathbb{E}\,|X|^p$, $p \geq 0$, and explicitly for $p = 0, 1$ and 2.

7.16. ★ Random variable X has a *generalized exponential distribution*, denoted $X \sim \text{GED}\,(p)$, if X has p.d.f.

$$f_X\,(x;\,p) = \frac{p}{2\Gamma\,\left(p^{-1}\right)}\exp\left(-\,|x|^p\right), \qquad p \in \mathbb{R}_{>0}. \qquad (7.73)$$

Let $Y = \sigma X$ for $\sigma > 0$, i.e. σ is a scale parameter.

(a) Verify that $f_X\,(x;\,p)$ integrates to one.

(b) Derive an expression for the c.d.f. of X in terms of the incomplete gamma function.

(c) Give the density of Y and parameter values p and σ such that the Laplace and normal distributions arise as special cases.

(d) Compute $\mathbb{V}\,(Y)$ and verify that it agrees with the special cases of Laplace and normal.

(e) Show that, if $X \sim \text{GED}\,(p)$, then $Y = |X|^p \sim \text{Gam}\,(1/p)$. This fact can be used for simulating GED r.v.s (which is not common to most software), via simulation of gamma r.v.s (for which most statistical programming software packages do have routines). In particular, generate $Y \sim \text{Gam}\,(1/p)$, set $X = Y^{1/p}$, and then attach a random sign to X, i.e. with probability 1/2, multiply X by negative one.

7.17. ★ Let $Z \sim \text{N}\,(0, 1)$ and $X \sim \text{N}\,(\mu, \sigma^2)$. Compute $M_Z\,(t) := \mathbb{E}\left[e^{tZ}\right]$ and $M_X\,(t) = \mathbb{E}\left[e^{tX}\right]$.

7.18. ★ ★ Let $Z \sim N(0, 1)$. Compute $I = \mathbb{E}\big[\exp\big(t\,(Z+b)^2\big)\big]$.

7.19. ★ Consider p.d.f. $f_X(x; n, m) = k\big(1 + |x|^{2n}\big)^{-m}$ for $n \in \mathbb{N}$, $m \in \mathbb{R}_{>0}$ and k is the constant of integration.

(a) Derive k. (Hint: try substituting $y = x^{2n}$ followed by $z = (1 + y)^{-1}$ (or combine the two).)

(b) Derive the absolute moments $\mu_r := \mathbb{E}\big[|X|^r\big]$ of $f_X(x; n, m)$ and specify when they exist.

7.20. ★ Let $X \sim F(n_1, n_2)$ and define

$$B = \frac{\frac{n_1}{n_2} X}{1 + \frac{n_1}{n_2} X}.$$

Show that $B \sim \text{Beta}(n_1/2, n_2/2)$ and $(1 - B) \sim \text{Beta}(n_2/2, n_1/2)$.

7.21. ★ Show that

$$(1 - a)^{-1/2} = \frac{1}{\sqrt{\pi}} \sum_{i=0}^{\infty} \frac{\Gamma(i + 1/2)}{i!} a^i.$$

7.22. ★ Let $\overline{\Phi}$ be the survivor function of a standard normal random variable. As in Smith (2004, p. 12), show that, for $k > 0$,

$$\lim_{x \to \infty} \frac{\overline{\Phi}(x + k/x)}{\overline{\Phi}(x)} = e^{-k}.$$

This result arises in the study of *extreme value theory* (see Smith (2004) and Coles (2001) for detailed, accessible accounts).

7.23. ★ Consider the c.d.f. of the beta distribution. Several numerical methods have been developed specifically for its evaluation (see, for example, Kennedy and Gentle, 1980; and Press, Teukolsky, Vetterling and Flannery, 1989), but here we consider two 'direct' ways: (i) direct numeric integration, and (ii) use of the identity

$$\sum_{k=j}^{n} \binom{n}{k} y^k (1 - y)^{n-k} = \frac{n!}{(j - 1)!\,(n - j)!} \int_0^y x^{j-1} (1 - x)^{n-j}\,dx \qquad (7.74)$$

(this is proved in Volume II in the context of order statistics; see also Example 1.28). Construct a program which compares the accuracy and computation time of the above two methods, along with the pre-programmed method `betainc` in Matlab (which uses one of the specific numeric algorithms). For the numerical integration, use the `quadl` function and, for the combinatoric exact solution, use the `binocdf` function, which evaluates the c.d.f. of the binomial distribution.

7.24. ★ There have been many numerical methods proposed for the fast approximate evaluation of the standard normal c.d.f. $\Phi(x)$ for $x > 0$. The references to the following cited methods (and further details) can be found in Johnson, Kotz and Balakrishnan (1994). From Pólya (1945),

$$\Phi(x) \approx \frac{1}{2}\left\{1 + \left[1 - \exp\left(-\frac{2x^2}{\pi}\right)\right]^{1/2}\right\}. \tag{7.75}$$

From Hastings (1955),

$$\Phi(x) \approx 1 - \left(a_1 t + a_2 t^2 + a_3 t^3\right)\varphi(x), \tag{7.76}$$

where $\varphi(x)$ denotes the standard normal p.d.f., $a_1 = 0.4361836$, $a_2 = -0.1201676$, $a_3 = 0.937298$ and $t = (1 + 0.33267x)^{-1}$. From Hart (1966),

$$\Phi(x) \approx 1 - \frac{c_x}{x\sqrt{2\pi}}\left(1 - \frac{f_x}{d_x + \sqrt{d_x^2 + f_x c_x}}\right), \tag{7.77}$$

where $c_x = \exp\left(-x^2/2\right)$, $d_x = x\sqrt{\frac{\pi}{2}}$, $f_x = \left(1 + bx^2\right)^{1/2}\left(1 + ax^2\right)^{-1}$, $a = p/(2\pi)$, $b = p^2/(2\pi)$ and $p = 1 + \sqrt{1 + 6\pi - 2\pi^2}$. Burr (1967) proposed

$$\Phi(x) \approx \frac{1}{2}[G(x) + 1 - G(-x)] \tag{7.78}$$

where $G(x) = 1 - \left[1 + (\alpha + \beta x)^c\right]^k$ with $\alpha = 0.644693$, $\beta = 0.161984$, $c = 4.874$ and $k = -6.158$, while Derenzo (1977) derived

$$1 - \Phi(x) \approx \begin{cases} \frac{1}{2}\exp\left[-\frac{(83x+351)x+562}{(703/x)+165}\right], & \text{if} \quad 0 \le x \le 5.5, \\ \frac{1}{x\sqrt{2\pi}}\exp\left(-\frac{x^2}{2} - \frac{0.94}{x^2}\right), & \text{if} \quad x > 5.5. \end{cases} \tag{7.79}$$

Page (1977) considered the simple function

$$\Phi(x) \approx \frac{e^{2y}}{1 + e^{2y}} \tag{7.80}$$

where $y = a_1 x\left(1 + a_2 x^2\right)$, $a_1 = 0.7988$ and $a_2 = 0.04417$ and Moran (1980) gave the approximation

$$\Phi(x) \approx \frac{1}{2} + \pi^{-1}\sum_{n=0}^{12} k_n \sin\left(\frac{\sqrt{2}}{3}mx\right) \tag{7.81}$$

where $k_n = \exp\left\{-m^2/9\right\}/m$ and $m = n + 0.5$.

Each method varies in terms of complexity and numerical accuracy. Construct a program, say cdfnorm(type,x), which plots the relative error compared to the 'exact' algorithm used in Matlab, `normcdf(x)`. Variable `type` contains any of the numbers 1 to 7 corresponding to the above methods and `x` is a vector of x coordinates with default 0.01:0.01:5. For each plot, report the largest discrepancy, and for which x it occurs. Example output is shown in Figure 7.15.

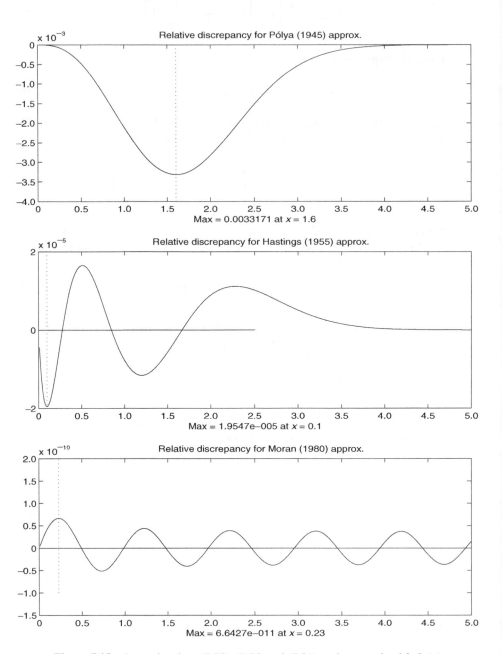

Figure 7.15   Approximations (7.75), (7.76) and (7.81) to the normal c.d.f. $\Phi(x)$

8

Joint and conditional random variables

Common sense is the collection of prejudices acquired by age eighteen.

(Albert Einstein)

Reproduced by permission of Mathematical Association of America

A great many people think they are thinking when they are merely re-arranging their prejudices.　(William James)

When dealing with people, remember you are not dealing with creatures of logic, but with creatures of emotion, creatures bristling with prejudice, and motivated by pride and vanity.　(Dale Carnegie)

Reproduced by permission of Mathematical Association of America

Some basic aspects of multivariate random variables were introduced in Chapter 5; in this chapter, we present additional concepts which arise in the multivariate setting, most notably conditioning. Similar to the case for univariate distributions, the monographs by Johnson, Kotz and Balakrishnan (1997) for discrete multivariate distributions and Kotz, Balakrishnan and Johnson (2000) for continuous multivariate distributions are indispensable works of reference containing a vast amount of information and cataloging most of the important multivariate distributions in use. The books by Fang, Kotz and Ng (1990) and Krzanowski and Marriott (1994) are also highly recommended.

8.1　Review of basic concepts

In some statistical applications, the (continuous) joint distribution of several parameters can be obtained, but interest centers only on a subset of them. Those which are not desired get *integrated out* to obtain the marginal density of the parameters of interest. Recall that, for bivariate density $f_{X,Y}$, the marginal densities of X and Y are

$$f_X(x) = \int_{-\infty}^{\infty} f_{X,Y}(x, y)\, \mathrm{d}y, \quad f_Y(y) = \int_{-\infty}^{\infty} f_{X,Y}(x, y)\, \mathrm{d}x,$$

with marginal c.d.f.s

$$F_X(x) = \int_{-\infty}^{x} \int_{-\infty}^{\infty} f_{X,Y}(x, y)\, dy\, dx = \int_{-\infty}^{x} f_X(x)\, dx,$$

$$F_Y(y) = \int_{-\infty}^{y} \int_{-\infty}^{\infty} f_{X,Y}(x, y)\, dx\, dy = \int_{-\infty}^{y} f_Y(y)\, dy.$$

The following two examples illustrate how one might start with a joint distribution of two r.v.s and then obtain their marginals.

⊖ **Example 8.1**   Assume you are confronted with the joint density

$$f_{X,Y}(x, y) = k\left(x^2 + y^2\right) \mathbb{I}_{(0,1)}(x) \mathbb{I}_{(0,1)}(y), \tag{8.1}$$

where k is a constant which needs to be determined.

Imagine, perhaps, that a sports researcher is concerned with modeling the success rate, based on his newly-designed fitness program, of people who are attempting to simultaneously gain a certain percentage increase in muscle strength and lose a certain percentage of body fat. Let X be the probability of reaching the strength goal, and let Y be the probability of reaching the fat-loss goal. He has postulated that their relationship can be expressed by (8.1).

Value k is given by the reciprocal of

$$\frac{1}{k} = \int_{-\infty}^{\infty} \int_{-\infty}^{\infty} \left(x^2 + y^2\right) \mathbb{I}_{(0,1)}(x) \mathbb{I}_{(0,1)}(y)\, dx\, dy$$

$$= \int_0^1 \int_0^1 x^2\, dx\, dy + \int_0^1 \int_0^1 y^2\, dx\, dy = 2 \int_0^1 \int_0^1 x^2\, dx\, dy = \frac{2}{3},$$

i.e. $k = 3/2$.

The marginal p.d.f. of X is computed as

$$f_X(x) = \frac{3}{2} \int_{-\infty}^{\infty} \left(x^2 + y^2\right) \mathbb{I}_{(0,1)}(x) \mathbb{I}_{(0,1)}(y)\, dy = \left(\frac{1}{2} + \frac{3}{2}x^2\right) \mathbb{I}_{(0,1)}(x)$$

which, in this case, does not depend on y. By symmetry,

$$f_Y(y) = \left(\frac{1}{2} + \frac{3}{2}y^2\right) \mathbb{I}_{(0,1)}(y).$$

The lower-order moments are straightforward to compute, with

$$\mathbb{E}[Y] = \mathbb{E}[X] = \int_0^1 x f_X(x)\, dx = \int_0^1 x \left(\frac{1}{2} + \frac{3}{2}x^2\right) dx = \frac{5}{8},$$

$$\mathbb{E}[Y^2] = \mathbb{E}[X^2] = \int_0^1 x^2 \left(\frac{1}{2} + \frac{3}{2}x^2\right) dx = \frac{7}{15}$$

and, recalling the definition of covariance from (5.24),

$$\mathrm{Cov}\,(X,\,Y) = \mathbb{E}\,[(X - \mathbb{E}\,[X])\,(Y - \mathbb{E}\,[Y])]$$

$$= \frac{3}{2} \int_0^1 \int_0^1 \left(x - \frac{5}{8}\right)\left(y - \frac{5}{8}\right)(x^2 + y^2)\,dx\,dy = -\frac{1}{64}.$$

Thus,

$$\sigma_X^2 = \sigma_Y^2 = \frac{7}{15} - \left(\frac{5}{8}\right)^2 = \frac{73}{960} \approx 0.076$$

and

$$\mathrm{Corr}\,(X,\,Y) = \frac{\mathrm{Cov}\,(X,\,Y)}{\sigma_X \sigma_Y} = -\frac{15}{73} \approx -0.2055.$$

As this is negative, one would say that the (simplistic) model used to describe the success rates X and Y captures the (say, empirically observed) fact that it is difficult simultaneously to reach both goals of gaining muscle strength and losing body fat.

Assume that the researcher is interested in knowing $\mathrm{Pr}\,(X \geq Y/3)$. This entails a double integral, and thus has 2! ways of being calculated; recall the discussion on multiple integrals in Section 5.1.2. Based on a sketch of the region of integration in the plane (which the reader is strongly encouraged to construct on his or her own), we can write

$$\mathrm{Pr}\,(X \geq Y/3) = \int_0^1 \int_{y/3}^1 f_{X,Y}\,(x,\,y)\,dx\,dy = \frac{47}{54}$$

or

$$\mathrm{Pr}\,(X \geq Y/3) = \int_0^{1/3} \int_0^{3x} f_{X,Y}\,(x,\,y)\,dy\,dx + \int_{1/3}^1 \int_0^1 f_{X,Y}\,(x,\,y)\,dy\,dx$$

$$= \frac{1}{18} + \frac{22}{27} = \frac{47}{54}.$$

Observe that integrating first with respect to y causes the range of integration to change, so that two integrals (and more work) are necessary. ∎

⊖ **Example 8.2** Let X and Y be continuous r.v.s such that, jointly, their coordinates are uniformly distributed in the triangular region given by the points (0,0), (0,1) and (1,0). The support of X and Y is shown in Figure 8.1 as the large triangle; a point in the plane $(X,\,Y)$ is equally likely to be anywhere in the triangular region. The dashed line going through (0, 1) and (1, 0) is $y = 1 - x$, showing that the sum of $X + Y$ cannot exceed one. From this description, the joint density of X and Y can be written as

$$f_{X,Y}\,(x,\,y) = 2\,\mathbb{I}_{(0,1)}\,(x)\,\mathbb{I}_{(0,1)}\,(y)\,\mathbb{I}_{(0,1)}\,(x + y).$$ (8.2)

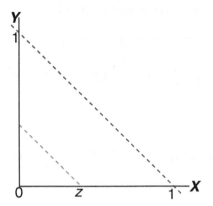

Figure 8.1 $\Pr(X + Y \leq z)$ as in (8.3)

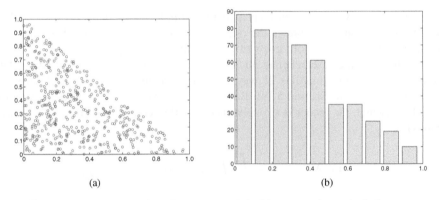

(a)   (b)

Figure 8.2 (a) Scatterplot of X and Y and (b) histogram of marginal of X

It is instructive in this case to simulate r.v.s which follow distribution (8.2). To do so, we use the symmetry of X and Y and the uniformity of the density in the triangular region. First let X and Y be independent standard uniform r.v.s, and accept the pair X, Y if $X + Y < 1$. A set of approximately $n/2$ pairs is generated and plotted with the code:

```
n=1000; xy=[rand(n,1) , rand(n,1)]; good=find( sum(xy')<1 );
use=xy(good,:); plot(use(:,1),use(:,2),'ro'), axis([0 1 0 1])
```

as shown in Figure 8.2(a).

In addition to enhancing our understanding of what it means for two r.v.s to be jointly uniformly distributed in the triangle (0,0), (0,1) and (1,0), the ability to simulate them also allows us easily to look at the marginal distributions. The marginal density of X is plotted as a histogram by calling hist(use(:,1),10), as shown in Figure 8.2(b) (and having used the code in Listing 2.2 for altering the appearance of the histogram). It indicates that $f_X(x)$ should be linear in x with negative slope. This is algebraically confirmed and made precise as follows. Combining $\mathbb{I}_{(0,1)}(y)$ and $\mathbb{I}_{(0,1)}(x+y)$ imply $0 < y < 1 - x$, so that

$$f_X(x) = 2\,\mathbb{I}_{(0,1)}(x) \int_0^{1-x} dy = 2\,(1-x)\,\mathbb{I}_{(0,1)}(x)\,.$$

Now consider computing an expression for the c.d.f. of $Z = X + Y$. This can be accomplished from geometric arguments: Figure 8.1 shows the line $y = z - x$, $0 < z < 1$, which forms a 'small triangle' with the origin. Then

$$\Pr(Z \le z) = \frac{\text{area of little triangle}}{\text{area of big triangle}} = 2 \cdot \frac{z^2/2}{1/2} = z^2\,\mathbb{I}_{(0,1)}(z)\,. \qquad (8.3)$$

This can, of course, be confirmed algebraically: From the joint density (8.2), for $0 \le z \le 1$,

$$\Pr(Z \le z) = \int_0^z \int_0^{z-x} f_{X,Y}(x, y)\,dy\,dx = 2 \int_0^z (z-x)\,dx = z^2\,\mathbb{I}_{(0,1)}(z),$$

yielding the same answer. ∎

Writing the marginal density functions as integral expressions involving the joint density is obviously a trivial act, while resolving the integrals into closed-form expressions can, in general, be quite challenging. The next example shows a case where the integrals initially look difficult but are not.

Example 8.3 (Pierce and Dykstra, 1969) From the bivariate distribution

$$f_{X_1,X_2}(x_1, x_2) = \frac{1}{2\pi} e^{-\frac{1}{2}\left(x_1^2 + x_2^2\right)} \left[1 + x_1 x_2 e^{-\frac{1}{2}\left(x_1^2 + x_2^2\right)}\right],$$

depicted in Figure 8.3, the marginals are computed as follows. Write

$$f_{X_2}(x_2) = \int_{-\infty}^{\infty} f_{X_1,X_2}(x_1, x_2)\,dx_1 = I_1 + I_2,$$

where

$$I_1 = \int_{-\infty}^{\infty} \frac{1}{2\pi} e^{-\frac{1}{2}\left(x_1^2 + x_2^2\right)}\,dx_1 = \frac{1}{2\pi} e^{-\frac{1}{2}x_2^2} \int_{-\infty}^{\infty} e^{-\frac{1}{2}x_1^2}\,dx_1 = \phi(x_2)\,,$$

the density of the normal distribution, recalling that the latter integral equals $\sqrt{2\pi}$. Next,

$$I_2 = \frac{1}{2\pi} x_2 e^{-x_2^2} \int_{-\infty}^{\infty} x_1 e^{-x_1^2}\,dx_1 = 0$$

because of symmetry. Clearly, $f_{X_1}(x_1)$ will also be normal. ∎

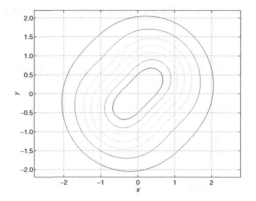

Figure 8.3 Contour plot of density $f_{X_1,X_2}(x_1, x_2)$ in Example 8.3

An important fact to keep in mind is that knowledge of all the marginals of a distribution does not necessarily allow the joint distribution to be determined. A simple example in the bivariate case is given by Mood, Graybill and Boes (1974, pp. 142–143); we will see others later.

8.2 Conditional distributions

The importance of conditional probability was illustrated in Section 3.2; the notion of conditional random variables is no less relevant. We illustrate these in the bivariate case, with more general multivariate extensions following analogously. Let $F_{X,Y}$ denote a bivariate c.d.f. with corresponding p.m.f. or p.d.f. $f_{X,Y}$ and define A and B to be the events $\{(x, y) : x \in A_0\}$ and $\{(x, y) : y \in B_0\}$ respectively where, for both events, x and y are in $\mathcal{S} \subseteq \mathbb{R}^2$, the support of (X, Y). From basic principles, the conditional probability $\Pr(x \in A_0 \mid y \in B_0)$ is given by

$$\frac{\Pr(x \in A_0, \, y \in B_0)}{\Pr(y \in B_0)} = \frac{\Pr(A \cap B)}{\Pr(B)}, \tag{8.4}$$

assuming $\Pr(B) > 0$, where

$$\Pr(B) = \Pr(x \in \mathbb{R}, \, y \in B_0) = \int_{y \in B_0} dF(y)$$

$$= \begin{cases} \int_{y \in B_0} f_Y(y)\,dy, & \text{if } (X, Y) \text{ is continuous}, \\ \sum_{y \in B_0} f_Y(y), & \text{if } (X, Y) \text{ is discrete}, \end{cases}$$

is evaluated from the marginal p.m.f. or p.d.f. of Y.

Most often, the event B_0 is not a particular range but rather a point y_0 in $\mathcal{S}_Y$. For this it is helpful to develop the discrete and continuous cases separately.

8.2.1 Discrete case

From (8.4), $\Pr(x \in A_0 \mid y \in B_0)$ is given by

$$\Pr(x \in A_0 \mid y \in B_0) = \frac{\sum_{x \in A_0} \sum_{y \in B_0} f_{X,Y}(x, y)}{\sum_{x \in \mathbb{R}} \sum_{y \in B_0} f_{X,Y}(x, y)}$$

$$= \sum_{x \in A_0} \frac{\sum_{y \in B_0} f_{X,Y}(x, y)}{\sum_{y \in B_0} f_Y(y)} =: \sum_{x \in A_0} f_{X \mid Y \in B_0}(x \mid B_0),$$

where $f_{X \mid Y \in B_0}$ is defined to be the *conditional p.m.f. given* $y \in B_0$.

Now let event $B = \{(x, y) : y = y_0\}$. If event $A = \{(x, y) : x \le x_0\}$, then the conditional c.d.f. of X given $Y = y_0$ is given by

$$\Pr(A \mid B) = \frac{\Pr(X \le x, Y = y_0)}{\Pr(Y = y_0)} = \sum_{i=-\infty}^{x} \frac{f_{X,Y}(i, y_0)}{f_Y(y_0)} =: F_{X \mid Y = y_0}(x \mid y_0)$$

and, likewise, if A is the event $\{(x, y) : x = x_0\}$, then the conditional p.m.f. of X given $Y = y_0$ is

$$\Pr(A \mid B) = \frac{\Pr(X = x, Y = y_0)}{\Pr(Y = y_0)} = \frac{f_{X,Y}(x, y_0)}{f_Y(y_0)} =: f_{X \mid Y}(x \mid y_0). \qquad (8.5)$$

ⓞ **Example 8.4** Let X_1 and X_2 be independently distributed r.v.s. Interest centers on the conditional distribution of X_1 given that their sum, $S = X_1 + X_2$, is some particular value, say s. From (8.5) with $X_1 = X$ and $S = Y$,

$$\Pr(X_1 = x \mid S = s) = \frac{\Pr(X_1 = x, S = s)}{\Pr(S = s)} = \frac{\Pr(X_1 = x, X_2 = s - x)}{\Pr(S = s)}.$$

Using the fact that X_1 and X_2 are independent, this is

$$\Pr(X_1 = x \mid S = s) = \frac{\Pr(X_1 = x) \Pr(X_2 = s - x)}{\Pr(S = s)}.$$

Now consider the following special cases.

1. For $X_i \overset{\text{i.i.d.}}{\sim} \text{Bin}(n, p)$. From the discussion in Section 6.4, $(X_1 + X_2) \sim \text{Bin}(2n, p)$ so that

$$
\frac{\Pr(X_1 = x)\Pr(X_2 = s - x)}{\Pr(X_1 + X_2 = s)} = \frac{\binom{n}{x}p^x(1 - p)^{n-x}\binom{n}{s-x}p^{s-x}(1 - p)^{n-s+x}}{\binom{2n}{s}p^s(1 - p)^{2n-s}}
$$

$$
= \frac{\binom{n}{x}\binom{n}{s-x}}{\binom{2n}{s}},
$$

i.e. $X_1 \mid (X_1 + X_2)$ is hypergeometric.

2. For $X_i \overset{\text{i.i.d.}}{\sim} \text{Geo}(p)$ using density (4.16). From the discussion in Section 6.4, $(X_1 + X_2) \sim \text{NBin}(r = 2, p)$ so that

$$
\frac{\Pr(X_1 = x)\Pr(X_2 = s - x)}{\Pr(X_1 + X_2 = s)} = \frac{p(1 - p)^x\, p(1 - p)^{s-x}}{\binom{r+s-1}{s}p^r(1 - p)^s} = \frac{1}{(1 + s)}\mathbb{I}_{(0,1,\dots,s)}(x),
$$

i.e. $X_1 \mid (X_1 + X_2)$ is discrete uniform.

3. For $X_i \overset{\text{ind}}{\sim} \text{Poi}(\lambda_i)$. Recall from Example 6.10 that $(X_1 + X_2) \sim \text{Poi}(\lambda_1 + \lambda_2)$ so that

$$
\frac{\Pr(X_1 = x)\Pr(X_2 = s - x)}{\Pr(X_1 + X_2 = s)} = \frac{e^{-\lambda_1}\lambda_1^x}{x!}\frac{e^{-\lambda_2}\lambda_2^{s-x}}{(s - x)!}\bigg/ \frac{e^{-\lambda_1 - \lambda_2}(\lambda_1 + \lambda_2)^s}{s!}
$$

$$
= \frac{s!}{x!\,(s - x)!}\frac{\lambda_1^x\lambda_2^{s-x}}{(\lambda_1 + \lambda_2)^s}
$$

$$
= \binom{s}{x}p^x(1 - p)^{s-x}, \tag{8.6}
$$

for $p = \lambda_1/(\lambda_1 + \lambda_2)$ and $x = 0, 1, \dots, s$, i.e. $X_1 \mid (X_1 + X_2)$ is binomial. ∎

8.2.2 Continuous case

Let continuous r.v.s X and Y have a joint p.d.f. $f = f_{X,Y}$ and marginals f_X and f_Y. Then, for a measurable event $B_0 \subset S_Y$ such that $\int_{y \in B_0} f_Y(y)\,dy > 0$, $\Pr(x \in A_0 \mid y \in B_0)$ is given by

$$
\frac{\int_{x \in A_0}\int_{y \in B_0} f_{X,Y}(x, y)\,dy\,dx}{\int_{x \in \mathbb{R}}\int_{y \in B_0} f_{X,Y}(x, y)\,dy\,dx} = \int_{x \in A_0}\frac{\int_{y \in B_0} f_{X,Y}(x, y)\,dy}{\int_{y \in B_0} f_Y(y)\,dy}\,dx
$$

$$
=: \int_{x \in A_0} f_{X \mid Y \in B_0}(x \mid B_0)\,dx
$$

and $f_{X \mid Y \in B_0}$ is referred to as the *conditional p.d.f. given* $y \in B_0$.

If B_0 is just a point in the support $\mathcal{S}_Y$, some care is required. In particular, writing

$$\Pr(X \leq x \mid Y = y_0) = \frac{\Pr(X \leq x, Y = y_0)}{\Pr(Y = y_0)}$$

yields the meaningless $0/0$, so that alternative definitions are needed. These are given by

$$F_{X|Y=y_0}(x \mid y_0) := \int_{-\infty}^{x} \frac{f(t, y_0)}{f(y_0)} dt,$$

$$f_{X|Y=y_0}(x \mid y_0) := \frac{\partial}{\partial x} F_{X|Y}(x \mid y_0) = \frac{f(x, y_0)}{f(y_0)},$$

which should appear quite natural in the light of the results for the discrete case. The equation for $F_{X|Y=y_0}$ can be justified by expressing $\Pr(x \in A_0 \mid Y = y_0)$ as

$$\lim_{h \to 0^+} \int_{x \in A_0} \frac{\int_{y_0}^{y_0+h} f_{X,Y}(t, y) dy}{\int_{y_0}^{y_0+h} f_Y(y) dy} dt = \int_{x \in A_0} \frac{h f_{X,Y}(x, y_0)}{h f_Y(y_0)} dx = \int_{x \in A_0} \frac{f_{X,Y}(x, y_0)}{f_Y(y_0)} dx,$$

$$(8.7)$$

using the mean value theorem for integrals (see Section A.2.3.1) and assuming $f_Y(y_0) > 0$ and that the order of operations $\lim_h$ and $\int_x$ can be exchanged. The c.d.f. $F_{X|Y}(x_0 \mid y_0)$ follows with $A = \{(x, y) : x \leq x_0\}$.

Remark: The construction of conditional p.d.f.s, c.d.f.s and conditional expectations (in Section 8.2.3 below) when conditioning on events of measure zero is more delicate than it appears and gives rise to what is called the *Borel paradox*. Although we will not encounter its effects in subsequent chapters, nor have we developed the required (measure–theoretic) mathematics to fully appreciate it, the interested reader is encouraged to have a look at some very simple demonstrations of it (see DeGroot (1986, pp. 171–4), Poirier (1995, p. 55(2.5.7)), Proschan and Presnell (1998), Casella and Berger (2002, p. 202(4.61), pp. 204–5, 358(7.17)), and the references therein). ∎

Usually $f_{X|Y=y_0}$ and $F_{X|Y=y_0}$ are not given for a specific value y_0 but rather as a general function of y, denoted respectively as

$$f_{X|Y}(x \mid y) := \frac{f_{X,Y}(x, y)}{f_Y(y)}, \qquad (8.8a)$$

and

$$F_{X|Y}(x \mid y) := \int_{-\infty}^{x} f_{X|Y}(t \mid y) dt = \frac{\int_{-\infty}^{x} f_{X,Y}(t, y) dt}{f_Y(y)}, \qquad (8.8b)$$

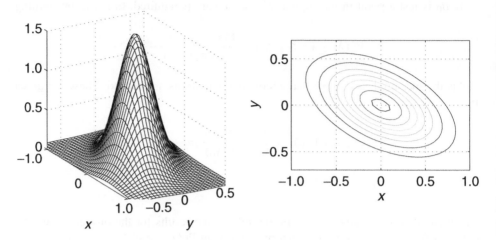

Figure 8.4 Mesh and contour plot of density $f_{X,Y}(x, y)$ in Example 8.5

which can then be evaluated for any particular $y = y_0$ of interest. In this case, we speak simply of the *conditional p.d.f.* and *conditional c.d.f.*, given as a function of y. Similar expressions arise in the discrete case.

⊖ **Example 8.5** Let the joint density of (X, Y) be given by

$$f_{X,Y}(x, y) = \frac{3\sqrt{3}}{\pi} \exp\left(-4x^2 - 6xy - 9y^2\right) \tag{8.9}$$

and depicted in Figure 8.4.

The marginal p.d.f. of Y is

$$f_Y(y) = \int_{-\infty}^{\infty} f_{X,Y}(x, y)\, dx = \frac{3\sqrt{3}}{\pi} \exp\left(-9y^2\right) \int_{-\infty}^{\infty} \exp\left[-\left(4x^2 + 6xy\right)\right] dx$$

$$= \frac{3\sqrt{3}}{\pi} \exp\left(-9y^2 + \frac{9}{4}y^2\right) \int_{-\infty}^{\infty} \exp\left[-\left(2x + \frac{3}{2}y\right)^2\right] dx,$$

obtained by the handy trick of *completing the square*, i.e.

$$4x^2 + 6xy + \frac{9}{4}y^2 - \frac{9}{4}y^2 = \left(2x + \frac{3}{2}y\right)^2 - \frac{9}{4}y^2.$$

However, with $\sigma^2 = 1/8$,

$$\int_{-\infty}^{\infty} \exp\left[-\left(2x + \frac{3}{2}y\right)^2\right] dx = \int_{-\infty}^{\infty} \exp\left[-\frac{1}{2\sigma^2}\left(x + \frac{3}{4}y\right)^2\right] dx = \sqrt{2\pi/8} = \frac{\sqrt{\pi}}{2},$$

which follows because the integrand is the $N\left(-\frac{3}{4}y, \sigma^2\right)$ kernel. Thus

$$f_Y(y) = \frac{3\sqrt{3}}{2\sqrt{\pi}} \exp\left(-\frac{27}{4}y^2\right). \tag{8.10}$$

Similarly, $f_X(x) = \sqrt{3/\pi} \exp\left(-3x^2\right)$ so that, using (8.8a) and (8.10),

$$f_{X|Y}(x \mid y) = \frac{f_{X,Y}(x, y)}{f_Y(y)} = \frac{2}{\sqrt{\pi}} \exp\left[-\frac{1}{4}(4x + 3y)^2\right],$$

$$f_{Y|X}(y \mid x) = \frac{f_{X,Y}(x, y)}{f_X(x)} = \frac{3}{\sqrt{\pi}} \exp\left[-(x + 3y)^2\right].$$

These are shown in Figures 8.5 and 8.6. ■

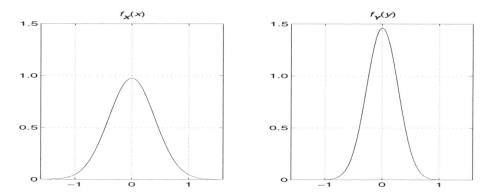

Figure 8.5 Marginal densities in Example 8.5

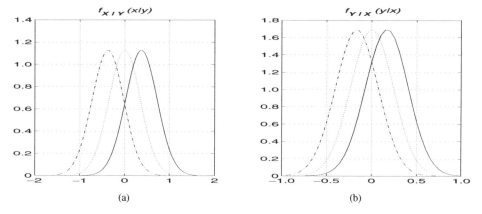

Figure 8.6 Conditional densities in Example 8.5: (a) $y = -0.5$ (solid), $y = 0$ (dotted), $y = 0.5$ (dashdot); (b) $x = -0.5$ (solid), $x = 0$ (dotted), $x = 0.5$ (dashdot)

Density (8.9) is a special case of the *bivariate normal*, which will be detailed in Volume II in the larger context of the multivariate normal, where all our essential results will be proven. We now just state the p.d.f. and some basic facts. If random variables X and Y are bivariate normally distributed, then we write

$$\begin{pmatrix} X \\ Y \end{pmatrix} \sim N\left(\begin{bmatrix} \mu_1 \\ \mu_2 \end{bmatrix}, \begin{bmatrix} \sigma_1^2 & \rho\sigma_1\sigma_2 \\ \rho\sigma_1\sigma_2 & \sigma_2^2 \end{bmatrix} \right), \tag{8.11}$$

where $\text{Corr}(X, Y) = \rho$. Their density is

$$f_{X,Y}(x, y) = \frac{1}{2\pi\sigma_1\sigma_2 (1 - \rho^2)^{1/2}} \exp\left[-\frac{N}{2(1 - \rho^2)} \right], \tag{8.12}$$

where

$$N = \left(\frac{x - \mu_1}{\sigma_1} \right)^2 - 2\rho \left(\frac{x - \mu_1}{\sigma_1} \right) \left(\frac{y - \mu_2}{\sigma_2} \right) + \left(\frac{y - \mu_2}{\sigma_2} \right)^2.$$

The marginal distributions are

$$X \sim N\left(\mu_1, \sigma_1^2\right), \quad Y \sim N\left(\mu_2, \sigma_2^2\right), \tag{8.13}$$

and the conditional distributions are

$$\begin{aligned} X \mid Y &\sim N\left(\mu_1 + \rho\sigma_1\sigma_2^{-1}(y - \mu_2), \sigma_1^2(1 - \rho^2) \right), \\ Y \mid X &\sim N\left(\mu_2 + \rho\sigma_2\sigma_1^{-1}(x - \mu_1), \sigma_2^2(1 - \rho^2) \right). \end{aligned} \tag{8.14}$$

We see that our special case (8.9) is the same as (8.12) when taking $\mu_1 = \mu_2 = 0$, $\sigma_1^2 = 1/6$, $\sigma_2^2 = 2/27$ and $\rho = -1/2$. Thus, from (8.13) the marginals are $X \sim N(0, 1/6)$ and $Y \sim N(0, 2/27)$, or, simplifying,

$$f_X(x) = \sqrt{3/\pi} \exp\left(-3x^2\right), \quad f_Y(y) = \frac{3\sqrt{3}}{2\sqrt{\pi}} \exp\left(-\frac{27}{4}x^2\right),$$

which agree with those obtained directly. With

$$\rho = \frac{-1/18}{\sqrt{(1/6)(2/27)}} = -\frac{1}{2},$$

the distribution of $X \mid (Y = y)$ is, from (8.14),

$$N\left(0 + \rho\frac{\sqrt{1/6}}{\sqrt{2/27}}(y - 0), \frac{1}{6}(1 - \rho^2)\right) \text{ or } N(-3y/4, 1/8),$$

i.e. simplifying,

$$f_{X|Y}(x \mid y) = \frac{2}{\sqrt{\pi}} \exp\left\{-4\left(x + \frac{3}{4}y\right)^2\right\},$$

as obtained directly.

Remark: In matrix notation, the bivariate normal density is written as

$$f_{\mathbf{Y}}(\mathbf{y}) = \frac{1}{2\pi\sqrt{|\Sigma|}} \exp\left[-\frac{1}{2}(\mathbf{y} - \boldsymbol{\mu})'\Sigma^{-1}(\mathbf{y} - \boldsymbol{\mu})\right],$$

where $\mathbf{Y} = (Y_1, Y_2)'$, $\mathbf{y} = (y_1, y_2)'$, and

$$\boldsymbol{\mu} = \begin{bmatrix} \mu_1 \\ \mu_2 \end{bmatrix}, \quad \Sigma = \begin{bmatrix} \sigma_1^2 & \rho\sigma_1\sigma_2 \\ \rho\sigma_1\sigma_2 & \sigma_2^2 \end{bmatrix}.$$

Thus, for (8.9), as

$$4x^2 + 6xy + 9y^2 = \begin{bmatrix} x & y \end{bmatrix} \begin{bmatrix} 4 & 3 \\ 3 & 9 \end{bmatrix} \begin{bmatrix} x \\ y \end{bmatrix},$$

we have

$$\Sigma^{-1} = 2\begin{bmatrix} 4 & 3 \\ 3 & 9 \end{bmatrix} = \begin{bmatrix} 8 & 6 \\ 6 & 18 \end{bmatrix}, \quad \text{or} \quad \Sigma = \begin{bmatrix} 1/6 & -1/18 \\ -1/18 & 2/27 \end{bmatrix},$$

with $\sqrt{|\Sigma|} = \sqrt{3}/18$. Volume II will have much more to say about the multivariate normal distribution. ∎

Rearranging the c.d.f. expression (8.8b) gives

$$F_{X|Y}(x \mid y) f_Y(y) = \int_{-\infty}^{x} f_{X,Y}(t, y) \, dt$$

or, integrating both sides,

$$\int_{-\infty}^{y} F_{X|Y}(x \mid w) f_Y(w) \, dw = \int_{-\infty}^{y} \int_{-\infty}^{x} f_{X,Y}(t, w) \, dt \, dw = F_{X,Y}(x, y),$$

i.e. the joint bivariate c.d.f. can be expressed as a function of a single integral. Similarly,

$$\boxed{F_{X,Y}(x, y) = \int_{-\infty}^{x} F_{Y|X}(y \mid t) f_X(t) \, dt}.$$

In words, the joint c.d.f. of X and Y can be interpreted as a weighted average of the conditional c.d.f. of Y given X, weighted by the density of X.

More generally, for vector $\mathbf{X} = (X_1, \ldots, X_n)$, its c.d.f. $F_\mathbf{X}$ can be expressed as an $(n-1)$-dimensional integral by conditioning on $n-1$ of the components and integrating them out, i.e.

$$F_\mathbf{X}(\mathbf{x}) = \int_{-\infty}^{x_{n-1}} \int_{-\infty}^{x_{n-2}} \cdots \int_{-\infty}^{x_1} F_{X_n | \mathbf{Y}}(x_n \mid \mathbf{y}) f_\mathbf{Y}(\mathbf{y}) \, dx_1 \cdots dx_{n-1}, \tag{8.15}$$

where $\mathbf{Y} = (X_1, \ldots, X_{n-1})$ and $\mathbf{y} = (x_1, \ldots, x_{n-1})$.

By multiplying both sides of the conditional p.d.f. (8.8a) with $f_Y(y)$ and integrating with respect to y, we obtain an expression for the marginal of X as

$$\boxed{f_X(x) = \int_{-\infty}^{\infty} f_{X|Y}(x \mid y) \, dF_Y(y)}. \tag{8.16}$$

In words, the p.m.f. or p.d.f. of X can be interpreted as being a weighted average of the conditional density of X given Y, weighted by the density of Y. The analogy to the law of total probability (3.10) is particularly clear in the discrete case, for which (8.16) states that

$$\Pr(X = x) = \sum \Pr(X = x \mid Y = y) \Pr(Y = y)$$

for all $x \in S_X$. Furthermore, Bayes' rule can be generalized as

$$\boxed{f_{X|Y=y}(x \mid y) = \frac{f_{Y|X}(y \mid x) f_X(x)}{\int_{-\infty}^{\infty} f_{Y|X}(y \mid x) \, dF_X(x)}}, \tag{8.17}$$

which provides an expression for conditional $X \mid Y$ in terms of that for $Y \mid X$ and the marginal of X.

⊖ **Example 8.6** (Example 8.5 cont.) Given $f_{X|Y}$ and f_Y, the p.d.f. of f_X can be elicited from (8.16), while that for $f_{X|Y=y}$ can be obtained from (8.17) if we additionally know $f_{Y|X}$. Similar calculations and use of completing the square yield

$$f_X(x) = \int_{-\infty}^{\infty} f_{X|Y}(x \mid y) f_Y(y) \, dy$$

$$= \frac{3\sqrt{3}}{\pi} \int_{-\infty}^{\infty} \exp\left[-\frac{1}{4}(4x + 3y)^2 - \frac{27}{4}y^2\right] dy = \frac{\sqrt{3}}{\sqrt{\pi}} \exp(-3x^2)$$

and

$$f_{X|Y=y}(x \mid y) = \frac{f_X(x) f_{Y|X}(y \mid x)}{\int_{-\infty}^{\infty} f_{Y|X}(y \mid x) \, dF_X(x)} = \frac{\exp\left[-3x^2 - (x + 3y)^2\right]}{\int_{-\infty}^{\infty} \exp\left[-3x^2 - (x + 3y)^2\right] dx}$$

$$= \frac{\exp\left[-3x^2 - (x + 3y)^2\right]}{\frac{\sqrt{\pi}}{2} \exp\left(-\frac{27}{4}y^2\right)} = \frac{2}{\sqrt{\pi}} \exp\left[-\frac{1}{4}(4x + 3y)^2\right],$$

which agree with the expressions in Example 8.5. ∎

◎ **Example 8.7** The so-called *exchange paradox*, or *wallet game* provides a nice example of the value and 'naturalness' of Bayesian arguments, and goes as follows. There are two sealed envelopes, the first with m dollars inside, and the second with $2m$ dollars inside, and they are otherwise identical (appearance, thickness, weight, etc.). You and your opponent are told this, but the envelopes are mixed up so neither of you know which contains more money. You randomly choose an envelope, and your opponent receives the other. You open yours, and find x dollars inside. Your opponent opens hers, and finds out the contents; call it Y dollars. The two players are then given the opportunity to trade envelopes. You quickly reason: with equal probability, her envelope contains either $Y = x/2$ or $2x$ dollars. If you trade, you thus expect to get

$$\frac{1}{2}\left(\frac{x}{2} + 2x\right) = \frac{5x}{4}, \tag{8.18}$$

which is greater than x, so you express your interest in trading. Your opponent, of course, made the same calculation, and is just as eager to trade.

The paradox is that, while the rule which led to the decision to trade seems simple and correct, the result that one should *always* trade seems intuitively unreasonable. The paradox appears to date back to 1953 (see Christensen and Utts (1992) for several earlier references, and the source of the resolution which we present).

The Bayesian resolution of the problem involves treating the amount of money in the first envelope, m, as a random variable, M, and with a prior probability distribution. (Recall Example 3.12 and the notion of prior probability before information is gathered.) To believe that no prior information exists on M violates all common sense. For example, if the game is being played at a party, the $3m$ dollars probably derived from contributions from a few people, and you have an idea of their disposable income and their willingness to part with their money for such a game. Let g_M be the p.d.f. of M, and let X be the amount of money in the envelope you chose. Then

$$\Pr(X = m \mid M = m) = \Pr(X = 2m \mid M = m) = \frac{1}{2}.$$

As X is either M or $2M$, it follows that, on observing $X = x$, M is either x or $x/2$. From Bayes' rule (3.11) and (8.17), $\Pr(M = x \mid X = x)$ is

$$\frac{\Pr(X = x \mid M = x)\, g_M(x)}{\Pr(X = x \mid M = x)\, g_M(x) + \Pr(X = x \mid M = x/2)\, g_M(x/2)}$$

$$= \frac{\frac{1}{2} g_M(x)}{\frac{1}{2} g_M(x) + \frac{1}{2} g_M(x/2)} = \frac{g_M(x)}{g_M(x) + g_M(x/2)},$$

and likewise,

$$\Pr(M = x/2 \mid X = x) = \frac{g_M(x/2)}{g_M(x) + g_M(x/2)}.$$

Thus, the expected amount after trading is

$$\mathbb{E}[Y \mid X = x] = \frac{g_M(x/2)}{g_M(x) + g_M(x/2)} \cdot \frac{x}{2} + \frac{g_M(x)}{g_M(x) + g_M(x/2)} \cdot 2x, \tag{8.19}$$

and when $g(x/2) = 2g(x)$, $E[Y \mid X = x] = x$. Thus, the decision rule is:

$$\text{trade if } g(x/2) < 2g(x),$$

and keep it otherwise. For example, if $g_M \sim \text{Exp}(\lambda)$, i.e. $g_M(m; \lambda) = \lambda e^{-\lambda m} \mathbb{I}_{(0,\infty)}(m)$, then

$$g(x/2) < 2g(x) \Rightarrow \lambda e^{-\lambda x/2} < 2\lambda e^{-\lambda x} \Rightarrow x < \frac{2\ln 2}{\lambda},$$

i.e. you should trade if $x < (2\ln 2)/\lambda$.

Note that (8.18) and (8.19) coincide when $g_M(x/2) = g_M(x)$ for all x, i.e. $g_M(x)$ is a constant, which implies that the prior distribution on M is a 'noninformative', improper uniform density on $(0, \infty)$. As mentioned above, it would defy logic to believe that, *in a realistic situation which gives clues about the amount of money involved*, someone would place equal probability on M being 10 dollars and M being 1000 dollars.

Thus, the paradox is resolved by realizing that prior information cannot be ignored if one wishes to make a rational decision. See Christensen and Utts (1992) for further calculations involving the 'frequentist perspective' of the problem. ∎

⊛ ***Example 8.8*** Let $X \sim \text{Exp}(\lambda)$. Then, for $x, s \in \mathbb{R}_{>0}$,

$$\Pr(X > s + x \mid X > s) = \frac{\Pr(X > s + x)}{\Pr(X > s)}$$

$$= \frac{\exp[-\lambda(s + x)]}{\exp[-\lambda s]} = e^{-\lambda x} = \Pr(X > x).$$

The fact that this conditional probability is not a function of s is referred to as the *memoryless property*. It is worth emphasizing with some concrete numbers what the memoryless property means, and what it does not. If, say, the lifetime of an electrical device, measured in days, follows an exponential distribution, and we have observed that it is still functioning at time $t = 100$ (the event $X > 100$), then the probability that it will last at least an additional 10 days (i.e. conditional on our observation) is the same as the *unconditional* probability that the device will last at least 10 days. What is *not* true is that the probability that the device, conditional on our observation, will last an additional 10 days, is the same as the unconditional probability of the device lasting at least 110 days. That is, it is *not* true that

$$\Pr(X > s + x \mid X > s) = \Pr(X > s + x).$$

It can be shown that the memoryless property *characterizes* the distribution, i.e. the exponential distribution is the only continuous distribution which has this property. Proofs can be found in many textbooks, for example Grimmett and Stirzaker (2001a, p. 210(5)). ∎

⊖ ***Example 8.9*** As with most mathematical objects, there are numerous ways in which a univariate distribution can be generalized to the multivariate case, and often the choice is dictated by an area of application. We illustrate some ways of constructing a bivariate density such that, when the two component r.v.s X and Y are independent, they are marginally exponentially distributed (but not necessarily the other way around).

The trivial case is when X and Y are independent exponential r.v.s, so that, for $\theta_1, \theta_2 \in \mathbb{R}_{>0}$, their joint density is

$$f_{X,Y}(x, y; \theta_1, \theta_2) = \theta_1 \theta_2 \exp\left(-\theta_1 x - \theta_2 y\right) \mathbb{I}_{(0,\infty)}(x) \, \mathbb{I}_{(0,\infty)}(y). \tag{8.20}$$

Figure 8.7(a) shows a contour plot of (8.20) for $\theta_1 = 2/5$ and $\theta_2 = 1/5$. That the contour lines are straight with slope proportional to the ratio of the two scale parameters is just seen by solving $f_{X,Y}(x, y; \theta_1, \theta_2) = k$ to get

$$y = \frac{\ln(\theta_1 \theta_2 / k)}{\theta_2} - \frac{\theta_1}{\theta_2}x.$$

We now present three nontrivial bivariate exponential distributions; the motivations for their construction, as well as further examples, information and references can

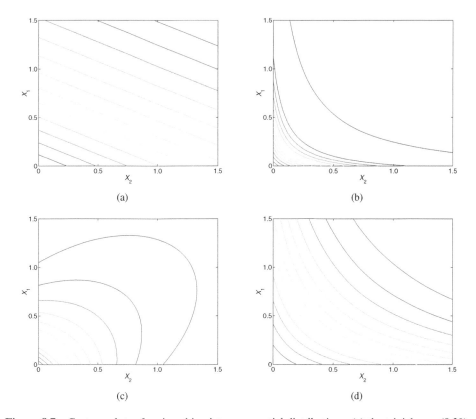

(a)

(b)

(c)

(d)

Figure 8.7 Contour plots of various bivariate exponential distributions: (a) the trivial case (8.20) for $\theta_1 = 2/5$ and $\theta_2 = 1/5$; (b) the Gumbel density (8.21) for $\theta = 10$; (c) the Moran and Downton density (8.23) for $\theta_1 = \theta_2 = 2/5$ and $\rho = 4/5$; (d) the Arnold and Strauss density (8.24) for $\beta_1 = \beta_2 = 2/5$ and $\beta_3 = 10$

be found in the references given at the beginning of this chapter, in particular Kotz, Balakrishnan and Johnson (2000).

(1) Gumbel (1960) proposed the density

$$f_{X,Y}(x, y; \theta) = \left[(1 + \theta x)(1 + \theta y) - \theta\right] \exp\left[-x - y - \theta xy\right] \mathbb{I}_{(0,\infty)}(x) \mathbb{I}_{(0,\infty)}(y),$$
(8.21)

for $\theta \geq 0$. When $\theta = 0$, X and Y are i.i.d. exp(1) r.v.s.[1] This density is plotted in Figure 8.7(b) for $\theta = 10$. The distribution has the property that, irrespective of θ, the marginal distributions are exponential, i.e. $X \sim \exp(1)$ and $Y \sim \exp(1)$. To see this for X, let $\lambda = 1 + \theta x$ so that

$$f_X(x; \theta) = \int_0^\infty f(x, y; \theta) \, dy = \int_0^\infty ((1 + \theta x)(1 + \theta y) - \theta) e^{-x} e^{-y} e^{-\theta xy} \, dy$$

$$= e^{-x} \left[\int_0^\infty (1 + \theta y) \lambda e^{-y\lambda} \, dy - \theta \int_0^\infty e^{-y\lambda} \, dy\right]$$

$$= e^{-x} \left(\int_0^\infty \lambda e^{-y\lambda} \, dy + \theta \int_0^\infty y\lambda e^{-y\lambda} \, dy - \frac{\theta}{\lambda}\right)$$

$$= e^{-x} \left(1 + \frac{\theta}{\lambda} - \frac{\theta}{\lambda}\right) = e^{-x},$$

where use was made of the expected value of an exponential random variable from (7.5), i.e.

$$\int_0^\infty z\lambda e^{-z\lambda} \, dz = 1/\lambda.$$

From the symmetry of x and y in (8.21), it follows immediately that $Y \sim \exp(1)$. The conditional distribution of X given Y easily follows as

$$f_{X|Y}(x \mid y) = \frac{f_{X,Y}(x, y)}{f_Y(y)} = \left[(1 + \theta x)(1 + \theta y) - \theta\right] e^{-x(1+\theta y)} \mathbb{I}_{(0,\infty)}(x), \quad y > 0.$$

Problem 8.12 shows that Cov (X, Y) can be expressed as

$$\text{Cov}(X, Y) = \int_0^\infty y e^{-y} \frac{1 + \theta y + \theta}{(1 + \theta y)^2} \, dy - 1,$$
(8.22)

which can be numerically evaluated. As the variance of each of the marginals is one, it follows that Cov $(X, Y) = $ Corr (X, Y). Figure 8.8 plots the correlation of X and Y as a function of θ. It is zero for $\theta = 0$ and decreases to -1 as θ increases.

---

[1] Two additional scale parameters, say β_1 and β_2, could also be introduced such that, when $\theta = 0$, $X \sim$ Exp(β_1) independent of $Y \sim$ Exp(β_2).

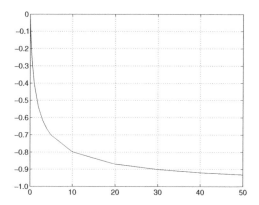

Figure 8.8 Correlation of the Gumbel bivariate exponential distribution (8.21) as a function of parameter θ

(2) One drawback of the Gumbel bivariate exponential (8.21) is that the correlation cannot be positive. A distribution with positive (but not negative) correlation was introduced by Moran in 1967, popularized by Downton in 1970, and further studied by Nagao and Kadoya in 1971 (see Kotz, Balakrishnan and Johnson, 2000, and Yue, 2001, for references). The density is

$$f_{X_1, X_2}(x_1, x_2; \theta_1, \theta_2, \rho) = \frac{\theta_1 \theta_2}{1 - \rho} I_0 \left(\frac{2\sqrt{\rho \theta_1 \theta_2 x_1 x_2}}{1 - \rho} \right)$$

$$\exp\left(-\frac{\theta_1 x_1 + \theta_2 x_2}{1 - \rho} \right) \times \mathbb{I}_{(0,\infty)}(x_1) \mathbb{I}_{(0,\infty)}(x_2)$$

(8.23)

for $\theta_1, \theta_2 \in \mathbb{R}_{>0}$ and $0 \leq \rho < 1$. The kernel of the density contains the term $I_v(\cdot)$, which is the modified Bessel function of the first kind, of order v (with $v = 0$ in this case). We encountered this already in (6.34), and with $v = 0$,[2]

$$I_0(x) = \sum_{j=0}^{\infty} \frac{(x^2/4)^j}{(j!)^2}.$$

With the usual conventions that $0^0 = 1$ and $0! = 1$, it follows that $I_0(0) = 1$, so that X_1 and X_2 are independent exponential r.v.s when $\rho = 0$. It can be shown that $\text{Corr}(X_1, X_2) = \rho$ and $\mathbb{E}[X_2 \mid X_1 = x_1] = [1 - \rho(1 - \theta_1 x_1)]/\theta_2$, with a similar expression for $\mathbb{E}[X_1 \mid X_2 = x_2]$.
Figure 8.7(c) shows (8.23) with $\theta_1 = \theta_2 = 2/5$ and $\rho = 4/5$.

(3) In both previous cases, the *conditional* distributions were not exponential. A bivariate distribution which has exponential conditionals was introduced by Arnold and

---

[2] There is an apparent typographical error in the discussion of this distribution in Kotz, Balakrishnan and Johnson (2000). The expression for $I_0(x)$ given on p. 350 is incorrect.

Strauss (1988), given by

$$f_{X,Y}(x, y) = C(\beta_3) \beta_1 \beta_2 \exp(-\beta_1 x - \beta_2 y - \beta_1 \beta_2 \beta_3 xy) \mathbb{I}_{(0,\infty)}(x) \mathbb{I}_{(0,\infty)}(y),$$

(8.24)

where its three parameters satisfy $\beta_1, \beta_2 > 0$, $\beta_3 \geq 0$ and

$$C(\beta_3) = \int_0^\infty \frac{e^{-u}}{1 + \beta_3 u} du$$

(8.25)

is the constant of integration. Using $v = 1/(1 + u)$, this constant can be written as

$$C(\beta_3) = \int_0^1 \frac{e^{-((1/v)-1)}}{v^2 + \beta_3 v(1 - v)} dv,$$

which is more suitable for numeric evaluation. Note the similarity of the exponential term in (8.24) and (8.21). The constraint $\beta_3 \geq 0$ also ensures that only negative correlation is possible. From (8.24) and (8.25), it is clear that, for $\beta_3 = 0$, X and Y are independent. The density with $\beta_1 = \beta_2 = 2/5$ and $\beta_3 = 10$ is plotted in Figure 8.7(d). ∎

Remark:

(a) Just as knowledge of all n marginal distributions of an n-variate random variable $\mathbf{X}$ does not completely determine (or *characterize*) the distribution of $\mathbf{X}$, knowledge of all the conditional distributions does not either (a simple example of this will be given in Volume II in the context of the multivariate normal distribution).

(b) It is also worth mentioning that a bivariate joint density involving discrete and continuous r.v.s can be obtained by use of the marginal and conditional distributions. In particular, let D be a discrete and C a continuous random variable. Then, from (8.17),

$$f_{C|D}(c \mid d) = \frac{f_{D|C}(d \mid c) f_C(c)}{f_D(d)} \quad \text{and} \quad f_{D|C}(d \mid c) = \frac{f_{C|D}(c \mid d) f_D(d)}{f_C(c)}.$$

For example, C could represent income (considered continuous) and D is profession (taking on a finite number of values). ∎

8.2.3   Conditional moments

Moments and expectations of other functions of a conditional distribution are defined in a natural way as follows. Letting $\mathbf{Y} = (X_{m+1}, \ldots, X_n)$ so that $\mathbf{X} = (\mathbf{X}_m, \mathbf{Y})$, the expected value of function $g(\mathbf{X}_m)$ conditional on $\mathbf{Y}$ is given by

$$\mathbb{E}[g(\mathbf{X}_m) \mid \mathbf{Y} = \mathbf{y}] = \int_{\mathbf{x} \in \mathbb{R}^m} g(\mathbf{x}) \, dF_{\mathbf{X}_m | \mathbf{Y}}(\mathbf{x} \mid \mathbf{y}).$$

(8.26)

The function $g(\mathbf{X}_m)$ in (8.26) could be replaced by $g(\mathbf{X})$ instead, i.e. letting g be a function of the entire random vector, but any occurrences of $\mathbf{Y}$ in g are just replaced by their nonstochastic counterparts $\mathbf{y}$ and are treated as constants. As $F_{\mathbf{X}|\mathbf{Y}}$ is degenerate in some dimensions, the conditional density used in the integral in (8.26) is still $F_{\mathbf{X}_m|\mathbf{Y}}$. To illustrate, assume the desired conditional expectation is $\mathbb{E}[(X+Y) \mid Y = y]$. Because we condition on Y, this is equivalent to $\mathbb{E}[X \mid Y = y] + y$.

A very important special case of (8.26) for random variables X, Y with bivariate density $f_{X,Y}$ is $\mathbb{E}[X \mid Y = y]$, the *conditional expectation of X given Y = y*,

$$\mathbb{E}\left[X \mid Y = y\right] = \int_{-\infty}^{\infty} x f_{X|Y}(x \mid y) \, dx \, . \tag{8.27}$$

There are two interpretations of $\mathbb{E}\left[X \mid Y = y\right]$. The first is *as univariate function of y*, i.e. as, say, $g : \mathcal{S} \subset \mathbb{R} \to \mathbb{R}$, with $g(y) := \mathbb{E}\left[X \mid Y = y\right]$ and $\mathcal{S}$ is the support of Y. The second interpretation, sometimes emphasized by using the shorter notation $\mathbb{E}[X \mid Y]$, is *as a random variable*. This follows because Y is an r.v. and, from the first interpretation, $\mathbb{E}[X \mid Y]$ is a function of Y. Thus, for example, one could compute its expectation and variance, as will be done shortly below.

These two interpretations of course hold in the more general case of (8.26). It is clear that $\mathbb{E}\left[g(\mathbf{X}_m) \mid \mathbf{Y} = \mathbf{y}\right]$ is a function of $\mathbf{y}$, which represents a particular value in the support of $\mathbf{Y}$. As such, it also makes sense to treat $\mathbb{E}\left[g(\mathbf{X}_m) \mid \mathbf{Y}\right]$ as a random variable, and we could take, say, its expectation.

Example 8.10 (Example 8.5 cont.) The conditional expectation of $X \mid Y$ could be computed from (8.27), but it is easier to note that, with $\sigma^2 = 1/8$,

$$\frac{2}{\sqrt{\pi}} \exp\left[-\frac{1}{4}(4x + 3y)^2\right] = \frac{1}{\sigma\sqrt{2\pi}} \exp\left[-\frac{1}{2\sigma^2}\left(x + \frac{3}{4}y\right)^2\right],$$

i.e. $X \mid Y \sim N\left(-3y/4, \sigma^2\right)$ with expected value $-3y/4$. From this and (4.40), i.e. that $\mathbb{V}(X) = \mathbb{E}[X^2] - (\mathbb{E}[X])^2$, it also follows that

$$\int_{-\infty}^{\infty} x^2 \frac{2}{\sqrt{\pi}} \exp\left[-\frac{1}{4}(4x + 3y)^2\right] dx = \frac{9}{16}y^2 + \frac{1}{8},$$

which would otherwise appear to be a difficult integral to resolve. A similar calculation shows that $Y \mid X \sim N(-x/3, 1/18)$. ∎

Example 8.11 (Example 8.1 cont.) We wish to compute $\mathbb{E}[X \mid Y]$ based on the joint density $f_{X,Y}(x, y) = (3/2)\left(x^2 + y^2\right)\mathbb{I}_{(0,1)}(x)\,\mathbb{I}_{(0,1)}(y)$.

From (8.8a) with the marginal f_Y as obtained in Example 8.1,

$$f_{X|Y}(x \mid y) = \frac{f_{X,Y}(x, y)}{f_Y(y)} = \frac{\frac{3}{2}\left(x^2 + y^2\right)\mathbb{I}_{(0,1)}(x)\,\mathbb{I}_{(0,1)}(y)}{\left(\frac{1}{2} + \frac{3}{2}y^2\right)\mathbb{I}_{(0,1)}(y)}$$

$$= \frac{x^2 + y^2}{\frac{1}{3} + y^2}\mathbb{I}_{(0,1)}(x)\,\mathbb{I}_{(0,1)}(y),$$

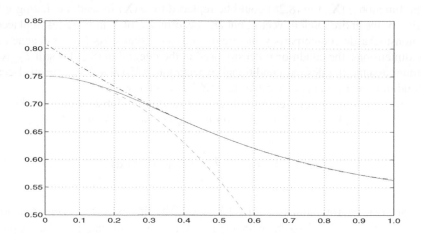

Figure 8.9   Conditional expected value $\mathbb{E}[X \mid Y]$ in (8.28) as a function of y, and Taylor series approximations (truncated)

so that, from (8.27)

$$\mathbb{E}[X \mid Y] = \int_0^1 x \frac{x^2 + y^2}{\frac{1}{3} + y^2} \, dx = \frac{\frac{1}{2}y^2 + \frac{1}{4}}{\frac{1}{3} + y^2} = \frac{3}{4} \frac{1 + 2y^2}{1 + 3y^2} \mathbb{I}_{(0,1)}(y), \qquad (8.28)$$

which is shown in Figure 8.9 as the solid line.

This is obviously a simple function of y, but let us imagine that the resulting conditional expectation is not practical for computation, and that an approximation is desired which is numerically cheaper to evaluate. From Taylor's formula (A.132), straightforward calculation shows that a quadratic Taylor series approximation to (8.28) is $(3/4) - (3/4)y^2$, which is plotted as the dashed line in Figure 8.9. It is adequate for $0 \le y \le 0.2$, but continues to worsen as $y \to 1$. This is not too surprising, because it contains no linear term. The Taylor series approximation around $y - 1/2$, i.e. of

$$\mathbb{E}[X \mid Y] = \frac{3}{4} \frac{1 + 2(q + 1/2)^2}{1 + 3(q + 1/2)^2}, \qquad q = y - 1/2,$$

might be better; it is

$$\frac{9}{14} - \frac{12}{49}q + \frac{60}{343}q^2$$

and is shown as the dash–dot line. It is very accurate for $0.3 \le y \le 1$ and not disastrous as $y \to 0$. ∎

As mentioned above after (8.27), we can take the expectation of a conditional expectation. Let X and Y have the continuous joint density $f_{X,Y}$. Then, subscripting the expectation operators for clarity,

$$\mathbb{E}_Y \mathbb{E}_{X|Y} \big[g\,(X) \mid Y \big] = \int_{-\infty}^{\infty} f_Y\,(y) \int_{-\infty}^{\infty} g\,(x)\,\frac{f_{X,Y}\,(x,\,y)}{f_Y\,(y)}\,dx\,dy$$

$$= \int_{-\infty}^{\infty} \int_{-\infty}^{\infty} g\,(x)\,f_{X,Y}\,(x,\,y)\,dx\,dy$$

$$= \int_{-\infty}^{\infty} g\,(x) \int_{-\infty}^{\infty} f_{X,Y}\,(x,\,y)\,dy\,dx$$

$$= \int_{-\infty}^{\infty} g\,(x)\,f_X\,(x)\,dx = \mathbb{E}_X \big[g\,(X) \big]. \qquad (8.29)$$

The same result holds if Y is discrete, in which case we can write

$$\mathbb{E}_X \big[g\,(X) \big] = \mathbb{E}_Y \mathbb{E}_{X|Y} \big[g\,(X) \mid Y \big] = \sum_{y=-\infty}^{\infty} \mathbb{E}_{X|Y} \big[g\,(X) \mid Y = y \big] \Pr(Y = y). \quad (8.30)$$

We write both results (8.29) and (8.30) as

$$\boxed{\mathbb{E}\mathbb{E} \big[g\,(X) \mid Y \big] = \mathbb{E} \big[g\,(X) \big]}, \qquad (8.31)$$

and refer to it as the *law of the iterated expectation*. This seemingly basic relation is of extreme importance; its use leads in many cases to simple solutions which, otherwise, would be far more challenging.

⊖ *Example 8.12* (Example 5.3 cont.) Consider the bivariate density

$$f_{X,Y}\,(x,\,y) = e^{-y} \mathbb{I}_{(0,\infty)}\,(x)\,\mathbb{I}_{(x,\infty)}\,(y).$$

The marginal distribution of X is $f_X\,(x) = \mathbb{I}_{(0,\infty)}\,(x) \int_x^\infty e^{-y} dy = e^{-x} \mathbb{I}_{(0,\infty)}\,(x)$, so that $X \sim \mathrm{Exp}\,(1)$ and $\mathbb{E}\,[X] = 1$. For Y, note that the range of x is 0 to y so that

$$f_Y\,(y) = \int_0^y e^{-y}\,dx = y e^{-y}$$

so that $Y \sim \mathrm{Gam}\,(2,\,1)$ and $\mathbb{E}\,[Y] = 2$. The conditional density of $Y \mid X$ is

$$f_{Y|X}\,(y \mid X = x) = \frac{f_{X,Y}\,(x,\,y)}{f_X\,(x)} = e^{x-y} \mathbb{I}_{(0,\infty)}\,(x)\,\mathbb{I}_{(x,\infty)}\,(y) = e^{x-y} \mathbb{I}\,(0 < x < y)$$

with expected value

$$\mathbb{E}\,[Y \mid X] = \int_x^\infty y f_{Y|X}\,(y \mid x)\,dy = e^x \int_x^\infty y e^{-y}\,dy = x + 1,$$

using the substitution $u = y$ and $dv = e^{-y}dy$. From (8.31),

$$\mathbb{E}[Y] = \mathbb{E}[\mathbb{E}[Y \mid X]] = \mathbb{E}[X + 1] = 2$$

as above.   ∎

◎ ***Example 8.13*** (Example 2.5 cont.) A random number, say N_m, Bernoulli i.i.d. trials with success probability p are performed until there is a run of m successes, i.e. until there are m *consecutive* successes. We denote this as $N_m \sim \text{Consec}(m, p)$; its p.m.f. is given in (2.7).

Define $E_m := \mathbb{E}[N_m]$, and from (8.31), $E_m = \mathbb{E}\left[\mathbb{E}\left[N_m \mid N_{m-1}\right]\right]$. For convenience, let $X = N_m \mid (N_{m-1} = n)$ and let Y be the $(n+1)$th trial. Then, from (8.30),

$$\mathbb{E}[X] = \mathbb{E}[X \mid Y = 0] \cdot \Pr(Y = 0) + \mathbb{E}[X \mid Y = 1] \cdot \Pr(Y = 1)$$
$$= (1 - p) \cdot \mathbb{E}[X \mid Y = 0] + p \cdot \mathbb{E}[X \mid Y = 1].$$

From the definitions of X and Y, it follows that $\mathbb{E}[X \mid Y = 1] = n + 1$, because the $(n+1)$th trial is a success ($Y = 1$) and the previous $m - 1$ trials were all successes ($N_{m-1} = n$). Similarly, $\mathbb{E}[X \mid Y = 0] = E_m + (n + 1)$, because the $(n+1)$th trial is a failure, and the experiment essentially 'starts over'. Thus,

$$\mathbb{E}\left[N_m \mid N_{m-1} = n\right] = (1 - p) \cdot (E_m + n + 1) + p \cdot (n + 1).$$

Now taking expectations of both sides of this expression with respect to N_{m-1} (which amounts to replacing n on the r.h.s. with $\mathbb{E}\left[N_{m-1}\right] = E_{m-1}$) gives

$$E_m = (1 - p) \cdot (E_m + E_{m-1} + 1) + p \cdot (E_{m-1} + 1)$$
$$= 1 + (1 - p) E_m + E_{m-1},$$

and solving yields

$$E_m = \frac{1}{p} + \frac{E_{m-1}}{p}. \tag{8.32}$$

However, $N_1 \sim \text{Geo}(p)$ with $E_1 = \mathbb{E}[N_1] = 1/p$. Thus, $E_2 = 1/p + 1/p^2$, etc., and

$$E_m = \frac{1}{p} + \frac{1}{p^2} + \cdots + \frac{1}{p^m} = \sum_{i=1}^{m} p^{-i} = \frac{p^{-m} - 1}{1 - p}. \tag{8.33}$$

Now let $p = 1/2$ so that $E_m = 2^{m+1} - 2$. If a fair coin is flipped E_m times, it seems natural to think that the probability of getting a run of m heads (successes) is $1/2$. As in Example 4.19, this event is not well-defined, as the number of runs is not specified. Instead, consider the probability of getting at least one such run. This can be computed with Gani's (1998) result (6.15) and (6.18) for $k = 2$. Figure 8.10 shows the resulting probability as a function of m. It appears to converge to $1 - e^{-1}$, similar to the results in Example 4.19.

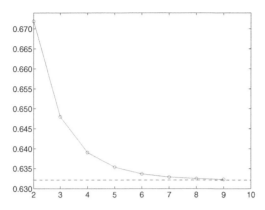

Figure 8.10 The probability of getting at least one m-length run of heads versus m, when flipping a fair coin n times, where $n = E_m = 2^{m+1} - 2$ is the expected number of flips required to get one such sequence; the dashed line is $1 - e^{-1}$

Problem 8.13 shows that

$$V(N_m) = \frac{1 - (1 + 2m)(1 - p) p^m - p^{2m+1}}{(1 - p)^2 p^{2m}},$$

which will be seen to take a bit more work. ∎

Analogous to the conditional expectation of X given Y, the *conditional variance of X given Y* is given by (8.26) with $g(X, Y) = (X - \mathbb{E}[X \mid Y])^2$, i.e.

$$\text{Var}(X \mid Y) = \mathbb{E}\left[(X - \mathbb{E}[X \mid Y])^2 \mid Y\right] = \mathbb{E}\left[X^2 \mid Y\right] - (\mathbb{E}[X \mid Y])^2, \qquad (8.34)$$

where the latter expression is similar to (4.40). Taking expectations with respect to Y and using (8.31) yields

$$\mathbb{E}[V(X \mid Y)] = \mathbb{E}\left[\mathbb{E}[X^2 \mid Y]\right] - \mathbb{E}\left[(\mathbb{E}[X \mid Y])^2\right]$$
$$= \mathbb{E}[X^2] - \mathbb{E}\left[(\mathbb{E}[X \mid Y])^2\right]. \qquad (8.35)$$

Now using (4.40) with (8.31) gives

$$V(\mathbb{E}[X \mid Y]) = \mathbb{E}\left[(\mathbb{E}[X \mid Y])^2\right] - (\mathbb{E}[X])^2. \qquad (8.36)$$

Finally, adding (8.35) and (8.36) results in

$$\boxed{V(X) = \mathbb{E}[V(X \mid Y)] + V(\mathbb{E}[X \mid Y])}, \qquad (8.37)$$

which is referred to as the *conditional variance formula*.

⊖ ***Example 8.14***   Independent Bernoulli trials are repeatedly performed, with the same success probability p, but the number of trials, N, is not fixed, but random, with $N \sim \mathrm{Poi}(\lambda)$. Let X be the number of successes. From (8.31),

$$\mathbb{E}[X] = \mathbb{E}_N[\mathbb{E}[X \mid N = n]] = \mathbb{E}_N[np] = p\,\mathbb{E}_N[n] = \lambda p,$$

and, from (8.37),

$$
\begin{aligned}
\mathbb{V}(X) &= \mathbb{E}_N[\mathbb{V}(X \mid N = n)] + \mathbb{V}_N(\mathbb{E}[X \mid N = n]) \\
&= \mathbb{E}_N[np(1-p)] + \mathbb{V}_N(np) \\
&= p(1-p)\lambda + p^2\lambda = \lambda p.
\end{aligned}
$$

See Problem 8.10 for derivation of f_X. ∎

8.2.4 Expected shortfall

The superior man, when resting in safety, does not forget that danger may come. When in a state of security he does not forget the possibility of ruin. When all is orderly, he does not forget that disorder may come. Thus his person is not endangered, and his States and all their clans are preserved.

(Confucius)

It is often of interest to calculate moments of an r.v. X, given that it exceeds (or is less than) a particular value. For example, a common calculation involved in empirical finance is the *expected shortfall*, defined as $\mathbb{E}[R \mid R < c]$, whereby r.v. R is the % return on the financial asset and c is usually taken to be a value far in the left tail, typically the 1% quantile of R. Denoting the p.d.f. and c.d.f. of R as f_R and F_R respectively, the expected value of measurable function $g(R)$, given that $R < c$, is

$$\boxed{\mathbb{E}[g(R) \mid R < c] = \frac{\int_{-\infty}^{c} g(r)\,f_R(r)\,dr}{F_R(c)}}. \tag{8.38}$$

⊖ ***Example 8.15***   Let $R \sim \mathrm{N}(0, 1)$ with p.d.f. ϕ and c.d.f. Φ, and take $g(r) = r$, so that

$$
\begin{aligned}
\mathbb{E}[R \mid R < c] &= \frac{1}{\Phi(c)} \int_{-\infty}^{c} r\phi(r)\,dr = \frac{1}{\Phi(c)} \frac{1}{\sqrt{2\pi}} \int_{-\infty}^{c} r \exp\left(-\frac{1}{2}r^2\right) dr \\
&= \frac{1}{\Phi(c)} \frac{1}{\sqrt{2\pi}} \left[-\exp\left(-\frac{1}{2}c^2\right)\right] = -\frac{\phi(c)}{\Phi(c)}.
\end{aligned}
$$

Notice that $\mathbb{E}[R \mid R < c]$ is negative for all $c < \infty$, which is intuitive, and approaches zero as $c \to \infty$. This also illustrates a case for which the integral in (8.38) can be algebraically solved. For the second moment, with $u = r^2/2$ and assuming $c < 0$, $r = -\sqrt{2u}$, $dr = -2^{-1/2}u^{-1/2}du$ and

$$
\begin{aligned}
\mathbb{E}\left[R^2 \mid R < c\right] &= \frac{1}{\Phi(c)} \frac{1}{\sqrt{2\pi}} \int_{-\infty}^{c} r^2 \exp\left(-\frac{1}{2}r^2\right) dr \\
&= \frac{1}{\Phi(c)} \frac{1}{\sqrt{\pi}} \int_{c^2/2}^{\infty} u^{1/2} \exp(-u)\, du \\
&= \frac{1}{\Phi(c)} \frac{1}{\sqrt{\pi}} \left[\Gamma(3/2) - \Gamma_{c^2/2}(3/2)\right] \qquad c < 0.
\end{aligned}
$$

For $c \to 0^-$, $\mathbb{E}\left[R^2 \mid R < 0\right] = 1$. This agrees, of course, with the result in Problem 7.5 for the second raw moment of the folded standard normal random variable. If $c > 0$, then split the integral at zero, use the previous result and $r = +\sqrt{2u}$ to get

$$
\begin{aligned}
\mathbb{E}\left[R^2 \mid R < c\right] &= \frac{1}{\Phi(c)} \left(\frac{\Gamma(3/2)}{\sqrt{\pi}} + \frac{1}{\sqrt{2\pi}} \int_{0}^{c} r^2 \exp\left(-\frac{1}{2}r^2\right) dr\right) \\
&= \frac{1}{\Phi(c)} \left(\frac{1}{2} + \frac{1}{\sqrt{\pi}} \int_{0}^{c^2/2} u^{1/2} \exp(-u)\, du\right) \\
&= \frac{1}{\Phi(c)} \left[\frac{1}{2} + \frac{1}{\sqrt{\pi}} \Gamma_{c^2/2}(3/2)\right] \qquad c > 0.
\end{aligned}
$$

As $c \to \infty$, $\mathbb{E}\left[R^2 \mid R < c\right] \to \left[1/2 + \Gamma(3/2)/\sqrt{\pi}\right] = 1$, as it should. ∎

Problem 8.1 asks the reader to calculate the expected shortfall when using a Student's t distribution, which usually provides a better description of certain financial data. It is easy to express the expected shortfall for a general location–scale family. Let $Y = \mu + \sigma Z$ and define $k = (c - \mu)/\sigma$. Then $F_Y(c) = \Pr(Y \leq c) = \Pr(Z \leq k) = F_Z(k)$ and

$$
\begin{aligned}
\mathbb{E}[Y \mid Y < c] &= \frac{1}{F_Y(c)} \int_{-\infty}^{c} y f_Y(y)\, dy = \frac{1}{F_Y(c)} \int_{-\infty}^{c} (\mu + \sigma z) \frac{1}{\sigma} f_Z\left(\frac{y - \mu}{\sigma}\right) dy \\
&= \frac{1}{F_Z(k)} \int_{-\infty}^{k} (\mu + \sigma z) f_Z(z)\, dz = \mu + \frac{\sigma}{F_Z(k)} \int_{-\infty}^{k} z f_Z(z)\, dz \\
&= \mu + \sigma \mathbb{E}[Z \mid Z < k], \qquad k = \frac{c - \mu}{\sigma}.
\end{aligned}
$$

8.2.5 Independence

It's déjà vu all over again.   (Yogi Berra)
Reproduced by permission of Workman Publishing Company

Our first formal encounter with independence was in Section 5.2.2 where we stated that r.v.s $X_1, \ldots, X_n$ are independent iff their joint density can be factored into the marginals as

$$f_{\mathbf{X}}(\mathbf{x}) = \prod_{i=1}^{n} f_{X_i}(x_i).$$

An equivalent statement in the bivariate case can be given in terms of conditional r.v.s: X and Y are independent when $f_{X|Y}(x \mid y) = f_X(x)$ or

$$f_{X,Y}(x, y) = f_{X|Y}(x \mid y) f_Y(y) = f_X(x) f_Y(y). \tag{8.39}$$

This can be generalized to n r.v.s by requiring that $f_{\mathbf{Z}|(\mathbf{X}\backslash\mathbf{Z})} = f_{\mathbf{Z}}$ for all subsets $\mathbf{Z} = (X_{j_1}, \ldots X_{j_k})$ of $\mathbf{X}$. Thus, a set of r.v.s are mutually independent if their marginal and conditional distributions coincide.

Let $\{\mathbf{X}_i, i = 1, \ldots, n\}$ be a set of independent d-dimensional r.v.s. It seems intuitive that all functions of disjoint subsets of the $\mathbf{X}_i$ should be independent. This is indeed the case. For example, if a random sample of n people are drawn from a certain population and $\mathbf{X}_i = (H_i, W_i)$ is the vector of their heights and weights, then $g_i(\mathbf{X}_i) = (2H_i, W_i^2)$, $i = 1, \ldots, n$, are independent. If the $\mathbf{X}_i$ are now ordered such that $\{\mathbf{X}_i, i = 1, \ldots, m\}$ and $\{\mathbf{X}_i, i = m+1, \ldots, n\}$ respectively correspond to the males and females, then $m^{-1} \sum_{i=1}^{m} g_i(\mathbf{X}_i)$ and $(n-m)^{-1} \sum_{i=1}^{n-m} g_i(\mathbf{X}_i)$ are also independent. As a second example, let $\mathbf{X} = \{X_i, i = 1, \ldots, n\}$ be an i.i.d. sample such that the r.v. $S = g(\mathbf{X}) := n^{-1} \sum_{i=1}^{n} X_i$ is independent of each $D_i = h_i(\mathbf{X}) := X_i - S$, $i = 1, \ldots, n$. Then S is also independent of any function of the D_i, such as $\sum_{i=1}^{n} D_i^2$. (This holds for $X_i \overset{\text{i.i.d.}}{\sim} \mathrm{N}(\mu, \sigma^2)$ as will be demonstrated in Volume II.)

8.2.6 Computing probabilities via conditioning

If A is an event of interest involving two or more r.v.s, the calculation of $\Pr(A)$ can sometimes be facilitated by conditioning on one or more r.v.s related to A. In particular, using the notation in (4.31),

$$\boxed{\Pr(A) = \int_{-\infty}^{\infty} \Pr(A \mid X = x) \, dF_X(x)}. \tag{8.40}$$

Observe that this is a natural generalization of the law of total probability (3.10). They coincide if X is discrete and we assign exclusive and exhaustive events B_i to all possible outcomes of X. Observe what happens in (8.40) if A and X are independent.

If X and Y are continuous random variables and event $A = \{X < aY\}$, then, conditioning on Y,

$$\Pr(A) = \Pr(X < aY) = \int_{-\infty}^{\infty} \Pr(X < aY \mid Y = y) f_Y(y) \, dy$$

$$= \int_{-\infty}^{\infty} F_{X|Y}(ay) f_Y(y) \, dy. \tag{8.41}$$

More generally, if event $B = \{X - aY < b\}$, then

$$\Pr(B) = \int_{-\infty}^{\infty} \Pr(X - aY < b \mid Y = y) \, f_Y(y) \, dy = \int_{-\infty}^{\infty} F_{X\mid Y}(b + ay) \, f_Y(y) \, dy.$$

As $\Pr(B)$ is the c.d.f. of $X - aY$, differentiating with respect to b gives the p.d.f. of $X - aY$ at b,

$$f_{X-aY}(b) = \int_{-\infty}^{\infty} \frac{d}{db} F_{X\mid Y}(b + ay) \, f_Y(y) \, dy = \int_{-\infty}^{\infty} f_{X\mid Y}(b + ay) \, f_Y(y) \, dy,$$

assuming that we can differentiate under the integral. From (8.8a), this can also be written

$$f_{X-aY}(b) = \int_{-\infty}^{\infty} f_{X,Y}(b + ay, y) \, dy$$

and, with $a = -1$ as an important special case,

$$f_{X+Y}(b) = \int_{-\infty}^{\infty} f_{X,Y}(b - y, y) \, dy \quad, \tag{8.42}$$

which is referred to as the *convolution* of X and Y. In many applications of (8.42), X and Y are independent, so that

$$f_{X+Y}(b) = \int_{-\infty}^{\infty} f_X(b - y) \, f_Y(y) \, dy. \tag{8.43}$$

⊙ **Example 8.16** The p.d.f. of the sum S of two independent standard (i.e. location zero and scale one) Cauchy r.v.s is given by (8.43) with

$$f_X(x) = f_Y(x) = \pi^{-1} \left(1 + x^2\right)^{-1},$$

or

$$f_S(s) = \frac{1}{\pi^2} \int_{-\infty}^{\infty} \frac{1}{1 + (s - x)^2} \frac{1}{1 + x^2} \, dx.$$

The calculation of this integral was shown in Example A.29, and results in

$$f_S(s) = \frac{1}{\pi} \frac{1}{2} \frac{1}{1 + (s/2)^2},$$

or $S \sim \text{Cau}(0, 2)$. A similar calculation shows that, if $X_i \overset{\text{ind}}{\sim} \text{Cau}(0, \sigma_i)$, then

$$S = \sum_{i=1}^{n} X_i \sim \text{Cau}(0, \sigma), \qquad \sigma = \sum_{i=1}^{n} \sigma_i. \tag{8.44}$$

An important special case of (8.44) is for $\sigma = 1/n$, so that the sample mean of n i.i.d. standard Cauchy r.v.s, $\overline{X} = n^{-1}(X_1 + \cdots + X_n)$, also follows a standard Cauchy distribution. ∎

⊙ **Example 8.17**   Let $X_i \overset{\text{ind}}{\sim} \text{Exp}(\lambda_i)$, $i = 1, 2$ and set $R = X_1/X_2$. As $\Pr(X_2 > 0) = 1$, the distribution of R is

$$F_R(r) = \Pr(R \leq r) = \Pr(X_1 \leq rX_2) = \Pr(X_1 - rX_2 \leq 0)$$

which, from (8.41), is

$$F_R(r; \lambda_1, \lambda_2) = \int_0^\infty F_{X_1}(rx_2) f_{X_2}(x_2)\, dx_2 = \int_0^\infty \left(1 - e^{-\lambda_1 r x_2}\right) \lambda_2 e^{-\lambda_2 x_2}\, dx_2$$

$$= 1 - \lambda_2 \int_0^\infty e^{-x_2(\lambda_1 r + \lambda_2)}\, dx_2 = 1 - \frac{\lambda_2}{\lambda_1 r + \lambda_2} = \frac{\lambda_1 r}{\lambda_1 r + \lambda_2}.$$

The density is

$$f_R(r; \lambda_1, \lambda_2) = \frac{\partial F_R(r)}{\partial r} = \frac{\lambda_1 \lambda_2}{(\lambda_1 r + \lambda_2)^2} \mathbb{I}_{(0,\infty)}(r)$$

$$= \frac{c}{(r+c)^2} \mathbb{I}_{(0,\infty)}(r), \qquad c = \frac{\lambda_2}{\lambda_1} = \frac{\mathbb{E}[X_1]}{\mathbb{E}[X_2]}, \tag{8.45}$$

which is similar to that in Example 7.2 with $a = 1$. In this case too, the mean (and higher moments) of R do not exist. ∎

⊛ **Example 8.18**   We wish to show that, if X and Y are independent $N(0, 1)$ r.v.s, then $Z = X/Y$ follows a Cauchy distribution with c.d.f. (7.23), i.e. $F_Z(z) = 1/2 + (\arctan z)/\pi$. To see this, write

$$F_Z(z) = \Pr(X/Y < z) = \Pr(X - zY < 0,\ Y > 0) + \Pr(X - zY > 0,\ Y < 0),$$

which is obtained by integrating the bivariate normal c.d.f. as

$$F_Z(z) = \int_0^\infty \int_{-\infty}^{zy} \frac{1}{2\pi} \exp\left[-\frac{1}{2}(x^2 + y^2)\right] dx\, dy$$

$$+ \int_{-\infty}^0 \int_{zy}^\infty \frac{1}{2\pi} \exp\left[-\frac{1}{2}(x^2 + y^2)\right] dx\, dy$$

$$:= I_1 + I_2.$$

Consider the case $0 < z < \infty$ or $0 < z^{-1} < \infty$ so that the range of integration is given in Figure 8.11(a), where $I_1 = (a) + (b)$ and $I_2 = (c) + (d)$. From the spherical

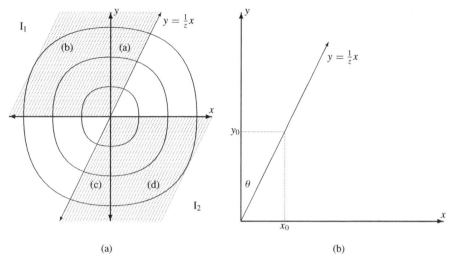

Figure 8.11 (a) The Cauchy c.d.f. is given by $I_1 + I_2$; (b) $\tan \theta = x_0/y_0 = z$

symmetry of the (zero-correlated) bivariate normal and the fact that it integrates to one, $(b) = (d) = 1/4$ and $(a) = (c) = \theta/2\pi$ so that

$$F_Z(z) = \frac{1}{2} + \frac{\theta}{\pi} = \frac{1}{2} + \frac{\arctan z}{\pi},$$

from Figure 8.11(b) with $\tan \theta = x$. A similar analysis holds for $z < 0$.

As an aside, there are distributions other than the normal for X and Y such that the ratio $Z = X/Y$ follows a Cauchy distribution (see Laha (1958), Kotlarski (1964), Popescu and Dumitrescu (1999) and the references therein for further details). The Student's t distribution is an example of a distribution which does not work; Figure 8.12 illustrates the joint density of X and Y when both are i.i.d. $t(v)$. ∎

⊖ ***Example 8.19*** Let the lifetime of two electrical devices, X_1 and X_2, be r.v.s with joint p.d.f. f_{X_1,X_2}. Interest centers on the distribution of the first device, given the information that the first device lasted longer than the second. The conditional c.d.f. of X_1 given that $X_1 > X_2$ is

$$F_{X_1|(X_1>X_2)}(t) = \Pr(X_1 \le t \mid X_1 > X_2) = \frac{\Pr((X_1 \le t) \cap (X_1 > X_2))}{\Pr(X_1 > X_2)}$$

$$= \frac{\iint_{y<x,\ x\le t} f_{X_1,X_2}(x, y)\ \mathrm{d}y\ \mathrm{d}x}{\iint_{y<x} f_{X_1,X_2}(x, y)\ \mathrm{d}y\ \mathrm{d}x}$$

$$= \frac{\int_{-\infty}^{t} \int_{-\infty}^{x} f_{X_1,X_2}(x, y)\ \mathrm{d}y\ \mathrm{d}x}{\int_{-\infty}^{\infty} \int_{-\infty}^{x} f_{X_1,X_2}(x, y)\ \mathrm{d}y\ \mathrm{d}x} = \frac{I(t)}{I(\infty)}, \qquad (8.46)$$

where $I(t)$ is defined as the numerator expression in (8.46). Depending on f_{X_1,X_2}, the value of $I(t)$ might need to be computed numerically.

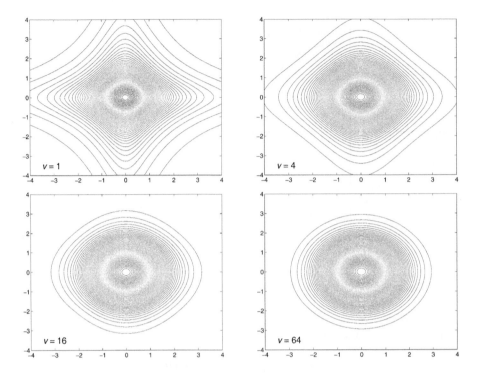

Figure 8.12 Bivariate distribution of i.i.d. Student's t r.v.s with v degrees of freedom

If we now assume that X_1 and X_2 are independent, then

$$I(t) = \int_{-\infty}^{t} \int_{-\infty}^{x} f_{X_1}(x) f_{X_2}(y) \, \mathrm{d}y \, \mathrm{d}x = \int_{-\infty}^{t} f_{X_1}(x) F_{X_2}(x) \, \mathrm{d}x, \qquad (8.47)$$

and further assuming that X_1 and X_2 are i.i.d. with common p.d.f. f_X, integrating by parts with $u = F_X(x)$ and $\mathrm{d}v = f_X(x) \, \mathrm{d}x$ gives

$$I(t) = [F_X(t)]^2 - \int_{-\infty}^{t} F_X(x) f_X(x) \, \mathrm{d}x = [F_X(t)]^2 - I(t),$$

or

$$I(t) = \frac{1}{2} [F_X(t)]^2, \qquad (8.48)$$

so that

$$F_{X_1|(X_1 > X_2)}(t) = \frac{[F_X(t)]^2 / 2}{[F_X(\infty)]^2 / 2} = [F_X(t)]^2.$$

For example, if $X_i \overset{\text{i.i.d.}}{\sim} \text{Exp}(\lambda)$, $i = 1, 2$, then $F_{X_1|(X_1 > X_2)}(t; \lambda) = \left[1 - e^{-\lambda t}\right]^2$.

If interest instead centers on the conditional c.d.f. of X_1 given that $X_1 < X_2$, then, similar to the above derivation and for $t < y$,

$$F_{X_1|(X_1<X_2)}(t) = \frac{\int_{-\infty}^t \int_x^\infty f_{X_1,X_2}(x,y)\, dy\, dx}{\int_{-\infty}^\infty \int_x^\infty f_{X_1,X_2}(x,y)\, dy\, dx} = \frac{J(t)}{J(\infty)},$$

and in the i.i.d. case, using (8.47) and (8.48) gives

$$J(t) = \int_{-\infty}^t f_X(x)[1 - F_X(x)]\, dx = \int_{-\infty}^t f_X(x)\, dx - \int_{-\infty}^t f_X(x) F_X(x)\, dx$$

$$= F_X(t) - \frac{1}{2}[F_X(t)]^2,$$

so that

$$F_{X_1|(X_1<X_2)}(t) = \frac{J(t)}{J(\infty)} = \frac{F_X(t) - \frac{1}{2}[F_X(t)]^2}{1 - \frac{1}{2}} = 2F_X(t) - [F_X(t)]^2$$

$$= 1 - [1 - F_X(t)]^2.$$

If $X_i \overset{\text{i.i.d.}}{\sim} \text{Exp}(\lambda)$, $i = 1, 2$, then

$$F_{X_1|(X_1<X_2)}(t) = 2\left(1 - e^{-\lambda t}\right) - \left(1 - e^{-\lambda t}\right)^2 = 1 - e^{-2\lambda t},$$

showing that $X_1 \mid (X_1 < X_2) \sim \exp(2\lambda)$. The Matlab code

```
lam1=1/10; lam2=1/10; s=10000; X=exprnd(1,s,1)/lam1;
Y=exprnd(1,s,1)/lam2; xly=X(find(X<Y)); hist(xly,40)
```

can be used to simulate the density of $X_1 \mid (X_1 < X_2)$ when X_1 and X_2 are independent. In Volume II, these results in the i.i.d. case will be generalized in the context of order statistics. See also Problem 8.14. ∎

8.3 Problems

There are two kinds of pain in life – the pain of discipline and the pain of regret. The pain of discipline weighs ounces and lasts for moments. The pain of regret weighs tons and lasts forever. (Jim Rohn)

Forget past mistakes. Forget failures. Forget everything except what you are going to do now and do it. (William Durant, founder of General Motors)

8.1. Let $R \sim t(\nu)$ for $\nu > 1$ and let $\Phi_\nu(z)$ denote its c.d.f. Compute $\mathbb{E}[R \mid R < c]$.

8.2. ★ ★ In the beginning of a game of bridge, 52 shuffled cards are dealt to four people, so that each gets 13 cards.

(a) What is the probability that two bridge players are dealt v_1 and v_2 hearts, respectively, where $v_1 + v_2 \leq 13$? Denote the two players of interest as A_1 and A_2, and the other two as B_1 and B_2.

(b) What is the probability that between two players, say A_1 and A_2, they have exactly $v = v_1 + v_2$ hearts, $v \leq 13$?

(c) Let C be the event that, between two players, say A_1 and A_2, one has v_1 hearts and the other v_2 hearts; and let D be the event that between these two, they have exactly $v = v_1 + v_2$ hearts, $v \leq 13$. Calculate $\Pr(C \mid D)$. Evaluate for $v_1 = 3$ and $v_2 = 2$.

(d) (Continuation of part (c)) Let $f(v_1 \mid v) = \Pr(C \mid D)$ in a sloppy r.v. notation. What is $\mathbb{E}[v_1 \mid v]$ and $\mathbb{V}(v_1 \mid v)$?

8.3. If $X_i \overset{\text{ind}}{\sim} \text{Exp}(\lambda_i)$, $i = 1, 2$, compute $\Pr(X_1 < aX_2 \mid X_1 < bX_2)$ for $0 < a \leq b < \infty$. Show algebraically and argue informally what happens in the limit as $b \to \infty$.

8.4. If X and Y are jointly distributed uniform over the unit square, determine $F_Z(z)$ and $f_Z(z)$, where $Z = XY$.

8.5. ★ Let N_i be independent $\text{Poi}(\lambda_i)$ processes, $i = 1, 2$ and let S_n^1 be the time of the nth event of the first process, and S_m^2 be the time of the mth event of the second process.

(a) What is $\Pr(S_1^1 < S_1^2)$?

(b) Using the memoryless property of the exponential, what is $\Pr(S_n^1 < S_1^2)$, i.e. n events from process 1 occur before any from process 2?

(c) At any given time, what is the probability that the next event which occurs will come from process i, $i = 1, 2$?

(d) In general, what is $\Pr(S_n^1 < S_m^2)$?

8.6. ★ ★ The number of winning tickets in a lottery is given by random variable $Y \sim \text{Bin}(N, \theta)$, whereby N, the number of total tickets, is assumed known. If a person buys n tickets, let X be the number of winning tickets among the n.

(a) Specify (without calculation) the conditional density, support, and expected value of $X \mid Y$.

(b) Calculate the joint density from X and Y.

(c) With respect to the definitions of N, n, y and x, show that

$$\sum_{y=0}^{N} \binom{N-n}{y-x} \theta^y (1-\theta)^{N-y} = \theta^x (1-\theta)^{n-x}.$$

(d) Calculate $f_X(x)$ and the conditional density $f_{Y|X}(y \mid x)$.

(e) Interpret your answer to part **(d)**. Could you have specified the distribution $f_{Y|X}(y \mid x)$ without having explicitly calculated the density?

(f) Calculate $\mathbb{E}[Y \mid X]$ and discuss the result for $n = 0$ and $n = N$.

(g) Calculate $\mathbb{E}[X]$:
   (i) directly using $f_X(x)$,
   (ii) using the relationship $\mathbb{E}[Y] = \mathbb{E}[\mathbb{E}[Y \mid X]]$
   (iii) and using the relationship $\mathbb{E}[X] = \mathbb{E}[\mathbb{E}[X \mid Y]]$.

8.7. ★ ★ Let the random variable $X \sim \text{Poi}(\lambda)$ describe the number of people that enter an elevator on the ground floor of a building with $N + 1$ floors. Let Y be the number of stops that the elevator needs to make, assuming that the X people independently choose one of the N floors and each floor is equally likely to be chosen.

(a) Show that $\mathbb{E}[Y] = N\left(1 - e^{-\lambda/N}\right)$. (Hint: for $i = 1, \ldots, N$, define

$$I_i = \begin{cases} 1, & \text{if the elevator stops on floor } i, \\ 0, & \text{otherwise,} \end{cases}$$

so that $Y = \sum_{i=1}^{N} I_i$.)

(b) Interpret the result for $N = 1$ and explain what happens as λ increases.

(c) Explain intuitively what $\lim_{N \to \infty} \mathbb{E}[Y]$ should be.

(d) Prove your result from the previous question. (Hint: either try a direct Taylor series approach, or consider the substitution $M = 1/n$.)

8.8. Calculate f_X, f_Y, $f_{X|Y}$ and $f_{Y|X}$ from the bivariate density

$$f_{X,Y}(x, y) = 2y\left(2 - x^2 - y\right)\mathbb{I}_{(0,1)}(x)\,\mathbb{I}_{(0,1)}(y),$$

plot all the densities and algebraically verify formulae (8.16) and (8.17).

8.9. From the joint density

$$f_{X,Y}(x, y) = cx^{a-1}(y - x)^{b-1}e^{-y}\mathbb{I}_{(0,\infty)}(x)\,\mathbb{I}_{(x,\infty)}(y),$$

derive c, f_X, f_Y, $f_{X|Y}$ and $f_{Y|X}$.   (Cacoullos, 1989, p. 44(174))

8.10. ★ Recall Example 8.14, in which independent Bernoulli trials are repeatedly performed with the same success probability p, and the number of trials is $N \sim \text{Poi}(\lambda)$. Let X be the number of successes. Derive the mass function $f_X(x) = \Pr(X = x)$.

8.11. ★ Consider the joint density function

$$f_{X,Y}(x, y) = e^{-x/y}e^{-y}y^{-1}\mathbb{I}_{(0,\infty)}(x)\,\mathbb{I}_{(0,\infty)}(y).$$

(a) Compute the marginal density of Y.

(b) Compute the conditional density of $X \mid Y$.

(c) Write two integral expressions for $\Pr(X > Y)$ and compute it using one of them.

8.12. ★ Derive (8.22).

8.13. ★ ★ As in Example 8.13, let $N_m \sim \mathrm{Consec}(m, p)$ and $E_m := \mathbb{E}[N_m]$.

(a) Show that

$$V(N_m) = \frac{1 - (1 + 2m)(1 - p)p^m - p^{2m+1}}{(1 - p)^2 p^{2m}}. \tag{8.49}$$

(Hint: compute $G_m := \mathbb{E}[N_m^2] = \mathbb{E}[\mathbb{E}[N_m^2 \mid N_{m-1}]]$, from which

$$V(N_m) = G_m - E_m^2$$

can be obtained. To do this, first show that

$$G_m = \frac{1}{p}\left[G_{m-1} + 2E_{m-1} + 1 + 2E_m(1 - p)(E_{m-1} + 1)\right]. \tag{8.50}$$

Then simplify for a few values of m and show that

$$G_m = \frac{p^{2m} - (3 + 2m)p^m - p^{2m+1} + (1 + 2m)p^{m+1} + 2}{(1 - p)^2 p^{2m}}, \tag{8.51}$$

from which (8.49) follows. Finally, to verify that (8.51) and, hence, (8.49) hold for all m, substitute the identity $E_{m-1} = pE_m - 1$ from (8.32) into (8.50), and recursively substitute to get

$$G_m = \left(\frac{2}{1 - p}\right)\left(\frac{p^{-2m} - p^{-m}}{1 - p} - mp^{-m}\right) - \frac{p^{-m} - 1}{1 - p},$$

and show that this is equivalent to (8.51).)
Based on the program in Listing 2.1, simulation can be used to corroborate the expected value (8.33) and variance (8.49). For example, these formulae with $p = 0.6$ and $m = 3$ give $\mathbb{E}[N_3] = 9.0741$ and $V(N_3) = 49.1907$. Simulation based on two million replications yields 9.0764 and 49.075, respectively.
The mean and variance of N_m can also be found by the use of *generating functions*; see, for example, Port (1994, p. 257(21.10))).

(b) Let $M_m := \mathbb{E}\left[e^{tN_m}\right]$ and show that

$$M_m = \frac{p\,e^t\,M_{m-1}}{1 - q\,e^t\,M_{m-1}}.$$

(8.52)

8.14. A red dot is placed randomly on a circle of unit perimeter. If you cut the perimeter of the circle twice, you obtain two pieces, one of which will contain the dot. What is the expected value of the length of the piece with the dot?

9

Multivariate transformations

In Section 7.3, we derived an expression for the pdf of $Y = g(X)$ for continuous random variable X as $f_Y(y) = f_X(x) |dx/dy|$, where $x = g^{-1}(y)$. The generalization to the multivariate case is more difficult to derive, but the result is straightforward to implement. The technique is relevant in numerous cases of important interest, some of which will be explored in this chapter, and others in Volume II.

9.1 Basic transformation

Let $\mathbf{X} = (X_1, \ldots, X_n)$ be an n-dimensional continuous r.v. and $\mathbf{g} = (g_1(\mathbf{x}), \ldots, g_n(\mathbf{x}))$ a continuous bijection which maps $\mathcal{S}_\mathbf{X} \subset \mathbb{R}^n$, the support of $\mathbf{X}$, onto $\mathcal{S}_\mathbf{Y} \subset \mathbb{R}^n$. Then the p.d.f. of $\mathbf{Y} = (Y_1, \ldots, Y_n) = \mathbf{g}(\mathbf{X})$ is given by

$$\boxed{f_\mathbf{Y}(\mathbf{y}) = f_\mathbf{X}(\mathbf{x}) |\det \mathbf{J}|}, \tag{9.1}$$

where $\mathbf{x} = \mathbf{g}^{-1}(\mathbf{y}) = \left(g_1^{-1}(\mathbf{y}), \ldots, g_n^{-1}(\mathbf{y}) \right)$ and

$$
\mathbf{J} = \begin{pmatrix}
\dfrac{\partial x_1}{\partial y_1} & \dfrac{\partial x_1}{\partial y_2} & \cdots & \dfrac{\partial x_1}{\partial y_n} \\
\dfrac{\partial x_2}{\partial y_1} & \dfrac{\partial x_2}{\partial y_2} & & \dfrac{\partial x_2}{\partial y_n} \\
\vdots & \vdots & \ddots & \vdots \\
\dfrac{\partial x_n}{\partial y_1} & \dfrac{\partial x_n}{\partial y_2} & \cdots & \dfrac{\partial x_n}{\partial y_n}
\end{pmatrix}
= \begin{pmatrix}
\dfrac{\partial g_1^{-1}(\mathbf{y})}{\partial y_1} & \dfrac{\partial g_1^{-1}(\mathbf{y})}{\partial y_2} & \cdots & \dfrac{\partial g_1^{-1}(\mathbf{y})}{\partial y_n} \\
\dfrac{\partial g_2^{-1}(\mathbf{y})}{\partial y_1} & \dfrac{\partial g_2^{-1}(\mathbf{y})}{\partial y_2} & & \dfrac{\partial g_2^{-1}(\mathbf{y})}{\partial y_n} \\
\vdots & \vdots & \ddots & \vdots \\
\dfrac{\partial g_n^{-1}(\mathbf{y})}{\partial y_1} & \dfrac{\partial g_n^{-1}(\mathbf{y})}{\partial y_2} & \cdots & \dfrac{\partial g_n^{-1}(\mathbf{y})}{\partial y_n}
\end{pmatrix}
\tag{9.2}
$$

is the Jacobian of $\mathbf{g}$. See Section A.3.3 for a discussion of the Jacobian matrix and its interpretation in terms of tangent maps. Notice that (9.1) reduces to the equation used in the univariate case.

A straightforward example should help illustrate the procedure.

⊛ ***Example 9.1*** Let X and Y be continuous r.v.s with joint distribution $f_{X,Y}$ and define $S = X + Y$. The density f_S is given in (8.42), where it was derived using a conditioning argument. To derive it using a bivariate transformation, a second, 'dummy' variable is required, which can often be judiciously chosen so as to simplify the calculation. In this case, we take $T = Y$, which is both simple and such that $(s, t) = (g_1(x, y), g_2(x, y)) = (x + y, y)$ is a bijection. In particular, the inverse transformation is easily seen to be $(x, y) = \left(g_1^{-1}(s, t), g_2^{-1}(s, t) \right) = (s - t, t)$, so that, from (9.1),

$$f_{S,T}(s, t) = |\det \mathbf{J}| \, f_{X,Y}(x, y),$$

where

$$\mathbf{J} = \begin{bmatrix} \partial x / \partial s & \partial x / \partial t \\ \partial y / \partial s & \partial y / \partial t \end{bmatrix} = \begin{bmatrix} 1 & -1 \\ 0 & 1 \end{bmatrix}, \quad |\det \mathbf{J}| = 1,$$

or $f_{S,T}(s, t) = f_{X,Y}(s - t, t)$. Thus,

$$f_S(s) = \int_{-\infty}^{\infty} f_{X,Y}(s - t, t) \, dt, \tag{9.3}$$

as in (8.42) for X and Y independent. ∎

In stark comparison to the univariate result in (7.65), relation (9.1) does not seem to exhibit an abundance of intuition. However, it actually is just an application of the multidimensional change of variable formula (A.165). Let $h : \mathbb{R}^n \to \mathbb{R}$ be a bounded, measurable function so that, from (5.18) and (A.165),

$$\mathbb{E}\left[h\left(\mathbf{Y}\right)\right] = \mathbb{E}\left[h\left(\mathbf{g}\left(\mathbf{X}\right)\right)\right] = \int_{S_{\mathbf{X}}} h\left(\mathbf{g}\left(\mathbf{x}\right)\right) f_{\mathbf{X}}\left(\mathbf{x}\right) d\mathbf{x} = \int_{S_{\mathbf{Y}}} h\left(\mathbf{y}\right) f_{\mathbf{X}}\left(\mathbf{g}^{-1}\left(\mathbf{y}\right)\right) |\det \mathbf{J}| \, d\mathbf{y}.$$

In particular, let $h = \mathbb{I}_B(\mathbf{y})$ for a Borel set $B \in \mathcal{B}^n$, so that

$$\mathbb{E}\left[h\left(\mathbf{Y}\right)\right] = \Pr\left(\mathbf{Y} \in B\right) = \int_{S_{\mathbf{Y}}} \mathbb{I}_B(\mathbf{y}) f_{\mathbf{X}}\left(\mathbf{g}^{-1}\left(\mathbf{y}\right)\right) |\det \mathbf{J}| \, d\mathbf{y}.$$

As this holds for all $B \in \mathcal{B}^n$, $f_{\mathbf{X}}\left(\mathbf{g}^{-1}\left(\mathbf{y}\right)\right) |\det \mathbf{J}|$ is a probability density function for $\mathbf{Y}$.

It is sometimes computationally advantageous to use the fact that $|\mathbf{J}| = 1/\left|\mathbf{J}^{-1}\right|$, where

$$\mathbf{J}^{-1} = \begin{pmatrix} \dfrac{\partial y_1}{\partial x_1} & \dfrac{\partial y_1}{\partial x_2} & \cdots & \dfrac{\partial y_1}{\partial x_n} \\ \dfrac{\partial y_2}{\partial x_1} & \dfrac{\partial y_2}{\partial x_2} & \cdots & \dfrac{\partial y_2}{\partial x_n} \\ \vdots & \vdots & \ddots & \vdots \\ \dfrac{\partial y_n}{\partial x_1} & \dfrac{\partial y_n}{\partial x_2} & \cdots & \dfrac{\partial y_n}{\partial x_n} \end{pmatrix} = \begin{pmatrix} \dfrac{\partial g_1(\mathbf{x})}{\partial x_1} & \dfrac{\partial g_1(\mathbf{x})}{\partial x_2} & \cdots & \dfrac{\partial g_1(\mathbf{x})}{\partial x_n} \\ \dfrac{\partial g_2(\mathbf{x})}{\partial x_1} & \dfrac{\partial g_2(\mathbf{x})}{\partial x_2} & \cdots & \dfrac{\partial g_2(\mathbf{x})}{\partial x_n} \\ \vdots & \vdots & \ddots & \vdots \\ \dfrac{\partial g_n(\mathbf{x})}{\partial x_1} & \dfrac{\partial g_n(\mathbf{x})}{\partial x_2} & \cdots & \dfrac{\partial g_n(\mathbf{x})}{\partial x_n} \end{pmatrix},$$

$$\tag{9.4}$$

which, in the univariate case, reduces to

$$\frac{dy}{dx} = 1/(dx/dy).\tag{9.5}$$

⊛ ***Example 9.2*** Let X and Y be continuous random variables with joint distribution $f_{X,Y}$ and define $P = XY$. Let $Q = Y$ so that the inverse transformation of $\{p = xy, q = y\}$ is $\{x = p/q, y = q\}$, and $f_{P,Q}(p, q) = |\det J| f_{X,Y}(x, y)$, where

$$\mathbf{J} = \begin{bmatrix} \partial x/\partial p & \partial x/\partial q \\ \partial y/\partial p & \partial y/\partial q \end{bmatrix} = \begin{bmatrix} 1/q & -pq^{-2} \\ 0 & 1 \end{bmatrix}, \quad |\det \mathbf{J}| = \frac{1}{|q|},$$

or $f_{P,Q}(p, q) = |q|^{-1} f_{X,Y}(p/q, q)$. Thus,

$$f_P(p) = \int_{-\infty}^{\infty} |q|^{-1} f_{X,Y}\left(\frac{p}{q}, q\right) dq.$$

Notice also that

$$\mathbf{J}^{-1} = \begin{bmatrix} \partial p/\partial x & \partial p/\partial y \\ \partial q/\partial x & \partial q/\partial y \end{bmatrix} = \begin{bmatrix} y & x \\ 0 & 1 \end{bmatrix} = \begin{bmatrix} q & p/q \\ 0 & 1 \end{bmatrix}, \quad |\det \mathbf{J}^{-1}| = |q|$$

and $\mathbf{J}\mathbf{J}^{-1} = \mathbf{I}$. ∎

It should be noted that relation (9.5) does not apply to the partial derivative elements in $\mathbf{J}$. For instance, in the previous example, regarding the diagonal elements of $\mathbf{J}$ and $\mathbf{J}^{-1}$,

$$\frac{1}{\partial p/\partial x} = \frac{1}{y} = \frac{1}{q} = \partial x/\partial p \quad \text{and} \quad \frac{1}{\partial q/\partial y} = \frac{1}{1} = \partial y/\partial q,$$

so that (9.5) indeed holds but, because $\partial y/\partial p = \partial q/\partial x = 0$, this clearly does not work for the off-diagonal elements.

The next two examples further illustrate the technique and also show that some care needs to be exercised when examining the support of the transformed r.v.s.

↺ ***Example 9.3*** Let $X_i \overset{\text{i.i.d.}}{\sim} \text{Exp}(\lambda)$, $i = 1, \ldots, n$, and define $Y_i = X_i$, $i = 2, \ldots, n$, and $S = \sum_{i=1}^{n} X_i$. This is a 1–1 transformation with $X_i = Y_i$, $i = 2, \ldots, n$, and $X_1 = S - \sum_{i=2}^{n} Y_i$. In this case, both (9.2) or (9.4) are easy to compute. Using the latter, the inverse Jacobian is

$$\mathbf{J}^{-1} = \begin{bmatrix} \frac{\partial S}{\partial x_1} & \frac{\partial S}{\partial x_2} & \frac{\partial S}{\partial x_3} & \cdots & \frac{\partial S}{\partial x_n} \\ \frac{\partial y_2}{\partial x_1} & \frac{\partial y_2}{\partial x_2} & \frac{\partial y_2}{\partial x_3} & \cdots & \frac{\partial y_2}{\partial x_n} \\ \frac{\partial y_3}{\partial x_1} & \frac{\partial y_3}{\partial x_2} & \frac{\partial y_3}{\partial x_3} & \cdots & \frac{\partial y_3}{\partial x_n} \\ \vdots & \vdots & \vdots & \ddots & \\ \frac{\partial y_n}{\partial x_1} & \frac{\partial y_n}{\partial x_2} & \frac{\partial y_n}{\partial x_3} & \cdots & \frac{\partial y_n}{\partial x_n} \end{bmatrix} = \begin{bmatrix} 1 & 1 & 1 & \cdots & 1 \\ 0 & 1 & 0 & \cdots & 0 \\ 0 & 0 & 1 & \cdots & 0 \\ \vdots & \vdots & \vdots & \ddots & \vdots \\ 0 & 0 & 0 & \cdots & 1 \end{bmatrix},$$

with determinant 1, so that

$$f_{S,Y}(s, y) = 1 \cdot f_X(x) = \lambda^n e^{-\lambda s} \mathbb{I}_{(0,\infty)} \left(s - \sum_{i=2}^n y_i\right) \prod_{i=2}^n \mathbb{I}_{(0,\infty)}(y_i)$$

$$= \lambda^n e^{-\lambda s} \mathbb{I}_{(0,\infty)}(s) \mathbb{I}_{(0,s)} \left(\sum_{i=2}^n y_i\right) \prod_{i=2}^n \mathbb{I}_{(0,\infty)}(y_i).$$

By expressing the indicator functions as

$$\mathbb{I}_{(0,\infty)}(s) \mathbb{I}_{(0,s)}(y_n) \mathbb{I}_{(0,s-y_n)}(y_{n-1}) \mathbb{I}_{(0,s-y_n-y_{n-1})}(y_{n-2}) \cdots \mathbb{I}_{(0,s-y_n-\cdots-y_3)}(y_2),$$

the Y_i can be integrated out to give the density of S:

$$f_S(s) = \lambda^n e^{-\lambda s} \int_0^s \int_0^{s-y_n} \int_0^{s-y_n-y_{n-1}}$$

$$\cdots \int_0^{s-y_n-\cdots-y_3} dy_2 \cdots dy_{n-2} dy_{n-1} dy_n \, \mathbb{I}_{(0,\infty)}(s).$$

Taking $n = 6$ for illustration and treating quantities in square brackets as constants (set them to t, say, to help see things), we get

$$\int_0^s \int_0^{s-y_6} \int_0^{s-y_6-y_5} \int_0^{s-y_6-y_5-y_4} \int_0^{s-y_6-y_5-y_4-y_3} dy_2 \, dy_3 \, dy_4 \, dy_5 \, dy_6$$

$$= \int_0^s \int_0^{s-y_6} \int_0^{s-y_6-y_5} \int_0^{[s-y_6-y_5-y_4]} \left([s - y_6 - y_5 - y_4] - y_3\right) dy_3 \, dy_4 \, dy_5 \, dy_6$$

$$= \frac{1}{2} \int_0^s \int_0^{s-y_6} \int_0^{[s-y_6-y_5]} \left([s - y_6 - y_5] - y_4\right)^2 dy_4 \, dy_5 \, dy_6$$

$$= \frac{1}{2}\frac{1}{3} \int_0^s \int_0^{[s-y_6]} \left([s - y_6] - y_5\right)^3 dy_5 \, dy_6$$

$$= \frac{1}{2}\frac{1}{3}\frac{1}{4} \int_0^s (s - y_6)^4 \, dy_6 = \frac{1}{2}\frac{1}{3}\frac{1}{4}\frac{1}{5} s^5,$$

which is just a series of simple integrals. We conclude without a formal proof by induction that

$$f_S(s) = \frac{\lambda^n}{(n-1)!} e^{-\lambda s} s^{n-1} \mathbb{I}_{(0,\infty)}(s),$$

or that $S \sim \text{Gam}(n, \lambda)$ (see also Problem 9.3). ∎

Notice that the above example also implies that, if $S_i \overset{\text{ind}}{\sim} \text{Gam}(n_i, \lambda)$, then

$$\sum_{i=1}^k S_i \sim \text{Gam}(n_\bullet, \lambda), \text{ where } n_\bullet = \sum_{i=1}^k n_i.$$

This is a fundamental result useful in various probabilistic and statistical contexts. The following is one application of such.

⊖ **Example 9.4** (Dubey, 1965) Let $X_i \overset{ind}{\sim} \text{Gam}(a_i, c)$, $i = 1, 2$ and define S as $X_1 + X_2$ which, from the above result, follows a $\text{Gam}(a_1 + a_2, c)$ distribution. This implies that

$$\mathbb{E}[S^k] = \mathbb{E}[(X_1 + X_2)^k] = \sum_{i=0}^{k} \binom{k}{i} \mathbb{E}[X_1^i] \mathbb{E}[X_2^{k-i}]$$

or, using (7.9),

$$\frac{\Gamma(k + a_1 + a_2)}{c^k \Gamma(a_1 + a_2)} = \sum_{i=0}^{k} \binom{k}{i} \frac{\Gamma(i + a_1)}{c^i \Gamma(a_1)} \frac{\Gamma(k - i + a_2)}{c^{k-i} \Gamma(a_2)}. \tag{9.6}$$

That is,

$$\frac{(k + a_1 + a_2 - 1)!}{(a_1 + a_2 - 1)! k!} = \sum_{i=0}^{k} \frac{(i + a_1 - 1)! (k - i + a_2 - 1)!}{i! (a_1 - 1)! (k - i)! (a_2 - 1)!}$$

or

$$\binom{k + a_1 + a_2 - 1}{k} = \sum_{i=0}^{k} \binom{i + a_1 - 1}{i} \binom{k - i + a_2 - 1}{k - i},$$

which is precisely (1.58). Rearranging (9.6) gives

$$\frac{\Gamma(a_1) \Gamma(a_2)}{\Gamma(a_1 + a_2)} = \sum_{i=0}^{k} \binom{k}{i} \frac{\Gamma(i + a_1) \Gamma(k - i + a_2)}{\Gamma(k + a_1 + a_2)}$$

or, from (1.48),

$$B(a_1, a_2) = \sum_{i=0}^{k} \binom{k}{i} B(a_1 + i, a_2 + k - i). \tag{9.7}$$

Letting $X \sim \text{Beta}(a_1, a_2)$, this latter result also follows directly from

$$1 = \int_0^1 f_X(x)\, dx = \frac{1}{B(a_1, a_2)} \int_0^1 (x + 1 - x)^k x^{a_1 - 1} (1 - x)^{a_2 - 1}\, dx$$

$$= \frac{1}{B(a_1, a_2)} \sum_{i=0}^{k} \binom{k}{i} \int_0^1 x^{a_1 + i - 1} (1 - x)^{a_2 - 1 + k - i}\, dx$$

$$= \frac{1}{B(a_1, a_2)} \sum_{i=0}^{k} \binom{k}{i} B(a_1 + i, a_2 + k - i),$$

using the binomial theorem. ■

⊚ **Example 9.5** Consider calculating the joint distribution of $S = X + Y$ and $D = X - Y$ and their marginals for $X, Y \overset{\text{i.i.d.}}{\sim} \exp(\lambda)$. Adding and subtracting the two equations yields $X = (S + D)/2$ and $Y = (S - D)/2$. From these,

$$
\mathbf{J} = \begin{bmatrix} \partial x/\partial s & \partial x/\partial d \\ \partial y/\partial s & \partial y/\partial d \end{bmatrix} = \begin{bmatrix} 1/2 & 1/2 \\ 1/2 & -1/2 \end{bmatrix}, \quad \det \mathbf{J} = -\frac{1}{2},
$$

so that

$$
f_{S,D}(s, d) = |\det \mathbf{J}|\, f_{X,Y}(x, y) = \frac{\lambda^2}{2} \exp(-\lambda s)\, \mathbb{I}_{(0,\infty)}(s + d)\, \mathbb{I}_{(0,\infty)}(s - d).
$$

The constraints imply $s > d$ and $s > -d$; if $d < 0$ $(d > 0)$ then $s > -d$ $(s > d)$ is relevant. Thus,

$$
f_D(d) = \mathbb{I}_{(-\infty,0)}(d) \int_{-d}^{\infty} f_{S,D}(s, d)\, ds + \mathbb{I}_{(0,\infty)}(d) \int_{d}^{\infty} f_{S,D}(s, d)\, ds
$$

$$
= \mathbb{I}_{(-\infty,0)}(d)\, \frac{\lambda}{2} e^{\lambda d} + \mathbb{I}_{(0,\infty)}(d)\, \frac{\lambda}{2} e^{-\lambda d}
$$

$$
= \frac{\lambda}{2} e^{-\lambda |d|},
$$

i.e. $D \sim \text{Lap}(0, \lambda)$. Next, $s + d > 0$ and $s - d > 0$ imply that $-s < d < s$, giving

$$
f_S(s) = \int_{-s}^{s} \frac{\lambda^2}{2} \exp(-\lambda s)\, dd = \frac{\lambda^2}{2} \exp(-\lambda s) \int_{-s}^{s} dd = s\lambda^2 e^{-\lambda s} \mathbb{I}_{(0,\infty)}(s),
$$

which follows because $S > 0$ from its definition. Thus, $S \sim \text{Gam}(2, \lambda)$, which agrees with the more general result established in Example 9.3. ∎

⊛ **Example 9.6** Let $Z_i \overset{\text{i.i.d.}}{\sim} N(0, 1)$, $i = 1, 2$. The inverse transformation and Jacobian of $S = Z_1 + Z_2$ and $D = Z_1 - Z_2$ was derived in Example 9.5, so that

$$
f_{S,D}(s, d) = \frac{1}{2} f_{Z_1,Z_2}(z_1, z_2)
$$

$$
= \frac{1}{2} \frac{1}{\sqrt{2\pi}} \exp\left(-\frac{1}{2} z_1^2\right) \frac{1}{\sqrt{2\pi}} \exp\left(-\frac{1}{2} z_2^2\right)
$$

$$
= \frac{1}{2} \frac{1}{\sqrt{2\pi}} \exp\left[-\frac{1}{2}\left(\frac{s+d}{2}\right)^2\right] \frac{1}{\sqrt{2\pi}} \exp\left[-\frac{1}{2}\left(\frac{s-d}{2}\right)^2\right]
$$

$$
= \frac{1}{2} \frac{1}{\sqrt{2\pi}} \frac{1}{\sqrt{2\pi}} \exp\left[-\frac{1}{2}\left(\frac{1}{2}d^2 + \frac{1}{2}s^2\right)\right]
$$

$$
= \frac{1}{\sqrt{2}} \frac{1}{\sqrt{2\pi}} \exp\left[-\frac{1}{2}\left(\frac{d}{\sqrt{2}}\right)^2\right] \frac{1}{\sqrt{2}} \frac{1}{\sqrt{2\pi}} \exp\left[-\frac{1}{2}\left(\frac{s}{\sqrt{2}}\right)^2\right],
$$

i.e. S and D are independent, with $S \sim N(0, 2)$ and $D \sim N(0, 2)$. ∎

9.2 The *t* and F distributions

We introduced the *t* and F distributions in Chapter 7, and mentioned their great relevance in statistical inference. The following two examples show how these distributions arise; a chapter in Volume II will discuss important generalizations (noncentral *t* and F) and later chapters will illustrate their use in statistical inference.

⊛ **Example 9.7**   Let $G \sim N(0, 1)$ independent of $C \sim \chi_n^2$ with joint p.d.f.

$$f_{G,C}(g, c) = \frac{1}{\sqrt{2\pi}} \exp\left(-\frac{1}{2}g^2\right) \frac{1}{2^{n/2}\Gamma(n/2)} c^{n/2-1} \exp\left(-\frac{c}{2}\right) \mathbb{I}_{(0,\infty)}(c)$$

and define $T = G/\sqrt{C/n}$. With $Y = C$, the inverse transform is $G = T\sqrt{Y/n}$ and $C = Y$ so that

$$\mathbf{J} = \left(\begin{array}{cc} \partial G/\partial T & \partial G/\partial Y \\ \partial C/\partial T & \partial C/\partial Y \end{array} \right) = \left(\begin{array}{cc} \sqrt{Y/n} & \cdot \\ 0 & 1 \end{array} \right), \quad |\det \mathbf{J}| = \sqrt{\frac{Y}{n}}$$

and

$$f_{T,Y}(t, y) = |\det \mathbf{J}| \, f_{G,C}(g, c)$$

$$= \frac{1}{\sqrt{2\pi}} \frac{2^{-n/2}n^{-1/2}}{\Gamma(n/2)} y^{n/2-1+1/2} \exp\left[-\frac{1}{2}\left(t\sqrt{\frac{y}{n}}\right)^2 - \frac{y}{2}\right] \mathbb{I}_{(0,\infty)}(y).$$

With $u = y\left(1 + t^2/n\right)/2$, $dy = 2/\left(1 + t^2/n\right) du$ and

$$\int_0^\infty y^{n/2-1/2} \exp\left[-\frac{y}{2}\left(1 + \frac{t^2}{n}\right)\right] dy = \left(\frac{2}{1 + t^2/n}\right)^{n/2+1/2} \Gamma(n/2 + 1/2),$$

the marginal of T simplifies to

$$f_T(t) = \frac{1}{\sqrt{n\pi}} \frac{\Gamma\left[(n+1)/2\right]}{\Gamma(n/2)} \left(1 + t^2/n\right)^{-(n+1)/2} = \frac{n^{-1/2}}{B(n/2, 1/2)} \left(1 + t^2/n\right)^{-(n+1)/2},$$

i.e. $T \sim t_n$.   ∎

From the independence of G and C in the previous example, we can use (5.20) to calculate $\mathbb{E}[T]$ and $\mathbb{E}|T|$. The former is just

$$\mathbb{E}[T] = \mathbb{E}\left[\frac{G}{\sqrt{C/n}}\right] = \sqrt{n}\mathbb{E}[G]\mathbb{E}\left[C^{-1/2}\right] = 0, \quad n > 1.$$

Similarly,

$$\mathbb{E}\,|T| = \sqrt{n}\mathbb{E}\,|G|\,\mathbb{E}\left[C^{-1/2}\right] = \sqrt{n}\sqrt{2/\pi}\,\Gamma\left(\frac{n-1}{2}\right) \Big/ \sqrt{2}\,\Gamma\left(\frac{n}{2}\right)$$

$$= \frac{\sqrt{n}\,\Gamma\left(\frac{n-1}{2}\right)}{\sqrt{\pi}\,\Gamma\left(\frac{n}{2}\right)},\ n > 1,$$

from Examples 7.3 and 7.5.

⊛ **Example 9.8** Consider the distribution of the ratio of two independent χ^2 r.v.s divided by their respective degrees of freedom (usually integers, although they need only be positive real numbers). Let $X_i \overset{\text{ind}}{\sim} \chi^2_{n_i}$ with

$$f_{X_1,X_2(x_1,x_2)} = K\, x_1^{(n_1-2)/2}\, x_2^{(n_2-2)/2}\,\exp\left[-\frac{1}{2}(x_1+x_2)\right]\mathbb{I}_{(0,\infty)}(x_1)\,\mathbb{I}_{(0,\infty)}(x_2),$$

$K^{-1} = \Gamma(n_1/2)\,\Gamma(n_2/2)\,2^{(n_1+n_2)/2}$, and define $Y_1 = (X_1/n_1)/(X_2/n_2)$, $Y_2 = X_2$ with inverse bijective functions $x_2 = y_2$ and $x_1 = y_1 y_2\,(n_1/n_2)$. Then

$$\mathbf{J} = \begin{pmatrix} \partial x_1/\partial y_1 & \partial x_1/\partial y_2 \\ \partial x_2/\partial y_1 & \partial x_2/\partial y_2 \end{pmatrix} = \begin{pmatrix} y_2\,(n_1/n_2) & \cdot \\ 0 & 1 \end{pmatrix},\quad |\det \mathbf{J}| = y_2\frac{n_1}{n_2}$$

and $f_{Y_1,Y_2}(y_1, y_2)$ is given by

$$y_2\frac{n_1}{n_2}\,K\left(y_1 y_2\frac{n_1}{n_2}\right)^{(n_1-2)/2} y_2^{(n_2-2)/2}\,\exp\left[-\frac{1}{2}\left(y_1 y_2\frac{n_1}{n_2}+y_2\right)\right]\mathbb{I}_{(0,\infty)}(y_1 y_2)\,\mathbb{I}_{(0,\infty)}(y_2)$$

or

$$K\left(\frac{n_1}{n_2}\right)^{n_1/2} y_1^{(n_1-2)/2} y_2^{(n_1+n_2-2)/2}\,\exp\left[-\frac{1}{2}\left(y_1 y_2\frac{n_1}{n_2}+y_2\right)\right]\mathbb{I}_{(0,\infty)}(y_1)\,\mathbb{I}_{(0,\infty)}(y_2).$$

Observe that, with $u = \frac{1}{2}\left(y_1 y_2\frac{n_1}{n_2}+y_2\right)$ and $y_2 = 2u\,/\left(1+y_1\frac{n_1}{n_2}\right)$, the integral

$$I := \int_0^\infty y_2^{(n_1+n_2-2)/2}\,\exp\left[-\frac{1}{2}\left(y_1 y_2\frac{n_1}{n_2}+y_2\right)\right]dy_2$$

$$= \left(\frac{2}{1+y_1\frac{n_1}{n_2}}\right)^{(n_1+n_2-2)/2+1}\int_0^\infty u^{(n_1+n_2-2)/2}e^{-u}\,du$$

$$= \left(\frac{2}{1+y_1\frac{n_1}{n_2}}\right)^{(n_1+n_2)/2}\Gamma\left(\frac{n_1+n_2}{2}\right),$$

so that integrating out Y_2 from f_{Y_1,Y_2} gives

$$f_{Y_1}(y_1) = K \left(\frac{n_1}{n_2}\right)^{n_1/2} y_1^{(n_1-2)/2} \left(\frac{2}{1+y_1\frac{n_1}{n_2}}\right)^{(n_1+n_2)/2} \Gamma\left(\frac{n_1+n_2}{2}\right) \mathbb{I}_{(0,\infty)}(y_1)$$

$$= \frac{\Gamma\left((n_1+n_2)/2\right)}{\Gamma(n_1/2)\,\Gamma(n_2/2)} \left(\frac{n_1}{n_2}\right)^{n_1/2} y_1^{(n_1-2)/2} \left(1+y_1\frac{n_1}{n_2}\right)^{-(n_1+n_2)/2} \mathbb{I}_{(0,\infty)}(y_1),$$

showing that $Y_1 \sim F(n_1, n_2)$. ∎

Use of the previous derivation also simplifies the calculation of moments of $F \sim F(n_1, n_2)$. That is, with $X_i, \overset{\text{ind}}{\sim} \chi^2_{n_i}$,

$$\mathbb{E}[F] = \mathbb{E}\left[\frac{X_1/n_1}{X_2/n_2}\right] = \frac{n_2}{n_1}\mathbb{E}[X_1]\mathbb{E}\left[X_2^{-1}\right] = \frac{n_2}{n_1}n_1\frac{1}{n_2-2} = \frac{n_2}{n_2-2}, \quad n_2 > 2,$$

from Example 7.5. The first four central moments of F are given in the more general context of the noncentral F distribution in Volume II.

⊖ **Example 9.9**   Let $Y_i \overset{\text{ind}}{\sim} \chi^2(d_i)$, $i = 1, \ldots, n$, and define

$$S_k = \frac{\sum_{i=1}^k Y_i}{\sum_{i=1}^k d_i}, \quad k = 1, \ldots, n-1, \quad \text{and} \quad F_k = \frac{Y_k/d_k}{S_{k-1}}, \quad k = 2, \ldots, n.$$

First consider the marginal distribution of F_k. Recall from Chapter 7 that a $\chi^2(d)$ distribution is just a gamma distribution with shape $d/2$ and (inverse) scale $1/2$. Thus, Example 9.3 and the comment directly following it imply that $\sum_{i=1}^k Y_i \sim \chi^2\left(\sum_{i=1}^k d_i\right)$. Next, Example 9.8 showed that an F distribution arises as a ratio of independent χ^2 r.v.s each divided by their degrees of freedom, so that $F_k \sim F\left(d_k, \sum_{i=1}^{k-1} d_i\right)$.

Next, and more challenging, for $n = 3$, we wish to show that C, F_2 and F_3 are independent, where $C = \sum_{i=1}^n Y_i$. This is true for general n, i.e. random variables

$$C = Y_1 + \cdots Y_n, \quad F_2 = \frac{Y_2/d_2}{Y_1/d_1}, \quad F_3 = \frac{Y_3/d_3}{(Y_1+Y_2)/(d_1+d_2)}, \quad \cdots,$$

$$F_n = \frac{Y_n/d_n}{(Y_1+Y_2+\cdots+Y_{n-1})/(d_1+d_2+\cdots+d_{n-1})},$$

are independent.[1]

---

[1] The result is mentioned, for example, in Hogg and Tanis (1963, p. 436), who state that 'this result is, in essence, well known and its proof, which is a rather easy exercise, is omitted'. They use it in the context of sequential, or iterative, testing of the equality of exponential distributions. It is also used by Phillips and McCabe (1983) in the context of testing for structural change in the linear regression model.

For $n = 3$, with Maple's assistance, we find

$$Y_1 = \frac{C(d_1 + d_2)d_1}{T_2 T_3}, \quad Y_2 = \frac{C F_2 d_2 (d_1 + d_2)}{T_2 T_3}, \quad Y_3 = \frac{C F_3 d_3}{T_3},$$

where $T_2 = (F_2 d_2 + d_1)$ and $T_3 = F_3 d_3 + d_1 + d_2$, and

$$
\begin{bmatrix}
\dfrac{\partial Y_1}{\partial C} & \dfrac{\partial Y_1}{\partial F_2} & \dfrac{\partial Y_1}{\partial F_3} \\[2mm]
\dfrac{\partial Y_2}{\partial C} & \dfrac{\partial Y_2}{\partial F_2} & \dfrac{\partial Y_2}{\partial F_3} \\[2mm]
\dfrac{\partial Y_3}{\partial C} & \dfrac{\partial Y_3}{\partial F_2} & \dfrac{\partial Y_3}{\partial F_3}
\end{bmatrix}
$$

$$
=
\begin{bmatrix}
\dfrac{d_1(d_1 + d_2)}{T_3 T_2} & -\dfrac{d_1 d_2 (d_1 + d_2)C}{T_3 T_2^2} & -\dfrac{d_1 d_3 (d_1 + d_2)C}{T_3^2 T_2} \\[3mm]
\dfrac{F_2 d_2 (d_1 + d_2)}{T_3 T_2} & \dfrac{d_1 d_2 (d_1 + d_2)C}{T_3 T_2^2} & -\dfrac{F_2 d_2 d_3 (d_1 + d_2)C}{T_3^2 T_2} \\[3mm]
\dfrac{d_3 F_3}{T_3} & 0 & \dfrac{d_3 (d_1 + d_2)C}{T_3^2}
\end{bmatrix}
$$

with

$$\det \mathbf{J} = \frac{d_1 d_2 d_3 (d_1 + d_2)^2 C^2}{T_3^3 T_2^2}.$$

Thus, with $d = d_1 + d_2 + d_3$ (and, for simplicity, using C, F_2 and F_3 as both the names of the r.v.s and their arguments in the p.d.f.), the joint density $f_{C, F_2, F_3}(C, F_2, F_3)$ is given by

$$|\det \mathbf{J}| \prod_{i=1}^{3} \frac{1}{2^{d_i/2}\Gamma(d_i/2)} y_i^{d_i/2 - 1} e^{-y_i/2}$$

$$= \frac{d_1 d_2 d_3 (d_1 + d_2)^2 C^2}{T_3^3 T_2^2} \frac{\left[\frac{C(d_1 + d_2)d_1}{T_2 T_3}\right]^{d_1/2 - 1} \left[\frac{C F_2 d_2 (d_1 + d_2)}{T_2 T_3}\right]^{d_2/2 - 1} \left(\frac{C F_3 d_3}{T_3}\right)^{d_3/2 - 1}}{2^{d/2}\Gamma\left(\frac{d_1}{2}\right)\Gamma\left(\frac{d_2}{2}\right)\Gamma\left(\frac{d_3}{2}\right)}$$

$$\times \exp\left\{-\frac{1}{2}\left[\frac{C(d_1 + d_2)d_1}{T_2 T_3} + \frac{C F_2 d_2 (d_1 + d_2)}{T_2 T_3} + \frac{C F_3 d_3}{T_3}\right]\right\}$$

$$= \Gamma(d/2) \frac{d_1 d_2 d_3 (d_1 + d_2)^2}{T_3^3 T_2^2} \frac{\left[\frac{(d_1 + d_2)d_1}{T_2 T_3}\right]^{d_1/2 - 1} \left[\frac{F_2 d_2 (d_1 + d_2)}{T_2 T_3}\right]^{d_2/2 - 1} \left(\frac{F_3 d_3}{T_3}\right)^{d_3/2 - 1}}{\Gamma\left(\frac{d_1}{2}\right)\Gamma\left(\frac{d_2}{2}\right)\Gamma\left(\frac{d_3}{2}\right)}$$

$$\times \frac{1}{2^{d/2}\Gamma(d/2)} C^{d/2 - 1} e^{-C/2}. \tag{9.8}$$

The p.d.f. of C has been separated from the joint density in (9.8), showing that $C \sim \chi^2(d)$ and that C is independent of F_2 and F_3. It remains to simplify the joint p.d.f. of F_2 and F_3. In this case, we have been told that F_2 and F_3 are independent, with $F_k \sim F\left(d_k, \sum_{i=1}^{k-1} d_i\right)$, so that we wish to confirm that

$$
\frac{\Gamma\left(\frac{d_1+d_2+d_3}{2}\right) \frac{d_3}{d_1+d_2} \left(\frac{d_3}{d_1+d_2} F_3\right)^{d_3/2-1}}{\Gamma\left(\frac{d_3}{2}\right)\Gamma\left(\frac{d_1+d_2}{2}\right)\left(1+\frac{d_3}{d_1+d_2}F_3\right)^{(d_1+d_2+d_3)/2}} \times \frac{\Gamma\left(\frac{d_1+d_2}{2}\right)\left(\frac{d_2}{d_1}\right)\left(\frac{d_2}{d_1}F_2\right)^{d_2/2-1}}{\Gamma\left(\frac{d_2}{2}\right)\Gamma\left(\frac{d_1}{2}\right)\left(1+\frac{d_2}{d_1}F_2\right)^{(d_1+d_2)/2}}
$$

$$
\stackrel{?}{=} \Gamma(d/2)\, \frac{d_1 d_2 d_3\,(d_1+d_2)^2}{T_3^3 T_2^2} \frac{\left[\frac{(d_1+d_2)d_1}{T_2 T_3}\right]^{d_1/2-1}\left[\frac{F_2 d_2 (d_1+d_2)}{T_2 T_3}\right]^{d_2/2-1}\left(\frac{F_3 d_3}{T_3}\right)^{d_3/2-1}}{\Gamma\left(\frac{d_1}{2}\right)\Gamma\left(\frac{d_2}{2}\right)\Gamma\left(\frac{d_3}{2}\right)}.
$$

The reader should check that, as $d = d_1 + d_2 + d_3$, all the gamma terms can be cancelled from both sides. Verifying the equality of the remaining equation just entails simple algebra, and Maple indeed confirms that both sides are equal. ∎

9.3 Further aspects and important transformations

Recall that an analytic expression for the inverse of the normal c.d.f. is not available, but could be numerically approximated by, for example, Moro's method in Program Listing 7.1. That could be used with the probability integral transform (Section 7.4) and a uniform random number generator to simulate normal r.v.s. The next example shows a different way of doing this.

Example 9.10 (Box and Muller, 1958) Let $U_i \overset{\text{i.i.d.}}{\sim} \text{Unif}(0,1)$ and define

$$
\begin{aligned}
X_1 &= g_1(U_1, U_2) = \sqrt{-2\ln U_1}\,\cos(2\pi U_2),\\
X_2 &= g_2(U_1, U_2) = \sqrt{-2\ln U_1}\,\sin(2\pi U_2).
\end{aligned}
\tag{9.9}
$$

This is a bijection with inverse

$$
u_1 = g_1^{-1}(x_1, x_2) = \exp\left[-\frac{1}{2}\left(x_1^2 + x_2^2\right)\right], \qquad u_2 = g_2^{-1}(x_1, x_2) = \frac{1}{2\pi}\arctan\left(\frac{x_2}{x_1}\right)
$$

and Jacobian

$$
\mathbf{J} = \begin{pmatrix} -x_1 \exp\left[-\frac{1}{2}\left(x_1^2 + x_2^2\right)\right] & -x_2 \exp\left[-\frac{1}{2}\left(x_1^2 + x_2^2\right)\right] \\ -\dfrac{1}{2\pi}\dfrac{x_2}{x_1^2 + x_2^2} & \dfrac{1}{2\pi}\dfrac{x_1}{x_1^2 + x_2^2} \end{pmatrix}
$$

with

$$
\det \mathbf{J} = -\frac{1}{2\pi}\exp\left[-\frac{1}{2}\left(x_1^2 + x_2^2\right)\right],
$$

so that $f_{X_1, X_2}(x_1, x_2)$ is given by

$|\det \mathbf{J}| f_{U_1, U_2}(u_1, u_2)$

$= \dfrac{1}{2\pi} \exp\left[-\dfrac{1}{2}\left(x_1^2 + x_2^2\right)\right] \mathbb{I}_{(0,1)} \left\{\exp\left[-\dfrac{1}{2}\left(x_1^2 + x_2^2\right)\right]\right\} \mathbb{I}_{(0,1)} \left[\dfrac{1}{2\pi} \arctan\left(\dfrac{x_2}{x_1}\right)\right]$

$= \dfrac{1}{2\pi} \exp\left[-\dfrac{1}{2}\left(x_1^2 + x_2^2\right)\right],$

i.e. $X_i \stackrel{\text{i.i.d.}}{\sim} \mathrm{N}(0, 1)$, $i = 1, 2$ (see also Problem 9.7). ∎

The next example serves three purposes. It generalizes the result mentioned directly after Example 9.3 by allowing for non-integer shape parameters in the sums of gammas; it introduces the Dirichlet distribution (after Johann Peter Gustav Lejeune Dirichlet, 1805–1859) and it proves (1.48), i.e. that $B(a, b) = \Gamma(a)\,\Gamma(b)\,/\,\Gamma(a + b)$.

⊙ **Example 9.11**   Let $\mathbf{X} = (X_1, \ldots, X_{n+1})$ with $X_i \stackrel{\text{ind}}{\sim} \mathrm{Gam}(\alpha_i, \beta)$, $\alpha_i, \beta \in \mathbb{R}$, and define $S = \sum_{i=1}^{n+1} X_i$, $\mathbf{Y} = (Y_1, \ldots, Y_n)$, $Y_i = X_i/S$, so that $\mathcal{S}_\mathbf{X} = \mathbb{R}_{>0}^{n+1}$ and

$$\mathcal{S}_{\mathbf{Y}, S} = \left\{(y_1, \ldots, y_n, s) : 0 < y_i < 1,\ \sum_{i=1}^{n} y_i < 1,\ s \in \mathbb{R}_{>0}\right\}.$$

The inverse functions are $x_i = y_i s$, $i = 1, \ldots, n$, $x_{n+1} = s\left(1 - \sum_{i=1}^{n} y_i\right)$ and

$$\mathbf{J} = \begin{pmatrix} s & 0 & \cdots & 0 & y_1 \\ 0 & s & \cdots & 0 & y_2 \\ \vdots & \vdots & \ddots & \vdots & \vdots \\ 0 & 0 & \cdots & s & y_n \\ -s & -s & \cdots & -s & 1 - \sum_{i=1}^{n} y_i \end{pmatrix}, \quad \det \mathbf{J} = s^n,$$

so that

$$f_{\mathbf{Y}, S}(\mathbf{y}, s) = |\det \mathbf{J}|\, f_{\mathbf{X}}(\mathbf{x}) = s^n \prod_{i=1}^{n+1} \frac{\beta^{\alpha_i}}{\Gamma(\alpha_i)} x_i^{\alpha_i - 1} e^{-\beta x_i} \mathbb{I}_{(0,\infty)}(x_i)$$

$$= s^n \prod_{i=1}^{n} \frac{\beta^{\alpha_i}}{\Gamma(\alpha_i)} (y_i s)^{\alpha_i - 1} e^{-\beta y_i s} \mathbb{I}_{(0,\infty)}(y_i s) \frac{\beta^{\alpha_{n+1}}}{\Gamma(\alpha_{n+1})}$$

$$\times \left[s\left(1 - \sum_{i=1}^{n} y_i\right)\right]^{\alpha_{n+1} - 1} e^{-\beta s\left(1 - \sum_{i=1}^{n} y_i\right)} \mathbb{I}_{(0,\infty)}\left[s\left(1 - \sum_{i=1}^{n} y_i\right)\right]$$

$$= \frac{\beta^{\alpha_1 + \cdots \alpha_{n+1}} s^{\alpha_1 + \cdots \alpha_{n+1} - 1}}{\Gamma(\alpha_1) \cdots \Gamma(\alpha_{n+1})} e^{-\beta s} \prod_{i=1}^{n} y_i^{\alpha_i - 1} \left(1 - \sum_{i=1}^{n} y_i\right)^{\alpha_{n+1} - 1}$$

$$\times \mathbb{I}_{(0,1)}(y_i)\, \mathbb{I}_{(0,1)}\left(\sum_{i=1}^{n} y_i\right) \mathbb{I}_{(0,\infty)}(s).$$

The density factors into two parts, $\beta^\alpha s^{\alpha-1} e^{-\beta s} \mathbb{I}_{(0,\infty)}(s)$ for $\alpha := \sum_{i=1}^{n+1} \alpha_i$ and

$$\frac{1}{\Gamma(\alpha_1)\cdots\Gamma(\alpha_{n+1})} \prod_{i=1}^{n} y_i^{\alpha_i-1} \left(1 - \sum_{i=1}^{n} y_i\right)^{\alpha_{n+1}-1} \mathbb{I}_{(0,1)}(y_i) \, \mathbb{I}_{(0,1)}\left(\sum_{i=1}^{n} y_i\right).$$

If we divide the former by $\Gamma(\alpha)$ and multiply the latter by $\Gamma(\alpha)$, we find

$$f_S(s) = \frac{\beta^\alpha}{\Gamma(\alpha)} s^{\alpha-1} e^{-\beta s} \mathbb{I}_{(0,\infty)}(s),$$

i.e. $S \sim \text{Gam}(\alpha, \beta)$ and

$$f_Y(y) = \frac{\Gamma(\alpha)}{\Gamma(\alpha_1)\cdots\Gamma(\alpha_{n+1})} \prod_{i=1}^{n} y_i^{\alpha_i-1} \left(1 - \sum_{i=1}^{n} y_i\right)^{\alpha_{n+1}-1} \mathbb{I}_{(0,1)}(y_i) \, \mathbb{I}_{(0,1)}\left(\sum_{i=1}^{n} y_i\right),$$

which is referred to as the *Dirichlet* distribution and denoted $\mathbf{Y} \sim \text{Dir}(\alpha_1, \ldots, \alpha_{n+1})$. It follows from (5.9) that S and $\mathbf{Y}$ are independent.

An important special case is $n = 1$, yielding $S = X_1 + X_2 \sim \text{Gam}(\alpha_1 + \alpha_2, \beta)$ independent of $Y = X_1/(X_1 + X_2)$ with distribution

$$f_Y(y) = \frac{\Gamma(\alpha_1 + \alpha_2)}{\Gamma(\alpha_1)\Gamma(\alpha_2)} y^{\alpha_1-1} (1-y)^{\alpha_2-1} \mathbb{I}_{(0,1)}(y),$$

i.e. $Y \sim \text{Beta}(\alpha_1, \alpha_2)$. This also confirms that

$$\boxed{B(a,b) = \int_0^1 y^{a-1}(1-y)^{b-1}\,dy = \frac{\Gamma(a)\Gamma(b)}{\Gamma(a+b)}},$$

which was proven directly in Section 1.5. ∎

Remark: Recall from Problem 7.20 that, if $X \sim F(n_1, n_2)$, then

$$B = \frac{\frac{n_1}{n_2} X}{1 + \frac{n_1}{n_2} X} \sim \text{Beta}\left(\frac{n_1}{2}, \frac{n_2}{2}\right), \tag{9.10}$$

i.e.

$$\Pr(B \leq b) = \Pr\left(X \leq \frac{n_2}{n_1}\frac{b}{1-b}\right),$$

which was shown directly using a transformation. This relation can now easily be verified using the results from Examples 9.11 and 9.8. For $n_1, n_2 \in \mathbb{N}$, let $G_i \overset{\text{ind}}{\sim} \chi^2_{n_i}$ or, equivalently, $G_i \overset{\text{ind}}{\sim} \text{Gam}(n_i/2, 1/2)$. Then, from Example 9.8,

$$F = \frac{G_1/n_1}{G_2/n_2} \sim F(n_1, n_2).$$

We can then express B from (9.10) as

$$B = \frac{\frac{n_1}{n_2} F}{1 + \frac{n_1}{n_2} F} = \frac{\frac{n_1}{n_2} \frac{G_1/n_1}{G_2/n_2}}{1 + \frac{n_1}{n_2} \frac{G_1/n_1}{G_2/n_2}} = \frac{\frac{G_1}{G_2}}{1 + \frac{G_1}{G_2}} = \frac{G_1}{G_1 + G_2}.$$

This also holds for any scale factor common to the G_i, i.e.

$$\boxed{\text{if } G_i \overset{\text{ind}}{\sim} \text{Gam}(\alpha_i, \beta), \text{ then } \frac{G_1}{G_1 + G_2} \sim \text{Beta}(\alpha_1, \alpha_2)},$$

as in Example 9.11. ∎

As in the univariate case, it is sometimes necessary to partition $\mathcal{S}_{\mathbf{X}}$ into disjoint subsets, say $\mathcal{S}_{\mathbf{X}}^{(1)}, \ldots, \mathcal{S}_{\mathbf{X}}^{(d)}$ in order for $\mathbf{g}$ to be piecewise invertible. Letting $\mathbf{g}_i = (g_{1i}(\mathbf{x}), \ldots, g_{ni}(\mathbf{x}))$ be continuous bijections from $\mathcal{S}_{\mathbf{X}}^{(i)}$ to (a subset of) $\mathbb{R}^n$, the general transformation can be written as

$$f_{\mathbf{Y}}(\mathbf{y}) = \sum_{i=1}^{d} f_{\mathbf{X}}(\mathbf{x}_i) |\det \mathbf{J}_i|, \tag{9.11}$$

where $\mathbf{x}_i = \left(g_{1i}^{-1}(\mathbf{y}), \ldots, g_{ni}^{-1}(\mathbf{y}) \right)$ and

$$\mathbf{J}_i = \begin{pmatrix} \dfrac{\partial g_{1i}^{-1}(\mathbf{y})}{\partial y_1} & \dfrac{\partial g_{1i}^{-1}(\mathbf{y})}{\partial y_2} & \cdots & \dfrac{\partial g_{1i}^{-1}(\mathbf{y})}{\partial y_n} \\[2ex] \dfrac{\partial g_{2i}^{-1}(\mathbf{y})}{\partial y_1} & \dfrac{\partial g_{2i}^{-1}(\mathbf{y})}{\partial y_2} & & \dfrac{\partial g_{2i}^{-1}(\mathbf{y})}{\partial y_n} \\[2ex] \vdots & \vdots & \ddots & \vdots \\[2ex] \dfrac{\partial g_{ni}^{-1}(\mathbf{y})}{\partial y_1} & \dfrac{\partial g_{ni}^{-1}(\mathbf{y})}{\partial y_2} & \cdots & \dfrac{\partial g_{ni}^{-1}(\mathbf{y})}{\partial y_n} \end{pmatrix}.$$

Further discussion of multivariate many-to-one transformations can be found in Hogg and Craig (1994, p. 190) and Port (1994, p. 462).

⊘ **Example 9.12**   Let X_1 and X_2 be continuous r.v.s with joint p.d.f. f_{X_1,X_2}. To derive an expression for the density of $S = X_1^2 + X_2^2$, start by letting $S = X_1^2 + X_2^2$, $Z = X_1^2$, $X_1 = \pm\sqrt{Z}$ and $X_2 = \pm\sqrt{S - Z}$. Then, considering each of the four possible sign configurations on the x_i,

$$f_{S,Z}(s, z) = |\det \mathbf{J}| f_{X_1,X_2}(x_1, x_2) \mathbb{I}_{(-\infty,0)}(x_1) \mathbb{I}_{(-\infty,0)}(x_2) + \cdots$$

with

$$\mathbf{J} = \begin{pmatrix} \partial x_1/\partial z & \partial x_1/\partial s \\ \partial x_2/\partial z & \partial x_2/\partial s \end{pmatrix} = \begin{pmatrix} \pm\frac{1}{2}z^{-1/2} & \cdot \\ 0 & \pm\frac{1}{2}(s-z)^{-1/2} \end{pmatrix}$$

and, as z and $s - z$ are both positive, all $|\det \mathbf{J}|$ are the same, namely

$$|\det \mathbf{J}| = \frac{1}{4}z^{-1/2}(s-z)^{-1/2}.$$

Thus

$$f_{S,Z}(s,z) = \frac{1}{4}z^{-1/2}(s-z)^{-1/2} \times \left[f_{X_1,X_2}\left(-\sqrt{z}, -\sqrt{s-z}\right) + \cdots \right] \mathbb{I}_{(0,s)}(z), \quad (9.12)$$

where the term in brackets has the form $f(-,-) + f(-,+) + f(+,-) + f(+,+)$ and

$$f_S(s) = \int_0^s f_{S,Z}(s,z)\,\mathrm{d}z.$$

For the special case of i.i.d. standard normal r.v.s, $f_{X_1,X_2}(x_1, x_2) = e^{-\frac{1}{2}\left(x_1^2+x_2^2\right)}/2\pi$, so that all four terms in (9.12) are the same, yielding

$$f_{S,Z}(s,z) = \frac{1}{4}z^{-1/2}(s-z)^{-1/2}\frac{4}{2\pi}e^{-\frac{1}{2}[z+(s-z)]} = \frac{1}{2\pi}z^{-1/2}(s-z)^{-1/2}e^{-\frac{1}{2}s}\mathbb{I}_{(0,s)}(z).$$

From (1.52),

$$f_S(s) = \frac{1}{2\pi}e^{-\frac{1}{2}s}\int_0^s z^{-1/2}(s-z)^{-1/2}\,\mathrm{d}z = \frac{1}{2\pi}e^{-\frac{1}{2}s}B\left(\frac{1}{2},\frac{1}{2}\right) = \frac{1}{2}e^{-\frac{1}{2}s}\mathbb{I}_{(0,\infty)}(s),$$

so that $S \sim \mathrm{Exp}(1/2)$ and also $S \sim \chi_2^2$.

Remark: Problem 9.6 extends the previous example and shows that, if $X_i \overset{\text{i.i.d.}}{\sim} N(0, 1)$, then $\sum_{i=1}^n X_i^2 \sim \chi_n^2$. ∎

Example 9.13 Let $Z_i \overset{\text{i.i.d.}}{\sim} N(0, 1)$ and define $Y = Z_1 Z_2 + Z_3 Z_4$. It appears as though a four-dimensional multivariate transformation and evaluation of a triple integral is necessary to calculate the distribution of Y. Fortunately, there is an easier way. As in Mantel and Pasternack (1966), let

$$H_1 = \frac{Z_1 + Z_2}{2}, \quad H_2 = \frac{Z_1 - Z_2}{2}, \quad H_3 = \frac{Z_3 + Z_4}{2}, \quad H_4 = \frac{Z_3 - Z_4}{2}.$$

From Example 9.6 and (7.35), $H_i \overset{\text{i.i.d.}}{\sim} N(0, 1/2)$. The inverse transformation is just

$$Z_1 = H_1 + H_2, \quad Z_2 = H_1 - H_2, \quad Z_3 = H_3 + H_4, \quad Z_4 = H_3 - H_4,$$

from which it follows that

$$Y = H_1^2 - H_2^2 + H_3^2 - H_4^2.$$

Next, from (7.35), $\sqrt{2}H_i \overset{\text{i.i.d.}}{\sim} N(0, 1)$ so that, from Example 9.12, $C_1 = 2\left(H_1^2 + H_2^2\right)$ $\sim \chi_2^2$ or $C_1 \sim \text{Exp}(1/2)$. Similarly, with $C_2 = 2\left(H_3^2 + H_4^2\right)$, $C_i \overset{\text{i.i.d.}}{\sim} \text{Exp}(1/2)$, $i = 1, 2$. Finally, from Example 9.5, if $X, Y \overset{\text{i.i.d.}}{\sim} \text{Exp}(\lambda)$, then $D = X - Y \sim \text{Lap}(0, \lambda)$, from which it follows that

$$2\left(H_1^2 + H_2^2 - H_3^2 - H_4^2\right) = C_1 - C_2 \sim \text{Lap}(0, 1/2),$$

or, after a scale transformation, $Y = \left(H_1^2 + H_2^2 - H_3^2 - H_4^2\right) \sim \text{Lap}(0, 1)$. ∎

It was mentioned in Section 8.2.5 that functions of independent r.v.s are themselves independent. This can be confirmed using a multivariate transformation. Let X_1 and X_2 be independent continuous r.v.s with joint p.d.f. $f_1(x_1) f_2(x_2)$ and define $Y_i = g_i(X_i)$, where each g_i is a bijection mapping $\mathcal{S}_{X_i}$ to $\mathcal{S}_{Y_i}$, $i = 1, 2$. Clearly, $X_i = g_i^{-1}(Y_1, Y_2) = g_i^{-1}(Y_i)$ so that, from (9.1),

$$f_{Y_1,Y_2}(y_1, y_2) = f_{X_1}(x_1) f_{X_2}(x_2) \left| \det \begin{bmatrix} \partial x_1/\partial y_1 & 0 \\ 0 & \partial x_2/\partial y_2 \end{bmatrix} \right|$$

$$= f_{X_1}(x_1) \left| \frac{dx_1}{dy_1} \right| f_{X_2}(x_2) \left| \frac{dx_2}{dy_2} \right|,$$

so that f_{Y_1,Y_2} factors into two univariate transformations rendering Y_1 and Y_2 independent. The result hinges on the diagonality of $\mathbf{J}$, which arises because the g_i are functions only of X_i. This can easily be generalized as follows. Let $\{\mathbf{X}_i, i = 1, \ldots, n\}$ be a set of independent d-dimensional r.v.s with p.d.f. $\prod_{i=1}^n f_{\mathbf{X}_i}(\mathbf{x}_i)$ and define

$$\mathbf{Y}_i = \mathbf{g}_i(\mathbf{X}_i) = (g_{i1}(\mathbf{X}_i), \ldots, g_{id}(\mathbf{X}_i)), i = 1, \ldots, n.$$

Then the Jacobian takes the *block diagonal* form $\mathbf{J} = \text{diag}(\mathbf{J}_1, \ldots, \mathbf{J}_n)$ with

$$\mathbf{J}_i = [\partial g_{ij}^{-1}(\mathbf{y}_i)/\partial y_{ik}]_{j,k=1}^d \text{ and } \det \mathbf{J} = \prod_{i=1}^n \det \mathbf{J}_i.$$

As in the simpler case above, the resulting p.d.f. of $\{\mathbf{Y}_i, i = 1, \ldots, n\}$ factors into marginals

$$\prod_{i=1}^n f_{\mathbf{Y}_i}(\mathbf{y}_i) = \prod_{i=1}^n f_{\mathbf{X}_i}(\mathbf{x}_i) |\det \mathbf{J}_i|.$$

Remark: There is an alternative method for deriving the joint distribution of a function of several random variables which does not require computing the Jacobian or the use of 'dummy' variables. It uses the so-called Dirac generalized function (see Au and Tam (1999) for details). ∎

9.4 Problems

9.1. ★ Let X and Y be continuous r.v.s with joint density $f_{X,Y}$. Derive an expression for the density of $R = X/(X+Y)$. Use it for the following two cases.

(a) $X, Y \overset{\text{i.i.d.}}{\sim} \text{Gam}(a, b)$,

(b) $X, Y \overset{\text{i.i.d.}}{\sim} \text{N}(0, 1)$.

9.2. ★ Let
$$Y = \frac{w_1 X_1 + w_2 X_2}{X_1 + X_2},$$

where $X_i \overset{\text{i.i.d.}}{\sim} \chi^2(1)$, $i = 1, 2$.

(a) Describe Y in terms of a 'common' density. (Hint: use Example 9.11).

(b) Assuming $0 < w_1 < w_2 < \infty$, derive the density of Y and its support.

(c) Show that it integrates to one. (Hint: substitute $u = (y - w_1)/(w_2 - w_1)$).

Random variable Y is a special case of

$$W = \sum_{i=1}^{n} w_i X_i \Big/ \sum_{i=1}^{n} X_i,$$

with $n = 2$. The density of W for $n = 3$ is more complicated, though still tractable, and is given in Provost and Cheong (2000). The distribution of W and of the more general random variable $\mathbf{Y'AY}/\mathbf{Y'BY}$, where $\mathbf{Y} \sim \text{N}_n(\boldsymbol{\mu}, \boldsymbol{\Sigma})$ and $\mathbf{A}$ and $\mathbf{B} > 0$ are $n \times n$ matrices, will be considered in a chapter on quadratic forms in Volume II.

9.3. ★ In reference to Example 9.3, the density of S can also be arrived at by use of the $1-1$ transformation

$$Y_1 = X_1,$$
$$Y_2 = X_1 + X_2,$$
$$\vdots$$
$$Y_n = X_1 + X_2 + \cdots + X_n =: S.$$

Derive the joint density of the Y_i and, by integrating out $Y_{n-1}, Y_{n-2}, \ldots, Y_1$, the marginal of S.

9.4. ★ Assume $X_1, X_2, X_3 \overset{\text{i.i.d.}}{\sim} \text{Exp}(1)$, and define

$$Y_1 = \frac{X_1}{X_1 + X_2 + X_3}, \quad Y_2 = \frac{X_1 + X_2}{X_1 + X_2 + X_3} \quad \text{and} \quad Y_3 = X_1 + X_2 + X_3.$$

(a) Show that the joint distribution of Y_1, Y_2 and Y_3 is given by

$$f_{Y_1,Y_2,Y_3}(y_1, y_2, y_3) = y_3^2 e^{-y_3} \mathbb{I}_{(0,\infty)}(y_3), \quad 0 < y_1 < y_2 < 1.$$

(b) Derive all univariate marginal distributions.

(c) Is the joint distribution of Y_1 and Y_2 independent of Y_3?

(d) Derive the distribution of $Y_1 \mid Y_2 = y_2$ for $0 < y_2 < 1$.

9.5. ★ ★ Let $X, Y \overset{\text{i.i.d.}}{\sim} \text{Exp}(\lambda)$ and define $Z = X + Y$.

(a) Compute the joint density of X and Z and from this, the marginal density of Z.

(b) Compute the conditional density of X given $Z = z$.

(c) Compute the conditional probability that $Z/3 < X < Z/2$ given $Z = z$.

(d) Compute the density of $D = X - Y$ using the formula

$$f_D(d) = \int_{-\infty}^{\infty} f_X(d + y) f_Y(y) \, dy,$$

which can be obtained from a bivariate transformation as was done above for the sum and product.

(e) Compute the density of $W = |X - Y|$ and its expected value.

(f) First determine the density of the maximum, and the density of the minimum, of two (or more) i.i.d. r.v.s. Then compute the conditional density of $\max(X, Y) - \min(X, Y)$ conditional on $\min(X, Y) = a$.

9.6. ★ ★ Let $X_i \overset{\text{i.i.d.}}{\sim} N(0, 1)$. Interest centers on the distribution of

$$Y_1 = g_1(\mathbf{X}) = \sum_{i=1}^{n} X_i^2.$$

To derive it, let $Y_i = g_i(\mathbf{X}) = X_i$, $i = 2, 3, \ldots, n$, perform the multivariate transformation to get $f_{\mathbf{Y}}$, and then integrate out $Y_2, \ldots, Y_n$ to get the distribution of Y_1. The following steps can be used.

(a) First do the $n = 2$ case, for which the fact that

$$\int \frac{1}{\sqrt{y_1 - y_2^2}} dy_2 = c + \arcsin\left(\frac{y_2}{\sqrt{y_1}}\right)$$

can be helpful.

(b) Simplify the following integral, which is used in the general case:

$$J = \int_0^{y_0} u^{m/2-1} (y_0 - u)^{-1/2} \, du.$$

(c) For the general case, conduct the multivariate transformation to show that

$$f_{Y_1}(y_1) = \frac{1}{(2\pi)^{n/2}} e^{-\frac{1}{2}y_1} \int \cdots \int_S \left(y_1 - \sum_{i=2}^{n} y_i^2 \right)^{-\frac{1}{2}} dy_2 \cdots dy_n,$$

where

$$S = \left\{ (y_2, \ldots, y_n) \in \mathbb{R}^{n-1} : 0 < \sum_{i=2}^{n} y_i^2 < y_1 \right\}.$$

Then use the following identity (due to Joseph Liouville (1809–1882) in 1839, which extended a result from Dirichlet).
Let $\mathcal{V}$ be a volume consisting of (i) $x_i \geq 0$ and (ii) $t_1 \leq \sum (x_i/a_i)^{p_i} \leq t_2$, and let f be a continuous function on (t_1, t_2). Then, with $r_i = b_i/p_i$ and $R = \sum r_i$,

$$\int \cdots \int_{\mathcal{V}} x_1^{b_1-1} \cdots x_n^{b_n-1} f \left[\left(\frac{x_1}{a_1} \right)^{p_1} + \cdots + \left(\frac{x_n}{a_n} \right)^{p_n} \right] dx_1 \cdots dx_n$$

$$= \frac{\prod a_i^{b_i} p_i^{-1} \Gamma(r_i)}{\Gamma(R)} \int_{t_1}^{t_2} u^{R-1} f(u) \, du. \tag{9.13}$$

For details on this and similar results, see Andrews, Askey and Roy (1999, Section 1.8) and Jones (2001, Chapter 9).

9.7. ★ The Box and Muller transformation for simulating normal r.v.s given in Example 9.10 was easy to calculate once given the formulae for X_1 and X_2, yet it seems difficult to have actually figured out what these formulae should be.

(a) Let $X, Y \overset{\text{i.i.d.}}{\sim} N(0, 1)$ and define $R = \sqrt{X^2 + Y^2}$ and $\Theta = \tan^{-1}(Y/X)$ to be the polar coordinates of X and Y when they are plotted in the plane. Use (9.1) and (A.166) to show that

$$f_{R,\Theta}(r, \theta) = \frac{1}{2\pi} r e^{-r^2/2} \mathbb{I}_{(0,\infty)}(r) \mathbb{I}_{(0,2\pi)}(\theta). \tag{9.14}$$

Note that R and Θ are independent with $f_R(r) = r e^{-r^2/2} \mathbb{I}_{(0,\infty)}(r)$.

(b) Calculate the density of $S = R^2$.

(c) Let $U_1, U_2 \overset{\text{i.i.d.}}{\sim} \text{Unif}(0, 1)$. Show that $-2 \ln U_1$ has the same distribution as S.

(d) Now set $R^2 = S = -2 \ln U_1$ and $\Theta = 2\pi U_2$ and justify (9.9).

9.8. ★ Calculate the density, expected value and distribution function of $X_1 \mid S$, where $X_i \overset{\text{i.i.d.}}{\sim} \text{Exp}(\lambda)$, $i = 1, \ldots, n$, and $S = \sum_{i=1}^{n} X_i$.

Calculus review

I advise my students to listen carefully the moment they decide to take no more mathematics courses. They might be able to hear the sound of closing doors.

(James Caballero)
Reproduced by permission of John Wiley & Sons, Ltd

One difficult decision relates to how much of an effort one should make to acquire basic technique, for example in mathematics and probability theory. One does not wish to be always training to run the race but never running it; however, we do need to train.   (E. J. Hannan, 1992)
Reproduced by permission of Springer

The purpose of this appendix is to refresh some fundamental results from basic calculus, and introduce a few elements which are possibly new to the reader. The presentation is not designed to replace a full textbook on the subject matter, and the first section mentions some of the many useful textbooks which could serve as outside reading for the interested student. Nevertheless, some proofs are provided, and numerous examples are given, many of which relate to issues arising in probability theory throughout the book. Some fundamental topics are omitted which are either assumed to be intuitively understood by the reader (such as the least upper bound property of $\mathbb{R}$), or not explicitly used (such as compactness).

Tables B.1 and B.3 contain a list of mathematical abbreviations and the Greek letters, respectively.

A.0 Recommended reading

The current mathematics teaching trend treats the real number system as a given – it is defined axiomatically. Ten or so of its properties are listed, called axioms of a complete ordered field, and the game becomes: deduce its other properties from the axioms. This is something of a fraud, considering that the entire structure of analysis is built on the real number system. For what if a system satisfying the axioms failed to exist?   (Charles C. Pugh, 2002, p. 10)

Fundamental Probability: A Computational Approach   M.S. Paolella
© 2006 John Wiley & Sons, Ltd

There are numerous good introductory real analysis textbooks, each covering the standard set of core material with slightly different paces and didactic approaches, as well as including some less standard material. As such, if you use results of real analysis in your study or work, you will inevitably accumulate a set of favorite books which cover a variety of important topics. I personally prefer to stay clear of terse, austere presentations, and enjoy more readable, livelier accounts peppered with superfluous things like anecdotes, historical information, complete proofs, examples, etc. (the size of *this* book should attest to that). For a traditional, detailed, unpretentious account of the core univariate topics, I very much like Stoll (2001) (which includes scores of useful and highly accessible references, new and old, to articles in the American Mathematical Monthly and Mathematics Magazine) whose contents are very similar to that of Goldberg (1964), another fine book, though less popular as the years go by. Both of those books include an introduction to Fourier series and Lebesgue integration and touch upon other optional topics (e.g. Goldberg discusses Abel summability and Tauber's theorem; Stoll covers the Riemann–Stieltjes integral). A well-written book which just covers the essential univariate topics is Bartle and Sherbert (1982).[1]

In addition to a development of the core topics, Protter and Morrey (1991), Lang (1997) and Trench (2003) also cover multivariate calculus in a fashion well-suited for beginners in real analysis. Browder (1996), Hijab (1997) and Pugh (2002) 'start from scratch', but briskly cover the core material in order to move on to more advanced subject matter. Each of these has their own unique strengths and accents on specific topics, and so augment each other very well. As many of my students are native German speakers, I also recommend Königsberger (1999), which is very accessible and also covers several less standard topics.

As a preparation for those needing a calculus review, Estep (2002) provides a slow, well motivated transition from basic algebra and beginning calculus to real analysis, with emphasis on the numerical solution of real problems. A good source of reading previous to, or concurrently with, a course in analysis or finite mathematics is Lipschutz (1998). Ostaszewski (1990) is a great book to work through; it is not an analysis book per se, but covers linear algebra in its first half and some elements of advanced calculus in the second. Along with generous use of graphics, it is very enjoyably written (with explanations like Alice in Wonderland's cardboard mill for the Riemann–Stieltjes integral).

Some elements of multivariate calculus are also required; the undergraduate books of Lang (1987) and Flanigan and Kazdan (1990) are great places to start. More rigorous developments of multivariate analysis are provided by the real analysis books mentioned above, while still more depth is given by Munkres (1991), Pugh (2002, Chapter 5) (who acknowledges Munkres' book), the last five chapters of Browder (1996) and Hubbard and Hubbard (2002). The latter book is the largest of these, is quite detailed, and has terrific graphics and typesetting.

We do not use any measure theory in this book, but your next step in probability theory might require it. Highly readable accounts are provided in the one or two specific chapters of Goldberg (1964) and Stoll (2001); those of Browder (1996) and Pugh (2002) are also recommended but are somewhat more advanced. A pleasant,

---

[1] Bartle and Sherbert is now in a third edition, 1999, and includes an introduction to the gauge integral (see footnote 8).

rudimentary and didactically excellent book-length treatment is provided by Priestley (1997), while Adams and Guillemin (1996), Chae (1995) and Jones (2001) are also very accessible and go somewhat further than Priestley.

Only rudimentary notions in complex analysis are required; the introductory chapters in Bak and Newman (1997) and Palka (1991) are enough. Basic linear and matrix algebra, however, are crucial in several places throughout the text. For this, many excellent texts exist, such as the short and light Axler (1996) and the heavy and detailed Meyer (2001), while several books are available which emphasize matrix algebra and its application to statistics, e.g. Searle (1982), Graybill (1983), Schott (2005), and Rao and Rao (1998). Harville (1997), along with its accompanying exercise and solutions manual, is a sheer goldmine of knowledge, though somewhat dense and perhaps not the best place to start learning matrix algebra. Magnus and Neudecker (1999) is a unique and popular book containing elements of both matrix algebra and multivariate analysis; it can be used for both teaching and as a reference.

A.1 Sets, functions and fundamental inequalities

The point of view that 'natural number' cannot be defined would be contested by many mathematicians who would maintain that the concept of 'set' is more primitive than that of 'number' and who would use it to define 'number'. Others would contend that the idea of 'set' is not at all intuitive and would contend that, in particular, the idea of an *infinite* set is very nebulous. They would consider a definition of 'number' in terms of sets to be an absurdity because it uses a difficult and perhaps meaningless concept to define a simple one.

(Harold M. Edwards, 1994, p. 461)
Reproduced by permission of Birkhauser

It turns out that, mathematically speaking, a precise definition of *set* is problematic. For our purposes, it can be thought of simply as a well-defined collection of objects. This intuitive description cannot be a definition, because the word 'collection' is nothing but a synonym for the word set. Nevertheless, in all contexts considered herein, the notion of set will be clear. For example, if $A = \{n \in \mathbb{N} : n < 7\}$, then A is the set of positive integers less than 7, or $A = \{1, 2, \ldots, 6\}$. If a is contained in A, then we write $a \in A$; otherwise, $a \notin A$. A set without any objects is called the *empty set* and is denoted $\emptyset$. A set with exactly one element is a *singleton set*.

Let A and B be two sets. The following handful of basic set operations will be used repeatedly throughout:

- the *intersection* of two sets, 'A and B' (or 'A inter B'), denoted AB or $A \cap B$. Each element of $A \cap B$ is contained in A, and contained in B; $A \cap B = \{x : x \in A, x \in B\}$.

- the *union* of two sets, 'A or B' (or 'A union B'), denoted $A \cup B$ or $A + B$. An element of $A \cup B$ is either in A, or in B, or in both.

- set *subsets*, 'A is a subset of B' or 'A is contained in B' or 'B contains A', denoted $A \subset B$ or $B \supset A$. If every element contained in A is also in B, then $A \subset B$. Like the ordering symbols $\leq$ and $<$ for real numbers, it is sometimes useful (if not more correct) to use the notation $A \subseteq B$ to indicate that A and B could be equal, and reserve $A \subset B$ to indicate that A is a *proper subset* of B, i.e. $A \subseteq B$ but $A \neq B$; in words, that there is at least one element in B which is not in A. Only when this distinction is important will we use $\subseteq$. Also, $\emptyset$ is a subset of every set.

- set *equality*, '$A = B$', which is true if and only if $A \subset B$ and $B \subset A$.

- the *difference*, or *relative complement*, 'B minus A', denoted $B \setminus A$ or $B - A$. It is the set of elements contained in B but not in A.

- If the set B is clear from the context, then it need not be explicitly stated, and the set difference $B \setminus A$ is written as A^c, which is the *complement* of A. Thus, we can write $B \setminus A = B \cap A^c$.

- the *product* of two sets, A and B, consists of all ordered pairs (a, b), such that $a \in A$ and $b \in B$; it is denoted $A \times B$.

The first four of the set operations above are extended to more than two sets in a natural way, i.e. for intersection, if $a \in A_1 \cap A_2 \cap \cdots \cap A_n$, then a is contained in each of the A_i, and is abbreviated by $a \in \bigcap_{i=1}^{n} A_i$. A similar notation is used for union. To illustrate this for subsets, let $A_n = [1/n, 1]$, $n \in \{1, 2, \ldots\}$, i.e. $A_1 = \{1\}$ and $A_2 = [1/2, 1] = \{x : 1/2 \leq x \leq 1\}$. Then $A_1 \subset A_2 \subset \cdots$, and $\bigcup_{n=1}^{\infty} = (0, 1] = \{x : 0 < x \leq 1\}$. In this case, the A_n are said to be *monotone increasing*. If sets A_i are monotone increasing, then

$$\lim_{i \to \infty} A_i = \bigcup_{i=1}^{\infty} A_i. \tag{A.1}$$

Similarly, the sets A_i are *monotone decreasing* if $A_1 \supset A_2 \supset \cdots$, in which case

$$\lim_{i \to \infty} A_i = \bigcap_{i=1}^{\infty} A_i. \tag{A.2}$$

We will also need basic familiarity with the following sets: $\mathbb{N} = \{1, 2, \ldots\}$ is the set of all natural numbers; $\mathbb{Z} = \{0, 1, -1, 2, -2, \ldots\}$ is the set of all integers or Zahlen (German for number); $\mathbb{Q} = \{m/n, m \in \mathbb{Z}, n \in \mathbb{N}\}$ is the set of all rational numbers (quotients); $\mathbb{R}$ is the set of all real numbers; $\mathbb{C}$ is the set of complex numbers, and $\mathbb{N} \subset \mathbb{Z} \subset \mathbb{Q} \subset \mathbb{R} \subset \mathbb{C}$. For convenience and clarity, we also define $\mathbb{R}_{>0} = \{x : x \in \mathbb{R}, x > 0\}$, $\mathbb{R}_{\geq 1} = \{x : x \in \mathbb{R}, x \geq 1\}$, etc.; if only a range is specified, then the real numbers are assumed, e.g. $x > 0$ is the same as $x \in \mathbb{R}_{>0}$. Also, we take $\mathbb{X} := \mathbb{R} \cup \{-\infty, \infty\}$, which is the *extended real line*. Letting $a \in \mathbb{R}$, properties of $\mathbb{X}$ include $\infty + \infty = \infty + a = \infty$, $a \cdot \infty = \text{sgn}(a) \cdot \infty$, but $\infty - \infty$, ∞/∞, etc., are undefined, as remains $0/0$.

We make use of the common abbreviations $\exists$ ('there exists'), $\nexists$ ('there does not exist'), $\Rightarrow$ ('implies'), iff (if and only if) and $\forall$ ('for all' or, better, 'for each'; see

Pugh, 2002, p. 5). As an example, $\forall x \in (0, 1), \exists y \in (x, 1)$. Also, the notation '$A := B$' means that A, or the l.h.s. (left-hand side) of the equation, is defined to be B, or the r.h.s. (right-hand side).

Sets obey certain rules, such as $A \cup A = A$ (idempotent); $(A \cup B) \cup C = A \cup (B \cup C)$ and $(A \cap B) \cap C = A \cap (B \cap C)$ (associative); $A \cup B = B \cup A$ and $A \cap B = B \cap A$ (commutative);

$$A \cup (B \cap C) = (A \cup B) \cap (A \cup C) \quad \text{and} \quad A \cap (B \cup C) = (A \cap B) \cup (A \cap C)$$

(distributive); $A \cup \emptyset = A$ and $A \cap \emptyset = \emptyset$ (identity); and $(A^c)^c = A$ (involution). Less obvious are De Morgan's laws, after Augustus De Morgan (1806–1871), which state that $(A \cup B)^c = A^c \cap B^c$ and $(A \cap B)^c = A^c \cup B^c$. More generally,

$$\left(\bigcup_{n=1}^{\infty} A_n \right)^c = \bigcap_{n=1}^{\infty} A_n^c \quad \text{and} \quad \left(\bigcap_{n=1}^{\infty} A_n \right)^c = \bigcup_{n=1}^{\infty} A_n^c. \tag{A.3}$$

⊖ **Example A.1**   Let $B_i := A_i \setminus \left[A_i \cap (A_1 \cup A_2 \cup \cdots \cup A_{i-1}) \right]$. To show that

$$B_i = A_i \setminus (A_1 \cup \cdots \cup A_{i-1}),$$

use the above rules for sets to get

$$\begin{aligned}
B_i &= A_i \setminus \left[A_i \cap (A_1 \cup A_2 \cup \cdots \cup A_{i-1}) \right] = A_i \cap \left[A_i \cap (A_1 \cup A_2 \cup \cdots \cup A_{i-1}) \right]^c \\
&= A_i \cap \left[(A_i \cap A_1) \cup (A_i \cap A_2) \cup \cdots \cup (A_i \cap A_{i-1}) \right]^c \\
&= A_i \cap \left[(A_i \cap A_1)^c \cap (A_i \cap A_2)^c \cap \cdots \cap (A_i \cap A_{i-1})^c \right] \\
&= A_i \cap \left(A_i^c \cup A_1^c \right) \cap \left(A_i^c \cup A_2^c \right) \cap \cdots \cap \left(A_i^c \cup A_{i-1}^c \right) \\
&= \left[A_i \cap \left(A_i^c \cup A_1^c \right) \right] \cap \left[A_i \cap \left(A_i^c \cup A_2^c \right) \right] \cap \cdots \cap \left[A_i \cap \left(A_i^c \cup A_{i-1}^c \right) \right] \\
&= \left[(A_i \cap A_i^c) \cup (A_i \cap A_1^c) \right] \cap \cdots \cap \left[(A_i \cap A_i^c) \cup (A_i \cap A_{i-1}^c) \right] \\
&= (A_i \cap A_1^c) \cap \cdots \cap (A_i \cap A_{i-1}^c) = A_i \cap \left(A_1^c \cap \cdots \cap A_{i-1}^c \right) \\
&= A_i \cap (A_1 \cup \cdots \cup A_{i-1})^c = A_i \setminus (A_1 \cup \cdots \cup A_{i-1}).
\end{aligned}$$ ∎

Two sets are *disjoint*, or *mutually exclusive*, if $A \cap B = \emptyset$, i.e. they have no elements in common. A set J is an *indexing set* if it contains a set of indices, usually a subset of $\mathbb{N}$, and is used to work with a group of sets A_i, where $i \in J$. If A_i, $i \in J$, are such that $\bigcup_{i \in J} A_i \supset \Omega$, then they are said to *exhaust*, or (form a) *cover* (for) the set Ω. If sets A_i, $i \in J$, are nonempty, mutually exclusive and exhaust Ω, then they (form a) *partition* (of) Ω.

⊙ **Example A.2**   Let A_i be monotone increasing sets, i.e. $A_1 \subset A_2 \subset \cdots$. Define $B_1 := A_1$ and $B_i := A_i \setminus (A_1 \cup A_2 \cup \cdots \cup A_{i-1})$. We wish to show that, for $n \in \mathbb{N}$,

$$\bigcup_{i=1}^{n} A_i = \bigcup_{i=1}^{n} B_i. \tag{A.4}$$

The B_i are clearly disjoint from their definition and such that B_i is the 'marginal contribution' of A_i over and above that of $(A_1 \cup A_2 \cup \cdots \cup A_{i-1})$, which follows because the A_i are monotone increasing. Thus, $B_i = A_i \setminus A_{i-1} = A_i \cap A_{i-1}^c$. If $\omega \in \bigcup_{i=1}^n A_i$, then, because the A_i are increasing, either $\omega \in A_1$ (and, thus, in all the A_i) or there exists a value $j \in \{2, \ldots, n\}$ such that $\omega \in A_j$ but $\omega \notin A_i$, $i < j$. It follows from the definition of the B_i that $\omega \in B_j$ and thus in $\bigcup_{i=1}^n B_i$, so that (i) $\bigcup_{i=1}^n A_i \subset \bigcup_{i=1}^n B_i$.

Likewise, if $\omega \in \bigcup_{i=1}^n B_i$ then, as the B_i are disjoint, ω is in exactly one of the B_i, say B_j, $j \in \{1, 2, \ldots, n\}$. From the definition of B_j, $\omega \in A_j$, so $\omega \in \bigcup_{i=1}^n A_i$, so that (ii) $\bigcup_{i=1}^n B_i \subset \bigcup_{i=1}^n A_i$. Together, (i) and (ii) imply that $\bigcup_{i=1}^n A_i = \bigcup_{i=1}^n B_i$.

Also, for $i > 1$,

$$B_i = A_i \setminus (A_1 \cup A_2 \cup \cdots \cup A_{i-1}) = A_i \cap (A_1 \cup A_2 \cup \cdots \cup A_{i-1})^c$$
$$= A_i A_1^c A_2^c \cdots A_{i-1}^c = A_i A_{i-1}^c,$$

where the last equality follows from $A_j = \bigcup_{n=1}^j A_n$ (because the A_i are monotone increasing) and thus $A_j^c = \bigcap_{n=1}^j A_n^c$. ∎

For $a, b \in \mathbb{R}$ with $a \leq b$, the interval $(a, b) = \{x \in \mathbb{R} : a < x < b\}$ is said to be an *open interval*,[2] while $[a, b] = \{x \in \mathbb{R} : a \leq x \leq b\}$ is a *closed interval*. In both cases, the interval has *length* $b - a$. For a set $S \subset \mathbb{R}$, the set of open intervals $\{O_i\}$, for $i \in J$ with J an indexing set, is an *open cover* of S if $\bigcup_{i \in J} O_i$ covers S, i.e. if $S \subset \bigcup_{i \in J} O_i$. Let $S \subset \mathbb{R}$ be such that there exists an open cover $\bigcup_{i \in N} O_i$ of S with a finite or countably infinite number of intervals. Denote the length of each O_i as $\ell(O_i)$. If $\forall \epsilon > 0$, there exists a cover $\bigcup_{i \in N} O_i$ of S such that $\sum_{i=1}^\infty \ell(O_i) < \epsilon$, then S is said to have *measure zero*.

For our purposes, the most important set with measure zero is any set with a finite or countable number of points. For example, if f and g are functions with domain $I = (a, b) \in \mathbb{R}$, where $a < b$, and such that $f(x) = g(x)$ for all $x \in I$ except for a finite or countably infinite number of points in I, then we say that f *and* g *differ on* I *by a set of measure zero*. As an example from probability, if U is a continuous uniform random variable on $[0, 1]$, then the event that $U = 1/2$ is not impossible, but it has probability zero, because the point $1/2$ has measure zero, as does any finite collection of points, or any countably infinite set of points on $[0, 1]$, e.g. $\{1/n, n \in \mathbb{N}\}$.

Let S be a nonempty subset of $\mathbb{R}$. We say S has an *upper bound* M iff $x \leq M \ \forall x \in S$, in which case S is *bounded above* by M. Note that, if S is bounded above, then it has infinitely many upper bounds. A fundamental property of $\mathbb{R}$ not shared by $\mathbb{Q}$ is that, if S has an upper bound M, then S possesses a unique *least upper bound*, or *supremum*, denoted sup S. That is, $\exists U \in \mathbb{R}$ such that U is an upper bound of S, and such that, if V is also an upper bound of S, then $V \geq U$. If S is not bounded above, then sup $S = \infty$. Also, sup $\emptyset = -\infty$. Similar terminology applies to the *greatest lower bound*, or *infimum* of S, denoted inf S. For example, let $S = \{1/n : n \in \mathbb{N}\}$. Then max S = sup S = 1 and inf $S = 0$, but S has no minimum value. Next, let S consist of the truncated values

---

[2] A popular notation for open sets is $]a, b[$, but we will not use this.

of $\sqrt{2}$ with $n \in \mathbb{N}$ decimal places, i.e. $S = \{1.4, 1.41, 1.414, 1.4142, \ldots\}$. Then $S \subset \mathbb{Q}$ but $\sup S = \sqrt{2} \notin \mathbb{Q}$.[3]

A *relation* between A and B is a subset of $A \times B$. If a relation f is such that, for each $a \in A$, there is one and only one $b \in B$ such that $(a, b) \in f$, then f is also a *mapping*; one writes $f : A \to B$ and $b = f(a)$, with A referred to as the *domain* and B as the *codomain* or *target*. When f is plotted on the plane in the standard fashion, i.e. with A on the horizontal axis and B on the vertical axis, then a mapping satisfies the 'vertical line test'. The subset of the codomain given by $\{b \in B : \exists a \in A \text{ with } f(a) = b\}$ is the *range* or *image* of f. For some subset $C \subset B$, the *pre-image* of C is the subset of the domain given by $\{a \in A : f(a) \in C\}$.

A mapping with codomain $B = \mathbb{R}$ is a *function*. Let f be a function with domain A and let $I \in A$ be an interval. If f is such that, $\forall a, b \in A, a < b \Rightarrow f(a) < f(b)$, then f is *strictly increasing* on I. Likewise, if $a < b \Rightarrow f(a) \leq f(b)$, then f is *(weakly) increasing*. The terms *strictly decreasing* and *(weakly) decreasing* are similarly defined. A function which is either increasing or decreasing is said to be *monotone*, while a function which is either strictly increasing or strictly decreasing is *strictly monotone*.

The mapping $f : A \to B$ is *injective* or *one-to-one* if $f(a_1) = f(a_2)$ implies $a_1 = a_2$ (that is, if a plot of f satisfies the 'horizontal line test'). A mapping is *surjective* or *onto* if the range is the (whole) codomain, and *bijective* if it is injective and surjective. If $f : A \to B$ is bijective, then the *inverse mapping* $f^{-1} : B \to A$ is bijective such that $f^{-1}(b)$ is the (unique) element in A such that $f(a) = b$. For mappings $f : A \to B$ and $g : B \to C$, the *composite mapping*, denoted $g \circ f : A \to C$, maps an element $a \in A$ to $g(f(a))$. Some thought confirms that, if f and g are injective, then so is $g \circ f$, and if f and g are surjective, then so is $g \circ f$; thus, if f and g are bijective, then so is $g \circ f$.

If $a \in \mathbb{R}$, then the *absolute value* of a is denoted by $|a|$, and is equal to a if $a \geq 0$, and $-a$ if $a < 0$. Clearly, $a \leq |a|$ and, $\forall a, b \in \mathbb{R}, |ab| = |a| \, |b|$. Observe that, for $b \in \mathbb{R}_{\geq 0}$, the inequality $-b < a < b$ is equivalent to $|a| < b$ and, similarly,

$$-b \leq a \leq b \quad \Leftrightarrow \quad |a| \leq b. \tag{A.5}$$

The *triangle inequality* states that

$$|x + y| \leq |x| + |y|, \quad \forall \, x, y \in \mathbb{R}. \tag{A.6}$$

This is seen by squaring both sides to get

$$|x + y|^2 = (x + y)^2 = x^2 + 2xy + y^2 \quad \text{and} \quad (|x| + |y|)^2 = x^2 + 2|x| \, |y| + y^2,$$

and noting that $xy \leq |xy| = |x| \, |y|$. Alternatively, note that, $\forall \, a \in \mathbb{R}, -|a| \leq a \leq |a|$, so adding $-|x| \leq x \leq |x|$ to $-|y| \leq y \leq |y|$ gives $-(|x| + |y|) \leq x + y \leq |x| + |y|$, which, from (A.5) with $a = x + y$ and $b = |x| + |y|$, is equivalent to $|x + y| \leq |x| + |y|$. Also, with $z = -y$, the triangle inequality states that, $\forall \, x, z \in \mathbb{R}, |x - z| \leq |x| + |z|$.

The *Cauchy–Schwarz inequality*, after Augustin Louis Cauchy (1789–1857) and Hermann Schwarz (1843–1921), also referred to as Cauchy's inequality or the Schwarz

---

[3] Observe that any element in $\mathbb{R}$ can be arbitrarily closely approximated by an element in $\mathbb{Q}$, which is the informal description of saying that $\mathbb{Q}$ is *dense* in $\mathbb{R}$. This is of enormous importance when actually working with numerical values in an (unavoidably) finite precision computing world.

inequality, is

$$|x_1 y_1 + \cdots + x_n y_n| \le (x_1^2 + \cdots x_n^2)^{1/2}(y_1^2 + \cdots y_n^2)^{1/2}, \tag{A.7}$$

for any points $\mathbf{x} = (x_1, \ldots, x_n)$ and $\mathbf{y} = (y_1, \ldots, y_n)$ in $\mathbb{R}^n$, $n \in \mathbb{N}$. It was first published by Cauchy in 1821. This is proved by letting $f(r) = \sum_{i=1}^n (r x_i + y_i)^2 = Ar^2 + Br + C$, where $A = \sum_{i=1}^n x_i^2$, $B = 2 \sum_{i=1}^n x_i y_i$ and $C = \sum_{i=1}^n y_i^2$. As $f(r) \ge 0$, the quadratic $Ar^2 + Br + C$ has one or no real roots, so that its discriminant $B^2 - 4AC \le 0$, i.e. $B^2 \le 4AC$ or, substituting, $\left(\sum_{i=1}^n x_i y_i\right)^2 \le \left(\sum_{i=1}^n x_i^2\right)\left(\sum_{i=1}^n y_i^2\right)$, which is (A.7) after taking square roots.

The Cauchy–Schwarz inequality is used to show the generalization of (A.6), which is also referred to as the *triangle inequality*:

$$\|\mathbf{x} + \mathbf{y}\| \le \|\mathbf{x}\| + \|\mathbf{y}\|, \tag{A.8}$$

where

$$\|\mathbf{x}\| = \sqrt{x_1^2 + \cdots + x_n^2} \tag{A.9}$$

is the *norm* of $\mathbf{x} = (x_1, \ldots, x_n) \in \mathbb{R}^n$. In particular, using the above notation for A, B and C,

$$\|\mathbf{x} + \mathbf{y}\|^2 = \sum_{i=1}^n (x_i + y_i)^2 = \sum_{i=1}^n x_i^2 + 2 \sum_{i=1}^n x_i y_i + \sum_{i=1}^n y_i^2 = A + B + C$$

and, as $B^2 \le 4AC$, $A + B + C \le A + 2\sqrt{AC} + C = \left(\sqrt{A} + \sqrt{C}\right)^2$. Taking square roots gives $\|\mathbf{x} + \mathbf{y}\| = \sqrt{A + B + C} \le \sqrt{A} + \sqrt{C} = \|\mathbf{x}\| + \|\mathbf{y}\|$.[4]

A.2 Univariate calculus

Leibniz never married; he had considered it at the age of fifty; but the person he had in mind asked for time to reflect. This gave Leibniz time to reflect, too, and so he never married. (Bernard Le Bovier Fontenelle)

---

[4] The Cauchy–Schwarz inequality can be generalized to *Hölder's inequality*: let $p, q \in \mathbb{R}_{>1}$ be such that $(p-1)(q-1) = 1$ (or, equivalently $p^{-1} + q^{-1} = 1$). Then

$$|x_1 y_1 + \cdots + x_n y_n| \le \left(|x_1|^p + \cdots + |x_n|^p\right)^{1/p} \left(|y_1|^q + \cdots + |y_n|^q\right)^{1/q},$$

while the triangle inequality can be generalized to *Minkowski's inequality*:

$$\left(\sum_{i=1}^n |x_i + y_i|^p\right)^{1/p} \le \left(\sum_{i=1}^n |x_i|^p\right)^{1/p} + \left(\sum_{i=1}^n |y_i|^p\right)^{1/p}, \quad p \in \mathbb{R}_{\ge 1},$$

(see, for example, Edwards, 1994, pp. 174–5, 185; Bachman, Narici and Beckenstein, 2000, Section 1.6; and Trench, 2003, p. 521). These can be further generalized to infinite sums and integral expressions (see, for example, Jones, 2001, Chapter 10; Browder, 1996, Section 10.4).

A.2.1 Limits and continuity

Informally, the limit of a function at a particular point, say x, is the value that $f(x)$ approaches, but need not assume at x. For example, $\lim_{x \to 0} (\sin x) / x = 1$, even though the ratio is not defined at $x = 0$. Formally, as instigated in 1821 by Cauchy, the function $f : A \to \mathbb{R}$ with $A \subset \mathbb{R}$ has the *right-hand limit* L at c, if, $\forall \epsilon > 0$, $\exists \delta > 0$ such that $|f(x) - L| < \epsilon$ whenever $x \in (c, c + \delta)$, for which we write $\lim_{x \to c^+} f(x)$. Likewise, f has the *left-hand limit* L at c, if, $\forall \epsilon > 0$, $\exists \delta > 0$ such that $|f(x) - L| < \epsilon$ whenever $x \in (c - \delta, c)$; we write $\lim_{x \to c^-} f(x)$. If $\lim_{x \to c^-} f(x)$ and $\lim_{x \to c^+} f(x)$ exist and coincide, then L is the *limit of* f *at* c, and write $\lim_{x \to c} f(x)$.

Of course, not all limits are finite. We write $\lim_{x \to c^+} f(x) = \infty$ if, $\forall M \in \mathbb{R}$, $\exists \delta > 0$ such that $f(x) > M$ for every $x \in (c, c + \delta)$; and $\lim_{x \to c^-} f(x) = \infty$ if, $\forall M \in \mathbb{R}$, $\exists \delta > 0$ such that $f(x) > M$ for every $x \in (c - \delta, c)$. Similar definitions hold for $\lim_{x \to c^+} f(x) = -\infty$ and $\lim_{x \to c^-} f(x) = -\infty$. As with a finite limit, if $\lim_{x \to c^+} f(x) = \lim_{x \to c^-} f(x) = \pm\infty$, then we write $\lim_{x \to c} f(x) = \pm\infty$. Lastly, we write $\lim_{x \to \infty} f(x) = L$ if, for each $\epsilon > 0$, $\exists x_0$ such that $|f(x) - L| < \epsilon$ for all $x > x_0$, and $\lim_{x \to -\infty} f(x) = L$ if, for each $\epsilon > 0$, $\exists x_0$ such that $|f(x) - L| < \epsilon$ for all $x < x_0$. As a shorthand, let $f(\infty) := \lim_{x \to \infty} f(x)$ and $f(-\infty) := \lim_{x \to -\infty} f(x)$. If $f(\infty) = f(-\infty)$, then we take $f(\pm\infty) := f(\infty) = f(-\infty)$.

Some important operations with limits include the following, which we state without complete rigor. Let f and g be functions whose domain contains the point c and such that $\lim_{x \to c} f(x) = L$ and $\lim_{x \to c} g(x) = M$. Then, for constant values $k_1, k_2 \in \mathbb{R}$,

$$\lim_{x \to c} [k_1 f(x) + k_2 g(x)] = k_1 L + k_2 M. \tag{A.10}$$

Also,

$$\lim_{x \to c} f(x) g(x) = LM, \tag{A.11}$$

$\lim_{x \to c} f(x) / g(x) = L/M$, if $M \neq 0$, and, if $g(x) \leq f(x)$, then $M \leq L$. Finally, for the limit of a composition of functions, let $b = \lim_{x \to a} f(x)$ and $L = \lim_{y \to b} g(y)$. Then $\lim_{x \to a} g(f(x)) = L$ (see, for example, Lang, 1997, Sections 2.2 and 2.3, for details and proofs).

Let f be a function with domain A and $a \in A$. If $a \in A$ and $\lim_{x \to a^+} f(x) = f(a)$, then f is said to be *continuous on the right at* a; and if $a \in A$ and $\lim_{x \to a^-} f(x) = f(a)$, then f is *continuous on the left at* a. If both of these conditions hold, then $\lim_{x \to a} f(x) = f(a)$, and f is said to be *continuous at* a. If f is continuous at each point $a \in S \subset A$, then f is *continuous on* S, in which case we also say that f is of *class* C^0 *on* S, or $f \in C^0(S)$. Often, subset S will be an interval, say (a, b) or $[a, b]$, in which case we write $f \in C^0(a, b)$ and $f \in C^0[a, b]$, respectively. If f is continuous on (its whole domain) A, then we say f is continuous, or that f is of class C^0, or $f \in C^0$.

An important result is the continuity of composite functions. Let $f : A \to B$ and $g : B \to C$ be continuous. Then $g \circ f : A \to C$ is continuous. More precisely, if f is continuous at $a \in A$, and g is continuous at $b = f(a) \in B$, then

$$\lim_{x \to a} g(f(x)) = g\left(\lim_{x \to a} f(x)\right). \tag{A.12}$$

Let f be a continuous function on the closed, bounded interval $I = [a, b]$, i.e. $f \in C^0[a, b]$. Then:

- $\forall x \in I, \exists m, M$ such that $m \leq f(x) \leq M$ (the image of f on I forms a bounded subset of $\mathbb{R}$);

- for some $x_0, x_1 \in I$ and $\forall x \in I, f(x_0) \leq f(x) \leq f(x_1)$ (f assumes minimum and maximum values). (A.13)

- (*Intermediate Value Theorem*) Let $\alpha = f(a)$ and $\beta = f(b)$. Given a number γ with $\alpha < \gamma < \beta, \exists c \in (a, b)$ such that $f(c) = \gamma$.

- f is uniformly continuous on I. (A.14)

These four facts together constitute what Pugh (2002, p. 39) argues could rightfully be called the *Fundamental Theorem of Continuous Functions*.

Above, we said that f is continuous at a if $\lim_{x \to a} f(x) = f(a)$. An equivalent, but seemingly more complicated, definition of continuity at a is: for a given $\epsilon > 0, \exists \delta > 0$ such that, if $|x - a| < \delta$ and $x \in A$, then $|f(x) - f(a)| < \epsilon$. Its value is seen when contrasting it with the definition of *uniform continuity*: Let f be a function with domain A and let $[a, b] \subset A$ be a closed, finite interval. Function f is uniformly continuous on $[a, b]$ if the condition holds: for a given $\epsilon > 0, \exists \delta > 0$ such that, if $x, y \in [a, b]$, and $|x - y| < \delta$, then $|f(x) - f(y)| < \epsilon$. Note that, with uniform continuity, δ does not depend on the choice of $x \in [a, b]$. The notions of uniform continuity and uniform convergence (discussed in Section A.2.4 below) play a major role in analysis and need to be well understood!

A.2.2   Differentiation

Definitions and techniques

Let $f \in C^0(I)$, where I is an interval of nonzero length. If the *Newton quotient*

$$\lim_{h \to 0} \frac{f(x + h) - f(x)}{h} \tag{A.15}$$

exists for $x \in I$, then f is *differentiable* at x, the limit is the *derivative* of f at x, and is denoted $f'(x)$ or df/dx. Similar to the notation for continuity, if f is differentiable at each point in I, then f is differentiable on I, and if f is differentiable on its domain, then f is differentiable. If f is differentiable and $f'(x)$ is a continuous function of x, then f is *continuously differentiable*, and is of *class* C^1.

Observe that, for h small,

$$f'(x) \approx \frac{f(x + h) - f(x)}{h} \quad \text{or} \quad f(x + h) \approx f(x) + hf'(x), \tag{A.16}$$

which, for constant x, is a linear function in h. By letting $h = y - x$, (A.15) can be equivalently written as

$$\lim_{y \to x} \frac{f(y) - f(x)}{y - x} \tag{A.17}$$

which is sometimes more convenient to work with.

The *Fundamental lemma of differentiation* makes the notion more precise that a differentiable function can be approximated at each point in (the interior of) its domain by a linear function whose slope is the derivative at that point. As in Protter and Morrey (1991, p. 85), let f be differentiable at the point x. Then there exists a function η defined on an interval about zero such that

$$f(x + h) - f(x) = \left[f'(x) + \eta(h) \right] \cdot h, \tag{A.18}$$

and η is continuous at zero, with $\eta(0) = 0$. The proof follows by solving (A.18) for $\eta(h)$ and defining $\eta(0) = 0$, i.e.

$$\eta(h) := \frac{1}{h} \left[f(x + h) - f(x) \right] - f'(x), \quad h \neq 0, \qquad \eta(0) := 0.$$

As f is differentiable at x, $\lim_{h \to 0} \eta(h) = 0$, so that η is continuous at zero.

Example A.3   From the definition of limit, $\lim_{h \to 0} 0/h = 0$, so that the derivative of $f(x) = k$ for some constant k is zero. For $f(x) = x$, it is easy to see from (A.15) that $f'(x) = 1$. For $f(x) = x^2$,

$$\lim_{h \to 0} \frac{(x + h)^2 - x^2}{h} = \lim_{h \to 0} (2x + h) = 2x.$$

For $n \in \mathbb{N}$, the binomial theorem (1.18) implies

$$f(x + h) = (x + h)^n = \sum_{i=0}^{n} \binom{n}{i} x^{n-i} h^i = x^n + nhx^{n-1} + \cdots + h^n,$$

so that

$$\lim_{h \to 0} \frac{f(x + h) - f(x)}{h} = \lim_{h \to 0} \left(nx^{n-1} + \cdots + h^{n-1} \right) = nx^{n-1}. \tag{A.19}$$

Now let $f(x) = x^{-n}$ for $n \in \mathbb{N}$. It is easy to verify that, for any $x \neq 0$ and $y \neq 0$,

$$f(y) - f(x) = y^{-n} - x^{-n} = \frac{x^n - y^n}{x^n y^n} = (y - x) \left(-\frac{y^{n-1} + y^{n-2}x + \cdots + x^{n-1}}{x^n y^n} \right),$$

so that (A.17) implies

$$\lim_{y \to x} \frac{f(y) - f(x)}{y - x} = -\lim_{y \to x} \left(\frac{y^{n-1} + y^{n-2}x + \cdots + x^{n-1}}{x^n y^n} \right) = -\frac{nx^{n-1}}{x^{2n}} = -nx^{-n-1}.$$

Thus, for $f(x) = x^n$ with $n \in \mathbb{Z}$, $f'(x) = nx^{n-1}$. ∎

Assume for functions f and g defined on I that $f'(x)$ and $g'(x)$ exist on I. Then

(*sum rule*) $(f+g)'(x) = f'(x) + g'(x)$,

(*product rule*) $(fg)'(x) = f(x)g'(x) + g(x)f'(x)$,   (A.20)

(*quotient rule*) $(f/g)'(x) = \dfrac{g(x)f'(x) - f(x)g'(x)}{[g(x)]^2}$, $g(x) \neq 0$,

(*chain rule*) $(g \circ f)'(x) = g'[f(x)]f'(x)$.   (A.21)

With $y = f(x)$ and $z = g(y)$, the usual mnemonic for the chain rule is $\dfrac{dz}{dx} = \dfrac{dz}{dy}\dfrac{dy}{dx}$.
The chain rule is simple to prove using the fundamental lemma of differentiation (A.18) (see, for example, Protter and Morrey (1991, p. 85)).

⊖ **Example A.4**   Result (A.19) for $n \in \mathbb{N}$ could also be established by using an induction argument. Let $f(x) = x^n$ and assume $f'(x) = nx^{n-1}$. It holds for $n = 1$; induction and the product rule imply for $f(x) = x^{n+1} = x^n \cdot x$ that $f'(x) = x^n \cdot 1 + x \cdot nx^{n-1} = (n+1)x^n$.   ∎

If f is differentiable at a, then f is continuous at a. This is seen by taking limits of

$$f(x) = \frac{f(x) - f(a)}{x - a}(x - a) + f(a),$$

which, using (A.10) and (A.11), gives

$$\lim_{x \to a} f(x) = f'(a) \cdot 0 + f(a) = f(a),$$ (A.22)

and recalling the definition of continuity. The function $f(x) = |x|$ at $x = 0$ is the showcase example that continuity does not imply differentiability. Next, differentiability of f does not imply that f' is continuous, as demonstrated via the popular example with $f(x) = x^2 \sin(1/x)$ for $x \neq 0$ and $f(0) = 0$. Then

$$f'(x) = 2x \sin\left(\frac{1}{x}\right) - \cos\left(\frac{1}{x}\right), \quad x \neq 0$$

and $\lim_{x \to 0} f'(x)$ does not exist. However, from the Newton quotient at $x = 0$,

$$\lim_{h \to 0} \frac{f(0+h) - f(0)}{h} = \lim_{h \to 0} h \sin(1/h) = 0,$$

so that $f'(0) = 0$, showing that $f'(x)$ is not continuous. What is true is that *uniform differentiability* implies uniform continuity of the derivative. As in Estep (2002, Section 32.4), a function f is said to be uniformly differentiable on an interval $[a, b]$ if, $\forall \epsilon > 0$, $\exists \delta > 0$ such that

$$\left| \frac{f(y) - f(x)}{y - x} - f'(x) \right| < \epsilon, \quad \forall x, y \in [a, b] \quad \text{with} \quad |x - y| < \delta.$$

If f is uniformly differentiable, then for $x, y \in [a, b]$ and $\epsilon > 0$, we can find a $\delta > 0$ such that, for $|x - y| < \delta$,

$$\left| f'(y) - f'(x) \right| = \left| f'(y) - \frac{f(y) - f(x)}{y - x} + \frac{f(y) - f(x)}{y - x} - f'(x) \right|$$

$$\leq \left| f'(y) - \frac{f(y) - f(x)}{y - x} \right| + \left| \frac{f(y) - f(x)}{y - x} - f'(x) \right| < 2\epsilon.$$

Thus, $f'(x)$ is uniformly continuous on $[a, b]$.

Of great use is *l'Hôpital's rule*[5] for evaluating *indeterminate forms or ratios*. Let f and g, and their first derivatives, be continuous functions on (a, b). If $\lim_{x \to a^+} f(x) = \lim_{x \to a^+} g(x) = 0$ and $\lim_{x \to a^+} f'(x) / g'(x) = L$, then $\lim_{x \to a^+} f(x) / g(x) = L$. Most students remember this very handy result, but few can intuitively justify it. Here is one way. Assume f and g are continuous at a, so that $f(a) = g(a) = 0$. Using (A.16) gives

$$\lim_{x \to a^+} \frac{f(x)}{g(x)} = \lim_{h \to 0} \frac{f(a + h)}{g(a + h)} \approx \lim_{h \to 0} \frac{f(a) + hf'(a)}{g(a) + hg'(a)} = \frac{f'(a)}{g'(a)} = \lim_{x \to a^+} \frac{f'(x)}{g'(x)}.$$

A different 'rough proof' of l'Hôpital's rule is given by Pugh (2002, p. 143). A similar result holds for $x \to b^-$, and for $x \to \infty$, and also for the case when $\lim_{x \to a^+} f(x) = \lim_{x \to a^+} g(x) = \infty$.

The following example illustrates some of aforementioned results and also serve to refresh some basic facts from trigonometry.

◎ **Example A.5** Let $f(x) = \sin(x)$ and recall the relations

$$\sin(x + y) = \sin x \cos y + \cos x \sin y, \tag{A.23a}$$

$$\cos(x + y) = \cos x \cos y - \sin x \sin y, \tag{A.23b}$$

which are straightforward to confirm geometrically (see, for example, Stillwell, 1998, Section 5.3). Using (A.23a), the derivative of f is

$$\frac{d \sin(x)}{dx} = \lim_{h \to 0} \frac{\sin(x + h) - \sin(x)}{h} = \lim_{h \to 0} \frac{\sin x \cos h - \sin x}{h} + \lim_{h \to 0} \frac{\cos x \sin h}{h}$$

$$= \sin(x) \lim_{h \to 0} \frac{\cos(h) - 1}{h} + \cos(x) \lim_{h \to 0} \frac{\sin(h)}{h}$$

$$= \cos(x), \tag{A.24}$$

where

$$L_s := \lim_{h \to 0} \frac{\sin(h)}{h} = 1 \quad \text{and} \quad \lim_{h \to 0} \frac{\cos(h) - 1}{h} = 0. \tag{A.25}$$

---

[5] Named after Guillaume François Antoine Marquis de l'Hôpital (1661–1704), who was taught calculus by Johann Bernoulli (for a high price), and wrote the first calculus textbook (1696) based on Bernoulli's notes, in which the result appeared. Not surprisingly, l'Hôpital's rule was also known to Bernoulli (confirmed in Basel, 1922, with the discovery of certain written documents).

Both limits in (A.25) need to be justified. If we *assume* that L_s is not infinite, then the second limit in (A.25) is easy to prove. Write

$$\frac{\cos{(h)} - 1}{h} = \frac{h\left[\cos{(h)} + 1\right]}{h\left[\cos{(h)} + 1\right]} \frac{\cos{(h)} - 1}{h} = \frac{h\left(\cos^2 h - 1\right)}{h^2\left[\cos{(h)} + 1\right]}$$

$$= -\left(\frac{\sin h}{h}\right)^2 \frac{h}{\cos{(h)} + 1}$$

using $\cos^2{(x)} + \sin^2{(x)} = 1$, so that, from (A.11) and because we assumed that $L_s \in \mathbb{R}$,

$$\lim_{h \to 0} \frac{\cos{(h)} - 1}{h} = -\lim_{h \to 0} \left(\frac{\sin h}{h}\right)^2 \lim_{h \to 0} \frac{h}{\cos{(h)} + 1} = 0.$$

Now let $f(x) = \cos{(x)}$. Using (A.24) and the relations

$$\cos{(x)} = \sin{(x + \pi/2)}, \qquad \sin{(x)} = -\cos{(x + \pi/2)} \qquad (A.26)$$

(which follow geometrically from the unit circle), the chain rule gives

$$\frac{d\cos x}{dx} = \frac{d\sin{(x + \pi/2)}}{dx} = \cos{(x + \pi/2)} = -\sin{(x)}.$$

Students remember that the derivatives of sine and cosine involve, respectively, cosine and sine, but some forget the signs. To recall them, just think of the unit circle at angle $\theta = 0$ and the geometric definition of sine and cosine. A slight increase in θ increases the vertical coordinate (sine) and decreases the horizontal one (cosine).

The easiest way of proving the former limit in (A.25) is using (A.24); it follows trivially by using the derivative of $\sin x$, i.e.[6]

$$\lim_{h \to 0} \frac{\sin{(h)}}{h} = \lim_{h \to 0} \frac{\sin h - \sin 0}{h} = \frac{d\sin x}{dx}\bigg|_{x=0} = \cos 0 = 1.$$

The circular logic between (A.24) and (A.25) is obviously not acceptable![7] The properties of the sine and cosine functions can be correctly, elegantly and easily derived from an algebraic point of view by starting with functions s and c such that

$$s' = c, \ c' = -s, \ s(0) = 0 \ \text{and} \ c(0) = 1 \qquad (A.27)$$

---

[6] The limits in (A.25) also follow by applying l'Hôpital's rule. For the latter,

$$\lim_{h \to 0} \frac{\cos{(h)} - 1}{h} = \frac{-\sin{(h)}}{1} = -\sin{(0)} = 0.$$

[7] Of course, from a geometric point of view, it is essentially obvious that $\lim_{h \to 0} h^{-1} \sin{(h)} = 1$. Let θ be the angle in the first quadrant of the unit circle, measured in radians. Recall that θ then represents the length of the arc on the unit circle, of which the total length is 2π. Then it seems apparent that, as θ decreases, the arc length coincides with $\sin(\theta)$.

(see, for example, Lang, 1997, Section 4.3). As definitions one takes

$$\cos(z) = \sum_{k=0}^{\infty} (-1)^k \frac{z^{2k}}{(2k)!} \quad \text{and} \quad \sin(z) = \sum_{k=0}^{\infty} (-1)^k \frac{z^{2k+1}}{(2k+1)!}, \quad \text{(A.28)}$$

which converge for all $z \in \mathbb{R}$ (see Section A.2.3 below for details). From (A.28), the properties of the trigonometric functions can be inferred, such as $\cos^2(x) + \sin^2(x) = 1$, (A.23), (A.26) and

$$\cos(-x) = \cos(x) \quad \text{and} \quad \sin(-x) = -\sin(x) \quad \text{(A.29)}$$

(see, for example, Browder (1996, Section 3.6) or Hijab (1997, Section 3.5) for details). From (A.29) and (A.23a),

$$\sin(x - y) + \sin(x + y) = \sin x \cos y - \cos x \sin y + \sin x \cos y + \cos x \sin y$$

$$= 2 \sin x \cos y.$$

Now let $b = x + y$ and $c = y - x$, so that $x = (b - c)/2$ and $y = (b + c)/2$. It follows that

$$\sin(b) - \sin(c) = 2 \sin\left(\frac{b-c}{2}\right) \cos\left(\frac{b+c}{2}\right), \quad \text{(A.30)}$$

which we need when working with characteristic functions of random variables.

Finally, let $f(x) = \tan(x) := \sin(x) / \cos(x)$ so that

$$f'(x) = \frac{\cos(x)\cos(x) - \sin(x)(-\sin(x))}{\cos^2(x)} = 1 + \frac{\sin^2(x)}{\cos^2(x)} = 1 + \tan^2(x) \quad \text{(A.31)}$$

from the quotient rule. ∎

The second derivative of f, if it exists, is the derivative of f', and denoted by f'' or $f^{(2)}$, and likewise for higher-order derivatives. If $f^{(r)}$ exists, then f is said to be rth *order differentiable*, and if $f^{(r)}$ exists for all $r \in \mathbb{N}$, then f is *infinitely differentiable*, or *smooth* (see, for example, Pugh, 2002, p. 147). Let $f^{(0)} \equiv f$. As differentiability implies continuity, it follows that, if f is rth order differentiable, then $f^{(r-1)}$ is continuous, and that smooth functions and all their derivatives are continuous. If f is rth order differentiable and $f^{(r)}$ is continuous, then f is *continuously rth order differentiable*, and f is of *class C^r*. An infinitely differentiable function is of *class C^∞*.

Let f be a strictly increasing continuous function on a closed interval I. Then the image of f is also a closed interval (from the intermediate value theorem). The *inverse function* g is defined as the function such that $g \circ f(x) = x$ and $f \circ g(y) = y$. It is also continuous and strictly increasing. If f is also differentiable in the interior of I with $f'(x) > 0$, then a fundamental result is that

$$g'(y) = \frac{1}{f'(x)} = \frac{1}{f'[g(y)]}, \quad \text{(A.32)}$$

the proof of which can be found in virtually all real analysis books.

⊙ ***Example A.6***   Let $f(x) = \sin(x)$ for $-\pi/2 < x < \pi/2$, with derivative $f'(x) = \cos x$ from (A.24). From (A.32) and relation $\cos^2(x) + \sin^2(x) = 1$, the inverse function $g(y) = \arcsin(y)$ has derivative

$$g'(y) = \frac{1}{\cos\left[\arcsin(y)\right]} = \frac{1}{\sqrt{1 - \sin^2\left[\arcsin(y)\right]}} = \frac{1}{\sqrt{1 - y^2}}.$$

Similarly, let $f(x) = \tan(x)$ for $-\pi/2 < x < \pi/2$ with inverse function $g(y) = \arctan(y)$ so that, from (A.31),

$$g'(y) = \frac{1}{1 + \tan^2\left[\arctan(y)\right]} = \frac{1}{1 + y^2}. \tag{A.33}$$

Now let z be a constant. Using (A.33) and the chain rule gives

$$\frac{d}{dx}\arctan(z - x) = -\frac{1}{1 + (z - x)^2}, \tag{A.34}$$

which we will use below in Example A.28.   ■

Mean value theorem

Let f be a function continuous on its domain $[a, b]$ and differentiable on (a, b). Of great importance is the *mean value theorem (of the differential calculus)*, which states that there exists a number $\xi \in (a, b)$ such that $f(b) - f(a) = f'(\xi)(b - a)$, more easily remembered as

$$\frac{f(b) - f(a)}{b - a} = f'(\xi), \tag{A.35}$$

for $b - a \neq 0$. The algebraic proof is easy, but informally, this is graphically 'obvious' for a differentiable (and thus continuous) function, as illustrated in Figure A.1.

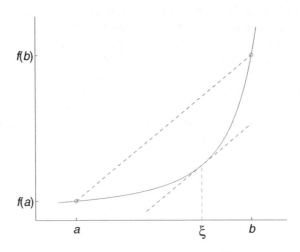

Figure A.1   The mean value theorem of the differential calculus

Remark: The need occasionally arises to construct simple graphics like Figure A.1, and it is often expedient to use the plotting and graphics generation capabilities of Matlab or other such software. In this case, the graph was constructed using the function $f(x) = 1/(1-x)^2$ with endpoints $a = 0.6$ and $b = 0.9$.

This is also a good excuse to illustrate Matlab's symbolic toolbox (which uses the Maple computing engine). The top third of the code in Program Listing A.1 uses some basic commands from the symbolic toolbox to compute ξ based on our choice of f, a and b. The rest of the code constructs Figure A.1.

While Matlab supports interactive graphics editing, use of the native graphics commands ('batch code') in Program Listing A.1 is not only faster the first time around (once you are familiar with them of course), but ensures that the picture can be identically and instantly reproduced. ∎

Let I be the open interval (a, b). If f is differentiable on I and, $\forall\, x \in I$, $\left| f'(x) \right| \leq M$, then the mean value theorem implies that $|f(y) - f(x)| \leq M|y - x|$ for all $x, y \in I$. This is referred to as the *(global) Lipschitz condition*. Note that, if $\forall\, x \in I$, $f'(x) = 0$, then $|f(y) - f(x)| \leq 0 \,\,\forall\, x, y \in I$, i.e. f is constant.

The mean value theorem is mainly used for proving other results, including the fundamental theorem of calculus (see Section A.2.3), the validity of interchanging derivative and integral (Section A.3.4), and the facts that, if $\forall x \in I$, $f'(x) \geq 0$ $(f'(x) > 0)$, then f is increasing (strictly increasing) on I, and if $\forall x \in I$, $f'(x) \leq 0$ $(f'(x) < 0)$, then f is decreasing (strictly decreasing) on I. These latter facts are of great use for deriving the *first derivative test*: Let f be differentiable on $I = (a, b)$ with $f'(c) = 0$ for $c \in I$. If $\forall x \in (a, c)$, $f'(x) < 0$, and $\forall x \in (c, b)$, $f'(x) > 0$ then f has a local minimum at c. Note that, if $f'(c) > 0$ for some point $c \in I$, *and* f' is continuous at c, then $\exists\, \delta > 0$ such that, $\forall x \in (c - \delta, c + \delta)$, $f'(x) > 0$, i.e. f is increasing on that interval (see, for example, the discussion in Stoll (2001, p. 182–3) for more detail).

The mean value theorem can be generalized to the *Cauchy* or *ratio mean value theorem*. If f and g are continuous functions on $[a, b]$ and differentiable on $I = (a, b)$, then $\exists\, c \in I$ such that $[f(b) - f(a)]g'(c) = [g(b) - g(a)]f'(c)$ or, easier to remember, if $g(b) - g(a) \neq 0$,

$$\frac{f(b) - f(a)}{g(b) - g(a)} = \frac{f'(c)}{g'(c)}.$$

This reduces to the usual mean value theorem when $g(x) = x$. It is used, for example, in proving l'Hôpital's rule. Given the prominence of the mean value theorem, it could justifiably called the *fundamental theorem of differential calculus* (Stoll, 2001, p. 204).

Let $I \subset \mathbb{R}$ be an interval and let $f : I \to \mathbb{R}$ be differentiable on I. If f' is continuous on I, then the intermediate value theorem applied to f' implies that, for $a, b \in I$ with $a < b$, $\alpha = f'(a)$, $\beta = f'(b)$ and a value $\gamma \in \mathbb{R}$ with either $\alpha < \gamma < \beta$ or $\alpha > \gamma > \beta$, $\exists\, c \in (a, b)$ such that $f'(c) = \gamma$. More interesting is the fact that this still holds *even if f' is not continuous*, a result attributed to Jean Gaston Darboux (1842–1917) (see Stoll (2001, p. 184), Browder (1996, Theorem 4.25) or Pugh (2002, p. 144)). It is referred to as the *intermediate value theorem for derivatives*.

Exponential and logarithm

> For the sake of brevity, we will always represent this number 2.718281828459...
> by the letter e.   (Leonhard Euler)

```
function meanvaluetheorem
%%%%%%%%%%%%%%%%%%%%%%%%%%%%%%%%%%%%%%%%%%%%%%%%%%%%%%%%%%%%%%%%%%%%%%%%%
%%%%%          Use the symbolic toolbox to compute xi          %%%%%
 syms x xi real  % declare x and xi to be real symbolic variables  %
 f=1/(1-x)^2     % our function                                    %
 a=0.6; b=0.9;   % use these two end points                        %
 fa=subs(f,'x',a); fb=subs(f,'x',b); % evaluate f at a and b       %
 ratio=(fb-fa)/(b-a)  % slope of the line                          %
 df=diff(f)            % first derivative of f                     %
 xi = solve(df-ratio) % find x such that f'(x) = ratio             %
 xi=eval(xi(1))  % there is only one real solution                 %
 subs(df,'x',xi) % just check if equals ratio                      %
%%%%%%%%%%%%%%%%%%%%%%%%%%%%%%%%%%%%%%%%%%%%%%%%%%%%%%%%%%%%%%%%%%%%%%%%%

% Plot function and the slope line
xx=0.57:0.002:0.908; ff=1./(1-xx).^2; h=plot(xx,ff)
hold on
  h=plot([a ; b],[fa ; fb],'go'); set(h,'linewidth',28)
hold off
set(h,'LineWidth',1.5), bot=-6; axis([0.53 0.96 bot 125])
set(gca,'fontsize',21,'Box','off', ...
        'YTick',[fa fb], 'YtickLabel',{'f(a)' ; 'f(b)'}, ...
        'XTick',[a b],   'XTickLabel',{'a' ; 'b'})
h=line([a b],[fa fb]);
set(h,'linestyle','-- ','color',[1 0 0],'linewidth',0.8)

% plot line y-y0 = m(x-x0) where m is slope and goes through (x0,y0)
x0=xi; y0=subs(f,'x',x0);
xa=a+0.4*(b-a);        xb=b-0.0*(b-a);
ya = y0+ratio*(xa-x0); yb = y0+ratio*(xb-x0);
h=line([xa xb],[ya yb]);
set(h,'linestyle','-- ','color',[1 0 0],'linewidth',0.8)

% vertical line at xi with label at xi
h=line([xi xi],[bot y0]);
set(h,'linestyle','-- ','color',[1 0.4 0.6],'linewidth',1.2)
text(xi-0.005,-13,'\xi','fontsize',24)
% Text command (but not the XTickLabel) supports use of
%  LaTeX-like text strings
```

Program Listing A.1 Computes ξ in the mean value theorem, and creates Figure A.1

The *exponential function* arises ubiquitously in the study of probability (and elsewhere of course), and so it is worth spending some time understanding it. As in Lang (1997, Sections 4.1 and 4.2), consider a function f such that

$$\text{(i) } f'(x) = f(x) \quad \text{and} \quad \text{(ii) } f(0) = 1.$$

From the product and chain rules, and (i),

$$[f(x) f(-x)]' = -f(x) f'(-x) + f(-x) f'(x)$$
$$= -f(x) f(-x) + f(-x) f(x) = 0,$$

so that $f(x) f(-x)$ is constant, and from (ii), equals 1. Thus, $f(x) \neq 0$ and $f(-x) = 1/f(x)$. From (ii), $f'(0) > 0$ and, as $f(x) \neq 0$, it follows that $f(x) > 0$ for all x and is strictly increasing. It is also straightforward to see that

$$f(x + y) = f(x) f(y) \quad \text{and} \quad f(nx) = [f(x)]^n, \quad n \in \mathbb{N}.$$

The latter also holds for $n \in \mathbb{R}$. Function f is the exponential, written $\exp(\cdot)$. Also, defining $e = f(1)$, we can write $f(x) = \exp(x) = e^x$. As $f(x)$ is strictly increasing and $f(0) = 1$, $f(1) = e > 1$.

It follows from (1.19) and the fact that $e > 1$ that $\lim_{n \to \infty} e^n / n^k = \infty$ for $n \in \mathbb{N}$ and all $k \in \mathbb{N}$. Now replace n by x, for $x \in \mathbb{R}_{>0}$. Use of the quotient rule gives

$$\frac{d}{dx} \left(\frac{e^x}{x^k} \right) = \frac{x^k e^x - e^x k x^{k-1}}{x^{2k}} = \frac{e^x}{x^k} (1 - k/x),$$

which is positive for $k < x$, i.e. for x large enough, e^x / x^k is increasing. This and the limit result for $n \in \mathbb{N}$ implies that

$$\lim_{x \to \infty} \frac{e^x}{x^k} = \infty, \quad \text{for all } k \in \mathbb{N}. \tag{A.36}$$

As $f(x)$ is strictly increasing, the inverse function g exists; and as $f(x) > 0$, $g(y)$ is defined for $y > 0$. From (ii), $g(1) = 0$; from (A.32) and (i),

$$g'(y) = \frac{1}{f'[g(y)]} = \frac{1}{f[g(y)]} = \frac{1}{y}. \tag{A.37}$$

For $a > 0$, the sum and chain rules imply

$$[g(ax) - g(x)]' = ag'(ax) - g'(x) = \frac{a}{ax} - \frac{1}{x} = 0,$$

so $g(ax) - g(x) = c$ or $g(ax) = c + g(x)$. Letting $x = 1$ gives $g(a) = c + g(1) = c$, which then implies

$$g(ax) = g(a) + g(x). \tag{A.38}$$

By induction, $g(x^n) = ng(x)$. Function g is the *natural logarithm*, denoted $\log(y)$, $\log y$ or, from the French *logarithm natural*, $\ln y$. Thus, $\ln x^n = n \ln x$. As $\ln 1 = 0$, write $0 = \ln 1 = \ln(x/x) = \ln x + \ln(1/x)$ from (A.38), so that $\ln x^{-1} = -\ln x$. The last two results generalize to (see Stoll, 2001, p. 234)

$$\ln(x^p) = p \cdot \ln(x), \quad p \in \mathbb{R}, \ x \in \mathbb{R}_{>0}. \tag{A.39}$$

See also Example A.20 below regarding (A.39). Based on their properties, the exponential and logarithmic functions are also used in the following way:

$$\text{for } r \in \mathbb{R} \text{ and } x \in \mathbb{R}_{>0}, \ x^r \text{ is defined by } x^r := \exp(r \ln x). \tag{A.40}$$

⊙ **Example A.7**   To evaluate $\lim_{x \to 0^+} x^x$, use l'Hôpital's rule to see that

$$\lim_{x \to 0^+} x \ln x = \lim_{x \to 0^+} \frac{\ln x}{1/x} = \lim_{x \to 0^+} \frac{1/x}{-x^{-2}} = - \lim_{x \to 0^+} x = 0.$$

Then, by the continuity of the exponential function,

$$\lim_{x \to 0^+} x^x = \lim_{x \to 0^+} \exp\left(\ln x^x\right) = \lim_{x \to 0^+} \exp(x \ln x) = \exp\left(\lim_{x \to 0^+} x \ln x\right) = \exp 0 = 1. \ ■$$

⊙ **Example A.8**   From the continuity of the exponential function and use of l'Hôpital's rule,

$$\lim_{k \to \infty} k^{1/k} = \lim_{k \to \infty} \exp\left(\ln k^{1/k}\right) = \lim_{k \to \infty} \exp\left(\frac{\ln k}{k}\right)$$

$$= \exp \lim_{k \to \infty} \frac{\ln k}{k} = \exp \lim_{k \to \infty} (1/k) = 1.$$

Also, for any $x \in \mathbb{R}_{>0}$,

$$\lim_{k \to \infty} x^{1/k} = \lim_{k \to \infty} \exp\left(\frac{\ln x}{k}\right) = \exp \lim_{k \to \infty} \left(\frac{\ln x}{k}\right) = \exp(0) = 1.$$

(Enter a positive number in your calculator, repeatedly press the $\sqrt{\ }$ key, and see what happens – either the key will break, or a 1 will result). ■

⊛ **Example A.9**   Let $f(x) = x^r$, for $r \in \mathbb{R}$ and $x \in \mathbb{R}_{>0}$. From (A.40) and the chain rule,

$$f'(x) = \exp(r \ln x) \frac{r}{x} = x^r \frac{r}{x} = r x^{r-1},$$

which extends the results in Example A.3 in a natural way. ■

⊙ **Example A.10**   For $x > 0$ and $p \in \mathbb{R}$, the chain rule implies

$$\frac{d}{dx} (\ln x)^p = \frac{p (\ln x)^{p-1}}{x},$$

so that, integrating both sides (and using the fundamental theorem of calculus – see Section A.2.3 below),

$$\frac{(\ln x)^p}{p} = \int \frac{dx}{x (\ln x)^{1-p}}. \tag{A.41}$$

Also, from (A.37) and the chain rule,

$$\frac{d}{dx} \ln (\ln x) = \frac{1}{\ln x} \frac{d}{dx} \ln x = \frac{1}{x \ln x}, \tag{A.42}$$

which we will use below in Example A.36. ∎

⊙ **Example A.11** For $y > 0$ and $k, p \in \mathbb{R}$,

$$\frac{d}{dp} y^{kp} = \frac{d}{dp} \exp (kp \ln y) = \exp (kp \ln y) k \ln y = y^{kp} k \ln y. \tag{A.43}$$

With $x > 1$ and $y = \ln x$, this implies

$$\frac{d}{dp} \left[(\ln x)^{-p} \right] = - (\ln x)^{-p} \ln (\ln x) . \tag{∎}$$

⊖ **Example A.12** In microeconomics, a *utility function*, $U (\cdot)$, is a preference ordering for different goods of choice ('bundles' of goods and services, amount of money, etc.). For example, if bundle A is preferable to bundle B, then $U (A) > U (B)$. Let $U : A \rightarrow \mathbb{R}$, $A \subset \mathbb{R}_{>0}$, be a continuous and twice differentiable utility function giving a preference ordering for overall wealth, W. Not surprisingly, one assumes that $U' (W) > 0$, i.e. people prefer more wealth to less, but also that $U'' (W) < 0$, i.e. the more wealth you have, the less additional utility you reap upon obtaining a fixed increase in wealth. (In this case, U is a concave function and the person is said to be *risk-averse* see Example 4.30.) A popular choice of U is $U (W; \gamma) = W^{1-\gamma} / (1 - \gamma)$ for a fixed parameter $\gamma \in \mathbb{R}_{>0} \setminus 1$ and $W > 0$. (Indeed, an easy calculation verifies that $U' (W) > 0$ and $U'' (W) < 0$.) Interest centers on the limit of U as $\gamma \rightarrow 1$. In this case, $\lim_{y \rightarrow 1} W^{1-\gamma} = 1$ and $\lim_{y \rightarrow 1} (1 - \gamma) = 0$ so that l'Hôpital's rule is not applicable. However, as utility is a relative measure, we can let $U (W; \gamma) = (W^{1-\gamma} - 1) / (1 - \gamma)$ instead. Then, from (A.43), $(d/d\gamma) W^{1-\gamma} = -W^{1-\gamma} \ln W$, so that

$$\lim_{\gamma \rightarrow 1} U (W; \gamma) = \lim_{\gamma \rightarrow 1} \frac{W^{1-\gamma} - 1}{1 - \gamma} = \lim_{\gamma \rightarrow 1} \frac{(d/d\gamma) \left(W^{1-\gamma} - 1 \right)}{(d/d\gamma) (1 - \gamma)}$$

$$= \lim_{\gamma \rightarrow 1} W^{1-\gamma} \ln W = \ln W.$$

(see also Example 4.21). ∎

A useful fact is that $\ln (1 + x) < x$, for all $x \in \mathbb{R}_{>0}$, easily seen as follows. With $f (x) = \ln (1 + x)$ and $g (x) = x$, note that f and g are continuous and differentiable, with $f (0) = g (0) = 0$, but their slopes are such that $f' (x) = (1 + x)^{-1} < 1 = g' (x)$, so that $f (x) < g (x)$ for all $x \in \mathbb{R}_{>0}$.

For any $k \in \mathbb{N}$, letting $x = e^z$ shows that

$$\lim_{x \rightarrow \infty} \frac{(\ln x)^k}{x} = \lim_{z \rightarrow \infty} \frac{z^k}{e^z} = 0 \tag{A.44}$$

from (A.36). Also, for some $p > 0$, with $z = x^p$, (A.44) implies

$$\lim_{x \to \infty} \frac{\ln x}{x^p} = p^{-1} \lim_{z \to \infty} \frac{(\ln z)}{z} = 0. \tag{A.45}$$

As the derivative of $\ln x$ at $x = 1$ is $1/x = 1$, the Newton quotient and (A.39) imply

$$1 = \lim_{h \to 0} \frac{\ln(1+h) - \ln 1}{h} = \lim_{h \to 0} \frac{\ln(1+h)}{h}$$

$$= \lim_{h \to 0} \left[\ln(1+h)^{1/h}\right] = \ln\left[\lim_{h \to 0} (1+h)^{1/h}\right].$$

Taking the inverse function (exponential) gives

$$e = \lim_{h \to 0} (1+h)^{1/h} = \lim_{n \to \infty} \left(1 + \frac{1}{n}\right)^n.$$

Going the other way, to evaluate $\lim_{n \to \infty} (1 + \lambda/n)^n$, take logs and use l'Hôpital's rule to get

$$\ln \lim_{n \to \infty} (1 + \lambda/n)^n = \lim_{n \to \infty} \ln(1 + \lambda/n)^n = \lim_{n \to \infty} n \ln(1 + \lambda/n)$$

$$= \lim_{n \to \infty} \frac{\frac{d}{dn} \ln(1 + \lambda/n)}{\frac{d}{dn} n^{-1}} = \lim_{n \to \infty} \frac{-\frac{\lambda}{n\lambda + n^2}}{-\frac{1}{n^2}} = \lambda \lim_{n \to \infty} \left(\frac{n}{n + \lambda}\right) = \lambda,$$

i.e.

$$\lim_{n \to \infty} (1 + \lambda/n)^n = e^\lambda. \tag{A.46}$$

A.2.3   Integration

If we evolved a race of Isaac Newtons, that would not be progress. For the price Newton had to pay for being a supreme intellect was that he was incapable of friendship, love, fatherhood, and many other desirable things. As a man he was a failure; as a monster he was superb.   (Aldous Huxley)

Every schoolchild knows the formula for the area of a rectangle. Under certain conditions, the area under a curve can be approximated by summing the areas of adjacent rectangles with heights coinciding with the function under study and ever-decreasing widths. Related concepts go back at least to Archimedes. This idea was of course known to Gottfried Leibniz (1646–1716) and Isaac Newton (1642–1727), though they viewed the integral as an antiderivative (see below) and used it as such. Augustin Louis Cauchy (1789–1857) is credited with using limits of sums as in the modern approach of integration, which led him to prove the fundamental theorem of calculus. Building on the work of Cauchy, Georg Bernhard Riemann (1826–1866) entertained working with discontinuous functions, and ultimately developed the modern definition of what is now called the Riemann integral in 1853, along with necessary and sufficient conditions for its existence. Contributions to its development were also made by Jean Gaston Darboux (1842–1917), while Thomas-Jean Stieltjes (1856–1894) pursued what is now referred to as the Riemann–Stieltjes integral.[8]

Definitions, existence and properties

The simplest schoolboy is now familiar with facts for which Archimedes would have sacrificed his life. (Ernest Renan)

To make precise the aforementioned notion of summing the area of rectangles, some notation is required. Let $A = [a, b]$ be a bounded interval in $\mathbb{R}$. A *partition* of A is a finite set $\pi = \{x_k\}_{k=0}^n$ such that $a = x_0 < x_1 < \cdots < x_n = b$, and its *mesh* (sometimes called the norm, or size), is given by $\mu(\pi) = \max\{x_1 - x_0, x_2 - x_1, \ldots, x_n - x_{n-1}\}$.

If π_1 and π_2 are partitions of I such that $\pi_1 \subset \pi_2$, then π_2 is a *refinement* of π_1. A *selection* associated to a partition $\pi = \{x_k\}_{k=0}^n$ is any set $\{\xi_k\}_{k=1}^n$ such that $x_{k-1} \leq \xi_k \leq x_k$ for $k = 1, \ldots, n$.

Now let $f : D \to \mathbb{R}$ with $A \subset D \subset \mathbb{R}$, $\pi = \{x_k\}_{k=0}^n$ be a partition of A, and $\sigma = \{\xi_k\}_{k=1}^n$ a selection associated to π. The *Riemann sum* for function f, with partition π and selection σ, is given by

$$S(f, \pi, \sigma) = \sum_{k=1}^n f(\xi_k)(x_k - x_{k-1}). \tag{A.47}$$

Observe how S is just a sum of areas of rectangles with heights dictated by f, π and σ. If the Riemann sum converges to a real number as the level of refinement increases, then f is *integrable*. Formally, function f is said to be *(Riemann) integrable* over $A = [a, b]$ if there is a number $I \in \mathbb{R}$ such that: $\forall \epsilon > 0$, there exists a partition π_0 of A such that, for every refinement π of π_0, and every selection σ associated to π, we have $|S(f, \pi, \sigma) - I| < \epsilon$. If f is Riemann integrable over $[a, b]$, then we write $f \in \mathcal{R}[a, b]$.

The number I is called the *integral* of f over $[a, b]$, and denoted by $\int_a^b f$ or $\int_a^b f(x) \, dx$. Observe how, in the latter notation, x is a 'dummy variable', in that it

---

[8] See Stoll (2001, Chapter 6) and Browder (1996, p. 121) for some historical commentary, and Hawkins (1970) for a detailed account of the development of the Riemann and Lebesgue integrals. The Riemann integral was fundamentally superseded and generalized by the work of Henri Léon Lebesgue (1875–1941) in 1902, as well as Émile Borel (1871–1956) and Constantin Carathéodory (1873–1950), giving rise to the Lebesgue integral. While it is considerably more complicated than the Riemann integral, it has important properties not shared by the latter, to the extent that the Riemann integral is considered by some to be just a historical relic. Somewhat unexpectedly, in the 1950s, Ralph Henstock and Jaroslav Kurzweil independently proposed an integral formulation which generalizes the Riemann integral, but in a more direct and much simpler fashion, without the need for notions of measurable sets and functions, or σ-algebras. It is usually referred to as the gauge integral, or some combination of the pioneers names. It not only nests the Lebesgue integral, but also the improper Riemann and Riemann–Stieltjes integrals. There are several textbooks which discuss it, such as those by or with Charles W. Swartz, and by or with Robert G. Bartle. See the web page by Eric Schechter, http://www.math.vanderbilt.edu/~schectex/ccc/gauge/ and the references therein for more information.

could be replaced by any other letter (besides, of course, f, a or b), and also how it mirrors the notation in (A.47), i.e. the term $\sum_{k=1}^{n}$ is replaced by $\int_{a}^{b}$, the term $f(\xi_k)$ is replaced by $f(x)$ and the difference $(x_k - x_{k-1})$ by dx. Indeed, the integral symbol $\int$ is an elongated letter S, for summation, introduced by Leibniz, and the word integral in this context was first used by Jakob Bernoulli.

For f to be integrable, it is necessary (but not sufficient) that f be bounded on $A = [a, b]$. To see this, observe that, if f were *not* bounded on A, then, for every given partition $\pi = \{x_k\}_{k=0}^{n}$, $\exists\, k \in \{1, \ldots, n\}$ and an $x \in [x_{k-1}, x_k]$ such that $|f(x)|$ is arbitrarily large. Thus, by varying the element ξ_k of the selection σ associated to π, the Riemann sum $S(f, \pi, \sigma)$ can be made arbitrarily large, and there can be no value I such that $|S(f, \pi, \sigma) - I| < \epsilon$.

⊖ **Example A.13**   Let $f(x) = x$. Then the graph of f from 0 to $b > 0$ forms a triangle with area $b^2/2$. For the equally-spaced partition $\pi_n = \{x_k\}_{k=0}^{n}$, $n \in \mathbb{N}$, with $x_k = kb/n$ and selection $\sigma = \{\xi_k\}_{k=1}^{n}$ with $\xi_k = x_k = kb/n$, the Riemann sum is

$$S(f, \pi, \sigma) = \sum_{k=1}^{n} f(\xi_k)(x_k - x_{k-1}) = \sum_{k=1}^{n} \frac{kb}{n}\left[\frac{kb}{n} - \frac{(k-1)b}{n}\right],$$

which simplifies to

$$S(f, \pi, \sigma) = \left(\frac{b}{n}\right)^2 \sum_{k=1}^{n} k = \left(\frac{b}{n}\right)^2 \left[\frac{n(n+1)}{2}\right] = \frac{b^2}{2}\frac{(n+1)}{n}. \tag{A.48}$$

This overestimates the area of the triangle because f is increasing and we took the selection $\xi_k = x_k$; likewise, choosing $\xi_k = x_{k-1}$ would underestimate it with $S = \frac{b^2}{2}\frac{(n-1)}{n}$; and, because of the linearity of f, choosing the midpoint $\xi_k = (x_{k-1} + x_k)/2$ gives exactly $b^2/2$. From the boundedness of f on $[a, b]$, the choice of selection will have vanishing significance as n grows so that, from (A.48), as $n \to \infty$, $S(f, \pi, \sigma) \to b^2/2 = I$. (Of course, to strictly abide by the definition, the partitions would have to be chosen as successive refinements, which is clearly possible.)[9]   ∎

Let $\pi = \{x_k\}_{k=0}^{n}$ be a partition of f. The *upper (Darboux) sum* of f for π is defined as $\overline{S}(f, \pi) = \sup_{\sigma}\{S(f, \pi, \sigma)\}$, i.e. the supremum of $S(f, \pi, \sigma)$ over all possible σ associated to π. Likewise, the *lower (Darboux) sum* is $\underline{S}(f, \pi) = \inf_{\sigma}\{S(f, \pi, \sigma)\}$, and $\underline{S}(f, \pi) \le \overline{S}(f, \pi)$. By defining

$$m_k = \inf\{f(t) : t \in [x_{k-1}, x_k]\} \quad \text{and} \quad M_k = \sup\{f(t) : t \in [x_{k-1}, x_k]\}, \tag{A.49}$$

we can write $\underline{S}(f, \pi) = \sum_{k=1}^{n} m_k(x_k - x_{k-1})$ and $\overline{S}(f, \pi) = \sum_{k=1}^{n} M_k(x_k - x_{k-1})$. Also, if $m \le f(x) \le M$ for $x \in [a, b]$, then $m(b - a) \le \underline{S}(f, \pi) \le \overline{S}(f, \pi) \le$

---

[9] The more general case with $f(x) = x^p$ for $x \ge 0$ and $p \ne -1$ is particularly straightforward when using a wise choice of nonequally-spaced partition, as was first shown by Pierre De Fermat before the fundamental theorem of calculus was known to him; see Browder (1996, pp. 102, 121) and Stahl (1999, p. 16) for details.

$M (b - a)$. It should be intuitively clear that, if π and π' are partitions of $[a, b]$ such that $\pi \subset \pi'$, then

$$\underline{S}(f, \pi) \leq \underline{S}(f, \pi') \leq \overline{S}(f, \pi') \leq \overline{S}(f, \pi). \tag{A.50}$$

Also, for *any* two partitions π_1 and π_2 of $[a, b]$, let $\pi_3 = \pi_1 \cup \pi_2$ be their *common refinement* (Pugh, 2002, p. 158), so that, from (A.50), $\underline{S}(f, \pi_1) \leq \underline{S}(f, \pi_3) \leq \overline{S}(f, \pi_3) \leq \overline{S}(f, \pi_2)$, i.e. the lower sum of any partition is less than or equal to the upper sum of any (other) partition. This fact is useful for proving the intuitively plausible result, due to Riemann, that, if f is a bounded function on $[a, b]$, then $\int_a^b f$ exists iff $\forall \epsilon > 0$, $\exists \pi$ of $[a, b]$ such that $\overline{S}(f, \pi) - \underline{S}(f, \pi) < \epsilon$. This, in turn, is used for proving the following important results.

- If f is a monotone, bounded function on $[a, b]$, then $\int_a^b f$ exists. (A.51)
  Let f be monotone increasing. Let $\pi = \{x_k\}_{k=0}^n$ be a partition of $[a, b]$ with $x_k = a + (k/n)(b - a)$. Then (A.49) implies that $m_k = f(x_{k-1})$ and $M_k = f(x_k)$, and

$$\overline{S}(f, \pi) - \underline{S}(f, \pi) = \sum_{k=1}^n \left[f(x_k) - f(x_{k-1}) \right] (x_k - x_{k-1})$$

$$= \frac{b-a}{n} \sum_{k=1}^n \left[f(x_k) - f(x_{k-1}) \right] = \frac{b-a}{n} \left[f(b) - f(a) \right].$$

  As f is bounded and increasing, $0 \leq f(b) - f(a) < \infty$, and n can be chosen such that the r.h.s. is less than any $\epsilon > 0$.

- If $f \in C^0[a, b]$, then $\int_a^b f$ exists. (A.52)
  Recalling (A.14), continuity of f on a closed interval I implies that f is uniformly continuous on I, so that, $\forall \epsilon > 0$, $\exists \delta > 0$ such that $|f(x) - f(y)| < \epsilon / (b - a)$ when $|x - y| < \delta$. Let $\pi = \{x_k\}_{k=0}^n$ be a partition of $[a, b]$ with mesh $\mu(\pi) < \delta$. Then, for any values $s, t \in [x_{k-1}, x_k]$, $|s - t| < \delta$ and $|f(s) - f(t)| < \epsilon / (b - a)$; in particular, from (A.49), $M_k - m_k \leq \epsilon / (b - a)$ (the strict equality is replaced with $\leq$ because of the nature of inf and sup) and

$$\overline{S}(f, \pi) - \underline{S}(f, \pi) = \sum_{k=1}^n (M_k - m_k)(x_k - x_{k-1}) \leq \frac{\epsilon}{b - a} \sum_{k=1}^n (x_k - x_{k-1}) = \epsilon.$$

- If f is a bounded function on $[a, b]$ whose set of discontinuities has measure zero, then $\int_a^b f$ exists. See Browder (1996, p. 104) for a short, easy proof when there exists a measure zero cover C for the set of discontinuity points, and C consists of a *finite* set of disjoint open intervals. This restriction to a finite set of intervals can be lifted, and is referred to as *Lebesgue's theorem*, given by him in 1902 (see, for example, Stoll (2001, Section 6.7) or Pugh (2002, pp. 165–7) for detailed proofs).

- Let f and ϕ be functions such that $f \in \mathcal{R}[a, b]$, with $m \leq f(x) \leq M$ for all $x \in [a, b]$, and $\phi \in C^0[m, M]$. Then $\phi \circ f \in \mathcal{R}[a, b]$. (A.53)
See, for example, Browder (1996, pp. 106–7) or Stoll (2001, pp. 217–8) for elementary proofs, and Stoll (2001, p. 220) or Pugh (2002, p. 168) for the extremely short proof using Lebesgue's theorem. Valuable special cases include $\phi(y) = |y|$ and $\phi(y) = y^2$, i.e. if f is integrable on $I = [a, b]$, then so are $|f|$ and f^2 on I. It is *not* necessarily true that the composition of two integrable functions is integrable.

With the existence of the integral established, we now state its major properties. With $f, g \in \mathcal{R}[a, b]$ and $I = [a, b]$,

$$\text{if } f(x) \leq g(x) \text{ for all } x \in I, \text{ then } \int_a^b f(x)\,dx \leq \int_a^b g(x)\,dx. \tag{A.54}$$

For any constants $k_1, k_2 \in \mathbb{R}$,

$$\int_a^b (k_1 f + k_2 g) = k_1 \int_a^b f + k_2 \int_a^b g. \tag{A.55}$$

These can be used to prove that $\left| \int_a^b f \right| \leq \int_a^b |f|$. If $f, g \in \mathcal{R}[a, b]$, then $fg \in \mathcal{R}[a, b]$ and

$$\left(\int_a^b fg \right)^2 \leq \left(\int_a^b f^2 \right) \left(\int_a^b g^2 \right), \tag{A.56}$$

known as the *Schwarz, Cauchy–Schwarz* or *Bunyakovsky–Schwarz* inequality.[10]
Let $a < c < b$ and let f be a function on $[a, b]$. Then, $\int_a^b f$ exists iff $\int_a^c f$ and $\int_c^b f$ exist, in which case

$$\int_a^b f = \int_a^c f + \int_c^b f. \tag{A.57}$$

Also, as definitions, $\int_a^a f := 0$ and $\int_b^a := -\int_a^b f$. The motivation for the former is obvious; for the latter definition, one reason is so that (A.57) holds even for $c < a < b$. That is,

$$\int_a^b f = \int_a^c f + \int_c^b f = -\int_c^a f + \int_c^b f = \int_c^b f - \int_c^a f,$$

which corresponds to our intuitive notion of working with pieces of areas.
The *mean value theorem for integrals* states the following. Let $f, g \in C^0(I)$ for $I = [a, b]$, with g nonnegative. Then $\exists\, c \in I$ such that

$$\int_a^b f(x)g(x)\,dx = f(c) \int_a^b g(x)\,dx. \tag{A.58}$$

---

[10] Bunyakovsky was first, having published the result in 1859, while Schwarz found it in 1885 (see Browder, 1996, p. 121). The finite-sum analog of (A.56) is (A.7).

A popular and useful form of the theorem just takes $g(x) \equiv 1$, so that $\int_a^b f = f(c)$ $(b - a)$. To prove this, use (A.13) to let $m = \min\{f(t) : t \in I\}$ and $M = \max\{f(t) : t \in I\}$. As $p(t) \geq 0$, $mp(t) \leq f(t) p(t) \leq Mp(t)$ for $t \in I$, so that

$$m \int_a^b p(t)\, \mathrm{d}t \leq \int_a^b f(t)\, p(t)\, \mathrm{d}t \leq M \int_a^b p(t)\, \mathrm{d}t$$

or, assuming $\int_a^b p(t)\, \mathrm{d}t > 0$, $m \leq \gamma \leq M$ where $\gamma = \int_a^b f(t)\, p(t)\, \mathrm{d}t / \int_a^b p(t)\, \mathrm{d}t$. From the intermediate value theorem, $\exists\, c \in I$ such that $f(c) = \gamma$.

The *Bonnet mean value theorem*, discovered in 1849 by Pierre Ossian Bonnet (1819–1892), states that, if $f, g \in C^0(I)$ and f is positive and decreasing, then $\exists\, c \in I$ such that

$$\int_a^b f(x) g(x)\, \mathrm{d}x = f(a) \int_a^c g(x)\, \mathrm{d}x. \tag{A.59}$$

See Lang (1997, p. 107) for proof, and Example 7.1 for an illustration of its use.

Fundamental theorem of calculus

Let $f : I \to \mathbb{R}$. The function $F : I \to \mathbb{R}$ is called an *indefinite integral*, or an *antiderivative*, or a *primitive* of f if, $\forall\, x \in I$, $F'(x) = f(x)$. The fundamental theorem of calculus (FTC) is the link between the differential and integral calculus, of which there are two forms.

- *For $f \in \mathcal{R}[a, b]$ with F a primitive of f, $\int_a^b f = F(b) - F(a)$.* $\qquad$ (A. 60)
  Let $\pi = \{x_k\}_{k=0}^n$ be a partition of $I = [a, b]$. As $F'(t) = f(t)$ for all $t \in I$, applying the mean value theorem to F implies that

$$F(x_k) - F(x_{k-1}) = F'(\xi_k)(x_k - x_{k-1}) = f(\xi_k)(x_k - x_{k-1})$$

for some $\xi_k \in (x_{k-1}, x_k)$. The set of ξ_k, $k = 1, \ldots n$ forms a selection, $\sigma = \{\xi_k\}_{k=1}^n$, associated to π, so that

$$S(f, \pi, \sigma) = \sum_{k=1}^n f(\xi_k)(x_k - x_{k-1}) = \sum_{k=1}^n \left[F(x_k) - F(x_{k-1}) \right] = F(b) - F(a).$$

This holds for any partition π, so that $\int_a^b f(t)\, \mathrm{d}t = F(b) - F(a)$.

- *For $f \in \mathcal{R}[a, b]$, define $F(x) = \int_a^x f$, $x \in I = [a, b]$. Then*

(1) $F \in C^0[a, b]$. $\qquad\qquad\qquad\qquad\qquad\qquad\qquad\qquad\qquad\qquad$ (A.61)

(2) *If f is continuous at $x \in I$, then $F'(x) = f(x)$.* $\qquad\qquad\qquad$ (A.62)

Let $M = \sup\{|f(t)| : t \in I\}$. Then, for any $x \in I$ and $h \in \mathbb{R}$ with $x + h \in I$,

$$|F(x+h) - F(x)| = \left| \int_a^{x+h} f(t)\,dt - \int_a^x f(t)\,dt \right|$$

$$= \left| \int_x^{x+h} f(t)\,dt \right| \leq \int_x^{x+h} |f(t)|\,dt \leq M|h|,$$

so that F is continuous at x, and thus on I.

If f is continuous at $x \in I$, then, $\forall \epsilon > 0$, $\exists \delta > 0$ such that, for all $t \in I$, $|f(t) - f(x)| < \epsilon$ when $|t - x| < \delta$. Then, for $h \in \mathbb{R}$ with $x + h \in I$ and $|h| < \delta$, use the fact that $\int_x^{x+h} dt = h$, or $h^{-1} \int_x^{x+h} f(y)\,dt = f(y)$, to get

$$\left| \frac{F(x+h) - F(x)}{h} - f(x) \right| = \left| \frac{1}{h} \int_x^{x+h} f(t)\,dt - f(x) \right|$$

$$= \left| \frac{1}{h} \int_x^{x+h} \left[f(t) - f(x) \right] dt \right|$$

$$\leq \frac{1}{|h|} \int_x^{x+h} \epsilon\,dt = \frac{|h|\,\epsilon}{|h|},$$

which shows that $F'(x) = f(x)$. It is also easy to show that (A. 62) implies (A.60) when f is continuous (see Browder (1996, p. 112) or Priestley (1997, Theorem 5.5)).

An informal, graphical proof of (A. 62) is of great value for remembering and understanding the result, as well as for convincing others. Figure A.2(a) shows a plot of part of a continuous function, $f(x) = x^3$, with vertical lines indicating $x = 3$ and $x = 3.1$. This is 'magnified' in Figure A.2(b), with a vertical line at $x = 3.01$, but keeping the same scaling on the y-axis to emphasize the approximate linearity of the function over a relatively small range of x-values. The rate of change, via the Newton quotient, of the area $A(t)$ under the curve from $x = 1$ to $x = t$ is

$$\frac{A(t+h) - A(t)}{h} = \frac{\int_1^{3+h} f - \int_1^3 f}{h} = \frac{1}{h} \int_3^{3+h} f \approx \frac{h\,f(t)}{h},$$

because, as $h \to 0$ the region under study approaches a rectangle with base h and height $f(t)$ (see Figure A.2(c)).

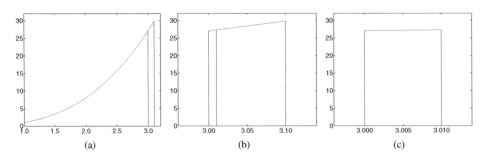

Figure A.2　Graphical illustration of the FTC (A. 62)

◎ **Example A.14** As in Hijab (1997, p. 103), let F and G be primitives of f on (a, b), so that $H = F - G$ is a primitive of zero, i.e. $\forall\, x \in (a, b)$, $H'(x) = [F(x) - G(x)]' = 0$. The mean value theorem implies that $\exists\, c \in (a, b)$ such that, for $a < x < y < b$,

$$H(x) - H(y) = H'(c)(x - y) = 0,$$

i.e. H is a constant. Thus, any two primitives of f differ by a constant. ∎

◎ **Example A.15** Observe that $F(x) = e^{kx}/k + C$ is a primitive of $f(x) = e^{kx}$ for $k \in \mathbb{R} \setminus 0$ and any constant $C \in \mathbb{R}$, because, via the chain rule, $dF(x)/dx = f(x)$. Thus, from (A.60),

$$\int_a^b f = F(b) - F(a) = k^{-1}(e^{kb} - e^{ka}), \tag{A.63}$$

a result we will make great use of. ∎

◎ **Example A.16** Let $I(x) = \int_0^{x^2} e^{-t} dt = 1 - e^{-x^2}$, so that $I'(x) = \frac{d}{dx}\left(1 - e^{-x^2}\right) = 2xe^{-x^2}$. Alternatively, let $G(y) = \int_0^y e^{-t} dt$, so that $I(x) = G(x^2) = G[f(x)]$, where $f(x) = x^2$. Then, from the chain rule and the FTC,

$$I'(x) = G'[f(x)]\, f'(x) = e^{-x^2} \cdot 2x,$$

as before, but without having to actually evaluate $I(x)$. ∎

◎ **Example A.17** Let $I = [a, b]$, $f : I \to \mathbb{R}$ be a continuous and integrable function on I, and $x \in I$. Differentiating the relation $\int_a^b f = \int_a^x f + \int_x^b f$ w.r.t. x and using (A.62) gives

$$0 = \frac{d}{dx}\left(\int_a^x f\right) + \frac{d}{dx}\left(\int_x^b f\right) = f(x) + \frac{d}{dx}\left(\int_x^b f\right),$$

which implies that

$$\frac{d}{dx}\left(\int_x^b f(t)\, dt\right) = -f(x), \quad x \in I. \quad\blacksquare$$

◎ **Example A.18** The FTC allows an easy proof of the mean value theorem for integrals. As f is integrable from (A.52), let $F(x) = \int_a^x f$. From (A.62), $F'(x) = f(x)$ for all $x \in I$. Then, from the mean value theorem (A.35), $\exists\, c \in I$ such that

$$\frac{F(b) - F(a)}{b - a} = F'(c),$$

i.e. $\int_a^b f(x) dx = F(b) - F(a) = F'(c)(b - a) = f(c)(b - a)$. ∎

The simple technique of *integration by parts* can be invaluable in many situations. Let $f, g \in \mathcal{R}[a, b]$ with primitives F and G, respectively. Then

$$\int_a^b F(t) g(t) \, dt = F(b) G(b) - F(a) G(a) - \int_a^b f(t) G(t) \, dt. \tag{A.64}$$

For proof, use the product rule to get $(FG)' = F'G + FG' = fG + Fg$. Integrating both sides of this and using (A.60) for the l.h.s. gives

$$F(b) G(b) - F(a) G(a) = \int \left[f(t) G(t) + F(t) g(t) \right] dt,$$

which, from (A.55), is equivalent to (A.64). Throughout, we will use the popular notation

$$\int_a^b u \, dv = uv \Big|_a^b - \int_a^b v \, du \qquad \text{or} \qquad \int u \, dv = uv - \int v \, du,$$

where $uv|_a^b := u(b) v(b) - u(a) v(a)$.

⊖ **Example A.19** Applying integration by parts to $\int_0^1 x^r (\ln x)^r \, dx$ for $r \in \mathbb{N}$ with $u = (\ln x)^r$ and $dv = x^r dx$ (so that $v = x^{r+1} / (r + 1)$ and $du = r (\ln x)^{r-1} x^{-1} dx$) gives

$$\int_0^1 x^r (\ln x)^r \, dx = (\ln x)^r \frac{x^{r+1}}{r+1} \Big|_0^1 - \int_0^1 \frac{x^{r+1}}{r+1} \frac{r}{x} (\ln x)^{r-1} \, dx$$

$$= -\frac{r}{r+1} \int_0^1 x^r (\ln x)^{r-1} \, dx.$$

Repeating this 'in a feast of integration by parts' (Havil, 2003, p. 44) leads to

$$\int_0^1 x^r (\ln x)^r \, dx = (-1)^r \frac{r!}{(r+1)^{r+1}}, \qquad r \in \mathbb{N}, \tag{A.65}$$

which is used in Example A.52 below. ∎

Another vastly useful technique is the *change of variables*: Let ϕ be differentiable on $I = [\alpha, \beta]$ and set $a = \phi(\alpha)$, $b = \phi(\beta)$. If f is continuous on $\phi(I)$ and $g = f \circ \phi$, then

$$\int_a^b f(t) \, dt = \int_\alpha^\beta g(u) \phi'(u) \, du. \tag{A.66}$$

To see this, note that (A.22) and (A.12) imply that g is continuous on I. From (A.52), f is integrable, so we can set $F(x) = \int_a^x f(t) \, dt$ for all $x \in I$, and from (A.62), $F'(x) = f(x)$ for all $x \in I$. Thus, from the chain rule, $G = F \circ \phi$ is differentiable on I, with

$$G'(u) = F'(\phi(u)) \phi'(u) = f(\phi(u)) \phi'(u) = g(u) \phi'(u)$$

for all $u \in I$. Thus,

$$\int_\alpha^\beta g(u)\,\phi'(u)\,du = \int_\alpha^\beta G'(u)\,du = G(\beta) - G(\alpha) = F(b) - F(a) = \int_a^b f(t)\,dt,$$

which is (A.66). The change of variables formula is often expressed as

$$\int f[g(x)]g'(x)\,dx = \int f(u)\,du, \quad \text{where} \quad u = g(x).$$

◎ **Example A.20** The natural logarithm is often defined as $\ln x = \int_1^x t^{-1}dt$, from which its most important properties follow, such as $\ln 1 = 0$ and $\ln(xy) = \ln x + \ln y$. For the latter, let $u = t/x$ so that $t = xu$, $dt = xdu$, so that

$$\int_x^{xy} t^{-1}dt = \int_1^y \frac{1}{u}\,du.$$

Thus,

$$\ln(xy) = \int_1^{xy} t^{-1}\,dt = \int_1^x t^{-1}\,dt + \int_x^{xy} t^{-1}\,dt$$

$$= \int_1^x t^{-1}\,dt + \int_1^y t^{-1}\,dt = \ln x + \ln y.$$

Similarly, $\ln(x/y) = \ln x - \ln y$.

Now let $y \in \mathbb{Z}$. From the integral representation, $\ln x^y = \int_1^{x^y} t^{-1}dt$, so that the FTC and the chain rule imply that

$$\frac{d}{dx}\ln x^y = \frac{1}{x^y}\frac{dx^y}{dx} = \frac{1}{x^y}yx^{y-1} = \frac{y}{x}.$$

Likewise,

$$\frac{d}{dx}y\ln x = y\frac{d}{dx}\int_1^x t^{-1}dt = \frac{y}{x},$$

so that $\ln x^y$ and $y\ln x$ have the same first derivative, and thus differ by a constant, $C = \ln x^y - y\ln x$. With $y = 0$, $C = 0 - \ln 1 = 0$, so that $\ln x^y = y\ln x$. This is easily extended to $y \in \mathbb{Q}$ (see, for example, Protter and Morrey, 1991, p. 118). The extension to $y \in \mathbb{R}$ follows by *defining* $x^r = \exp(r\ln x)$, as was given in Section A.2.2. ∎

◎ **Example A.21** Let $b, c \in \mathbb{R}$ and $T \in \mathbb{R}_{>0}$. From (A.29) and $u = -t$,

$$A = \int_{-T}^T t^{-1}\sin(bt)\sin(ct)dt$$

$$= \int_T^{-T} -u^{-1}\sin(-bu)\sin(-cu)(-1)du = -\int_{-T}^T u^{-1}\sin(bu)\sin(cu)du = -A,$$

so that $A = 0$. Similarly,

$$B = \int_0^T t^{-1} \sin(bt) \cos(ct) dt$$

$$= \int_0^{-T} t^{-1} \sin(-bu) \cos(-cu)(-1) du = \int_{-T}^0 u^{-1} \sin(bu) \cos(cu) du,$$

so that

$$\int_{-T}^T t^{-1} \sin(bt) \cos(ct) dt = 2 \int_0^T t^{-1} \sin(bt) \cos(ct) dt.$$

These results will be useful when working with characteristic functions of random variables.   ∎

Improper integrals

Recall that the Riemann integral is designed for bounded functions on a closed, bounded interval domain. An extension, credited to Cauchy, is to let $f : (a, b] \to \mathbb{R}$ such that, $\forall c \in (a, b)$, f is integrable on $[c, b]$, and define

$$\int_a^b f = \lim_{c \to a^+} \int_c^b f. \tag{A.67}$$

A similar definition holds when the limit is taken at the upper boundary. This is termed an *improper integral (of the second kind)*; if the limit in (A.67) exists, then $\int_a^b f$ is *convergent*, otherwise *divergent*.

⊚ **Example A.22**   (Stoll, 2001, p. 241) Let $f(x) = x^{-1/2}$, for $x \in (0, 1]$. Then

$$\int_0^1 f = \lim_{c \to 0^+} \int_c^1 x^{-1/2} dx = \lim_{c \to 0^+} \left(2 - 2\sqrt{c}\right) = 2,$$

and $\int_0^1 f$ is convergent. As $f \in \mathcal{R}[c, 1]$ for $c \in (0, 1)$, (A.53) implies that $f^2 \in \mathcal{R}[c, 1]$; but that does *not* imply that the improper integral $\int_0^1 f^2$ is convergent. Indeed,

$$\lim_{c \to 0^+} \int_c^1 x^{-1} dx = - \lim_{c \to 0^+} \ln c = \infty,$$

i.e. $\int_0^1 f^2$ is divergent.   ∎

Integrals can also be taken over infinite intervals, i.e. $\int_a^\infty f(x)\,dx$, $\int_{-\infty}^b f(x)\,dx$ and $\int_{-\infty}^\infty f(x)\,dx$; these are (also) referred to as *improper integrals (of the first kind)*. If function f is defined on $(-\infty, \infty)$, then $\int_{-\infty}^\infty f(x)\,dx$ is defined by

$$\int_{-\infty}^\infty f(x)\,dx = \lim_{a \to -\infty} \int_a^t f(x)\,dx + \lim_{b \to \infty} \int_t^b f(x)\,dx, \tag{A.68}$$

for any point $t \in \mathbb{R}$, when both limits on the r.h.s. exist. An example which we will use often is

$$\int_0^\infty e^{-u} du = \lim_{b \to \infty} \int_0^b e^{-u} du = \lim_{b \to \infty} \left(1 - e^{-b}\right) = 1, \qquad (A.69)$$

having used (A.63) with $k = -1$.

⊖ **Example A.23** The unpleasant looking integral

$$\int_0^1 \frac{e^{-((1/v)-1)}}{v^2} dv$$

is easily handled by using the substitution $u = (1/v) - 1$, so that $v = 1/(1+u)$ and $dv = -(1+u)^{-2} du$. Thus,

$$\int_0^1 \frac{e^{-((1/v)-1)}}{v^2} dv = -\int_\infty^0 \frac{e^{-u}}{(1+u)^{-2}} (1+u)^{-2} du = \int_0^\infty e^{-u} du = 1,$$

from (A.69). ∎

⊖ **Example A.24** Applying integration by parts to $\int e^{at} \cos(bt) dt$ with $u = e^{at}$ and $dv = \cos(bt) dt$ (so that $du = ae^{at} dt$ and $v = (\sin bt)/b$) gives

$$\int e^{at} \cos(bt) dt = e^{at} \frac{\sin bt}{b} - \frac{a}{b} \int e^{at} \sin(bt) dt.$$

Similarly,

$$\int e^{at} \sin(bt) dt = -e^{at} \frac{\cos bt}{b} + \frac{a}{b} \int e^{at} (\cos bt) dt,$$

so that

$$\int e^{at} \cos(bt) dt = e^{at} \frac{\sin bt}{b} + e^{at} \frac{a \cos bt}{b^2} - \frac{a^2}{b^2} \int e^{at} (\cos bt) dt$$

or

$$\int e^{at} \cos(bt) dt = \frac{e^{at}}{a^2 + b^2} (a \cos bt + b \sin bt), \qquad (A.70)$$

which is given, for example, in Abramowitz and Stegun (1972, Equation 4.3.137). For the special case with $a = -1$ and $b = s$, (A.70) reduces to

$$\int_0^\infty e^{-t} \cos(st) dt = \frac{1}{1+s^2}. \qquad (A.71)$$

A similar derivation confirms that

$$\int e^{at} \sin(bt)\, dt = \frac{e^{at}}{a^2 + b^2} (a \sin bt - b \cos bt),$$

with special case

$$\int_0^\infty e^{-t} \sin(st)\, dt = \frac{s}{1+s^2}, \tag{A.72}$$

for $a = -1$ and $b = s$. Results (A.71) and (A.72) can also be derived via the Laplace transform (see Volume II). ■

⊙ **Example A.25**   The integral $\int_{-\infty}^\infty \exp(-x^2)\, dx$ is of utmost importance, as it arises in conjunction with the normal distribution (they are related by substituting $y = x/\sqrt{2}$). We wish to verify that it is convergent. Via symmetry, it suffices to study $\int_0^\infty \exp(-x^2)\, dx$. Let $f(x) = e^{-x^2}$, $x \in \mathbb{R}_{\geq 0}$. As f is bounded and monotone on $[0, k]$ for all $k \in \mathbb{R}_{>0}$, it follows from (A.51) that $\int_0^k f$ exists. Alternatively, (A.12) and (A.52) also imply its existence. Thus, for examining the limit as $k \to \infty$, it suffices to consider $\int_1^k f$.

As $e^{x^2} = 1 + x^2 + \frac{1}{2}(x^2)^2 + \cdots > x^2$ (for $x \in \mathbb{R}$), it follows that, for $x > 0$,

$$e^{x^2} > x^2 \Rightarrow x^2 > 2\ln x \Rightarrow -x^2 < -2\ln x \Rightarrow e^{-x^2} < x^{-2}.$$

Thus, from (A.54),

$$\int_1^k e^{-x^2}\, dx < \int_1^k x^{-2}\, dx = \frac{k-1}{k} < 1, \quad \forall\, k > 1,$$

and $\int_1^\infty e^{-x^2}\, dx$ is convergent. Alternatively, for $x > 1$,

$$x^2 > x \Rightarrow -x^2 < -x \Rightarrow e^{-x^2} < e^{-x},$$

so that, from (A.54) and (A.69),

$$\int_1^k e^{-x^2}\, dx < \int_1^k e^{-x}\, dx = e^{-1} - e^{-k} < e^{-1} \approx 0.367879, \quad \forall\, k > 1. \tag{A.73}$$

Thus, $\int_1^\infty e^{-x^2} dx$ and, hence, $\int_0^\infty e^{-x^2} dx$ are convergent. Its value will be computed in Examples A.31 and A.79 below.[11] ■

---

[11] To see how close (A.73) is to the true value, use of (A.57), the result in Example A.31, and numeric integration gives

$$\int_1^\infty e^{-x^2} dx = \int_0^\infty e^{-x^2} dx - \int_0^1 e^{-x^2} dx \approx \frac{\sqrt{\pi}}{2} - 0.746824 \approx 0.1394.$$

⊖ **Example A.26** To show that

$$I := \int_0^\infty (1+t)\, e^{-t} \cos{(st)}\, dt = \frac{2}{\left(1+s^2\right)^2}, \tag{A.74}$$

let $C = \int_0^\infty t e^{-t} \cos{(st)}\, dt$ and $S = \int_0^\infty t e^{-t} \sin{(st)}\, dt$. Set $u = t e^{-t}$ and $dv = \cos{(st)}\, dt$ so that $du = (1-t)\, e^{-t} dt$ and $v = (\sin st)/s$. Then, from (A.72),

$$C = t e^{-t} \frac{\sin st}{s} \Big|_{t=0}^\infty - \int_0^\infty \frac{\sin st}{s} (1-t)\, e^{-t} dt$$

$$= 0 \quad - \frac{1}{s} \int_0^\infty e^{-t} \sin{(st)}\, dt + \frac{1}{s} \int_0^\infty t e^{-t} \sin{(st)}\, dt = -\frac{1}{1+s^2} + \frac{S}{s}.$$

Similarly, with $dv = \sin{(st)}\, dt$, $v = -(\cos st)/s$ and using (A.71),

$$S = -t e^{-t} \frac{\cos st}{s} \Big|_{t=0}^\infty + \int_0^\infty \frac{\cos st}{s} (1-t)\, e^{-t} dt$$

$$= 0 \quad + \frac{1}{s} \int_0^\infty e^{-t} \cos{(st)}\, dt - \frac{1}{s} \int_0^\infty t e^{-t} \cos{(st)}\, dt = \frac{1}{s\left(1+s^2\right)} - \frac{C}{s}.$$

Combining these yields

$$C = \int_0^\infty t e^{-t} \cos{(st)}\, dt = \frac{1-s^2}{\left(1+s^2\right)^2}, \tag{A.75}$$

so that, from (A.71) and (A.75),

$$I = \frac{1}{1+s^2} + \frac{1-s^2}{\left(1+s^2\right)^2} = \frac{2}{\left(1+s^2\right)^2},$$

which is (A.74). ∎

⊘ **Example A.27** Let $f(x) = x/(1+x^2)$. Then, using the substitution $u = 1+x^2$, a straightforward calculation yields

$$\int_0^c \frac{x}{1+x^2}\, dx = \frac{1}{2} \ln{(1+c^2)} \quad \text{and} \quad \int_{-c}^0 \frac{x}{1+x^2}\, dx = -\frac{1}{2} \ln{(1+c^2)},$$

which implies that

$$\lim_{c \to \infty} \int_{-c}^c \frac{x}{1+x^2}\, dx = \lim_{c \to \infty} 0 = 0. \tag{A.76}$$

This would seem to imply that

$$0 = \lim_{c \to \infty} \int_{-c}^c \frac{x}{1+x^2}\, dx = \int_{-\infty}^\infty \frac{x}{1+x^2}\, dx,$$

but the second equality is not true, because the limits in (A.68) do not exist. In (A.76), the order is conveniently chosen so that positive and negative terms precisely cancel, resulting in zero. ∎

Remark:

(a) A similar calculation shows that, for $c > 0$ and $k > 0$,

$$\lim_{c \to \infty} \int_{-c}^{kc} \frac{x}{1+x^2} \, dx = \ln k.$$

This expression could also be used for evaluating $\int_{-\infty}^{\infty} f(x) \, dx$, but results in a different value for each k. Thus, it also shows that $\int_{-\infty}^{\infty} f(x) \, dx$ does not exist.

(b) Notice that $f(x) = (1 + x^2)^{-1}$ is an *even function*, i.e. it satisfies $f(-x) = f(x)$ for all x (or is symmetric about zero). In this case, f is continuous for all x, so that, for any finite $c > 0$, $\int_0^c f(x) dx = \int_{-c}^0 f(x) dx$. On the other hand, $g(x) = x$ is an *odd function*, i.e. satisfies $g(-x) = -g(x)$, and, as g is continuous, for any finite $c > 0$, $\int_0^c g(x) dx = -\int_{-c}^0 g(x) dx$. Finally, as $h(x) = f(x)g(x)$ is also odd, $\int_{-c}^c h(x) dx = 0$. Thus, the result in (A.76) could have been immediately determined.

(c) The integral $\int_a^{\infty} \cos x \, dx$ also does not exist, because $\sin x$ does not have a limit as $x \to \infty$. Notice, however, that, for any value $t > 0$, the integral $\int_a^t \cos x \, dx$ is bounded. This shows that, if $\int_a^{\infty} f(x) dx$ does not exist, then it is not necessarily true that $\int_a^t f(x) dx$ increases as $t \to \infty$. ∎

⊚ **Example A.28**   (Example A.27 cont.) We have seen that the improper integral $\int_{-\infty}^{\infty} x (1 + x^2)^{-1} dx$ does not exist but, for any $z \in \mathbb{R}$,

$$\int_{-\infty}^{\infty} \left[\frac{x}{1+x^2} - \frac{x}{1+(z-x)^2} \right] dx = -\pi z. \tag{A.77}$$

This follows because

$$I = \int \left[\frac{x}{1+x^2} - \frac{x}{1+(z-x)^2} \right] dx = \frac{1}{2} \ln \left[\frac{1+x^2}{1+(z-x)^2} \right] + z \arctan (z - x),$$

seen by the FTC and (A.34). For fixed z,

$$\lim_{x \to \pm \infty} \ln \left[\frac{1+x^2}{1+(z-x)^2} \right] = \ln \left[\lim_{x \to \pm \infty} \frac{1+x^2}{1+(z-x)^2} \right] = \ln 1 = 0$$

and, from the graph of $\tan (x)$ for $-\pi/2 < x < \pi/2$, we see that, for fixed z,

$$\lim_{x \to \infty} \arctan (z - x) = -\frac{\pi}{2} \quad \text{and} \quad \lim_{x \to -\infty} \arctan (z - x) = \frac{\pi}{2}, \tag{A.78}$$

from which the value of $-\pi z$ for (A.77) follows. ∎

⊙ **Example A.29** (Example A.28 cont.) Consider the integral

$$I(s) = \int_{-\infty}^{\infty} \frac{1}{1+x^2} \frac{1}{1+(s-x)^2} \, dx.$$

To resolve this, the first step is to use a *partial fraction decomposition* for the integrand,

$$\frac{1}{1+x^2} \frac{1}{1+(s-x)^2} = \frac{Ax}{1+x^2} + \frac{B}{1+x^2} - \frac{Cx}{1+(s-x)^2} + \frac{D}{1+(s-x)^2},$$

where

$$A = \frac{2}{sR}, \quad B = \frac{1}{R}, \quad C = A, \quad D = \frac{3}{R},$$

and $R = s^2 + 4$, which can be easily verified with symbolic math software. As

$$\int_{-\infty}^{\infty} \frac{dx}{1+(s-x)^2} = \int_{-\infty}^{\infty} \frac{du}{1+u^2} = \pi,$$

integration of the B and D terms gives

$$\int_{-\infty}^{\infty} \left[\frac{B}{1+x^2} + \frac{D}{1+(s-x)^2} \right] dx = \pi (B+D) = \frac{4\pi}{R} = \frac{4\pi}{s^2+4}.$$

For the remaining two terms, use of (A.77) leads to

$$\int_{-\infty}^{\infty} \left[\frac{Ax}{1+x^2} - \frac{Cx}{1+(s-x)^2} \right] dx = A \int_{-\infty}^{\infty} \left[\frac{x}{1+x^2} - \frac{x}{1+(s-x)^2} \right] dx$$

$$= -\frac{2\pi}{s^2+4},$$

so that

$$I(s) = \frac{2\pi}{4+s^2} = \frac{\pi}{2} \frac{1}{1+(s/2)^2}.$$

This result is required in Example 8.16, which determines the density of the sum of two independent standard Cauchy random variables via the convolution formula given in (8.43).   ∎

⊙ **Example A.30** The integral

$$S = \int_0^{\infty} \frac{\sin x}{x} \, dx = \frac{\pi}{2} \tag{A.79}$$

arises in many contexts; we will require it when working with characteristic functions. To show that it converges, recall from (A.25) that $\lim_{x \to 0} x^{-1} \sin x = 1$, so that

$\int_0^1 x^{-1} \sin x \, dx$ is well defined, and it suffices to consider $\lim_{M \to \infty} \int_1^M x^{-1} \sin x \, dx$. Integration by parts with $u = x^{-1}$ and $dv = \sin x \, dx$ gives

$$\int_1^M x^{-1} \sin x \, dx = -x^{-1} \cos x \Big|_1^M - \int_1^M x^{-2} \cos x \, dx.$$

Clearly, $\lim_{M \to \infty} (\cos M) / M = 0$, so the first term is unproblematic. For the integral, note that

$$\left| \frac{\cos x}{x^2} \right| \le \frac{1}{x^2} \quad \text{and} \quad \int_1^\infty \frac{dx}{x^2} < \infty,$$

so that S converges.

Showing that the integral equals $\pi/2$ is a standard calculation via contour integration (see, for example, Bak and Newman, 1997, p. 134), though derivation without complex analysis is also possible. As in Hijab (1997, p. 197) and Jones (2001, p. 192), let $F(x) = \int_0^x r^{-1} \sin r \, dr$ for $x > 0$ and apply integration by parts to

$$I(b, s) := \int_0^b e^{-sx} \frac{\sin x}{x} \, dx$$

with $u = e^{-sx}$, $dv = x^{-1} \sin x \, dx$, $du = -se^{-sx} dx$ and $v = F(\cdot)$ to get

$$I(b, s) = e^{-sb} F(b) + s \int_0^b F(x) e^{-sx} dx = e^{-sb} F(b) + \int_0^{sb} F(y/s) e^{-y} \, dy,$$

with $y = sx$. Letting $b \to \infty$ and using the boundedness of $F(b)$ from (A.25) gives

$$I(b, s) \to I(\infty, s) = \int_0^\infty F(y/s) e^{-y} dy.$$

Now taking the limit in s,

$$\lim_{s \to 0^+} I(\infty, s) = \int_0^\infty F(\infty) e^{-y} dy = F(\infty) \int_0^\infty e^{-y} dy = F(\infty).$$

But using (A.78) and the fact, proven in (A.124), that

$$\int_0^\infty e^{-sx} \frac{\sin x}{x} dx = \arctan \left(s^{-1} \right),$$

we have

$$\lim_{s \to 0^+} I(\infty, s) = \lim_{s \to 0^+} \int_0^\infty e^{-sx} \frac{\sin x}{x} dx = \lim_{s \to 0^+} \left[\arctan \left(s^{-1} \right) \right] = \arctan(\infty) = \frac{\pi}{2},$$

i.e.

$$\frac{\pi}{2} = \lim_{s \to 0^+} I(\infty, s) = F(\infty) = \int_0^\infty \frac{\sin r}{r} dr,$$

as was to be shown. Other elementary proofs are outlined in Beardon (1997, p. 182) and Lang (1997, p. 343), while Goldberg (1964, p. 192) demonstrates that $\int_0^\infty \frac{\sin x}{x} dx$ does not converge absolutely, i.e. $\int_0^\infty \frac{|\sin x|}{x} dx$ does not exist. As an aside, by substituting $u = ax$, we find that $\int_0^\infty (\sin ax)/x \, dx = \text{sgn}(a)\pi/2$. ■

⊚ **Example A.31** Example A.25 showed that $I = \int_0^\infty \exp\left(-x^2\right) dx$ is convergent. Its value is commonly, if not almost exclusively, derived by use of *polar coordinates* (see Section A.3.4 below), but it can be done without them. As in Weinstock (1990), let

$$I = \int_0^\infty \exp\left(-x^2\right) dx$$

and

$$f(x) = \int_0^1 \frac{\exp\left[-x\left(1+t^2\right)\right]}{1+t^2} dt, \quad x > 0. \tag{A.80}$$

From (A.33) and the FTC,

$$f(0) = \int_0^1 (1+t^2)^{-1} dt = \arctan(1) - \arctan(0) = \pi/4 \tag{A.81}$$

and

$$0 < f(x) = e^{-x} \int_0^1 \frac{e^{-xt^2}}{1+t^2} dt < e^{-x} \int_0^1 \frac{1}{1+t^2} dt = \frac{\pi}{4} e^{-x},$$

so that $f(\infty) = 0$. Differentiating (A.80) with respect to x (and assuming we can interchange derivative and integral, see Section A.3.4), $f'(x)$ is given by

$$\int_0^1 \frac{d}{dx} \frac{\exp\left[-x\left(1+t^2\right)\right]}{1+t^2} dt = -\int_0^1 \exp\left[-x\left(1+t^2\right)\right] dt = -e^{-x} \int_0^1 \exp\left(-xt^2\right) dt.$$

Now, with $u = t\sqrt{x}$, $t = u/\sqrt{x}$ and $dt = x^{-1/2} du$,

$$f'(x) = -e^{-x} x^{-1/2} \int_0^{\sqrt{x}} \exp\left(-u^2\right) du = -e^{-x} x^{-1/2} g\left(\sqrt{x}\right), \tag{A.82}$$

where $g(z) := \int_0^z \exp\left(-u^2\right) du$. From (A.81) and that $f(\infty) = 0$, integrating both sides of (A.82) from 0 to ∞ and using the FTC gives, with $z = \sqrt{x}$, $x = z^2$ and $dx = 2z \, dz$,

$$0 - \frac{\pi}{4} = f(\infty) - f(0) = -\int_0^\infty e^{-x} x^{-1/2} g\left(\sqrt{x}\right) dx = -2 \int_0^\infty e^{-z^2} g(z) \, dz$$

or $\int_0^\infty \exp(-z^2) g(z) \, dz = \pi/8$. Again from the FTC and the definition of $g(z)$,

$$\frac{\pi}{8} = \int_0^\infty g'(z) g(z) \, dz \stackrel{u=g(z)}{=} \int_0^I u \, du = \frac{I^2}{2}.$$

or $I = \sqrt{\pi}/2$. Weinstock also gives similar derivations of the *Fresnel integrals* $\int_0^\infty \cos y^2 dy$ and $\int_0^\infty \sin y^2 dy$. Another way of calculating $\int_{-\infty}^\infty \exp(-x^2) dx$ without use of polar coordinates is detailed in Hijab (1997, Section 5.4). ■

A.2.4   Series

Of particular importance to us is differentiation and integration of infinite series, results for which are given below, and which the reader may skip if 'in a rush'. Their development, however, first requires establishing various notions and tools, some of which are also required in the text.

Definitions

> Mathematics is an experimental science, and definitions do not come first, but later on.   (Oliver Heaviside)

A *sequence* is a function $f : \mathbb{N} \to \mathbb{R}$, with $f(n)$, $n \in \mathbb{N}$, being the nth *term* of f. We often denote the sequence of f as $\{s_n\}$, where $s_n = f(n)$. Let $\{s_n\}$ be a sequence. If for any given $\epsilon > 0$, $\exists a \in \mathbb{R}$ and $\exists N \in \mathbb{N}$ such that, $\forall n \geq N$, $|a - s_n| < \epsilon$, then the sequence is *convergent*, and *converges to* (the unique value) a. If $\{s_n\}$ converges to a, then we write $\lim_{n \to \infty} s_n = a$. If $\{s_n\}$ does not converge, then it is said to *diverge*. Sequence $\{s_n\}$ is *strictly increasing* if $s_{n+1} > s_n$, and *increasing* if $s_{n+1} \geq s_n$. The sequence is *bounded from above* if $\exists c \in \mathbb{R}$ such that $s_n \leq c$ for all n. Similar definitions apply to *decreasing*, *strictly decreasing* and *bounded from below*. A simple but fundamental result is that, if $\{s_n\}$ is bounded from above and increasing, or bounded from below and decreasing, then it is convergent. Sequence $\{s_n\}$ is termed a *Cauchy sequence* if, for a given $\epsilon \in \mathbb{R}_{>0}$, $\exists N \in \mathbb{N}$ such that $\forall n, m \geq N$, $|s_m - s_n| < \epsilon$. An important result is that

$$a \text{ sequence } \{s_n\} \text{ converges iff } \{s_n\} \text{ is a Cauchy sequence.} \tag{A.83}$$

Let $\{s_n\}$ be a sequence. For each $k \in \mathbb{N}$, define the two sequences $a_k = \inf\{s_n : n \geq k\}$ and $b_k = \sup\{s_n : n \geq k\}$. A bit of thought confirms that a_k is increasing and b_k is decreasing. Thus, if they are bounded, then they converge to a value in $\mathbb{R}$, and if they are not bounded, then they diverge to plus or minus ∞. Either way, the two sequences have limits in $\mathbb{X}$. The *limit supremum* (or *limit superior*) of $\{s_n\}$, denoted $\limsup s_n$, is

$$\limsup_{n \to \infty} s_n = \lim_{k \to \infty} b_k = \lim_{k \to \infty} \left(\sup_{n \geq k} s_k \right),$$

and the *limit infimum* (or *limit inferior*), denoted $\liminf s_n$, is

$$\liminf_{n \to \infty} s_n = \lim_{k \to \infty} a_k = \lim_{k \to \infty} \left(\inf_{n \geq k} s_k \right).$$

Because a_k is increasing, and b_k is decreasing, it follows that

$$\liminf_{n \to \infty} s_n = \sup_{k \in \mathbb{N}} \inf_{n \geq k} s_k \quad \text{and} \quad \limsup_{n \to \infty} s_n = \inf_{k \in \mathbb{N}} \sup_{n \geq k} s_k.$$

The following facts are straightforward to prove, and should at least be informally thought through until they become intuitive:

- $\lim\sup s_n = -\infty$ iff $\lim_{n\to\infty} s_n = -\infty$;

- $\lim\sup s_n = \infty$ iff, $\forall M \in \mathbb{R}$ and $n \in \mathbb{N}$, $\exists k \in \mathbb{N}$ with $k \geq n$ such that $s_k \geq M$;[12]

- suppose $\lim\sup s_n \in \mathbb{R}$ (i.e. it is finite); then, $\forall \epsilon > 0$, $U = \lim\sup s_n$ iff:

  (1) $\exists N \in \mathbb{N}$ such that, $\forall n \geq N$, $s_n < U + \epsilon$, and   (A.84)

  (2) given $n \in \mathbb{N}$, $\exists k \in \mathbb{N}$ with $k \geq n$ such that $s_k > U - \epsilon$.   (A.85)

  (To verify these, think what would happen if they were not true.)

Similarly:

- $\lim\inf s_n = \infty$ iff $\lim_{n\to\infty} s_n = \infty$;

- $\lim\inf s_n = -\infty$ iff, $\forall M \in \mathbb{R}$ and $n \in \mathbb{N}$, $\exists k \in \mathbb{N}$ with $k \geq n$ such that $s_k \leq M$;

- suppose $\lim\inf s_n \in \mathbb{R}$; then, given any $\epsilon > 0$, $L = \lim\inf s_n$ iff:

  (1) $\exists N \in \mathbb{N}$ such that, $\forall n \geq N$, $s_n > L + \epsilon$, and

  (2) given $n \in \mathbb{N}$, $\exists k \in \mathbb{N}$ with $k \geq n$ such that $s_k < L - \epsilon$.

It is easy to see that, if $\ell = \lim_{n\to\infty} s_n$ for $\ell \in \mathbb{X}$, then $\lim\inf s_n = \lim\sup s_n = \ell$. The converse is also true: if $\lim\inf s_n = \lim\sup s_n = \ell$ for $\ell \in \mathbb{X}$, then $\ell = \lim_{n\to\infty} s_n$. To see this for $\ell \in \mathbb{R}$, note that, for a given $\epsilon > 0$, the above results imply that $\exists n_1, n_2 \in \mathbb{N}$ such that, $\forall n \geq n_1$, $s_n < \ell + \epsilon$ and, $\forall n \geq n_2$, $s_n > \ell - \epsilon$. Thus, with $N = \max\{n_1, n_2\}$, $\ell - \epsilon < s_n < \ell + \epsilon$ for all $n \geq N$, or $\lim_{n\to\infty} s_n = \ell$.

Let f_k be a sequence. The sum

$$S = \sum_{k=1}^{\infty} f_k = \lim_{n\to\infty} \sum_{k=1}^{n} f_k = \lim_{n\to\infty} s_n$$

is referred to as a *series* associated with the sequence $\{f_k\}$ and $s_n = \sum_{k=1}^{n} f_k$ is its *n*th partial sum. Series S *converges* if $\lim_{n\to\infty} s_n$ exists, i.e. if the limit is bounded, and *diverges* if the partial sums are not bounded. If S converges, then $\lim_{n\to\infty} f_n = 0$, but the converse is not true (use $\sum_{k=1}^{\infty} k^{-1}$ as an example). If $\sum_{k=1}^{\infty} |f_k|$ converges, then S is said to *converge absolutely*. If S is absolutely convergent, then S is convergent, but the converse is not true. For example, the *alternating (harmonic) series*

---

[12] Note the difference to saying that $\lim_{n\to\infty} s_n = \infty$: if $s_n \to \infty$, then $\lim\sup s_n = \infty$, but the converse need not be true.

$S = \sum_{k=1}^{\infty}(-1)^k k^{-1}$ is convergent, but not absolutely convergent. It is *conditionally convergent*. The *Cauchy criterion*, obtained by Cauchy in 1821, is:

$$\sum_{k=1}^{\infty} f_k \text{ converges} \Leftrightarrow \exists N \in \mathbb{N} \text{ such that, } \forall \, n, m \geq N, \left| \sum_{k=n+1}^{m} f_k \right| < \epsilon, \qquad (A.86)$$

for any given $\epsilon > 0$. This follows from (A.83) and writing $\left| \sum_{k=n+1}^{m} f_k \right| = |s_m - s_n|$.
The *geometric series*, and the *p-series* or *(Riemann's) zeta function*

$$\sum_{k=1}^{\infty} c^k , \quad c \in [0, 1) \qquad \text{and} \qquad \sum_{k=1}^{\infty} \frac{1}{k^p} , \quad p \in \mathbb{R}_{>0}$$

respectively, are important convergent series because they can often be used to help prove the convergence of other series via the tests outlined below. Indeed, for the geometric series $S_0 = S_0(c) := \sum_{k=1}^{\infty} c^k$, $S_0 = c + c^2 + c^3 + \cdots$, $cS_0 = c^2 + c^3 + \cdots$ and $S_0 - cS_0 = c - \lim_{k\to\infty} c^k = c$, for $c \in [0, 1)$. Solving $S_0 - cS_0 = c$ implies

$$S_0 = \frac{c}{1 - c}, \qquad c \in [0, 1). \qquad (A.87)$$

For the zeta function, writing

$$\zeta(p) = \sum_{k=1}^{\infty} \frac{1}{k^p} = 1 + \frac{1}{2^p} + \frac{1}{3^p} + \cdots$$

$$= 1 + \left(\frac{1}{2^p} + \frac{1}{3^p} \right) + \left(\frac{1}{4^p} + \frac{1}{5^p} + \frac{1}{6^p} + \frac{1}{7^p} \right) + \cdots$$

$$< 1 + \frac{2}{2^p} + \frac{4}{4^p} + \cdots = \sum_{i=0}^{\infty} \left(\frac{1}{2^{p-1}} \right)^i = \frac{1}{1 - \frac{1}{2^{p-1}}},$$

which is valid for $1/2^{p-1} < 1$ or $(p-1)\ln 2 > 0$ or $p > 1$.

◎ ***Example A.32*** Let $\zeta(p) = \sum_{r=1}^{\infty} r^{-p}$. The well-known result

$$\zeta(2) = \sum_{n=1}^{\infty} \frac{1}{n^2} = \frac{\pi^2}{6} \qquad (A.88)$$

is often proven via contour integration (Bak and Newman, 1997, p. 141) or in the context of Fourier analysis (Stoll, 2001, p. 413; Jones, 2001, p. 410). The first proof, by Euler in 1735, involves the use of infinite products; see Havil (2003, p. 39) for a simple account, or Hijab (1997, Section 5.6) and Bak and Newman (1997, p. 223) for a rigorous proof. Before Euler solved it, the problem had been unsuccessfully attempted by Wallis, Leibniz, Jakob Bernoulli, and others (Havil, 2003, p. 38). Today, there are many known methods of proof.[13] It can also be shown that $\zeta(4) = \pi^4/90$ and $\zeta(6) = \pi^6/945$. In general, expressions exist for even p. ∎

⊙ **Example A.33** Recall that $e^\lambda = \lim_{n\to\infty} (1 + \lambda/n)^n$ from (A.46). We wish to confirm that sequence $s_n := (1 + 1/n)^n$ converges. Applying the binomial theorem (1.18) to s_n gives $s_n = \sum_{i=0}^n \binom{n}{i}(\frac{1}{n})^i$, with $(i + 1)$th term

$$\binom{n}{i}n^{-i} = \frac{n_{[i]}}{i!}n^{-i} = \frac{n(n-1)\cdots(n-i+1)}{i!}n^{-i}$$

$$= \frac{1}{i!}(1)\left(1 - \frac{1}{n}\right)\left(1 - \frac{2}{n}\right)\cdots\left(1 - \frac{i-1}{n}\right). \tag{A.89}$$

Similarly, $s_{n+1} = \sum_{i=0}^{n+1} \binom{n+1}{i}(n+1)^{-i}$ with $(i + 1)$th term

$$\binom{n+1}{i}(n+1)^{-i} = \frac{(n+1)n(n-1)\cdots(n-i+2)}{i!}(n+1)^{-i}$$

$$= \frac{1}{i!}(1)\left(\frac{n}{n+1}\right)\left(\frac{n-1}{n+1}\right)\cdots\left(\frac{n-i+2}{n+1}\right)$$

$$= \frac{1}{i!}\left(1 - \frac{1}{n+1}\right)\left(1 - \frac{2}{n+1}\right)\cdots\left(1 - \frac{i-1}{n+1}\right). \tag{A.90}$$

As the quantity in (A.90) is larger than that in (A.89), it follows that $s_n \le s_{n+1}$, i.e. s_n is a nondecreasing sequence. Also, for $n \ge 2$,

$$s_n = \sum_{i=0}^n \binom{n}{i}\frac{1}{n^i} = \sum_{i=0}^n \frac{n(n-1)\cdots(n-i+1)}{n^i}\frac{1}{i!} \le \sum_{i=0}^n \frac{1}{i!}.$$

Note that $2! < 2^2$ and $3! < 2^3$, but $4! > 2^4$. Assume it holds for $k \ge 4$. It holds for $k+1$ because $(k+1)! = (k+1)k! > (k+1)2^k > 2 \cdot 2^k = 2^{k+1}$. Thus, $k! > 2^k$ for $k \ge 4$, and

$$\sum_{i=0}^n \frac{1}{i!} = \frac{8}{3} + \sum_{i=4}^n \frac{1}{i!} < \frac{8}{3} + \sum_{i=4}^n \frac{1}{2^i} = \frac{8}{3} + \sum_{i=0}^n \frac{1}{2^i} - \sum_{j=0}^3 \frac{1}{2^j} = \frac{19}{24} + \sum_{i=0}^n \frac{1}{2^i}$$

$$< \frac{19}{24} + \sum_{i=0}^\infty \frac{1}{2^i} = \frac{19}{24} + 2 < 2.8.$$

Thus, s_n is a nondecreasing, bounded sequence, and is thus convergent. ∎

---

[13] Matsuoka (1961) gives an elementary one, requiring two integration by parts of $\int_0^{\pi/2} \cos^{2n}(t)\,dt = (\pi/2)(2n-1)!!\,/\,(2n)!!$, where $(2n)!! = 2 \cdot 4 \cdots (2n)$, $0!! = 1$, $(2n+1)!! = 1 \cdot 3 \cdots (2n+1)$ and $(-1)!! = 1$. Kortam (1996) illustrates further simple proofs, Hofbauer (2002) and Harper (2003) each contribute yet another method, and Chapman (2003) provides 14 proofs (not including the previous two, but including that from Matsuoka).

Tests for Convergence and Divergence

With the exception of the geometric series, there does not exist in all of mathematics a single infinite series whose sum has been determined rigorously.

(Niels Abel)

The following are some of the many conditions, or 'tests', which can help determine if a series of nonnegative terms is convergent or divergent.

- *The Comparison Test* Let $S = \sum_{k=1}^{\infty} f_k$ and $T = \sum_{k=1}^{\infty} g_k$ with $0 \leq f_k, g_k < \infty$ and T convergent. If there exists a positive constant C such that $f_k \leq C g_k$ for all k sufficiently large (i.e. $\exists K \in \mathbb{N}$ such that it holds for all $k \geq K$), then S also converges. The proof when $f_k \leq C g_k$ for all $k \in \mathbb{N}$ is simply to note that

$$\sum_{k=1}^{n} f_k \leq \sum_{k=1}^{n} C g_k \leq C \sum_{k=1}^{\infty} g_k < \infty$$

is true for all n, so that the partial sum $\sum_{k=1}^{n} f_k$ is bounded. In a similar way, the comparison test can be used to show that a series diverges.

- *The Ratio Test* Let $S = \sum_{k=1}^{\infty} f_k$ with $0 \leq f_k < \infty$. If $\exists c \in (0, 1)$ such that $f_{k+1}/f_k \leq c$ for all k sufficiently large, then S converges. To prove this, let K be such that $f_{k+1} \leq c f_k$ for all $k \geq K$. Then $f_{K+1} \leq c f_K$, and $f_{K+2} \leq c f_{K+1} \leq c^2 f_K$, etc. and $f_{K+n} \leq c^n f_K$. Then

$$\sum_{n=1}^{\infty} f_{K+n} = f_{K+1} + f_{K+2} + \cdots$$

$$\leq c f_K + c^2 f_K + \cdots = \frac{c}{1-c} f_K,$$

which is finite for $c \in [0, 1)$. More generally, allow f_k to be negative or positive, and let $c = \lim_{k \to \infty} |f_{k+1}/f_k|$. If $c < 1$, then a similar argument shows that $S = \sum_{k=1}^{\infty} f_k$ converges absolutely. If $c > 1$ or ∞, then $\exists K \in \mathbb{N}$ such that $\forall k \geq K$, $|f_k| > |f_K|$, which implies that $\lim_{k \to \infty} |f_k| \neq 0$, and S diverges.

- *The Root Test* Let $S = \sum_{k=1}^{\infty} f_k$ and assume (for now) that $r = \lim_{k \to \infty} |f_k|^{1/k}$ exists. If $r < 1$, then $\exists \epsilon > 0$ such that $r + \epsilon < 1$, and $\exists K \in \mathbb{N}$ such that $|f_k|^{1/k} < r + \epsilon$, or $|f_k| < (r + \epsilon)^k$, $\forall k \geq K$. It follows by the comparison test with the geometric series $\sum_{k=1}^{\infty} (r + \epsilon)^k$ that $\sum_{k=1}^{\infty} |f_k|$ converges, i.e. S is absolutely convergent.

Similarly, if $r > 1$ or ∞, then $\exists \epsilon > 0$ such that $r - \epsilon > 1$, and $\exists K \in \mathbb{N}$ such that $|f_k|^{1/k} > r - \epsilon$, or $|f_k| > (r - \epsilon)^k$, $\forall k \geq K$. Thus, $\lim_{k \to \infty} |f_k| > 1$, and S diverges.

If $r = 1$, the test is inconclusive. To see this, take the zeta function, with $f_k = k^{-p}$, and observe that

$$\lim_{k \to \infty} f_k^{1/k} = \lim_{k \to \infty} \left(\frac{1}{k^p} \right)^{1/k} = \left(\frac{1}{\lim_{k \to \infty} k^{1/k}} \right)^p = 1$$

for any $p \in \mathbb{R}$; but we know that $\zeta(p)$ converges for $p > 1$ and diverges otherwise, so that the root test is inconclusive.

- *The Integral Test* Let $f(x)$ be a nonnegative, decreasing function for all $x \geq 1$. If $\int_1^\infty f(x)\, dx < \infty$, then $S = \sum_{k=1}^\infty f(k)$ exists. The proof rests upon the fact that

$$f(k) \leq \int_{k-1}^k f(x)\, dx,$$

which is graphically obvious from Figure A.3; the area of the rectangle from $n-1$ to n with height $f(n)$ is $1 \times f(n) = f(n)$, which is less than or equal to the area under $f(x)$ between $x = n-1$ and $x = n$. Thus,

$$f(2) + f(3) + \cdots + f(k) \leq \int_1^k f(x)\, dx \leq \int_1^\infty f(x)\, dx < \infty,$$

and the partial sums are bounded. To show divergence, note from Figure A.3 that $f(k) \geq \int_k^{k+1} f(x)\, dx$, so that

$$f(1) + f(2) + \cdots + f(k) \geq \int_1^{k+1} f(x)\, dx.$$

If the latter integral diverges as $k \to \infty$, then so does the partial sum.

- *The Dirichlet Test* Let $\{a_k\}$ and $\{b_k\}$ be sequences such that the partial sums of $\{a_k\}$ are bounded, $\{b_k\}$ is positive and decreasing, i.e. $b_1 \geq b_2 \geq \cdots \geq 0$, and $\lim_{k\to\infty} b_k = 0$. Then $\sum_{k=1}^\infty a_k b_k$ converges. See, for example, Stoll (2001, Section 7.2) for proof. As a special case, if f_k is a positive, decreasing sequence with $\lim_{k\to\infty} f_k = 0$, then $\sum_{k=1}^\infty (-1)^k f_k$ converges, which is often referred to as the *alternating series test*.

  If the sequence $\{b_k\}$ is positive and decreasing, and $\lim_{k\to\infty} b_k = 0$, then the Dirichlet test can also be used to prove that $\sum_{k=1}^\infty b_k \sin(kt)$ converges for all $t \in \mathbb{R}$,

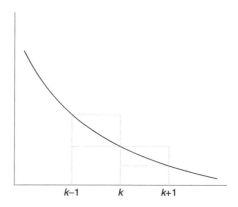

Figure A.3   For continuous, positive, decreasing function f, $f(k) \leq \int_{k-1}^k f(x)dx$

and that $\sum_{k=1}^{\infty} b_k \cos(kt)$ converges for all $t \in \mathbb{R}$, except perhaps for $t = 2z\pi$, $z \in \mathbb{Z}$ (see Stoll (2001, pp. 296–7) for proof).

⊚ **Example A.34** Let $f(x) = 1/x^p$ for $x \in \mathbb{R}_{\geq 1}$ and $p \in \mathbb{R}_{>0}$. As $f(x)$ is nonnegative and decreasing, the integral test implies that $\zeta(p) = \sum_{x=1}^{\infty} x^{-p}$ exists if

$$\int_1^{\infty} \frac{1}{x^p}\, dx = \lim_{x \to \infty} \left(\frac{x^{1-p}}{1-p} \right) - \frac{1}{1-p} < \infty,$$

which is true for $1 - p < 0$, i.e. $p > 1$, and does not exist otherwise. Thus, $\zeta(1)$ diverges, but $\zeta(p)$ converges for $p > 1$. ∎

⊚ **Example A.35** Let $S(p) = \sum_{k=1}^{\infty} (\ln k)/k^p$. For $p > 1$, use the 'standard trick' (Lang, 1997, p. 212) and write $p = 1 + \epsilon + \delta$, $\delta > 0$. From (A.45) $\lim_{k \to \infty} (\ln k)/k^{\delta} = 0$, which implies that, for large enough k, $(\ln k)/k^{\delta} \leq 1$. Thus, for k large enough and $C = 1$,

$$\frac{\ln k}{k^p} = \frac{\ln k}{k^{\delta} k^{1+\epsilon}} \leq C \frac{1}{k^{1+\epsilon}},$$

so that the comparison test and the convergence of the zeta function imply that $S(p)$ converges for $p > 1$. A similar analysis shows that $S(p)$ diverges for $p < 1$. For $p = 1$, as $\ln k > 1$ for $k \geq 3$, the comparison test with $\zeta(1)$ confirms that it also diverges. The integral test also works in this case (see Stoll (2001, p. 286)). ∎

⊚ **Example A.36** Continuing with the investigation of how inserting $\ln k$ affects convergence, consider now

$$S(q) = \sum_{k=2}^{\infty} \frac{1}{(k \ln k)^q}.$$

First, take $q = 1$. Let $f : D \to \mathbb{R}$ be given by $f(x) = (x \ln x)^{-1}$, with $D = \{x \in \mathbb{R} : x \geq 2\}$. It is straightforward to show that $f'(x) = -(1 + \ln x)(x \ln x)^{-2} < 0$ on D, so that, from the first derivative test, f is decreasing on D, and the integral test is applicable. However, from (A.42), the improper integral

$$\lim_{t \to \infty} \int_2^t \frac{dx}{x \ln x} = \lim_{t \to \infty} [\ln (\ln t) - \ln (\ln 2)]$$

diverges, so that $S(1)$ diverges.[14] For $q > 1$, let $q = 1 + m$, so that (A.41) with $p = -m$ implies that

$$\int_2^t \frac{dx}{x (\ln x)^{1+m}} = -\frac{(\ln x)^{-m}}{m} \Big|_2^t = \frac{(\ln 2)^{-m}}{m} - \frac{1}{m (\ln t)^m}, \quad m > 0,$$

---

[14] The divergence is clearly very slow. The largest number in Matlab is obtained by t=realmax, which is about 1.7977×10^{308}, and $\ln (\ln t) = 6.565$.

so that

$$\int_2^\infty \frac{dx}{x \, (\ln x)^{1+m}} = \frac{(\ln 2)^{-m}}{m} < \infty,$$

and $S(q)$ converges for $q > 1$. ∎

⊖ **Example A.37** Let $S = \sum_{k=2}^\infty (\ln k)^{-vk}$, for constant $v \in \mathbb{R}_{>0}$. As

$$\lim_{k \to \infty} \left| (\ln k)^{-vk} \right|^{1/k} = \lim_{k \to \infty} \left(\frac{1}{\ln k} \right)^v = 0,$$

the root test shows that S converges. ∎

⊚ **Example A.38** Let

$$\gamma_n = 1 + \frac{1}{2} + \frac{1}{3} + \cdots + \frac{1}{n} - \ln n,$$

which converges to *Euler's constant*, denoted γ, with $\gamma \approx 0.5772156649$. To see that γ_n is convergent, as in Beardon (1997, p. 176), let

$$a_n = \ln \left(\frac{n}{n-1} \right) - \frac{1}{n}.$$

As $\sum_{i=1}^n a_i = 1 - \gamma_n$, it suffices to show that $\sum_{i=1}^\infty a_i$ converges. Observe that

$$\int_0^1 \frac{t}{n \, (n-t)} \, dt = \int_0^1 \left(\frac{1}{n-t} - \frac{1}{n} \right) dt = a_n.$$

Next, let $f(t) = t / (n-t)$ for $n \geq 1$ and $t \in (0, 1)$. Clearly, $f(0) = 0$, $f(t)$ is increasing on $(0, 1)$ and, as $f''(t) = 2n \, (n-t)^{-3} > 0$, it is convex. Thus, the area under its curve on $(0, 1)$ is bounded by that of the right triangle with vertices $(0, 0)$, $(1, 0)$ and $(1, f(1))$, which has area $f(1)/2 = \frac{1}{2(n-1)}$. Thus

$$\int_0^1 \frac{t}{(n-t)} \, dt \leq \frac{1}{2 \, (n-1)}$$

or

$$0 \leq a_n = \frac{1}{n} \int_0^1 \frac{t}{(n-t)} \, dt \leq \frac{1}{2n \, (n-1)} \leq \frac{1}{n^2}.$$

By the comparison test with the zeta function, $\sum_{i=1}^\infty a_i$ converges; see also Example A.53. ∎

○ **Example A.39** To see that $\cos(z) = \sum_{k=0}^{\infty} (-1)^k z^{2k} / (2k)!$ converges, use the ratio test to see that

$$c = \lim_{k\to\infty} \left| \frac{(-1)^{k+1} \frac{z^{2(k+1)}}{[2(k+1)]!}}{(-1)^k \frac{z^{2k}}{(2k)!}} \right| = \lim_{k\to\infty} \left| \frac{z^2}{(2k+1)(2k+2)} \right| = 0,$$

for all $z \in \mathbb{R}$. ∎

○ **Example A.40** Consider the sum $S = \lim_{r\to\infty} \sum_{i=1}^{r-1} \frac{i}{(r-i)^2}$ and let $j = r - i$ so that

$$S = \lim_{r\to\infty} \sum_{j=1}^{r} \frac{r-j}{j^2} = \lim_{r\to\infty} \left(r \sum_{j=1}^{r} \frac{1}{j^2} - \sum_{j=1}^{r} \frac{1}{j} \right).$$

Then

$$\frac{S}{r} = \lim_{r\to\infty} \sum_{j=1}^{r} \frac{1}{j^2} - \lim_{r\to\infty} \frac{1}{r} \sum_{j=1}^{r} \frac{1}{j}. \tag{A.91}$$

From (A.88), the first sum in (A.91) converges to $\pi^2/6$. Using the comparison test, the second sum is bounded because, for $r \geq 2$,

$$\sum_{j=1}^{r} \frac{1}{rj} < \sum_{j=1}^{r} \frac{1}{r} = 1.$$

To see that it converges to zero, use the fact that $(rj)^{-1}$ is a positive, decreasing function in j, so that the conditions of the integral test hold. Then,

$$\sum_{j=2}^{r} \frac{1}{rj} \leq \int_{1}^{r} \frac{1}{rx} dx \quad \text{and} \quad \lim_{r\to\infty} \int_{1}^{r} \frac{1}{rx} dx = \lim_{r\to\infty} \frac{\ln r}{r} = \lim_{r\to\infty} \frac{r^{-1}}{1} = 0$$

from l'Hôpital's rule. Thus, S/r in (A.91) converges to $\pi^2/6$ and, as $r \to \infty$,

$$S \to \frac{r\pi^2}{6},$$

in the sense that the ratio of S to $r\pi^2/6$ converges to unity as r grows. We need this result in Example 6.8 for the calculation of the asymptotic variance of a particular random variable. ∎

In the root test, we assumed that $\lim_{k\to\infty} |f_k|^{1/k}$ exists, but this assumption can be relaxed by working with the *exponential growth rate* of the series $\sum_{k=1}^{\infty} f_k$, given by $L = \limsup |f_k|^{1/k}$, which always exists (in $\mathbb{X}$). The proof is as before. If $L < 1$, then $\exists \epsilon > 0$ such that $L + \epsilon < 1$, and, from (A.84), $\exists K \in \mathbb{N}$ such that, $\forall k \geq K$, $|f_k|^{1/k} < L + \epsilon$, or $|f_k| < (L + \epsilon)^k$. The comparison test is then used as before. A similar argument using (A.85) shows that the series diverges if $L > 1$.

⊖ **Example A.41** (Stoll, 2001, p. 289) Let $a_n = 2^{-k}$ if $n = 2k$ and $a_n = 3^{-k}$ if $n = 2k + 1$, so that

$$S = \sum_{n=1}^{\infty} a_n = 1 + \frac{1}{2} + \frac{1}{3} + \frac{1}{2^2} + \frac{1}{3^2} + \cdots,$$

and $a_n^{1/n} = 2^{-1/2}$ for $n = 2k$ and $a_n^{1/n} = 3^{-k/(2k+1)}$ for $n = 2k + 1$. As $\lim_{k \to \infty} 3^{-k/(2k+1)} = 3^{-1/2}$ and $\max \left(2^{-1/2}, 3^{-1/2} \right) = 2^{-1/2}$, we have $\limsup a_n^{1/n} = 2^{-1/2} < 1$ and thus, by the root test, S converges. ■

Remark:

(a) It turns out that the root test is 'more powerful' than the ratio test, in the sense that, if the ratio test proves convergence or divergence, then so does the root test, but the converse is not true. (Our comfortable statistics language is not actually used; one says that the root test has a *strictly wider scope* than the ratio test.) The reason for this is that, for the positive sequence $\{a_n\}$,

$$\liminf \frac{a_{n+1}}{a_n} \leq \liminf a_n^{1/n} \leq \limsup a_n^{1/n} \leq \limsup \frac{a_{n+1}}{a_n}. \tag{A.92}$$

The proof of this is straightforward; see, for example, Stoll (2001, p. 291). It is, however, not hard to construct series for which the ratio test is much easier to apply than the root test, so that both tests are indeed valuable.

(b) For the series $S = \sum_{k=1}^{\infty} f_k$, if $L = \limsup |f_k|^{1/k} < 1$, then S is absolutely convergent, and hence also convergent. If $L > 1$, then $\sum_{k=1}^{\infty} |f_k|$ diverges, but could it be the case that S converges? To answer this, recall from (A.85) that, if $L > 1$, then there are infinitely many k such that $|f_k| > 1$. Thus, whatever the sign of the terms, $\{f_k\}$ is not converging to zero, which implies that S diverges. ■

The tools for convergence of infinite series can be extended to convergence of *infinite products* by use of basic properties of the logarithm and continuity of the exponential function (see Section A.2.2). Let $a_k \geq 0$ such that $\lim_{k \to \infty} a_k = 0$. If $S = \sum_{k=1}^{\infty} a_k$ converges, then so does the infinite product $P = \prod_{k=1}^{\infty} (1 + a_k)$. Let $p_n = \prod_{k=1}^{n} (1 + a_k)$. As $\ln (1 + a_k) \leq a_k$,

$$\ln p_n = \sum_{k=1}^{n} \ln (1 + a_k) \leq \sum_{k=1}^{n} a_k \leq S,$$

so that, from the comparison test, $\ln P = \sum_{k=1}^{\infty} \ln (1 + a_k)$ converges. Taking exponents gives the result.

⊙ **Example A.42** *Wallis' formula*, or *Wallis' product* (John Wallis, 1616–1703), is

$$\lim_{n \to \infty} \frac{1}{\sqrt{n}} \frac{2 \cdot 4 \cdot 6 \cdots 2n}{1 \cdot 3 \cdot 5 \cdots (2n - 1)} = \sqrt{\pi} \tag{A.93}$$

or

$$\lim_{n\to\infty} \frac{2}{1}\frac{2}{3}\frac{4}{3}\frac{4}{5}\frac{6}{5}\frac{6}{7}\cdots\frac{2n}{2n-1}\frac{2n}{2n+1} = \frac{\pi}{2}.$$ (A.94)

To see that the product in (A.94) converges, use the previous result with

$$\frac{2n}{2n-1}\frac{2n}{2n+1} = \frac{4n^2}{4n^2-1} = 1 + \frac{1}{4n^2-1} =: 1 + a_k.$$

To show that $\sum_{k=1}^{\infty} a_k$ converges, use the comparison test with $\zeta\,(2)$. There are several ways of proving that the infinite product in (A.94) is $\pi/2$; see, for example, Keeping (1995, p. 392), Hijab (1997, p. 173) and Andrews, Askey and Roy (1999, p. 46) for three different proofs. We reproduce the former one in Example 1.27, after the gamma and beta functions are introduced. ∎

The Wallis product is important for deriving *Stirling's approximation* to $n!$ as given in (1.42); see, for example, Lang (1997, p. 120) or Andrews, Askey and Roy (1999, Section 1.4). The next example shows a less important application of it.

⊖ *Example A.43*   Let

$$w_n = \frac{2^n\,n!}{1\cdot 3\cdot 5\cdots(2n-1)} = \frac{2\cdot 4\cdot 6\cdots 2n}{1\cdot 3\cdot 5\cdots(2n-1)} \approx \sqrt{n\pi}$$

from (A.93). Then, using the generalized binomial coefficient (1.8),

$$\binom{N+1/2}{N} = \frac{\left(N+\frac{1}{2}\right)\left(N-\frac{1}{2}\right)\cdots\frac{3}{2}}{N!}$$

$$= \frac{(2N+1)\,(2N-1)\cdots 3}{2^N\,N!} = \frac{2N+1}{w_N} \approx \frac{2N+1}{\sqrt{n\pi}}.$$ (A.95)

From Figure A.4, we see that this approximation is very good even for modest values of N. This result is used in Problem 4.13. ∎

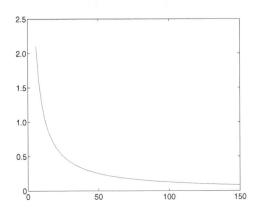

Figure A.4   The relative percentage error, $100\,(\text{approx.} - \text{true})\,/\text{true}$, for the approximation in (A.95) as a function of N

Cauchy product

> Cauchy is mad and there is nothing that can be done about him, although, right now, he is the only one who knows how mathematics should be done.
>
> (Niels Abel)

Consider the product of the two series $\sum_{k=0}^{\infty} a_k$ and $\sum_{k=0}^{\infty} b_k$. Multiplying their values out in tabular form

	b_0	b_1	b_2	$\cdots$
a_0	$a_0 b_0$	$a_0 b_1$	$a_0 b_2$	$\cdots$
a_1	$a_1 b_0$	$a_1 b_1$	$a_1 b_2$	
a_2	$a_2 b_0$	$a_2 b_1$	$a_2 b_2$	
$\vdots$	$\vdots$			$\ddots$

and summing the off-diagonals suggests that the product is given by the

$$a_0 b_0 + (a_0 b_1 + a_1 b_0) + (a_0 b_2 + a_1 b_1 + a_2 b_0) + \cdots,$$

which is referred to as the *Cauchy product*. It is a standard result in real analysis that, if $\sum_{k=0}^{\infty} a_k$ and $\sum_{k=0}^{\infty} b_k$ are absolutely convergent series with sums A and B, respectively, then their Cauchy product

$$\sum_{k=0}^{\infty} c_k, \qquad c_k = a_0 b_k + a_1 b_{k-1} + \cdots + a_k b_0, \tag{A.96}$$

is absolutely convergent with sum AB (see, for example, Trench, 2003, p. 227).

⊖ **Example A.44** Let $S = \sum_{k=0}^{\infty} a^k = (1-a)^{-1}$ for $a \in [0, 1)$. As this is absolutely convergent, (A.96) with $a_k = b_k = a^k$ implies that $c_k = a^0 a^k + a^1 a^{k-1} + \cdots + a^k a^0 = (k+1)\, a^k$ and $(1-a)^{-2} = S^2 = \sum_{k=0}^{\infty} c_k = 1 + 2a + 3a^2 + \cdots$. ∎

The Cauchy product result can be generalized. Let $x_{nm} := x\,(n, m)$ be a function of $n, m \in \mathbb{N}$ with $x_{nm} \in \mathbb{R}_{\geq 0}$. Then

$$\lim_{N \to \infty} \sum_{n=0}^{N} \sum_{m=0}^{N} x_{nm} = \lim_{N \to \infty} \sum_{m=0}^{N} \sum_{n=0}^{N} x_{nm} = \sum_{s=0}^{\infty} \sum_{\substack{m \geq 0, n \geq 0 \\ m+n=s}} x_{nm}, \tag{A.97}$$

if the unordered sum converges or, equivalently, if the terms are absolutely summable (see, for example, Beardon, 1997, Section 5.5; Browder, 1996, Section 2.5). As before, the values to be summed can be shown in a table as

		n		
		0	1	2	$\cdots$
	0	x_{00}	x_{01}	x_{02}	$\cdots$
m	1	x_{10}	x_{11}	x_{12}	
	2	x_{20}	x_{21}	x_{22}	
	$\vdots$	$\vdots$			$\ddots$

and, if the double sum is absolutely convergent, then the elements can be summed in any order, i.e. by columns, by rows, or by summing the off-diagonals.

⊚ **Example A.45**   It is instructive to show the properties of the exponential starting from its power series expression

$$f(x) = \exp(x) = 1 + x + \frac{x^2}{2!} + \frac{x^3}{3!} + \cdots. \tag{A.98}$$

Clearly, $f(0) = 1$, and observe that

$$\lim_{k \to \infty} \left| \frac{x^{k+1}/(k+1)!}{x^k/k!} \right| = \lim_{k \to \infty} \left| \frac{x}{k+1} \right| = 0$$

for all $x \in \mathbb{R}$, so that the ratio test shows absolute convergence of the series. Using (A.98), $f(0) = 1$, and differentiating (A.98) termwise[15] shows that $f(x) = f'(x)$. To show $f(x+y) = f(x)f(y)$ from (A.98), use the binomial theorem (1.18) to get

$$\exp(x+y) = \sum_{s=0}^{\infty} \frac{(x+y)^s}{s!} = \sum_{s=0}^{\infty} \frac{1}{s!} \sum_{n=0}^{s} \binom{s}{n} x^n y^{s-n}$$

$$= \sum_{s=0}^{\infty} \frac{1}{s!} \sum_{n=0}^{s} \frac{s!}{n!(s-n)!} x^n y^{s-n} = \sum_{s=0}^{\infty} \sum_{n=0}^{s} \frac{x^n}{n!} \frac{y^{s-n}}{(s-n)!}$$

$$= \sum_{s=0}^{\infty} \sum_{\substack{m \ge 0, n \ge 0 \\ m+n=s}} \frac{x^n y^m}{n! \, m!}.$$

It follows from (A.97) that

$$\exp(x+y) = \lim_{N \to \infty} \sum_{n=0}^{N} \sum_{m=0}^{N} \frac{x^n y^m}{n! \, m!} = \lim_{N \to \infty} \sum_{n=0}^{N} \frac{x^n}{n!} \sum_{m=0}^{N} \frac{y^m}{m!} = \exp(x)\exp(y).$$

To show $[\exp(x)]^n = \exp(nx)$, apply the multinomial theorem (1.34) to (A.98) to get

$$[f(x)]^n = \lim_{r \to \infty} \left(\sum_{s=0}^{r} \frac{x^s}{s!} \right)^n = \sum_{\mathbf{n}: n_\bullet = n, \, n_i \ge 0} \binom{n}{n_0, \ldots, n_r} 1^{n_0} \left(\frac{x}{1!}\right)^{n_1} \cdots \left(\frac{x^r}{r!}\right)^{n_r}$$

$$= \binom{n}{n, 0, \ldots, 0} 1^n \left(\frac{x}{1!}\right)^0 \cdots \left(\frac{x^r}{r!}\right)^0$$

$$+ \binom{n}{n-1, 1, 0 \ldots, 0} 1^{n-1} \left(\frac{x}{1!}\right)^1 \cdots \left(\frac{x^r}{r!}\right)^0$$

---

[15] See Example A.47 and result (A.2.3) for the justification of termwise differentiation.

$$+\binom{n}{n-1,0,1,0\ldots,0}\left(\frac{x}{1!}\right)^0\left(\frac{x^2}{2!}\right)^1\cdots\left(\frac{x^r}{r!}\right)^0$$

$$+\binom{n}{n-2,2,0,\ldots,0}\left(\frac{x}{1!}\right)^2\left(\frac{x^2}{2!}\right)^0\cdots\left(\frac{x^r}{r!}\right)^0$$

$$+\cdots+\sum_{n_1+2n_2+\cdots rn_r=k}\binom{n}{n_0,\ldots,n_r}\left(\frac{x}{1!}\right)^{n_1}\cdots\left(\frac{x^r}{r!}\right)^{n_r}+\cdots$$

$$=1+\frac{n!}{(n-1)!}x+\frac{n!}{(n-1)!}\frac{x^2}{2}+\frac{n!}{(n-2)!2!}x^2+\cdots+C_kx^k+\cdots$$

$$=1+nx+\frac{nx^2}{2}+\frac{n(n-1)}{2}x^2+\cdots$$

$$=1+(nx)+\frac{(nx)^2}{2}+\cdots$$

which indeed looks like $f(nx)$. Of course, more work is required to actually determine that

$$C_k=\sum_{n_1+2n_2+\cdots rn_r=k}\binom{n}{n_0,\ldots,n_r}\prod_{i=1}^r(i!)^{-n_i}=\frac{n^k}{k!},$$

which might be amenable to induction, which we do not attempt. Just to check for $k=3$, the solutions to $n_1+2n_2+\cdots rn_r=3$ (with n_0 such that $\sum_{i=0}^r n_r=n$) are $(n_0,\ldots,n_r)=(n_0,3,0,\ldots,0)$, $(n_0,1,1,0,\ldots,0)$, and $(n_0,0,0,1,0,\ldots,0)$. Thus,

$$C_3=\frac{n!}{(n-3)!3!}\frac{1}{1^3}+\frac{n!}{(n-2)!}\frac{1}{1^1}\frac{1}{2^1}+\frac{n!}{(n-1)!}\frac{1}{6^1}=\frac{n^3}{3!}.\quad\blacksquare$$

Sequences and series of functions

It is true that a mathematician who is not also something of a poet will never be a perfect mathematician.

(Karl Weierstrass)

Up to this point, f_k represented a series. Now let $\{f_n(x)\}$ be a *sequence of functions* with the same domain, say D.

The function f is the *pointwise limit* of sequence $\{f_n\}$, or $\{f_n\}$ *converges pointwise* to f, if, $\forall\ x\in D$, $\lim_{n\to\infty}f_n(x)=f(x)$. That is, $\forall\ x\in D$ and for every given $\epsilon>0$, $\exists\ N\in\mathbb{N}$ such that $|f_n(x)-f(x)|<\epsilon$, $\forall n>N$. We will also denote this by writing $f_n\to f$. It is helpful to read $\forall\ x$ in the previous sentence as 'for each', and not 'for all', to emphasize that N depends on both x and ϵ.

Just as (infinite) series can be associated with sequences, consider the (infinite) *series (of functions)*, $\sum_{n=0}^{\infty}f_n(x)$, $x\in D$. If, for each $x\in D$, the sequence $s_i=\sum_{n=0}^i f_n(x)$ converges pointwise, then the series is said to *converge pointwise (on D)* to *(the function) $S=\sum_{n=0}^{\infty}f_n(x)$, $x\in D$.*

⊖ **Example A.46** (Stoll, 2001, p. 320) Let $f_k(x) = x^2 (1 + x^2)^{-k}$, for $x \in \mathbb{R}$ and $k = 0, 1, \ldots$, and observe that $f_k(x)$ is continuous. The series

$$S(x) := \sum_{k=0}^{\infty} f_k(x) = x^2 \sum_{k=0}^{\infty} \frac{1}{(1 + x^2)^k} = 1 + x^2$$

for $x \neq 0$, and $S(0) = 0$. Thus, $S(x)$ converges pointwise on $\mathbb{R}$ to the function $f(x) = (1 + x^2)\mathbb{I}(x \neq 0)$, and is not continuous at zero.   ∎

The above example shows that the pointwise limit may not be continuous even if each element in the sequence is continuous. Similarly, differentiability or integrability of f_n does not ensure that the pointwise limit shares that property; a standard example for the latter is

$$f_n(x) = nx(1 - x^2)^n, \quad x \in [0, 1], \tag{A.99}$$

with $\lim_{n \to \infty} f_n(x) = 0$ for $x \in (0, 1)$ and $f_n(0) = f_n(1) = 0$, so that the pointwise limit is $f(x) = 0$, but $\frac{1}{2} = \lim_{n \to \infty} \int_0^1 f_n \neq \int_0^1 f$ (see, for example, Stoll, 2001, p. 321; Estep, 2002, Section 33.3).

The function f is the *uniform limit* on D of the sequence $\{f_n\}$, (or $\{f_n\}$ is *uniformly convergent* to f), if, for every given $\epsilon > 0$, $\exists N \in \mathbb{N}$ such that $|f_n(x) - f(x)| < \epsilon$ for every $x \in D$ whenever $n > N$. The difference between pointwise and uniform convergence parallels that between continuity and uniform continuity; in the uniform limit, N is a function of ϵ, but not of x. In this case, we write $f_n \rightrightarrows f$. Sequence $\{f_n\}$ is said to be *uniformly Cauchy* if, for every given $\epsilon > 0$, $\exists N \in \mathbb{N}$ such that $|f_n(x) - f_m(x)| < \epsilon$ for every $n, m > N$ and every $x \in D$. Let $\{f_n\}$ be a sequence of functions. Important results include:

• if $f_n \rightrightarrows f$, then $f_n \to f$;

• $\{f_n\}$ is uniformly convergent iff $\{f_n\}$ is uniformly Cauchy;   (A. 100)

• if $f_n \in C^0(D)$, $\forall n \in \mathbb{N}$, and $f_n \rightrightarrows f$, then $f \in C^0(D)$;   (A. 101)

• if $f, f_n \in C^0[a, b]$, $\forall n \in \mathbb{N}$, where $[a, b] \subset D$ and $m_n = \max_x |f_n(x) - f(x)|$, then $f_n(x) \rightrightarrows f(x)$ on $[a, b]$ iff $\lim_{n \to \infty} m_n = 0$.   (A.102)

To prove (A.102), first assume $f_n(x) \rightrightarrows f(x)$ so that, by definition, $\forall \epsilon > 0$, $\exists N \in \mathbb{N}$ such that, $\forall x \in [a, b]$ and $\forall n > N$, $D_n(x) := |f_n(x) - f(x)| < \epsilon$. By assumption, f_n and f are continuous on $[a, b]$, so that $D_n(x)$ is as well and, from (A.13), $\exists x_n \in [a, b]$ such that $x_n = \arg\max_x D_n(x)$. Thus, $\forall n > N$, $m_n = D_n(x_n) = \max_x D_n(x) < \epsilon$, i.e. $\lim_{n \to \infty} m_n = 0$. Now assume $m_n = \max_x D_n(x) \to 0$ as $n \to \infty$. Then, $\forall \epsilon > 0$, $\exists N \in \mathbb{N}$ such that, $\forall n > N$, $m_n < \epsilon$, so that, $\forall x \in [a, b]$ and $n > N$, $D_n(x) = |f_n(x) - f(x)| \leq m_n < \epsilon$.   ∎

The sequence $\{f_n\}$ is said to be *monotone decreasing (increasing)* if, $\forall n \in \mathbb{N}$ and $\forall x \in D$, $f_{n+1}(x) \leq f_n(x)$ ($f_{n+1}(x) \geq f_n(x)$), and *monotone* if it is either monotone decreasing or monotone increasing. The monotonicity of f_n is key for pointwise convergence to imply uniform convergence:

- if (i) the $f_n : D \to \mathbb{R}$ are continuous, (ii) the f_n are monotone, (iii) D is a closed, bounded interval, and (iv) $f_n \to f$, $f \in C^0$, then $f_n \rightrightarrows f$ on D.   (A.103)
  For proof, see Browder (1996, p. 64).

Example A.47 Let $f_n(x) = x^n/n!$, $n = 0, 1, \ldots$ and $s_n(x) = \sum_{k=0}^{n} f_k(x)$. Then (i) $f_n(x)$, $s_n(x) \in C^0$ for $x \in \mathbb{R}$ and all $n \in \mathbb{N}$, (ii) for each $x \geq 0$, $s_n(x)$ is monotone increasing in n, (iii) $\forall r \in \mathbb{R}_{>0}$, $D = [0, r]$ is a closed, bounded interval, and (iv) from Example A.45, $s_n \to \exp(x)$ for all $x \in \mathbb{R}$. Thus, from (A.103), $\forall r \in \mathbb{R}_{>0}$ and $\forall x \in [0, r]$, $\lim_{n \to \infty} s_n(x) = \sum_{k=0}^{\infty} x^n/n! \rightrightarrows \exp(x)$. ∎

Let $\{f_n\}$ be a sequence of functions, each with domain D. The series $\sum_{n=1}^{\infty} f_n(x)$ is said to *converge uniformly (on D) to the function* S if the associated sequence of partial sums converges uniformly on D. In this case, we write $\sum_{n=1}^{\infty} f_n(x) \rightrightarrows S(x)$. For sequence $\{f_n(x)\}$ with domain D:

- if, $\forall n \in \mathbb{N}$, $f_n \in C^0$ and $\sum_{n=1}^{\infty} f_n(x) \rightrightarrows S(x)$, then $S \in C^0$;   (A.104)

- (Weierstrass M-test) *if there exists a sequence of constants M_n such that, $\forall x \in D$ and $\forall n \in \mathbb{N}$, $|f_n(x)| \leq M_n$, and $\sum_{n=1}^{\infty} M_n < \infty$, then $\sum_{n=1}^{\infty} f_n(x)$ is uniformly convergent on D.*

The Weierstrass M-test, after Karl Theodor Wilhelm Weierstrass (1815–1897), is proved by letting $S_n(x) = \sum_{k=1}^{n} f_k(x)$ be the nth partial sum of sequence f_k. Then, for $n > m$,

$$|S_n(x) - S_m(x)| = \left| \sum_{k=m+1}^{n} f_k(x) \right| \leq \sum_{k=m+1}^{n} |f_k(x)| \leq \sum_{k=m+1}^{n} M_k. \qquad (A.105)$$

As the series M_n is convergent, the Cauchy criterion (A.86) implies that the r.h.s. of (A.105) can be made arbitrarily close to zero. The result now follows from (A.100).

Example A.48 Let $f_n(x) = (-1)^n x^{2n}/(2n)!$, $x \in \mathbb{R}$. Then, $\forall L \in \mathbb{R}_{>0}$,

$$|f_n(x)| = \left| \frac{x^{2n}}{(2n)!} \right| \leq \left| \frac{L^{2n}}{(2n)!} \right| =: M_n, \qquad x \in D = [-L, L].$$

By the ratio test, $M_{n+1}/M_n = L^2/(2n+1)(2n+2) \to 0$ as $n \to \infty$, so that $\sum_{n=0}^{\infty} M_n < \infty$, and, from the Weierstrass M-test, $\sum_{n=0}^{\infty} f_n(x)$ converges uniformly on $D = [-L, L]$. This justifies the definitions

$$\cos(x) = \sum_{k=0}^{\infty} (-1)^k \frac{x^{2k}}{(2k)!} \qquad \text{and} \qquad \sin(x) = \sum_{k=0}^{\infty} (-1)^k \frac{x^{2k+1}}{(2k+1)!},$$

as in (A.28).[16] Next, for $x \neq 0$, we wish to know if

$$\frac{\sin x}{x} = \frac{1}{x} \sum_{n=0}^{\infty} \frac{(-1)^n x^{2n+1}}{(2n+1)!} \overset{?}{=} \sum_{n=0}^{\infty} \frac{(-1)^n x^{2n}}{(2n+1)!}.$$

As the series for $\sin x$ is convergent, it follows from the definition of convergence that, for any $\epsilon > 0$ and $x \neq 0$, $\exists\, N = N(x, \epsilon) \in \mathbb{N}$ such that, $\forall\, n > N$,

$$\left| \sin x - \sum_{n=0}^{N} \frac{(-1)^n x^{2n+1}}{(2n+1)!} \right| < \epsilon |x| \quad \text{or} \quad \left| \frac{\sin x}{x} - \sum_{n=0}^{N} \frac{1}{x} \frac{(-1)^n x^{2n+1}}{(2n+1)!} \right| < \epsilon,$$

so that, for $x \neq 0$,

$$\frac{\sin x}{x} = \sum_{n=0}^{\infty} \frac{(-1)^n x^{2n}}{(2n+1)!}. \tag{A.106}$$

With $0^0 = 1$, the r.h.s. equals 1 for $x = 0$, which coincides with the limit as $x \to 0$ of the l.h.s. As above, the Weierstrass M-test shows that this series is uniformly convergent on $[-L, L]$ for any $L \in \mathbb{R}_{>0}$. ∎

⊖ **Example A.49** (Example A.47 cont.) Again let $f_n(x) = x^n/n!$, $n = 0, 1, \ldots$, and $s_n(x) = \sum_{k=0}^{n} f_n(x)$. For all $L \in \mathbb{R}_{>0}$,

$$|f_n(x)| = \frac{|x^n|}{n!} \le \frac{L^n}{n!} =: M_n, \quad x \in [-L, L],$$

and $\sum_{n=0}^{\infty} M_n < \infty$ from Example A.45, which showed that $s_n(L) \to \exp(L)$ absolutely. Thus, the Weierstrass M-test implies that $\sum_{k=0}^{\infty} f_n(x)$ converges uniformly on $[-L, L]$, where L is an arbitrary positive real number. It is, however, *not* true that $\sum_{k=0}^{\infty} x^n/n!$ converges uniformly on $(-\infty, \infty)$. Also,

$$\sum_{r=0}^{\infty} \frac{(-z \ln z)^r}{r!} \rightrightarrows e^{-z \ln z}, \tag{A.107}$$

by taking $x = z \ln z \in \mathbb{R}$. ∎

⊙ **Example A.50** Let $f_k(x) = (x/R)^k =: a^k$ for a fixed $R \in \mathbb{R}_{>0}$. Then $G(x) = \sum_{k=1}^{\infty} f_k(x)$ is a geometric series which converges to $S(x) = (x/R)/(1 - x/R)$ for $|x/R| < 1$, or $|x| < R$. Let S_n be the partial sum of sequence $\{f_k\}$. For $G(x)$ to converge

---

[16] Observe that, while $\lim_{n \to \infty} x^{2n}/(2n)! = 0$ for all $x \in \mathbb{R}$, for any fixed n, $\lim_{x \to \infty} x^{2n}/(2n)! = \infty$. This means that, although the series converges, evaluation of the truncated sum will be numerically problematic because of the limited precision with which numbers are digitally stored. Of course, the relations $\cos(x + \pi) = -\cos x$, $\cos(x + 2\pi) = \cos x$ and $\cos(-x) = \cos x$; and $\sin(x + \pi) = -\sin x$, $\sin(x + 2\pi) = \sin x$ and $\sin(-x) = -\sin x$; and $\sin x = -\cos(x + \pi/2)$, imply that only the series for cosine is required, with x restricted to $[0, \pi/2]$. For $x \in [0, \pi/2]$, it is enough to sum the $\cos x$ series up to $n = 10$ to ensure 15 digit accuracy.

uniformly to $S(x)$, it must be the case that, for any $\epsilon > 0$, there is a value $N \in \mathbb{N}$ such that, $\forall\, n > N$,

$$|S_n - G| = \left| \sum_{k=1}^{n} \left(\frac{x}{R}\right)^k - \frac{x/R}{1 - x/R} \right| = \left| \frac{a - a^{n+1}}{1 - a} - \frac{a}{1 - a} \right| = \left| \frac{a^{n+1}}{1 - a} \right| < \epsilon. \quad \text{(A.108)}$$

However, for any n,

$$\lim_{x \to R^-} \frac{a^{n+1}}{1 - a} = \lim_{a \to 1^-} \frac{a^{n+1}}{1 - a} = \infty,$$

so that the inequality in (A.108) cannot hold. Now choose a value b such that $0 < b < R$ and let $M_k = (b/R)^k$, so that $\sum_{k=1}^{\infty} M_k = (b/R)/(1 - b/R) < \infty$. Then, for $|x| \le b$, $\left|(x/R)^k\right| \le (b/R)^k = M_k$, and use of the Weierstrass M-test shows that the series $G(x)$ converges uniformly on $[-b, b]$ to $S(x)$ (see also Example A.61). ∎

Remark:
(a) When using the Maple engine which accompanies Scientific Workplace 4.0, evaluating $\lim_{n \to \infty} \sum_{k=1}^{n} (-1)^k$ yields the interval $-1..0$, yet evaluating $\sum_{k=1}^{\infty} (-1)^k$ produces $-1/2$. Presumably, the latter result is obtained because Maple computes $\lim_{x \to -1+} \sum_{k=1}^{\infty} x^k = \lim_{x \to -1+} x/(1 - x) = -1/2$, which is itself correct, but

$$\sum_{k=1}^{\infty} (-1)^k \ne \lim_{x \to -1+} \sum_{k=1}^{\infty} x^k.$$

From (A.104), this would be true if $\sum_{k=1}^{\infty} x^k$ were uniformly convergent for $x = -1$, which it is not. While this is probably a mistake in Maple, it need not be one, as the next remark shows.

(b) The series $\sum_{k=1}^{\infty} a_k$ is said to be *Abel summable* to L if $\lim_{x \to 1-} f(x) = L$, where $f(x) = \sum_{k=1}^{\infty} a_k x^k$ for $0 \le x < 1$ (after Neils Henrik Abel, 1802–1829; see Goldberg, 1964, p. 251). For example, with $a_k = (-1)^k$, the series $f(x) = \sum_{k=1}^{\infty} a_k x^k = -x + x^2 - x^3 + \cdots$ converges for $|x| < 1$ to $-x/(x + 1)$. Then the series $-1 + 1 - 1 + 1 - \cdots$ is clearly divergent, but is Abel summable to $\lim_{x \to 1-} f(x) = -1/2$. ∎

⊖ ***Example A.51*** The contrapositive of (A.104) implies that, $f_n \in C^0$, $f \notin C^0 \Rightarrow \sum_{n=1}^{\infty} f_n(x) \not\rightrightarrows f$. In words, if the f_n are continuous but f is not, then $\sum_{n=1}^{\infty} f_n(x)$ is not uniformly convergent to f. In Example A.46, $f(x)$ is not continuous at $x = 0$, so that $\sum_{k=0}^{\infty} f_k(x)$ is not uniformly convergent on any interval containing zero. ∎

Recall from (A.101) and (A.104) that uniform convergence of sequences and series of continuous functions implies continuity of the limiting function. A similar result holds for integrability of sequences and series:

• *if,* $\forall\, n \in \mathbb{N}$, $f_n \in \mathcal{R}[a, b]$ *and* $f_n(x) \rightrightarrows f(x)$ *on* $[a, b]$, *then* $f \in \mathcal{R}[a, b]$ *and*

$$\lim_{n \to \infty} \int_a^b f_n(x)\, dx = \int_a^b f(x)\, dx. \quad \text{(A.109)}$$

The proof is worth showing, as it clearly demonstrates why the uniform convergence assumption is necessary. Let $\epsilon_n = \max_{x \in [a,b]} |f_n(x) - f(x)|$. Uniform convergence implies that $\lim_{n \to \infty} \epsilon_n = 0$ using (A.102), and that $f_n(x) - \epsilon_n \leq f(x) \leq f_n(x) + \epsilon_n$, i.e. $\forall n \in \mathbb{N}$,

$$\int_a^b \left[f_n(x) - \epsilon_n \right] dx \leq \sup_\pi \underline{S}(f, \pi) \leq \inf_\pi \overline{S}(f, \pi) \leq \int_a^b \left[f_n(x) + \epsilon_n \right] dx, \quad (A.110)$$

recalling the definitions of $\underline{S}(f, \pi)$ and $\overline{S}(f, \pi)$ from Section A.2.3. This implies

$$0 \leq \inf_\pi \overline{S}(f, \pi) - \sup_\pi \underline{S}(f, \pi) \leq \int_a^b \left[f_n(x) + \epsilon_n \right] dx - \int_a^b \left[f_n(x) - \epsilon_n \right] dx$$

$$= \int_a^b 2\epsilon_n dx,$$

but $\lim_{n \to \infty} \epsilon_n = 0$, so that $\lim_{n \to \infty} \left[\inf_\pi \overline{S}(f, \pi) - \sup_\pi \underline{S}(f, \pi) \right] = 0$, and $f \in \mathcal{R}[a, b]$. Now we can write

$$\left| \int_a^b f_n(x) \, dx - \int_a^b f(x) \, dx \right| = \left| \int_a^b \left[f_n(x) - f(x) \right] dx \right| = \left| \int_a^b \epsilon_n dx \right| = \epsilon_n (b - a),$$

and taking the limit yields (A.109). ∎

- If $\forall n \in \mathbb{N}$, $f_n \in \mathcal{R}[a, b]$ and $\sum_{n=1}^\infty f_n(x) \rightrightarrows S(x)$ for $x \in [a, b]$, then $S \in \mathcal{R}[a, b]$ and

$$\int_a^b S(x) \, dx = \sum_{n=1}^\infty \int_a^b f_n(x) \, dx. \quad (A.111)$$

To see this, let $S_n(x) = \sum_{k=1}^n f_k(x)$, so that, $\forall n \in \mathbb{N}$, $S_n \in \mathcal{R}[a, b]$. From the previous result, $S \in \mathcal{R}[a, b]$ and, from (A.109) applied to $S(x)$ and $S_n(x)$,

$$\int_a^b \sum_{k=1}^\infty f_k(x) \, dx = \int_a^b S(x) \, dx = \lim_{n \to \infty} \int_a^b S_n(x) \, dx = \lim_{n \to \infty} \sum_{k=1}^n \int_a^b f_k(x) \, dx,$$

$$(A.112)$$

which is (A.111). ∎

⊖ **Example A.52** Recall from Example A.7 that $\lim_{x \to 0} x^x = 1$. The integral $I = \int_0^1 x^{-x} dx$ was shown to be equal to $\sum_{r=1}^\infty r^{-r}$ by Johann Bernoulli in 1697. To see this, as in Havil (2003, p. 44), write

$$I = \int_0^1 e^{-x \ln x} dx = \int_0^1 \sum_{r=0}^\infty \frac{(-x \ln x)^r}{r!} dx = \sum_{r=0}^\infty \frac{(-1)^r}{r!} \int_0^1 (x \ln x)^r \, dx,$$

where the exchange of sum and integral is justified by (A.107) and (A.111). The result now follows from (A.65), i.e. $\int_0^1 (x \ln x)^r \, dx = (-1)^r r! \, / \, (r+1)^{r+1}$, or

$$I = \sum_{r=0}^{\infty} \frac{(-1)^r}{r!} \frac{(-1)^r r!}{(r+1)^{r+1}} = \sum_{r=0}^{\infty} \frac{1}{(r+1)^{r+1}}. \qquad \blacksquare$$

◎ **Example A.53**   (Browder, 1996, p. 113) Let $f_n(x) = x/n(x+n)$ for $x \in [0,1]$ and $n \in \mathbb{N}$. Clearly, $0 \le f_n(x) \le 1/n(n+1)$ and, from the comparison test with $g_n = n^{-2}$, $\sum_{n=1}^{\infty} [n(n+1)]^{-1}$ converges. This series has the 'telescoping property', i.e.

$$\frac{1}{n(n+1)} = \frac{1}{n} - \frac{1}{n+1}, \qquad \text{so that} \qquad \sum_{k=1}^{n} \frac{1}{k(k+1)} = \frac{n}{n+1} \to 1.$$

Thus, from the Weierstrass M-test, $\sum_{n=1}^{\infty} f_n(x)$ converges uniformly on $[0,1]$ to a function, say $S(x)$, which, by (A.104), is continuous and, clearly, $0 \le S(x) \le 1$. From (A.111),

$$\sum_{n=1}^{\infty} \int_0^1 f_n(x) \, dx = \int_0^1 S(x) \, dx =: \gamma, \qquad (A.113)$$

and $\int_0^1 S(x) \, dx < \int_0^1 dx = 1$, so that $0 < \gamma < 1$. As

$$\int_0^1 f_n = \int_0^1 \left(\frac{1}{n} - \frac{1}{x+n} \right) dx = \frac{1}{n} - \ln \frac{n+1}{n},$$

(A.113) implies $\gamma = \lim_{N \to \infty} \sum_{n=1}^{N} \int_0^1 f_n(x) \, dx = \lim_{N \to \infty} \left[\sum_{n=1}^{N} n^{-1} - \ln(N+1) \right]$ or, as $\lim_{N \to \infty} [\ln(N+1) - \ln N] = 0$,

$$\gamma = \lim_{N \to \infty} \left(\sum_{n=1}^{N} \frac{1}{n} - \ln N \right),$$

which is referred to as *Euler's constant* (see also Example A.38). $\qquad \blacksquare$

◎ **Example A.54**   Let $S(x) = 1 - x + x^2 - x^3 + \cdots$. Then, for $-1 < x \le 0$, $S(x) = 1 + y + y^2 + \cdots$, where $y = -x$, and so converges to $1/(1-y) = 1/(1+x)$. For $0 \le x < 1$, the alternating series test shows that $S(x)$ converges, and it is easy to verify that it converges to $1/(1+x)$, so that $S(x)$ converges to $1/(1+x)$ for $|x| < 1$. Similar to the derivation in Example A.50, for every $b \in [0,1)$, $S(x)$ is uniformly convergent for $x \in [-b, b]$. So, from (A.111),

$$\int_0^b \frac{1}{1+x} \, dx = \int_0^b 1 \, dx - \int_0^b x \, dx + \int_0^b x^2 \, dx - \cdots.$$

For the first integral, let $u = 1 + x$ so that $\int_0^b (1 + x)^{-1} \, dx = \int_1^{b+1} u^{-1} \, du = \ln(1 + b)$. Thus,

$$\ln(1 + b) = b - \frac{b^2}{2} + \frac{b^3}{3} - \cdots, \qquad |b| < 1. \tag{A.114}$$

Example A.64 will show that (A.114) also holds for $b = 1$.   ∎

⊙ ***Example A.55***   Similar to Example A.54,

$$\frac{1}{1 + y^2} = 1 - y^2 + y^4 - y^6 + \cdots, \qquad |y| < 1, \tag{A.115}$$

and, for every $b \in [0, 1)$, the r.h.s. is uniformly convergent for $y \in [-b, b]$, so that termwise integration of the r.h.s. is permitted. From (A.33) and the FTC (A.60),

$$\int_0^t \frac{1}{1 + y^2} \, dy = \arctan(t) - \arctan(0) = \arctan(t), \tag{A.116}$$

so that

$$\arctan(t) = t - \frac{t^3}{3} + \frac{t^5}{5} - \cdots = \sum_{n=0}^{\infty} \frac{(-1)^n}{2n + 1} t^{2n+1}, \qquad t \in [0, 1), \tag{A.117}$$

from which $\arctan(t)$, $t \in [0, 1)$, can be computed to any degree of accuracy. Example A.65 below considers the case when $t = 1$.   ∎

Another useful result is the *bounded convergence theorem*, which involves the interchange of limit and integral using only pointwise convergence, but requires that f can also be integrable on $I = [a, b]$, and that the f_n are bounded for all n and all $x \in I$.

• *If, $\forall \, n \in \mathbb{N}$, $f_n \in \mathcal{R}[a, b]$ with $f_n(x) \to f(x)$, $f \in \mathcal{R}[a, b]$, and $\exists \, M \in \mathbb{R}_{>0}$ such that, $\forall \, x \in [a, b]$ and $\forall \, n \in \mathbb{N}$, $|f_n(x)| \le M$, then*

$$\lim_{n \to \infty} \int_a^b f_n(x) \, dx = \int_a^b f(x) \, dx. \tag{A.118}$$

This is explicitly proven in Stoll (2001, Section 10.6) and is a special case of *Lebesgue's dominated convergence theorem*, detailed in, for example, Browder (1996, Section 10.2), Stoll (2001, Section 10.7), and Pugh (2002, Section 6.4). Paralleling results (A.109) and (A.111), let $\{f_n(x)\}$ be a sequence of functions with nth partial sum $S_n(x)$, such that $S_n(x) \to S(x)$. If the conditions leading to (A.118) apply to $\{S_n(x)\}$ and $S(x)$, then (A.111) also holds, the proof of which is the same as (A.112).

⊖ ***Example A.56***   With $f_n(x) = nx(1 - x^2)^n$ as in (A.99), use of a symbolic software package easily shows that $f_n'(x) = -n(1 - x^2)^{n-1}(x^2(1 + 2n) - 1)$ and solving $f_n'(x_m) = 0$ yields $x_m = (1 + 2n)^{-1/2}$, so that

$$f_n(x_m) = \frac{n}{\sqrt{1 + 2n}}\left(\frac{2n}{1 + 2n}\right)^n \quad \text{and} \quad \lim_{n \to \infty} f_n(x_m) = \infty.$$

Thus, $\nexists M$ such that, $\forall\, x \in [0, 1]$ and $\forall\, n \in \mathbb{N}$, $|f_n(x)| \leq M$, and the contrapositive of the bounded convergence theorem implies that $\lim_{n\to\infty} \int_0^1 f_n \neq \int_0^1 f$, as was previously determined with a direct calculation of the integrals. ∎

⊖ **Example A.57** [17] For a fixed $x \in \mathbb{R}_{>0}$ and all $t \in \mathbb{R}$, define

$$h(t) := \frac{\exp\left[-x\left(1 + t^2\right)\right]}{1 + t^2} = e^{-x}\,\frac{e^{-xt^2}}{1 + t^2}, \qquad (A.119)$$

which is the integrand in (A.80). Interest centers on developing formulae for $\int_0^1 h$.
   Method 1. From (A.98) with x replaced by $-xt^2$,

$$h(t) = \frac{e^{-x}}{1 + t^2}\left(1 - xt^2 + \frac{x^2 t^4}{2!} - \cdots\right) = e^{-x}\sum_{k=0}^{\infty}(-1)^k\,\frac{x^k}{k!}\,\frac{t^{2k}}{1 + t^2},$$

and termwise integration gives

$$\int_0^1 h(t)\,dt = e^{-x}\sum_{k=0}^{\infty}(-1)^k\,\frac{x^k}{k!}\,J_k,$$

where

$$J_k := \int_0^1 \frac{t^{2k}}{1 + t^2}\,dt, \qquad k \in \mathbb{N},$$

which seems resilient to use of a transformation or integration by parts. It is, however, quite simple, with the following 'trick':

$$\int \frac{t^{2k}}{1 + t^2}\,dt = \int \frac{t^{2k-2}(t^2 + 1 - 1)}{1 + t^2}\,dt = \int \frac{t^{2k-2}(1 + t^2)}{1 + t^2}\,dt - \int \frac{t^{2k-2}}{1 + t^2}\,dt,$$

giving the recursion

$$\int_a^b \frac{t^{2k}}{1 + t^2}\,dt = \left.\frac{t^{2k-1}}{2k - 1}\right|_a^b - \int_a^b \frac{t^{2k-2}}{1 + t^2}\,dt.$$

With $a = 0$ and $b = 1$,

$$J_k = \frac{1}{2k - 1} - J_{k-1}. \qquad (A.120)$$

From (A.81), $J_0 = \pi/4$, and iterating (A.120) gives

$$J_1 = 1 - \pi/4, \quad J_2 = 1/3 - 1 + \pi/4, \quad J_3 = 1/5 - 1/3 + 1 - \pi/4$$

---

[17] This example was contributed by Walther Paravicini.

and the general formula

$$J_k = (-1)^k \left(\sum_{m=1}^{k} \frac{(-1)^m}{2m - 1} + \pi/4 \right),$$

so that

$$\int_0^1 h(t) \, dt = e^{-x} \sum_{k=0}^{\infty} \frac{x^k}{k!} \left(\sum_{m=1}^{k} \frac{(-1)^m}{2m - 1} + \pi/4 \right). \tag{A.121}$$

This sum converges very fast because of the $k!$ in the denominator. To illustrate, take $x = 0.3$. Accurate numeric integration of h, and also evaluation of (A.121) truncating the infinite sum at $U = 200$, gives 0.5378448777, which we will deem correct to 10 digits. With only $U = 4$, (A.121) yields 0.5378456, accurate to five digits.

   Method 2. We make the *ansatz* that $h(t)$ can be expressed as the series $h(t) = \sum_{k=0}^{\infty} a_k t^{2k}$, and calculate the a_k. With $j = k - 1$,

$$e^{-xt^2} = e^x(1 + t^2) \sum_{k=0}^{\infty} a_k t^{2k} = e^x \left(a_0 + \sum_{k=1}^{\infty} a_k t^{2k} + \sum_{k=0}^{\infty} a_k t^{2k+2} \right)$$

$$= e^x \left(a_0 + \sum_{j=0}^{\infty} a_{j+1} t^{2j+2} + \sum_{k=0}^{\infty} a_k t^{2k+2} \right) = e^x \left[a_0 + \sum_{j=0}^{\infty} (a_{j+1} + a_j) t^{2j+2} \right]$$

$$= e^x a_0 + e^x \sum_{k=1}^{\infty} (a_k + a_{k-1}) t^{2k}.$$

With $t = 0$, it follows immediately that $a_0 = e^{-x}$. By comparison with $\exp(-xt^2) = \sum_{k=0}^{\infty} (-xt^2)^k / k!$, we see that

$$e^x(a_k + a_{k-1}) = \frac{(-1)^k}{k!} x^k, \qquad k = 1, 2, \ldots .$$

Iterating on

$$e^x a_k = -e^x a_{k-1} + \frac{(-1)^k}{k!} x^k$$

with $a_0 = e^{-x}$ gives

$$e^x a_1 = -1 + \frac{(-1)^1}{1!} x^1 = -\left(1 + \frac{x^1}{1!} \right),$$

$$e^x a_2 = +\left(1 + \frac{x^1}{1!} \right) + \frac{(-1)^2}{2!} x^2 = +\left(1 + \frac{x^1}{1!} + \frac{x^2}{2!} \right),$$

and, in general,

$$e^x a_k = (-1)^k \left(1 + \frac{x^1}{1!} + \frac{x^2}{2!} + \cdots + \frac{x^k}{k!} \right) = (-1)^k \sum_{j=0}^{k} \frac{x^j}{j!}.$$

Thus,

$$h(t) = \sum_{k=0}^{\infty} a_k t^{2k} = e^{-x} \sum_{k=0}^{\infty} (-1)^k \left(\sum_{j=0}^{k} \frac{x^j}{j!} \right) t^{2k}$$

and, as $\int_0^1 t^{2k} dt = 1/(2k+1)$,

$$\int_0^1 h(t) \, dt = e^{-x} \sum_{k=0}^{\infty} (-1)^k \frac{1}{2k+1} \left(\sum_{j=0}^{k} \frac{x^j}{j!} \right). \tag{A.122}$$

Whereas (A.121) has a $k!$ in the denominator, (A.122) has only $2k+1$, so we expect it to converge much slower. Indeed, with $x = 0.3$, the use of 1000 terms in the sum results in 0.538 09, which is correct only to three digits. The formula is useless for numeric purposes.

Method 3. Expanding the enumerator of (A.119) as a power series in $-x \left(1 + t^2\right)$ gives

$$h(t) = \frac{1}{1+t^2} \sum_{k=0}^{\infty} (-1)^k \frac{x^k}{k!} \left(1 + t^2\right)^k = \sum_{k=0}^{\infty} (-1)^k \frac{x^k}{k!} \left(1 + t^2\right)^{k-1},$$

so that

$$\int_0^1 h(t) \, dt = \sum_{k=0}^{\infty} (-1)^k \frac{x^k}{k!} I_k,$$

where $I_k := \int_0^1 (1 + t^2)^{k-1} \, dt$. From (A.81), $I_0 = \pi/4$, and for $k > 0$, use of the binomial formula gives

$$I_k = \int_0^1 \sum_{m=0}^{k-1} \binom{k-1}{m} t^{2m} \, dt = \sum_{m=0}^{k-1} \binom{k-1}{m} \frac{1}{2m+1},$$

yielding

$$\int_0^1 h(t) \, dt = \frac{\pi}{4} + \sum_{k=1}^{\infty} (-1)^k \frac{x^k}{k!} \sum_{m=0}^{k-1} \binom{k-1}{m} \frac{1}{2m+1}. \tag{A.123}$$

Like (A.121), (A.123) converges fast: with $x = 0.3$, truncating the infinite sum at $U = 6$ gives 0.5378453, which is accurate to five digits. Based on this value of x, it appears that (A.121) converges the fastest. ∎

The following example makes use of the *gamma function*, which is described in Section 1.5.1.

⊖ **Example A.58**   Consider evaluating the improper integral $\int_0^\infty e^{-sx} x^{-1} \sin x \, dx$ for $s \in \mathbb{R}_{>1}$. From (A.106) and assuming that $x > 0$,

$$\frac{\sin x}{x} = \sum_{n=0}^\infty (-1)^n \frac{x^{2n}}{(2n+1)!} < \sum_{n=0}^\infty \frac{x^{2n}}{(2n+1)!} < \sum_{n=0}^\infty \frac{x^{2n}}{(2n)!} < \sum_{n=0}^\infty \frac{x^n}{(n)!} = e^x,$$

so that, as $x > 0$ and $s > 1$,

$$e^{-sx} \frac{\sin x}{x} < e^{-sx} e^x = e^{-x(s-1)} < e^0 = 1.$$

The conditions in the bounded convergence theorem are fulfilled, and termwise integration can be performed. Using (1.41), (1.38) and (A.117), this gives

$$\int_0^\infty e^{-sx} \frac{\sin x}{x} dx = \sum_{n=0}^\infty \frac{(-1)^n}{(2n+1)!} \int_0^\infty e^{-sx} x^{2n} dx = \sum_{n=0}^\infty \frac{(-1)^n}{(2n+1)!} \frac{\Gamma(2n+1)}{s^{2n+1}}$$

$$= \sum_{n=0}^\infty \frac{(-1)^n}{2n+1} \left(\frac{1}{s}\right)^{2n+1} = \arctan\left(s^{-1}\right), \tag{A.124}$$

which is required in Example A.30.   ■

We now turn to the conditions which allow for interchange of limits and differentiation, beginning with some illustrations of what conditions are *not* sufficient.

⊖ **Example A.59**   Let $f_n(x) = (\sin nx)/n$ so that, $\forall x \in \mathbb{R}$, $\lim_{n\to\infty} f_n(x) = 0 =: f(x)$. Then $f'(x) = 0$ but $f'_n(x) = \cos nx$ and, $\forall n \in \mathbb{N}$, $f'_n(0) = 1$, so that $\exists x \in \mathbb{R}$ such that $\lim_{n\to\infty} \frac{d}{dx} f_n(x) \neq \frac{d}{dx} \lim_{n\to\infty} f_n(x)$. Given the previous results on interchange of limit and integral, one might expect that uniform convergence is sufficient. However, $\forall x \in \mathbb{R}$,

$$|f_n(x) - f_m(x)| = \left| \frac{\sin nx}{n} - \frac{\sin mx}{m} \right| \leq \left| \frac{1}{n} - \frac{-1}{m} \right| = \left| \frac{1}{n} + \frac{1}{m} \right|,$$

so $\forall \epsilon > 0$, $\exists N \in \mathbb{N}$ such that, $\forall n, m > N$, $|f_n(x) - f_m(x)| < \epsilon$, i.e. f_n is uniformly Cauchy and, by (A.100), f_n is uniformly convergent. Thus, uniform convergence is not enough to ensure the interchange of limit and derivative.

Observe that $f'_n(x) = \cos nx$ is not convergent (pointwise or uniformly). It turns out that uniform convergence of $\{f'_n\}$ is necessary for interchange.   ■

⊖ **Example A.60**   Let $I = [-1, 1]$ and $f_n(x) = \left(x^2 + n^{-1}\right)^{1/2}$ for $x \in I$, so that $f_n \in \mathcal{C}^1$ with $f'_n(x) = x \left(x^2 + n^{-1}\right)^{-1/2}$. Figure A.5 shows f_n and f'_n for several n. In the limit, $f_n(x) \to f(x) := |x|$, which is not differentiable at $x = 0$. In fact, $f_n(x) \rightrightarrows f(x)$,

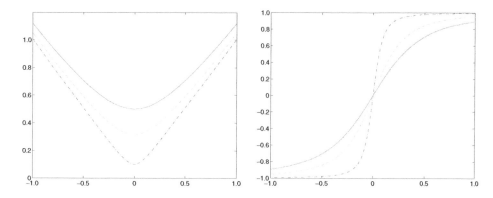

Figure A.5 Function (a) $f_n(x) = \left(x^2 + n^{-1}\right)^{1/2}$ and (b) f_n', for $n = 4$ (solid), $n = 10$ (dashed) and $n = 100$ (dash–dot)

because $m_n = \max_{x \in I} |f_n(x) - f(x)| = n^{-1/2}$ and[18] result (A.102). Also, $f_n'(x) \to x/|x|$ for $x \neq 0$, but the convergence cannot be uniform at $x = 0$, because, for any $n \in \mathbb{N}$,

$$\lim_{x \to 0^+} \left| f_n'(x) - \frac{x}{|x|} \right| = \lim_{x \to 0^+} \left| \frac{x}{\sqrt{x^2 + n^{-1}}} - \frac{x}{|x|} \right| = \lim_{x \to 0^+} \frac{x}{|x|} - \lim_{x \to 0^+} \frac{x}{\sqrt{x^2 + n^{-1}}}$$

$$= 1 - \lim_{x \to 0^+} \frac{x/x}{\sqrt{x^2/x^2 + 1/nx^2}} = 1 - 0 = 1. \qquad \blacksquare$$

Let $f : D \to \mathbb{R}$, where $I = [a, b] \subset D$. The following theorem gives the desired result, recalling that $\mathcal{C}^1$ is the class of continuously differentiable functions.

- Let $f_n \in \mathcal{C}^1(I)$ such that $f_n'(x) \rightrightarrows g(x)$ and $f_n(x) \to f(x)$ on I. Then $g \in \mathcal{C}^0(I)$ and, $\forall x \in I$, $f'(x) = g(x)$. (A.125)

Proof: That $g \in \mathcal{C}^0$ follows directly from (A.101). For $a, x \in I$, (A.52) and (A.60) imply that, $\forall n \in \mathbb{N}$, $\int_a^x f_n' = f_n(x) - f_n(a)$. As $f_n'(x) \rightrightarrows g(x)$, taking the limit as $n \to \infty$ and using (A.109) gives $\int_a^x g = f(x) - f(a)$. As $g \in \mathcal{C}^0$, differentiating this via (A.62) yields $g(x) = f'(x)$. $\blacksquare$

The assumptions in result (A.125) can be somewhat relaxed, though we will not require it. In particular, as proven, for example, in Stoll (2001, pp. 340–1):

---

[18] To derive m_n, first note that $f_n(x) - f(x)$ is symmetric in x, so we can restrict attention to $x \in [0, 1]$, in which case $d(x) = |f_n(x) - f(x)| = \left(x^2 + n^{-1}\right)^{1/2} - x > 0$, for $x \in [0, 1]$. Its first derivative is $d'(x) = x\left(x^2 + n^{-1}\right)^{-1/2} - 1$, which is strictly negative for all $x \in [0, 1]$ and $n \in \mathbb{N}$. (At $x = 1$, $d'(x) = \sqrt{n/(n+1)} - 1$.) Thus, $d(x)$ reaches its maximum on $[0, 1]$ at $x = 0$, so that

$$\max_{x \in [0,1]} |d(x)| = \max_{x \in I} |f_n(x) - f(x)| = d(0) = n^{-1/2}.$$

- Let $\{f_n\}$ *be a sequence of differentiable functions on* $I = [a, b]$. *If* $f_n'(x) \rightrightarrows g(x)$ *on* I *and* $\exists\, x_0 \in I$ *such that* $\{f_n(x_0)\}$ *converges, then* $f_n(x) \rightrightarrows f(x)$ *on* I, *and,* $\forall\, x \in I$, $f'(x) = g(x)$.

◎　***Example A.61***　Again consider the geometric series $\sum_{k=1}^{\infty} x^k$, which converges for $x \in (-1, 1)$ and then equals $x/(1-x)$. Let $S_n(x) = \sum_{k=1}^{n} x^k$ be the nth partial sum, with $S_n'(x) = \sum_{k=1}^{n} k x^{k-1}$. From the binomial theorem (1.18),

$$(r + \epsilon)^k = \sum_{i=0}^{k} \binom{k}{i} \epsilon^i r^{k-i} = r^k + k \epsilon r^{k-1} + \cdots + \epsilon^k, \tag{A.126}$$

and the positivity of the terms on the r.h.s. of (A.126) imply that $k \epsilon r^{k-1} < (r + \epsilon)^k$, so that the Weierstrass M-test implies that $\sum_{k=1}^{\infty} k \epsilon r^{k-1} = \epsilon \sum_{k=1}^{\infty} k r^{k-1}$ is also uniformly convergent for $x \in [-r, r]$. Thus, from (A.125) with $f_n = S_n$,

$$S_n'(x) = \sum_{k=1}^{n} k x^{k-1} \rightrightarrows \frac{\mathrm{d}}{\mathrm{d}x} \left(\lim_{n \to \infty} \sum_{k=1}^{n} x^k \right) = \frac{\mathrm{d}}{\mathrm{d}x} \left(\frac{x}{1-x} \right) = \frac{1}{(1-x)^2}. \tag{A.127}$$

As this holds $\forall\, x \in [-r, r]$, where r is an arbitrary number from the open interval $(0, 1)$, (A.127) holds for all $|x| < 1$. (If this were not true, then there would exist an $x \in (0, 1)$, say x_0, for which it were not true, but the previous analysis applies to all $x \in [0, x_0 + \epsilon]$, where ϵ is such that $x_0 + \epsilon < 1$, which always exists.)

To add some intuition and informality, let

$$S = \lim_{n \to \infty} \sum_{k=1}^{n} k x^{k-1} = 1 + 2x + 3x^2 + \cdots \quad \text{and} \quad xS = x + 2x^2 + 3x^3 + \cdots,$$

so that $S - xS = 1 + x + x^2 + \cdots = \sum_{k=0}^{\infty} x^k = (1-x)^{-1}$, which, for $|x| < 1$, converges, and $S = (1-x)^{-2}$ (also see the next section). ∎

Power and Taylor series

> I regard as quite useless the reading of large treatises of pure analysis: too large a number of methods pass at once before the eyes. It is in the works of applications that one must study them.
>
> (Joseph-Louis Lagrange)

A series of the form $\sum_{k=0}^{\infty} a_k x^k$ for sequence $\{a_k\}$ is a *power series* in x with coefficients a_k. More generally, $S(x) = \sum_{k=0}^{\infty} a_k (x-c)^k$ is a power series in $(x-c)$, where $c \in \mathbb{R}$. With $f_k = a_k (x-c)^k$, the exponential growth rate of S is $g(x) = \limsup |f_k|^{1/k} = |x-c| \limsup |a_k|^{1/k}$, and, from the root test, S converges absolutely if $g(x) < 1$ and diverges for $g(x) > 1$. The *radius of convergence* of S is $R = 1/\limsup |a_k|^{1/k}$, with $1/0 = \infty$ i.e. $R = \infty$ if $\limsup |a_k|^{1/k} = 0$, and $1/\infty = 0$. Thus, we can say that $S(x)$ converges if $g(x) = |x-c| R^{-1} < 1$, or $|x-c| < R$, and diverges if $|x-c| > R$.

⊙ **Example A.62** Particularly when working with series involving factorials, the following result is quite useful. If $\lim_{k \to \infty} |a_{k+1}/a_k|$ exists, then $\lim \inf |a_{k+1}/a_k| = \lim \sup |a_{k+1}/a_k|$ and, from (A.92), these equal $\lim \sup |a_k|^{1/k}$, so that the radius of convergence is $R = 1/\lim_{k \to \infty} |a_{k+1}/a_k|$. For example, consider the power series of the form

$$S(x) = \sum_{n=0}^{\infty} a_k x^k, \quad \text{where} \quad a_k = \frac{(-1)^k}{m^k (k!)^p} \quad \text{and} \quad m, p \in \mathbb{R}_{>0}.$$

As

$$\lim_{k \to \infty} \frac{|a_{k+1}|}{|a_k|} = \lim_{k \to \infty} \frac{m^k (k!)^p}{m^{k+1} [(k+1)!]^p} = \lim_{k \to \infty} \frac{1}{m (1+k)^p} = 0,$$

we have $R = \infty$. ∎

- *If power series S has radius of convergence $R > 0$, then, $\forall b \in (0, R)$, S converges uniformly for all x with $|x - c| \leq b$.* (A.128)

This is easily proved as follows. Choose $\epsilon > 0$ such that $b + \epsilon \in (b, R)$, which implies $\lim \sup |a_k|^{1/k} = R^{-1} < (b + \epsilon)^{-1}$. From (A.84), $\exists N \in \mathbb{N}$ such that, $\forall n \geq N, |a_k|^{1/k} < (b + \epsilon)^{-1}$, so that, $\forall n \geq N$ and $|x - c| \leq b$, $|a_k (x - c)^k| \leq |a_k| b^k < (b/(b + \epsilon))^k$. As $\sum_{k=1}^{\infty} [b/(b + \epsilon)]^k < \infty$, the result follows from the Weierstrass M-test. ∎

⊙ **Example A.63** In Example A.61, the uniform convergence of $\sum_{k=1}^{\infty} k x^{k-1}$ was shown via the binomial theorem and the Weierstrass M-test. The following way is easier. As $\sum_{k=1}^{\infty} k x^{k-1} = \sum_{j=0}^{\infty} (j+1) x^j$, let $a_j = j + 1$ so that, from Example A.8, $\lim \sup |a_j|^{1/j} = \lim_{j \to \infty} (j+1)^{1/j} = 1$, and $R = 1$. Thus, $\sum_{k=1}^{\infty} k x^{k-1}$ converges for $x \in (-1, 1)$, and (A.128) implies that $\sum_{k=1}^{\infty} k x^{k-1}$ is uniformly convergent on $[-r, r]$ for each $r \in (0, 1)$. ∎

- *(Abel's Theorem) Suppose $S(x) = \sum_{k=0}^{\infty} a_k x^k$ has radius of convergence $R = 1$. If $\sum_{k=0}^{\infty} a_k < \infty$, then $\lim_{x \to 1-} S(x) = S(1) = \sum_{k=0}^{\infty} a_k$* see, for example, Goldberg (1964, Section 9.6) or Stoll (2001, Section 8.7) for proof. Naturally, Abel's theorem can also be stated for general c and $R > 0$.

⊖ **Example A.64** It is easy to see that the radius of convergence of $S(x) = \sum_{k=1}^{\infty} (-1)^{k+1} x^k / k$ is $R = 1$ and, from the alternating series test (page 387), $S(1) = \sum_{k=1}^{\infty} (-1)^{k+1}/k$ is also convergent. Abel's theorem and (A.114) thus imply that $\ln 2 = 1 - \frac{1}{2} + \frac{1}{3} - \frac{1}{4} + \cdots$. ∎

⊖ **Example A.65** From Example A.55, $\arctan(t) = \sum_{n=0}^{\infty} \frac{(-1)^n}{2n+1} y^{2n+1} =: S(y)$, for $t \in [0, 1)$. From the alternating series test, $S(1)$ converges, so that

$$\frac{\pi}{4} = \arctan(1) = \sum_{n=0}^{\infty} \frac{(-1)^n}{2n+1} = 1 - \frac{1}{3} + \frac{1}{5} - \frac{1}{7} + \cdots.$$

from Abel's theorem. ∎

- Let $f(x) = \sum_{k=0}^{\infty} a_k (x - c)^k$ for $|x - c| < R$, where $R > 0$ is the radius of convergence of f. Then $d(x) = \sum_{k=1}^{\infty} k a_k (x - c)^{k-1}$ has radius of convergence R and $f'(x) = d(x)$ for x such that $|x - c| < R$. (A.129)

Proof: Using the limit results in Example A.8, the exponential growth rate of $d(x)$ is, for $x \neq c$,

$$\limsup \left| k a_k (x - c)^{k-1} \right|^{1/k} = \limsup |k|^{1/k} \limsup \left| (x - c)^{k-1} \right|^{1/k} \limsup |a_k|^{1/k}$$

$$= 1 \cdot \limsup \left| \frac{(x - c)^k}{x - c} \right|^{1/k} \limsup |a_k|^{1/k}$$

$$= \frac{|x - c|}{\lim_{k \to \infty} |x - c|^{1/k}} \limsup |a_k|^{1/k}$$

$$= |x - c| \limsup |a_k|^{1/k} ,$$

so that $d(x)$ has the same radius of convergence as does $f(r)$, namely R. That $f'(x) = d(x)$ for $|x - c| < R$ follows directly from the results in (A.128) and (A.125). ∎

Thus, the result in Example A.61 could have been obtained immediately via (A.129). It also implies that, if $f(x) = \sum_{k=0}^{\infty} a_k (x - c)^k$ and $g(x) = \sum_{k=0}^{\infty} b_k (x - c)^k$ are power series with radius of convergence R which are equal for $|x - c| < R$, then $a_k = b_k$ for $k = 0, 1, \ldots$. To see this, note that $f, g \in C^{\infty}$ so $f^{(n)}(x) = g^{(n)}(x)$ for $n \in \mathbb{N}$ and $|x - c| < R$. In particular,

$$f^{(n)}(x) = \sum_{k=n}^{\infty} k (k - 1) (k - n + 1) a_k (x - c)^{k-n}$$

and, as $0^0 = 1$, $f^{(n)}(c) = n! a_n$. Thus, $n! a_n = f^{(n)}(x) = g^{(n)}(x) = n! b_n$ for $n = 0, 1, \ldots$, which implies that $a_n = b_n$ for $n = 0, 1, \ldots$.

Repeated use of (A.129) implies that $f \in C^{\infty}(x - R, x + R)$, i.e. that f is infinitely differentiable on $(x - R, x + R)$. The converse, however, does not hold, i.e. there exist functions in $C^{\infty}(I)$ which cannot be expressed as a power series for particular $c \in I$. The ubiquitous example is to use $c = 0$ and the function given by $f(x) = \exp(-1/x^2)$ for $x \neq 0$ and $f(0) = 0$. Let I be an open interval. A function $f : I \to \mathbb{R}$ is said to be *analytic* in I if, $\forall c \in I$, there exists a sequence $\{a_k\}$ in $\mathbb{R}$ and a $\delta > 0$ such that, $\forall x$ with $|x - c| < \delta$, $f(x) = \sum_{k=0}^{\infty} a_k (x - c)^k$. Thus, the class of analytic functions is a proper subset of C^{∞}.

Recall from (A.16) that, for a differentiable function f, $f(x + h) \approx f(x) + h f'(x)$, accurate for h near zero, i.e. knowledge of a function and its derivative at a specified point, x, can be used to approximate the function at other points near x. By replacing x with c and then setting $h = x - c$, this can be written as

$$f(x) \approx f(c) + (x - c) f'(c) . \tag{A.130}$$

For example, with $f(x) = e^x$ and $c = 0$, (A.130) reads $e^x \approx e^0 + xe^0 = 1 + x$, which is accurate for $x \approx 0$. When evaluated at $x = c$, (A.130) is exact, and taking first derivatives of both sides w.r.t. x gives $f'(x) \approx f'(c)$, which is again exact at $x = c$. One might imagine that accuracy is improved if terms involving higher derivatives are taken into account. This is the nature of a *Taylor polynomial*, which was developed by Brooks Taylor, 1685–1731 (though variants were independently discovered by others, such as Gregory, Newton, Leibniz, Johann Bernoulli and de Moivre). It was only in 1772 that Joseph-Louis Lagrange (1736–1813) recognized the importance of the contribution, proclaiming it the basic principle of the differential calculus. Lagrange is also responsible for characterizing the error term. The first usage of the term Taylor series appears to be by Simon Lhuilier (1750–1840) in 1786.

Let $f : I \to \mathbb{R}$, where I is an open interval, and let $c \in I$. If $f^{(n)}(x)$ exists for all $x \in I$, then the nth order *Taylor polynomial* of f at c is

$$T_n(x; f, c) = T_n(x) = \sum_{k=0}^{n} \frac{f^{(k)}(c)}{k!} (x - c)^k,$$

and if $f \in C^\infty(I)$, then the *Taylor series* of f at c is

$$\sum_{k=0}^{\infty} \frac{f^{(k)}(c)}{k!} (x - c)^k. \tag{A.131}$$

When $c = 0$, (A.131) is also referred to as the *Maclaurin series*, after Colin Maclaurin (1698–1746). As in (A.130), observe that $T_n(c) = f(c)$, $T_n'(c) = T_n'(x)\big|_{x=c} = f'(c)$, up to $T_n^{(r)}(c) = f^{(r)}(c)$, so that locally (i.e. for x near c), $T_n(c)$ behaves similarly to $f(x)$ and could be used for effective approximation. The *remainder* between f and $T_n(x)$ is defined as $R_n(x) := f(x) - T_n(x)$, and *Taylor's formula with remainder* is given by

$$f(x) = T_n(x) + R_n(x) = \sum_{k=0}^{n} \frac{f^{(k)}(c)}{k!} (x - c)^k + R_n(x). \tag{A.132}$$

Clearly, $f(x) = T(x)$ iff $\lim_{n \to \infty} R_n(x) = 0$. If $f^{(n+1)}(x)$ exists for all $x \in I$, then the *Lagrange form of the remainder* is

$$R_n(x) = \frac{f^{(n+1)}(\zeta)}{(n+1)!} (x - c)^{n+1}, \quad \zeta \text{ between } x \text{ and } c.$$

To prove this, as in Bartle and Sherbert (1982, p. 222), assume $x \neq c$ and let $J = [x, c] \cup [c, x]$, i.e. $J = [x, c]$ if $x < c$, and $[c, x]$ if $c < x$. Then, for $t \in J$, let

$$P_n(t) := f(x) - f(t) - (x - t) f'(t) - \frac{(x - t)^2 f''(t)}{2!} \cdots - \frac{(x - t)^n f^{(n)}(t)}{n!}, \tag{A.133}$$

with $P_n(x) = 0$. Then $P_1'(t) = -f'(t) - [(x-t) f''(t) + f'(t)(-1)] = -(x-t)$ $f''(t)$, which can be written as $-(x-t)^n f^{(n+1)}(t)/n!$ for $n = 1$. Now use induction: assume this holds for $n - 1$; then

$$P_n'(t) = \frac{d}{dx}\left(P_{n-1}(t) - \frac{(x-t)^n f^{(n)}(t)}{n!}\right)$$

$$= -\frac{(x-t)^{n-1} f^{(n)}(t)}{(n-1)!} - \frac{(x-t)^n f^{(n+1)}(t) - f^{(n)}(t) n (x-t)^{n-1}(-1)}{n!}$$

$$= -\frac{(x-t)^n f^{(n+1)}(t)}{n!}.$$

Now let

$$G(t) := P_n(t) - \left(\frac{x-t}{x-c}\right)^{n+1} P_n(c), \quad t \in J,$$

so that $G(c) = 0$ and $G(x) = P_n(x) = 0$. The mean value theorem then implies that there exists a $\zeta \in J$ (actually, the interior of J) such that

$$\frac{G(c) - G(x)}{c - x} = G'(\zeta),$$

so that $0 = G'(\zeta) = P_n'(\zeta) + (n+1)\frac{(x-\zeta)^n}{(x-c)^{n+1}} P_n(c)$. Thus,

$$P_n(c) = -\frac{1}{n+1}\frac{(x-c)^{n+1}}{(x-\zeta)^n} P_n'(\zeta) = \frac{1}{n+1}\frac{(x-c)^{n+1}}{(x-\zeta)^n}\frac{(x-\zeta)^n f^{(n+1)}(\zeta)}{n!}$$

$$= \frac{f^{(n+1)}(\zeta)}{(n+1)!}(x-c)^{n+1},$$

and (A.133) reads

$$\frac{f^{(n+1)}(\zeta)}{(n+1)!}(x-c)^{n+1} = f(x) - f(c) - (x-c) f'(c)$$

$$- \frac{(x-c)^2 f''(c)}{2!} \cdots - \frac{(x-c)^n f^{(n)}(c)}{n!},$$

as was to be shown.[19]

---

[19] This proof, like other variants of it, are somewhat 'rabbit-out-of-the-hat', in the sense that it is not at all clear how one stumbles upon choosing $P_n(t)$ and $G(t)$. Such elegant proofs are just the result of concerted effort and much trial and error, and abound in mathematics, old and new. Indeed, referring to Gauss' style of mathematical proof, Niels Abel said that 'he is like the fox, who effaces his tracks in the sand with his tail'. In defense of his style, Gauss exclaimed that 'no self-respecting architect leaves the scaffolding in place after completing the building'. As encouragement, Gauss also said 'If others would but reflect on mathematical truths as deeply and continuously as I have, then they would also make my discoveries.'

The remainder $R_n(x)$ can be expressed in integral form, provided $f^{(n+1)}(x)$ exists for each $x \in I$ and, $\forall a, b \in I$, $f^{(n+1)} \in \mathcal{R}[a, b]$. Then

$$R_n(x) = \frac{1}{n!} \int_c^x f^{(n+1)}(t)(x - t)^n \, dt, \quad x \in I.$$

⊖ **Example A.66** Let $f(x) = \sin x$, so that, from the conditions in (A.27), $f'(x) = \cos x$ and $f''(x) = -\sin x$. Thus, $f^{(2n)}(x) = (-1)^n \sin x$ and $f^{(2n+1)}(x) = (-1)^n \cos x$, for $x \in \mathbb{R}$ and $n \in \mathbb{N} \cup 0$. As $\sin 0 = 0$ and $\cos 0 = 1$, the nth order Taylor polynomial for $c = 0$ is thus

$$T_n(x) = 0 + x - 0 - \frac{1}{6}x^3 + 0 + \frac{1}{120}x^5 + \cdots = \sum_{k=0}^n \frac{(-1)^k}{(2k+1)!} x^{2k+1}.$$

As $|\sin x| \le 1$ and $|\cos x| \le 1$, the remainder satisfies $|R_n(x)| \le |x|^{n+1} / (n+1)!$, which goes to zero as $n \to \infty$. Thus,

$$\sin x = \sum_{k=0}^\infty \frac{(-1)^k}{(2k+1)!} x^{2k+1},$$

which is its definition (see also Example A.48). ∎

A.3 Multivariate calculus

The discovery in 1846 of the planet Neptune was a dramatic and spectacular achievement of mathematical astronomy. The very existence of this new member of the solar system, and its exact location, were demonstrated with pencil and paper; there was left to observers only the routine task of pointing their telescopes at the spot the mathematicians had marked.

(James R. Newman)
Reproduced by permission of Simon and Schuster, NY

Here, as throughout the book, we use bold face to denote a point in $\mathbb{R}^n$, e.g. $\mathbf{x} = (x_1, x_2, \ldots, x_n)$, and also for multivariate functions, e.g. $\mathbf{f} : \mathbb{R} \to \mathbb{R}^m$, $m > 1$.

A.3.1 Neighborhoods and open sets

For any $\mathbf{x} \in \mathbb{R}^n$ and $r \in \mathbb{R}_{>0}$, the *open ball of radius r around* $\mathbf{x}$ is the subset $B_r(\mathbf{x}) \subset \mathbb{R}^n$ with $B_r(\mathbf{x}) := \{\mathbf{y} \in \mathbb{R}^n : \|\mathbf{x} - \mathbf{y}\| < r\}$ (note the strict inequality), where, recalling (A.9), $\|\mathbf{x}\|$ is the *norm* of $\mathbf{x}$. A *neighborhood* of a point $\mathbf{x} \in \mathbb{R}^n$ is a subset $A \subset \mathbb{R}^n$ such that there exists an $\epsilon > 0$ with $B_\epsilon(\mathbf{x}) \subset A$. If, for some $r \in \mathbb{R}_{>0}$, the set $A \subset \mathbb{R}^n$ is contained in the ball $B_r(\mathbf{0})$, then A is said to be *bounded*.

The subset $U \subset \mathbb{R}^n$ is *open in* $\mathbb{R}^n$ if, for every point $\mathbf{x} \in U$, $\exists r > 0$ such that $B_r(\mathbf{x}) \subset U$. It is easy to prove that, for every $\mathbf{x} \in \mathbb{R}^n$ and $r > 0$, the open ball $B_r(\mathbf{x})$ is open in $\mathbb{R}^n$.

For example, the open interval $\{x \in \mathbb{R} : |x - c| < r\} = (c - r, c + r)$, $c \in \mathbb{R}$, $r \in \mathbb{R}_{>0}$, is an open set in $\mathbb{R}$, but it is not open in the plane $\mathbb{R}^2$. Likewise, the square region $S_1 = \{\mathbf{y} = (y_1, y_2) \in \mathbb{R}^2 : |y_1| < 1, |y_2| < 1\}$ is open in $\mathbb{R}^2$, but not in $\mathbb{R}^3$.

A set $C \subset \mathbb{R}^n$ is *closed* if its complement, $\mathbb{R}^n \setminus C$ is open. By convention, the empty set $\emptyset$ is open (indeed, every point in $\emptyset$ satisfies the requirement), so that its complement, $\mathbb{R}^n$, is closed; but, from the definition, $\mathbb{R}^n$ is open, so that $\emptyset$ is closed. This is not incorrect: sets can be open and closed (or neither). The closed interval $[a, b]$ is a closed set, as is the square region $S_2 = \{\mathbf{y} = (y_1, y_2) \in \mathbb{R}^2 : |y_1| \leq 1, |y_2| \leq 1\}$. In Section A.3.2 below, after vector sequences are introduced, we state a definition of a closed set which is equivalent to its above definition in terms of open sets, but adds considerably more intuition into what a closed set represents.

The point $\mathbf{x} \in A \subset \mathbb{R}^n$ is an *interior point* of A if $\exists\, r > 0$ such that $B_r(\mathbf{x}) \subset A$. The *interior* of A is the set of all interior points of A, denoted A^o. Observe that the biggest open set contained in any set $A \subset \mathbb{R}^n$ is A^o. Likewise, the smallest closed set which contains A is the *closure* of A, denoted $\overline{A}$; it is the set of $\mathbf{x} \in \mathbb{R}^n$ such that, $\forall\, r > 0$, $B_r(\mathbf{x}) \cap A \neq \emptyset$. The closure of a set is closed. For example, the closure of (a, b), $a < b$, is $[a, b]$ (note that its complement, $(-\infty, a) \cup (b, \infty)$ is open). As an example in $\mathbb{R}^2$, using the sets S_1 and S_2 given above, $\overline{S_1} = S_2$. In words, the closure of a set includes its 'boundary', and indeed, the *boundary* of a set $A \subset \mathbb{R}^n$, denoted ∂A, is defined to be the difference between its closure and interior, i.e. $\partial A = \overline{A} - A^o$. (The notation ∂ is used because it signifies a line around a region, and has nothing to do with the symbol for the partial derivative.)

⊖ ***Example A.67*** For points $\mathbf{x}_1, \mathbf{x}_2 \in \mathbb{R}^n$, the *line segment* from $\mathbf{x}_1$ to $\mathbf{x}_2$ is the set of points

$$\mathbf{x}_1 + t(\mathbf{x}_2 - \mathbf{x}_1) = t\mathbf{x}_2 + (1 - t)\mathbf{x}_1, \quad 0 \leq t \leq 1.$$

For point $\mathbf{c} \in \mathbb{R}^n$ and $r > 0$, let $B_r(\mathbf{c})$ be the open ball of radius r around $\mathbf{c}$. It should be geometrically obvious that, if $\mathbf{x}_1, \mathbf{x}_2 \in B_r(\mathbf{c})$, then so are all the points on the line segment from $\mathbf{x}_1$ to $\mathbf{x}_2$. To see this algebraically, let $\mathbf{x} = \mathbf{x}(t) = t\mathbf{x}_2 + (1 - t)\mathbf{x}_1$ for $0 \leq t \leq 1$, and use the triangle inequality (A.8) to get

$$\begin{aligned} \|\mathbf{x} - \mathbf{c}\| &= \|t\mathbf{x}_2 + (1 - t)\mathbf{x}_1 - t\mathbf{c} - (1 - t)\mathbf{c}\| \\ &= \|t(\mathbf{x}_2 - \mathbf{c}) + (1 - t)(\mathbf{x}_1 - \mathbf{c})\| \\ &\leq t\|\mathbf{x}_2 - \mathbf{c}\| + (1 - t)\|\mathbf{x}_1 - \mathbf{c}\| \\ &< t \cdot r + (1 - t) \cdot r = r. \end{aligned}$$

As $B_r(\mathbf{c})$ is open, $\|\mathbf{x} - \mathbf{c}\|$ is strictly less than r, though $\sup \|\mathbf{x} - \mathbf{c}\| = r$. ∎

A.3.2 Sequences, limits and continuity

A (multivariate, or vector) sequence is a function $\mathbf{f} : \mathbb{N} \to \mathbb{R}^n$ with kth term $\mathbf{f}(k)$, $k \in \mathbb{N}$. As in the univariate case, the more common notation for sequence $\mathbf{a}_1 = \mathbf{f}(1)$, $\mathbf{a}_2 =$

$\mathbf{f}(2),\ldots$ is $\{\mathbf{a}_k\}$, and the ith component of $\mathbf{a}_k$ is denoted by $(\mathbf{a}_k)_i$, $i = 1,\ldots,n$. For sequence $\{\mathbf{a}_k\}$ and set $S \subset \mathbb{R}^n$, the notation $\mathbf{a}_k \in S$ indicates that $\mathbf{a}_k \in S$, $\forall\, k \in \mathbb{N}$.

The sequence $\{\mathbf{a}_k\}$, $\mathbf{a}_k \in \mathbb{R}^n$, converges to the unique point $\mathbf{a} \in \mathbb{R}^n$ if, $\forall\, \epsilon > 0$, $\exists\, K \in \mathbb{N}$ such that, $\forall\, k > K$, $\|\mathbf{a}_k - \mathbf{a}\| < \epsilon$. Point $\mathbf{a}$ is the *limit* of $\{\mathbf{a}_k\}$ if $\{\mathbf{a}_k\}$ converges to $\mathbf{a}$, in which case one writes $\lim_{k\to\infty} \mathbf{a}_k = \mathbf{a}$. In order for $\lim_{k\to\infty} \mathbf{a}_k = \mathbf{a} = (a_1,\ldots,a_n)$ to hold, it is necessary and sufficient that $\lim_{k\to\infty} (\mathbf{a}_k)_i = a_i$, $i = 1,\ldots,n$. As in the univariate case, $\{\mathbf{a}_k\}$ is a *Cauchy sequence* if, for a given $\epsilon > 0$, $\exists\, N \in \mathbb{N}$ such that $\forall\, n, m \geq N$, $\|\mathbf{a}_m - \mathbf{a}_n\| < \epsilon$. As expected, (A.83) generalizes: sequence $\{\mathbf{a}_k\}$ converges iff $\{\mathbf{a}_k\}$ is a Cauchy sequence.

The point $\mathbf{a} \in S \subset \mathbb{R}^n$ is said to be a *limit point* of S if $\exists\, \{\mathbf{a}_k\}$, $\mathbf{a}_k \in S$, such that $\lim_{k\to\infty} \mathbf{a}_k = \mathbf{a}$. In other words, $\mathbf{a}$ is a limit point of S if there exists a sequence with terms in S which converge to it.[20]

As in the univariate case, the series $\sum_{k=1}^{\infty} \mathbf{a}_k$ is convergent if the sequence of partial sums $\{\mathbf{s}_p\}$, $\mathbf{s}_p = \sum_{k=1}^{p} \mathbf{a}_k$, is convergent. Consider the function $f : \mathbb{N} \to I$, $I = (0,1]$, given by $f(k) = 1/k$. Observe that I is neither open nor closed. Clearly, $\lim_{k\to\infty} a_k = 0$, and $0 \notin I$. However, $\lim_{k\to\infty} a_k$ is contained in the closure of I, which is the closed set $[0,1]$. With this concept in mind, the following basic result of analysis should appear quite reasonable: The set $C \subset \mathbb{R}^n$ is *closed* iff it contains all its limit points.

The concepts of limits and continuity for univariate functions considered in Section A.2.1 generalize to the multivariate case. Let $\mathbf{f} : A \to \mathbb{R}^m$ with $A \subset \mathbb{R}^n$ and $\mathbf{x}_0 \in \bar{A}$ (the closure of A). Then $\lim_{\mathbf{x}\to\mathbf{x}_0} \mathbf{f}(\mathbf{x}) = \mathbf{b}$ if, $\forall\, \epsilon > 0$, $\exists\, \delta > 0$ such that, when $\|\mathbf{x} - \mathbf{x}_0\| < \delta$ and $\mathbf{x} \in A$, $\|\mathbf{f}(\mathbf{x}) - \mathbf{b}\| < \epsilon$. If the limit exists, then it is unique.

⊖ **Example A.68** Let $f(x) = 1/x$ for $x \neq 0$. It is easy to see that $\lim_{x\to 0} f(x)$ does not exist, though one-sided limits do exist. Similar phenomena exist in the multivariate case. Let $f : A \to \mathbb{R}$ with $A = \mathbb{R}^2 \setminus \mathbf{0}$ and $f(\mathbf{x}) = x_1 x_2/(x_1^2 + x_2^2)$. To see that $\lim_{\mathbf{x}\to\mathbf{0}} f(\mathbf{x})$ does not exist, set $x_2(x_1) = kx_1$ for some fixed $k \in \mathbb{R}$ so that $\lim_{x_1\to 0} x_2(x_1) = 0$ and $f(\mathbf{x}) = f(x_1, x_2(x_1)) = f(x_1, kx_1) = kx_1^2/(x_1^2 + k^2 x_1^2) = k/(1 + k^2)$. Thus, along the line $x_2 = kx_1$, $\lim_{\mathbf{x}\to\mathbf{0}} f(\mathbf{x}) = k/(1 + k^2)$, i.e. it depends on the choice of k, showing that $\lim_{\mathbf{x}\to\mathbf{0}} f(\mathbf{x})$ depends on the path which $\mathbf{x}$ takes towards zero. Thus, $\lim_{\mathbf{x}\to\mathbf{0}} f(\mathbf{x})$ cannot exist. ∎

The following results should not be surprising. Let $\mathbf{f}, \mathbf{g} : A \to \mathbb{R}^m$ and assume that $\lim_{\mathbf{x}\to\mathbf{x}_0} \mathbf{f}(\mathbf{x})$ and $\lim_{\mathbf{x}\to\mathbf{x}_0} \mathbf{g}(\mathbf{x})$ exist. Then, for constant values $k_1, k_2 \in \mathbb{R}$,

$$\lim_{\mathbf{x}\to\mathbf{x}_0} (k_1\mathbf{f} + k_2\mathbf{g})(\mathbf{x}) = k_1 \lim_{\mathbf{x}\to\mathbf{x}_0} \mathbf{f}(\mathbf{x}) + k_2 \lim_{\mathbf{x}\to\mathbf{x}_0} \mathbf{g}(\mathbf{x})$$

and

$$\lim_{\mathbf{x}\to\mathbf{x}_0} (\mathbf{f}\cdot\mathbf{g})(\mathbf{x}) = \lim_{\mathbf{x}\to\mathbf{x}_0} \mathbf{f}(\mathbf{x}) \cdot \lim_{\mathbf{x}\to\mathbf{x}_0} \mathbf{g}(\mathbf{x}),$$

---

[20] For this definition, some mathematicians will additionally require that the sequence consists of unique terms, i.e. $\mathbf{a}_k \in S$ with $\mathbf{a}_k \neq \mathbf{a}_h$, $h \neq k$, which, for example, precludes a finite set of points from having limit points. In this case, as in Stoll (2001, p. 69), a point $\mathbf{a} \in \mathbb{R}^n$ is a limit point of S if, $\forall\, r > 0$, $(B_r(\mathbf{a}) \setminus \mathbf{a}) \cap S \neq \emptyset$. In words, $\mathbf{a}$ is a limit point of S if, for any $r > 0$, the open ball of radius r around $\mathbf{a}$, but 'punctured' at $\mathbf{a}$ (i.e. excluding the point $\mathbf{a}$) contains at least one point which is in S. As in Pugh (2002, p. 58) and Hubbard and Hubbard (2002, p. 97), we will not require that $\mathbf{a}_k \neq \mathbf{a}_h$.

where, for $\mathbf{x}, \mathbf{y} \in \mathbb{R}^n$, $\mathbf{x} \cdot \mathbf{y} = \sum_{i=1}^{n} x_i y_i$ is the *dot product* of $\mathbf{x}$ and $\mathbf{y}$. Let $A \subset \mathbb{R}^n$ and $B \subset \mathbb{R}^m$. If $\mathbf{f} : A \to \mathbb{R}^m$ and $\mathbf{g} : B \to \mathbb{R}^p$ such that $\mathbf{f}(A) \subset B$, then the composite function $\mathbf{g} \circ \mathbf{f}$ is well-defined. If $\mathbf{y}_0 := \lim_{\mathbf{x} \to \mathbf{x}_0} \mathbf{f}(\mathbf{x})$ and $\lim_{\mathbf{y} \to \mathbf{y}_0} \mathbf{g}(\mathbf{y})$ both exist, then $\lim_{\mathbf{x} \to \mathbf{x}_0} (\mathbf{g} \circ \mathbf{f})(\mathbf{x}) = \lim_{\mathbf{y} \to \mathbf{y}_0} \mathbf{g}(\mathbf{y})$.

Let $\mathbf{f} : A \to \mathbb{R}^m$ with $A \subset \mathbb{R}^n$. Paralleling the univariate case discussed in Section A.2.1, the function $\mathbf{f}$ is *continuous* at $\mathbf{a} \in A$ if, for $\mathbf{x} \in A$, $\lim_{\mathbf{x} \to \mathbf{a}} \mathbf{f}(\mathbf{x}) = \mathbf{f}(\mathbf{a})$. Equivalently, $\mathbf{f}$ is continuous at $\mathbf{a} \in A$ if, for a given $\epsilon > 0$, $\exists\, \delta > 0$ such that, if $\|\mathbf{x} - \mathbf{a}\| < \delta$ and $\mathbf{x} \in A$, then $\|\mathbf{f}(\mathbf{x}) - \mathbf{f}(\mathbf{a})\| < \epsilon$. If $\mathbf{f}$ is continuous at every point in A, then $\mathbf{f}$ is said to be continuous, and we write $\mathbf{f} \in \mathcal{C}^0$. Function $\mathbf{f}$ is *uniformly continuous* on subset $S \subset A$ if: for a given $\epsilon > 0$, $\exists\, \delta > 0$ such that, if $\mathbf{x}, \mathbf{y} \in S$, and $\|\mathbf{x} - \mathbf{y}\| < \delta$, then $\|\mathbf{f}(\mathbf{x}) - \mathbf{f}(\mathbf{y})\| < \epsilon$. Similar to the previous results on limits, if $\mathbf{f}$ and $\mathbf{g}$ are continuous at $\mathbf{x}_0$, then so are $\mathbf{f} + \mathbf{g}$ and $\mathbf{f} \cdot \mathbf{g}$ at $\mathbf{x}_0$. For the composite function $\mathbf{g} \circ \mathbf{f}$, if $\mathbf{f}$ is continuous at $\mathbf{x}_0$, and $\mathbf{g}$ is continuous at $\mathbf{f}(\mathbf{x}_0)$, then $\mathbf{g} \circ \mathbf{f}$ is continuous at $\mathbf{x}_0$.

Similar to (A.13), if $A \subset \mathbb{R}^n$ is closed and bounded, and $f : A \to \mathbb{R}$ is continuous, then f takes on minimum and maximum values, i.e. $\exists\, \mathbf{a}, \mathbf{b} \in A$ such that, $\forall \mathbf{x} \in A$, $f(\mathbf{a}) \leq f(\mathbf{x}) \leq f(\mathbf{b})$. The intermediate value theorem given for univariate functions in Section A.2.1 also generalizes. Let $f : A \to \mathbb{R}$ be continuous on subset $S \subset A \subset \mathbb{R}^n$ and assume $\mathbf{a}, \mathbf{b} \in S$. Let $\alpha = f(\mathbf{a})$ and $\beta = f(\mathbf{b})$. Given a number γ with $\alpha < \gamma < \beta$, $\exists\, \mathbf{c} \in S$ such that $f(\mathbf{c}) = \gamma$. For proof see, for example, Trench (2003, p. 313).

A.3.3   Differentiation

Do not worry about your problems with mathematics, I assure you, mine are far greater.
(Albert Einstein)
Reproduced by permission of Princeton University Press

Partial derivatives and the gradient

Let $f : A \to \mathbb{R}$ with $A \subset \mathbb{R}^n$ an open set. For every $\mathbf{x} = (x_1, \ldots, x_n) \in A$ and for each $i = 1, 2, \ldots, n$, the *partial derivative* of f with respect to x_i is defined as

$$\frac{\partial f}{\partial x_i}(\mathbf{x}) = \lim_{h \to 0} \frac{f(x_1, \ldots, x_{i-1}, x_i + h, x_{i+1}, \ldots, x_n) - f(x_1, \ldots, x_n)}{h}, \qquad (A.134)$$

if the limit exists. Because the remaining $n - 1$ variables in $\mathbf{x}$ are held constant, the partial derivative is conceptually identical to the Newton quotient (A.15) for univariate functions. This can be more compactly written by defining $\mathbf{e}_i = (0, 0, \ldots, 0, 1, 0, \ldots, 0)$ to be the n-vector with a one in the ith position, and zero elsewhere, so that

$$\frac{\partial f}{\partial x_i}(\mathbf{x}) = \lim_{h \to 0} \frac{f(\mathbf{x} + h\mathbf{e}_i) - f(\mathbf{x})}{h}.$$

A popular and useful alternative notation for the partial derivative is $D_i f(\mathbf{x})$ or, better, $(D_i f)(\mathbf{x})$, with the advantage that the name of the ith variable (in this case x_i) does not need to be explicitly mentioned. This is termed the partial derivative of f with respect to the ith variable. The *gradient* of f, denoted $(\text{grad } f)(\mathbf{x})$ (and rhyming with sad and glad), or $(\nabla f)(\mathbf{x})$, is the *row vector* of all partial derivatives:

$$(\text{grad } f)(\mathbf{x}) = (D_1 f(\mathbf{x}), \ldots, D_n f(\mathbf{x})). \tag{A.135}$$

In the following, let $f : A \to \mathbb{R}$ with $A \subset \mathbb{R}^n$ an open set and such that $(\text{grad } f)(\mathbf{x})$ exists $\forall \mathbf{x} \in A$. Recall that, for $n = 1$, the *tangent to the curve* at the point (x_0, y_0), for $x_0 \in A \subset \mathbb{R}$ and $y_0 = f(x_0)$ is the (non-vertical) line $T(x) = y_0 + f'(x_0)(x - x_0)$. This is the best linear approximation to f in a neighborhood of x_0 such that $f(x_0) = T(x_0)$, and, from (A.15) and (A.16), satisfies

$$\lim_{x \to x_0} \frac{f(x) - T(x)}{x - x_0} = 0.$$

For $n = 2$, envisaging a thin, flat board resting against a sphere in three-space, we seek a (non-vertical) plane in $\mathbb{R}^3$ which is 'tangent' to f at a given point, say (x_0, y_0, z_0), for $(x_0, y_0) \in A$ and $z_0 = f(x_0, y_0)$. A plane is linear in both x and y, so its equation is $z = z_0 + s(x - x_0) + t(y - y_0)$, where s and t need to be determined. When restricted to the plane $y = y_0$, the surface f is just the curve $z = g(x) := f(x, y_0)$ in $\mathbb{R}^2$, and the plane we seek is just the line $z = z_0 + s(x - x_0)$. This is the $n = 1$ case previously discussed, so the tangent to the curve $g(x)$ at x_0 is the line $z = z_0 + g'(x_0)(x - x_0)$, i.e. $s = D_1 f(x_0, y_0)$. Similarly, $t = D_2 f(x_0, y_0)$, so that the *tangent plane* of f at (x_0, y_0, z_0) is $T(x, y) = z_0 + (D_1 f(x_0, y_0))(x - x_0) + (D_2 f(x_0, y_0))(y - y_0)$, which satisfies $f(x_0, y_0) = T(x_0, y_0)$ and

$$\lim_{(x,y) \to (x_0,y_0)} \frac{f(x, y) - T(x, y)}{\|(x, y) - (x_0, y_0)\|} = 0.$$

This motivates the following definition for general n. Let $f : A \to \mathbb{R}$ for $A \subset \mathbb{R}^n$ and let $\mathbf{x}_0 = (x_{01}, x_{02}, \ldots, x_{0n})'$ be an interior point of A. The function f is said to be *differentiable at* $\mathbf{x}_0$ if $(\text{grad } f)(\mathbf{x}_0)$ exists, and there exists a *tangent map* $T : \mathbb{R}^n \to \mathbb{R}$ of f at $\mathbf{x}_0$, such that $f(\mathbf{x}_0) = T(\mathbf{x}_0)$ and

$$\lim_{\mathbf{x} \to \mathbf{x}_0} \frac{f(\mathbf{x}) - T(\mathbf{x})}{\|\mathbf{x} - \mathbf{x}_0\|} = 0. \tag{A.136}$$

If the tangent map of f at $\mathbf{x}_0$ exists, then it is unique and, *with $\mathbf{x}$ and $\mathbf{x}_0$ column vectors*, is given by

$$T(\mathbf{x}) = f(\mathbf{x}_0) + (\text{grad } f)(\mathbf{x}_0)(\mathbf{x} - \mathbf{x}_0)$$

$$= f(\mathbf{x}_0) + \sum_{i=1}^{n} (D_i f)(\mathbf{x}_0)(x_i - x_{0i}) \tag{A.137}$$

$$=: f(\mathbf{x}_0) + \mathrm{d}f(\mathbf{x}_0, \mathbf{x} - \mathbf{x}_0), \tag{A.138}$$

where the term defined in (A.138) is the *total differential*, i.e.

$$\mathrm{d}f(\mathbf{x}, \mathbf{h}) = (\text{grad } f)(\mathbf{x}) \cdot \mathbf{h}.$$

⊖ *Example A.69*   Let $f : A \subset \mathbb{R}^2 \to \mathbb{R}$ be differentiable on the open set A. It will be shown below that f is also continuous on A. For $(x, y) \in A$, let $z = f(x, y)$ and let δ_x and δ_y represent very small, positive quantities such that $(x + \delta_x, y + \delta_y) \in A$. Let δ_z be such that $z + \delta_z = f(x + \delta_x, y + \delta_y)$, i.e.

$$
\begin{aligned}
\delta_z &= f(x + \delta_x, y + \delta_y) - f(x, y) \\
&= f(x + \delta_x, y + \delta_y) - f(x, y + \delta_y) + f(x, y + \delta_y) - f(x, y) \\
&= \frac{f(x + \delta_x, y + \delta_y) - f(x, y + \delta_y)}{\delta_x} \delta_x + \frac{f(x, y + \delta_y) - f(x, y)}{\delta_y} \delta_y .
\end{aligned}
$$

From the fundamental lemma of differentiation (A.18), this can be written as

$$
\begin{aligned}
\delta_z &= \left[\frac{\partial f(x, y + \delta_y)}{\partial x} + \eta_x \right] \delta_x + \left[\frac{\partial f(x, y)}{\partial y} + \eta_y \right] \delta_y \\
&= \frac{\partial f(x, y + \delta_y)}{\partial x} \delta_x + \frac{\partial f(x, y)}{\partial y} \delta_y \quad + \eta_x \delta_x + \eta_y \delta_y ,
\end{aligned}
$$

where $\lim_{\delta_x \to 0} \eta_x = 0$ and likewise for η_y. This coincides with the interpretation of (A.137) for $n = 2$.   ∎

The Jacobian and the chain rule

Now consider a multivariate function $\mathbf{f} : A \subset \mathbb{R}^n \to \mathbb{R}^m$ with A an open set, where $\mathbf{f}$ is such that $\mathbf{f}(\mathbf{x}) = (f_1(\mathbf{x}), \dots, f_m(\mathbf{x}))$, $f_i : A \to \mathbb{R}$, $i = 1, \dots, m$, for all $\mathbf{x} = (x_1, \dots, x_n)' \in A$.

If each partial derivative, $\partial f_i(\mathbf{x}_0)/\partial x_j$, $i = 1, \dots, m$, $j = 1, \dots, n$, exists, then the *total derivative* of $\mathbf{f}$ at $\mathbf{x}_0 \in A$ is the $m \times n$ matrix

$$
\mathbf{f}'(\mathbf{x}_0) := \mathbf{J}_{\mathbf{f}}(\mathbf{x}_0) := \begin{pmatrix} \dfrac{\partial f_1}{\partial x_1}(\mathbf{x}_0) & \cdots & \dfrac{\partial f_1}{\partial x_n}(\mathbf{x}_0) \\ \vdots & \ddots & \vdots \\ \dfrac{\partial f_m}{\partial x_1}(\mathbf{x}_0) & \cdots & \dfrac{\partial f_m}{\partial x_n}(\mathbf{x}_0) \end{pmatrix} = \begin{pmatrix} (\mathrm{grad}\, f_1)(\mathbf{x}_0) \\ \vdots \\ (\mathrm{grad}\, f_m)(\mathbf{x}_0) \end{pmatrix}, \quad \text{(A.139)}
$$

also referred to as the *Jacobian matrix* of $\mathbf{f}$ at $\mathbf{x}_0$.[21] When $m = 1$, the total derivative is just the gradient (A.135).

---

[21] After the prolific Carl Gustav Jacob Jacobi (1804–1851), who made contributions in several branches of mathematics, including the study of functional determinants. Though the theory goes back (at least) to Cauchy in 1815, Jacobi's 1841 memoir *De determinantibus functionalibus* had the first modern definition of determinant, and the first use of the word Jacobian was by Sylvester in 1853. Jacobi is also remembered as an excellent teacher who introduced the 'seminar method' for teaching the latest advances in math (whereby students present and discuss current articles and papers.)

In what follows, let $\mathbf{f} = (f_1(\mathbf{x}), \ldots, f_m(\mathbf{x})) : A \subset \mathbb{R}^n \to \mathbb{R}^m$, where A is an open set. Function $\mathbf{f}$ is said to be *differentiable at* $\mathbf{x}_0 \in A$ if each f_i, $i = 1, \ldots, m$, is differentiable at $\mathbf{x}_0 \in A$. Analogous to the previous case with $m = 1$, let

$$\mathbf{T}(\mathbf{x}) = \mathbf{f}(\mathbf{x}_0) + \mathbf{J_f}(\mathbf{x}_0)(\mathbf{x} - \mathbf{x}_0) = \mathbf{f}(\mathbf{x}_0) + d\mathbf{f}(\mathbf{x}_0, \mathbf{x} - \mathbf{x}_0),$$

where $d\mathbf{f}$ is the total differential.

- $\mathbf{f}$ *is differentiable at* $\mathbf{x}_0 \in A$ *iff the Jacobian* $\mathbf{J_f}(\mathbf{x}_0)$ *exists,* $\mathbf{f}(\mathbf{x}_0) = \mathbf{T}(\mathbf{x}_0)$, *and*

$$\lim_{\mathbf{x} \to \mathbf{x}_0} \frac{\|\mathbf{f}(\mathbf{x}) - \mathbf{T}(\mathbf{x})\|}{\|\mathbf{x} - \mathbf{x}_0\|} = 0. \tag{A.140}$$

The proof is easy and offers some practice with the notation.

($\Rightarrow$) Assume $\mathbf{f}$ is differentiable at $\mathbf{x}_0 \in A$. Then, by definition, $(\text{grad } f_i)(\mathbf{x}_0)$ exists, $i = 1, \ldots, m$, so that the form of (A.139) shows that $\mathbf{J_f}(\mathbf{x}_0)$ exists. Next, differentiability of $\mathbf{f}$ means that there exists a tangent map of each f_i at $\mathbf{x}_0$, given by, say, $T_i(\mathbf{x}) = f_i(\mathbf{x}_0) + (\text{grad } f_i)(\mathbf{x}_0)(\mathbf{x} - \mathbf{x}_0)$ and, for each i, $f_i(\mathbf{x}_0) = T_i(\mathbf{x}_0)$. Thus, taking

$$\mathbf{T}(\mathbf{x}) = \begin{bmatrix} T_1(\mathbf{x}) \\ \vdots \\ T_m(\mathbf{x}) \end{bmatrix} = \begin{bmatrix} f_1(\mathbf{x}_0) + (\text{grad } f_1)(\mathbf{x}_0)(\mathbf{x} - \mathbf{x}_0) \\ \vdots \\ f_m(\mathbf{x}_0) + (\text{grad } f_m)(\mathbf{x}_0)(\mathbf{x} - \mathbf{x}_0) \end{bmatrix} \tag{A.141}$$

$$= \mathbf{f}(\mathbf{x}_0) + \mathbf{J_f}(\mathbf{x}_0)(\mathbf{x} - \mathbf{x}_0),$$

it is clear that $\mathbf{T}(\mathbf{x}_0) = (f_1(\mathbf{x}_0), \ldots, f_m(\mathbf{x}_0))' = \mathbf{f}(\mathbf{x}_0)$. Lastly, from (A.136), differentiability of $\mathbf{f}$ implies that, for each $i = 1, \ldots, m$,

$$\lim_{\mathbf{x} \to \mathbf{x}_0} \frac{f_i(\mathbf{x}) - T_i(\mathbf{x})}{\|\mathbf{x} - \mathbf{x}_0\|} = 0 = \lim_{\mathbf{x} \to \mathbf{x}_0} \frac{|f_i(\mathbf{x}) - T_i(\mathbf{x})|}{\|\mathbf{x} - \mathbf{x}_0\|}. \tag{A.142}$$

Next note that, for any real vector $\mathbf{z} = (z_1, \ldots, z_m)$, $\|\mathbf{z}\| \le |z_1| + \cdots + |z_m|$ which is easily confirmed by squaring both sides. Thus, with

$$\mathbf{z} = \frac{\mathbf{f}(\mathbf{x}) - \mathbf{T}(\mathbf{x})}{\|\mathbf{x} - \mathbf{x}_0\|} = \frac{1}{\|\mathbf{x} - \mathbf{x}_0\|} \begin{bmatrix} f_1(\mathbf{x}) - T_1(\mathbf{x}) \\ \vdots \\ f_m(\mathbf{x}) - T_m(\mathbf{x}) \end{bmatrix},$$

it follows that

$$\|\mathbf{z}\| = \frac{\|\mathbf{f}(\mathbf{x}) - \mathbf{T}(\mathbf{x})\|}{\|\mathbf{x} - \mathbf{x}_0\|} \le \sum_{i=1}^{m} \frac{|f_i(\mathbf{x}) - T_i(\mathbf{x})|}{\|\mathbf{x} - \mathbf{x}_0\|},$$

i.e. (A.140) follows from (A.142).

($\Leftarrow$) If $\mathbf{J_f}(\mathbf{x}_0)$ exists, then $(\text{grad } f_i)(\mathbf{x}_0)$ exists, $i = 1, \ldots, m$. From (A.141), if $\mathbf{T}(\mathbf{x}_0) = \mathbf{f}(\mathbf{x}_0)$, then $f_i(\mathbf{x}_0) = T_i(\mathbf{x}_0)$, $i = 1, \ldots, m$. Lastly, it trivially follows from the definition of the norm that

$$\frac{|f_i(\mathbf{x}) - T_i(\mathbf{x})|}{\|\mathbf{x} - \mathbf{x}_0\|} \leq \frac{\|\mathbf{f}(\mathbf{x}) - \mathbf{T}(\mathbf{x})\|}{\|\mathbf{x} - \mathbf{x}_0\|},$$

so that (A.142) follows from (A.140).   ∎

We say that $\mathbf{f}$ is *differentiable on A* if $\mathbf{f}$ is differentiable at each $\mathbf{x}_0 \in A$. Furthermore, if $\mathbf{f}$ is differentiable and all the partial derivatives of each f_i are continuous, then $\mathbf{f}$ is *continuously differentiable*, and we write $\mathbf{f} \in \mathcal{C}^1$.

⊖ **Example A.70**   Let $\mathbf{f} : \mathbb{R}^2 \to \mathbb{R}^3$, $(x, y) \mapsto (ye^x, x^2y^3, -x)$. Then

$$\mathbf{f}'(x, y) = \mathbf{J_f}(x, y) = \begin{pmatrix} ye^x & e^x \\ 2xy^3 & x^23y^2 \\ -1 & 0 \end{pmatrix},$$

and $\mathbf{f}$ is continuously differentiable.   ∎

In Section A.2.2, we showed that, for $f : A \subset \mathbb{R} \to \mathbb{R}$, if f is differentiable at the point $a \in A$, then f is continuous at a. This extends as follows.

- *If $\mathbf{f}$ is differentiable at $\mathbf{x}_0 \in A$, then $\mathbf{f}$ is continuous at $\mathbf{x}_0$.*

The proof is instructive and hinges on the two most important inequalities in analysis. From (A.140), $\exists \delta^* > 0$ such that, for $\mathbf{x} \in A$, if $\|\mathbf{x} - \mathbf{x}_0\| < \delta^*$, then $\|\mathbf{f}(\mathbf{x}) - \mathbf{T}(\mathbf{x})\| < \|\mathbf{x} - \mathbf{x}_0\|$, where $\mathbf{T}(\mathbf{x}) = \mathbf{f}(\mathbf{x}_0) + \mathbf{J_f}(\mathbf{x}_0)(\mathbf{x} - \mathbf{x}_0)$. Let $\mathbf{K} = \mathbf{J_f}(\mathbf{x}_0)(\mathbf{x} - \mathbf{x}_0)$ so that $\mathbf{T}(\mathbf{x}) = \mathbf{f}(\mathbf{x}_0) + \mathbf{K}$. If $\|\mathbf{x} - \mathbf{x}_0\| < \delta^*$, then, from the triangle inequality (A.8),

$$\begin{aligned}
\|\mathbf{f}(\mathbf{x}) - \mathbf{f}(\mathbf{x}_0)\| &= \|\mathbf{f}(\mathbf{x}) - \mathbf{f}(\mathbf{x}_0) - \mathbf{K} + \mathbf{K}\| \\
&= \|\mathbf{f}(\mathbf{x}) - \mathbf{T}(\mathbf{x}) + \mathbf{K}\| \\
&\leq \|\mathbf{f}(\mathbf{x}) - \mathbf{T}(\mathbf{x})\| + \|\mathbf{K}\| \\
&< \|\mathbf{x} - \mathbf{x}_0\| + \|\mathbf{K}\|.
\end{aligned} \tag{A.143}$$

From (A.141), with row vector $\mathbf{w}_i = (w_{i1}, \ldots, w_{in}) := (\text{grad } f_i)(\mathbf{x}_0)$ and column vector $\mathbf{z}_i = (z_{i1}, \ldots, z_{in}) := (\mathbf{x} - \mathbf{x}_0)$,

$$\|\mathbf{K}\|^2 = \sum_{i=1}^{m} \left[(\text{grad } f_i)(\mathbf{x}_0)(\mathbf{x} - \mathbf{x}_0)\right]^2 = \sum_{i=1}^{m} \left[\mathbf{w}_i \mathbf{z}_i\right]^2 = \sum_{i=1}^{m} \left(\sum_{j=1}^{n} w_{ij} z_{ij}\right)^2.$$

For each $i = 1, \ldots, m$, the Cauchy–Schwarz inequality (A.7) implies

$$\left(\sum_{j=1}^{n} w_{ij} z_{ij}\right)^2 \leq \left(\sum_{j=1}^{n} w_{ij}^2\right)\left(\sum_{j=1}^{n} z_{ij}^2\right) = \|(\text{grad } f_i)(\mathbf{x}_0)\|^2 \|(\mathbf{x} - \mathbf{x}_0)\|^2,$$

so that

$$
\|\mathbf{K}\| \leq \left(\sum_{i=1}^{m} \|(\text{grad } f_i)(\mathbf{x}_0)\|^2 \|(\mathbf{x} - \mathbf{x}_0)\|^2 \right)^{1/2}
$$

$$
= \|(\mathbf{x} - \mathbf{x}_0)\| \left(\sum_{i=1}^{m} \|(\text{grad } f_i)(\mathbf{x}_0)\|^2 \right)^{1/2} =: \|(\mathbf{x} - \mathbf{x}_0)\| \, G. \tag{A.144}
$$

Thus, from (A.143) and (A.144),

$$
\|\mathbf{f}(\mathbf{x}) - \mathbf{f}(\mathbf{x}_0)\| < \|\mathbf{x} - \mathbf{x}_0\| + \|\mathbf{K}\| < \|\mathbf{x} - \mathbf{x}_0\| \, (1 + G). \tag{A.145}
$$

Because we assume that $\mathbf{f}$ is differentiable at $\mathbf{x}_0$, G is finite. Thus, for a given $\epsilon > 0$, we can find a $\delta > 0$ such that, if $\|\mathbf{x} - \mathbf{x}_0\| < \delta$ and $\mathbf{x} \in A$, then $\|\mathbf{f}(\mathbf{x}) - \mathbf{f}(\mathbf{x}_0)\| < \epsilon$. In particular, from (A.145), $\delta = \min(\delta^*, \epsilon/(1 + G))$. ∎

The following two results are of great practical importance; see, for example, Hubbard and Hubbard (2002, pp. 159 and 680) for detailed proofs.

- If all the partial derivatives of $\mathbf{f} = (f_1(\mathbf{x}), \ldots, f_m(\mathbf{x})) : A \subset \mathbb{R}^n \to \mathbb{R}^m$ exist and are continuous on A, then $\mathbf{f}$ is differentiable on A with derivative $\mathbf{J_f}$.

- Let $\mathbf{f} : A \subset \mathbb{R}^n \to \mathbb{R}^m$ and $\mathbf{g} : B \subset \mathbb{R}^m \to \mathbb{R}^p$ with A and B open sets and $\mathbf{f}(A) \subset B$. If $\mathbf{f}$ is differentiable at $\mathbf{x} \in A$ and $\mathbf{g}$ is differentiable at $\mathbf{f}(\mathbf{x})$, then the composite function $\mathbf{g} \circ \mathbf{f}$ is differentiable at $\mathbf{x}$, with derivative

$$
\mathbf{J}_{\mathbf{g} \circ \mathbf{f}}(\mathbf{x}) = \mathbf{J_g}(\mathbf{f}(\mathbf{x})) \, \mathbf{J_f}(\mathbf{x}). \tag{A.146}
$$

This generalizes the chain rule (A.21) for univariate functions.

⊖ **Example A.71** As in Example A.70, let $\mathbf{f} : \mathbb{R}^2 \to \mathbb{R}^3$, $(x, y) \mapsto (ye^x, x^2 y^3, -x)$, and also let $\mathbf{g} : (\mathbb{R}_{>0} \times \mathbb{R}) \to \mathbb{R}^2$, $(x, y) \mapsto (\ln x, x + 2y) = (g_1(x, y), g_2(x, y))$. The function $\mathbf{g}$ is continuously differentiable with derivative at $(x, y) \in (\mathbb{R}_{>0} \times \mathbb{R})$:

$$
\mathbf{g}'(x, y) = \mathbf{J_g}(x, y) = \begin{bmatrix} 1/x & 0 \\ 1 & 2 \end{bmatrix}.
$$

Let $\mathbf{h} = \mathbf{f} \circ \mathbf{g}$. The composition $\mathbf{h}$ is continuously differentiable and its derivative at $(x, y) \in (\mathbb{R}_{>0} \times \mathbb{R})$ is given by

$$
\mathbf{h}'(x, y) = \mathbf{J_h}(x, y) = \mathbf{J_f}(\mathbf{g}(x, y)) \cdot \mathbf{J_g}(x, y)
$$

$$
= \begin{bmatrix} (x + 2y)x & x \\ 2\ln(x)(x + 2y)^3 & [\ln(x)]^2 3(x + 2y)^2 \\ -1 & 0 \end{bmatrix} \cdot \begin{pmatrix} 1/x & 0 \\ 1 & 2 \end{pmatrix}
$$

$$
= \begin{bmatrix} (x + 2y) + x & 2x \\ (1/x)2\ln x(x + 2y)^3 + (\ln x)^2 3(x + 2y)^2 & 2(\ln x)^2 3(x + 2y)^2 \\ -1/x & 0 \end{bmatrix}.
$$

The reader is encouraged to calculate an expression for $\mathbf{h}(x, y)$ and compute $\mathbf{J_h}$ directly. ∎

For the chain rule (A.146), a special case of interest is $n = p = 1$, i.e. $\mathbf{f} = (f_1, \ldots, f_m) : A \subset \mathbb{R} \to \mathbb{R}^m$ and $g : B \subset \mathbb{R}^m \to \mathbb{R}$. Then for $x \in A$, $\mathbf{J}_g$ is a row vector and $\mathbf{J}_\mathbf{f}$ is a column vector, so that (A.146) simplifies to

$$\mathbf{J}_{g \circ \mathbf{f}}(x) = \sum_{i=1}^{m} \frac{\partial g}{\partial f_i}(\mathbf{f}(x)) \frac{\mathrm{d} f_i}{\mathrm{d} x}(x), \tag{A.147}$$

where $\partial g / \partial f_i$ denotes the ith partial derivative of g.[22] A mnemonic version of this formula is, with $h = g \circ \mathbf{f}$,

$$\frac{\mathrm{d} h}{\mathrm{d} x} = \sum_{i=1}^{m} \frac{\partial h}{\partial f_i} \frac{\mathrm{d} f_i}{\mathrm{d} x}.$$

⊖ **Example A.72** Assume that the United States GDP, denoted by P, is a continuously differentiable function of the capital, C, and the work force, W. Moreover, assume C and W are continuously differentiable functions of time, t. Then P is a continuously differentiable function of t and economists would write:

$$\frac{\mathrm{d} P}{\mathrm{d} t} = \frac{\partial P}{\partial C} \frac{\mathrm{d} C}{\mathrm{d} t} + \frac{\partial P}{\partial W} \frac{\mathrm{d} W}{\mathrm{d} t},$$

showing us how the change of P can be split into a part due to the decrease or increase of C and another due to the change of W. ∎

⊖ **Example A.73** (Counterexample to A.72) Consider the following functions:

$$g : \mathbb{R}^2 \to \mathbb{R}, (x, y) \mapsto \begin{cases} \dfrac{x^3 + y^3}{x^2 + y^2}, & \text{if } (x, y) \neq (0, 0), \\ 0, & \text{if } (x, y) = (0, 0), \end{cases}$$

and $\mathbf{f} : \mathbb{R} \to \mathbb{R}^2, t \mapsto (t, t)$. The partial derivatives of g exist on the whole domain, in particular, at $(x, y) = (0, 0)$:

$$\frac{\partial g}{\partial x}(0, 0) = \lim_{x \to 0} \frac{g(x, 0) - g(0, 0)}{x - 0} = \lim_{x \to 0} \frac{x^3 + 0}{x(x^2 + 0)} = \lim_{x \to 0} 1 = 1 = \frac{\partial g}{\partial y}(0, 0),$$

but the partial derivatives of g are not continuous. However, $\mathbf{f}$ is continuously differentiable with derivative

$$\mathbf{f}' : \mathbb{R} \to \mathbb{R}^2, t \mapsto \begin{pmatrix} 1 \\ 1 \end{pmatrix}.$$

---

[22] This notation is somewhat misleading, as f_i is a function itself. Keep in mind that, in '$\partial g / \partial f_i$', the f_i could be replaced by, say, y_i or any other name of variable as long as it can be easily inferred which partial derivative is meant.

So, if the chain rule were applicable here, then the derivative of $h := g \circ \mathbf{f}$ at $t = 0$ is calculated to be

$$h'(0) = \mathbf{J}_g(\mathbf{f}(0))\mathbf{J}_{\mathbf{f}}(0) = (1 \quad 1)\begin{pmatrix} 1 \\ 1 \end{pmatrix} = 2.$$

On the other hand, we can calculate $h(t)$ for $t \in \mathbb{R}$ directly as

$$h(t) = \frac{t^3 + t^3}{t^2 + t^2} = \frac{2t^3}{2t^2} = t,$$

so that $h'(0) = 1$. This demonstrates that the chain rule generally does not hold when one of the functions is not continuously differentiable. ∎

Higher-order derivatives and Taylor series

Let $f : A \to \mathbb{R}$ with $A \subset \mathbb{R}^n$ an open set such that the partial derivatives $D_i f(\mathbf{x})$ are continuous at point $\mathbf{x} = (x_1, \ldots, x_n) \in A$, $i = 1, 2, \ldots, n$. As $D_i f$ is a function, its partial derivative may be computed, if it exists, i.e. we can apply the D_j operator to $D_i f$ to get $D_j D_i f$, called the iterated partial derivative of f with respect to i and j. The following is a fundamental result, the proof of which can be found in most of the analysis books mentioned in Section A.0, for example, Lang (1997, p. 372) or Protter and Morrey (1991, p. 179).

- If $D_i f$, $D_j f$, $D_i D_j f$ and $D_j D_i f$ exist and are continuous, then

$$D_i D_j f = D_j D_i f. \tag{A.148}$$

This extends as follows. Let $D_i^k f$ denote k applications of D_i to f, e.g., $D_i^2 f = D_i D_i f$. Then, for nonnegative integers k_i, $i = 1, \ldots, n$, any iterated partial derivative operator can be written as $D_1^{k_1} \cdots D_n^{k_n}$, with $D_i^0 = 1$, i.e. the derivative with respect to the ith variable is not taken. The *order* of $D_1^{k_1} \cdots D_n^{k_n}$ is $\sum k_i$. Extending (A.148), if $D_i^j f$ is continuous for $i = 1, \ldots, n$ and $j = 0, \ldots, k_i$, then $D_1^{k_1} \cdots D_n^{k_n} f$ is the same for all possible orderings of the elements of $D_1^{k_1} \cdots D_n^{k_n}$.

Let $f : I \to \mathbb{R}$, with $I \subset \mathbb{R}$ an open interval which contains points x and $x + c$. From (A.132), the Taylor series expansion of $f(c + x)$ around c can be expressed as

$$f(c + x) = \sum_{k=0}^{r} \frac{f^{(k)}(c)}{k!} x^k + \text{remainder term}, \tag{A.149}$$

if $f^{(r+1)}$ exists, where the order of the expansion is r (before we used n, as is standard convention, but now n is the dimension of the domain of f, which is also standard). This can be extended to function $f : \mathbb{R}^n \to \mathbb{R}$. We first consider the case with $n = 2$ and $r = 2$, which suffices for the applications considered in the main text.

Let $f : A \to \mathbb{R}$ with $A \subset \mathbb{R}^2$ an open set, and let $\mathbf{x} = (x_1, x_2)$ and $\mathbf{c} = (c_1, c_2)$ be column vectors such that $(\mathbf{c} + t\mathbf{x}) \in A$ for $0 \le t \le 1$. Let $g : I \to \mathbb{R}$ be the univariate function defined by $g(t) = f(\mathbf{c} + t\mathbf{x})$ for $I = [0, 1]$, so that $g(0) = f(\mathbf{c})$ and $g(1) = f(\mathbf{c} + \mathbf{x})$. Applying (A.149) to g (with $c = 0$ and $x = 1$) gives

$$g(1) = g(0) + g'(0) + \frac{g''(0)}{2} + + \frac{g'''(0)}{6} + \cdots + \frac{g^{(k)}(0)}{k!} + \cdots. \tag{A.150}$$

From the chain rule (A.147),

$$g'(t) = (\text{grad } f)(\mathbf{c} + t\mathbf{x})\,\mathbf{x} = x_1 D_1 f(\mathbf{c} + t\mathbf{x}) + x_2 D_2 f(\mathbf{c} + t\mathbf{x}) =: f_1(\mathbf{c} + t\mathbf{x}), \tag{A.151}$$

where f_1 is the linear combination of differential operators applied to f given by $f_1 := (x_1 D_1 + x_2 D_2) f$. Again from the chain rule, now applied to f_1, and using (A.148),

$$\begin{aligned}
g''(t) &= (\text{grad } f_1)(\mathbf{c} + t\mathbf{x})\,\mathbf{x} \\
&= x_1 D_1 f_1(\mathbf{c} + t\mathbf{x}) + x_2 D_2 f_1(\mathbf{c} + t\mathbf{x}) \\
&= x_1^2 (D_1^2 f)(\mathbf{c} + t\mathbf{x}) + 2x_1 x_2 (D_1 D_2 f)(\mathbf{c} + t\mathbf{x}) + x_2^2 (D_2^2 f)(\mathbf{c} + t\mathbf{x}) \\
&= \left[(x_1 D_1 + x_2 D_2)^2 f \right](\mathbf{c} + t\mathbf{x}) =: f_2(\mathbf{c} + t\mathbf{x}).
\end{aligned}$$

This and (A.151) give $g'(0) = f_1(\mathbf{c})$ and $g''(0) = f_2(\mathbf{c})$, so that (A.150) yields

$$f(\mathbf{c} + \mathbf{x}) = f(\mathbf{c}) + f_1(\mathbf{c}) + \frac{1}{2} f_2(\mathbf{c}) + \cdots \tag{A.152}$$

$$= f(\mathbf{c}) + x_1 D_1 f(\mathbf{c}) + x_2 D_2 f(\mathbf{c})$$

$$+ \frac{x_1^2}{2}(D_1^2 f)(\mathbf{c}) + x_1 x_2 (D_1 D_2 f)(\mathbf{c}) + \frac{x_2^2}{2}(D_2^2 f)(\mathbf{c}) + \cdots. \tag{A.153}$$

If we 'remove' the second coordinate used in f, writing x instead of $\mathbf{x} = (x_1, x_2)$ and similar for $\mathbf{c}$, then (A.153) simplifies to

$$f(c + x) = f(c) + x D_1 f(c) + \frac{x^2}{2}(D_1^2 f)(c) + \cdots,$$

which agrees with (A.149).

From (A.150), expansion (A.152) can be continued with

$$\begin{aligned}
g'''(t) &= \frac{d}{dt} f_2 (\mathbf{c} + t\mathbf{x}) = (\text{grad } f_2)\,(\mathbf{c} + t\mathbf{x})\,\mathbf{x} = x_1 D_1 f_2 (\mathbf{c} + t\mathbf{x}) + x_2 D_2 f_2 (\mathbf{c} + t\mathbf{x}) \\
&= x_1 D_1 \left[x_1^2 (D_1^2 f)(\mathbf{c} + t\mathbf{x}) + 2x_1 x_2 (D_1 D_2 f)(\mathbf{c} + t\mathbf{x}) + x_2^2 (D_2^2 f)(\mathbf{c} + t\mathbf{x}) \right] \\
&\quad + x_2 D_2 \left[x_1^2 (D_1^2 f)(\mathbf{c} + t\mathbf{x}) + 2x_1 x_2 (D_1 D_2 f)(\mathbf{c} + t\mathbf{x}) + x_2^2 (D_2^2 f)(\mathbf{c} + t\mathbf{x}) \right] \\
&= x_1^3 (D_1^3 f)(\mathbf{c} + t\mathbf{x}) + 3x_1^2 x_2 (D_1^2 D_2 f)(\mathbf{c} + t\mathbf{x}) \\
&\quad + 3x_1 x_2^2 (D_1 D_2^2 f)(\mathbf{c} + t\mathbf{x}) + x_2^3 (D_2^3 f)(\mathbf{c} + t\mathbf{x}) \\
&= \left[(x_1 D_1 + x_2 D_2)^3 f \right](\mathbf{c} + t\mathbf{x}) \\
&=: f_3(\mathbf{c} + t\mathbf{x}),
\end{aligned}$$

and it seems natural to postulate that (A.152) takes the form

$$f(\mathbf{c} + \mathbf{x}) = \sum_{k=0}^{\infty} \frac{f_k(\mathbf{c})}{k!}, \tag{A.154}$$

where $f_k := (x_1 D_1 + x_2 D_2)^k f$, $k = 0, 1, \ldots$. This is indeed true if all the derivatives exist, and can be proven by induction. Note that f_k can be expanded by the binomial theorem (1.18).

Expression (A.154) is for $n = 2$, although the extension to the case of general n is the same, except that

$$f_k := (x_1 D_1 + \cdots + x_n D_n)^k f = (\nabla \mathbf{x})^k f,$$

where we use ∇ to represent the operator which, when applied to f, returns the gradient, i.e. $\nabla := (D_1, \ldots, D_n)$, which is a row vector, and $\mathbf{x} = (x_1, \ldots, x_n)$ is a column vector. Note that $f_1(\mathbf{c})$ is the total differential. Now, f_k is evaluated via the multinomial theorem (1.34), and each f_k will have $\binom{k+n-1}{k}$ terms. With this notation, and assuming all relevant partial derivatives exist,

$$f(\mathbf{c} + \mathbf{x}) = \sum_{k=0}^{r} \frac{\left[(\nabla \mathbf{x})^k f\right](\mathbf{c})}{k!} + \text{remainder term},$$

where it can be shown (see, for example, Lang, 1997, Section 15.5) that the

$$\text{remainder term} = \frac{\left[(\nabla \mathbf{x})^{r+1} f\right](\mathbf{c} + t\mathbf{x})}{(r+1)!}, \quad \text{for} \quad 0 \le t \le 1.$$

A.3.4 Integration

Mathematics is not a deductive science – that's a cliche. When you try to prove a theorem, you don't just list the hypotheses, and then start to reason. What you do is trial and error, experimentation, guesswork.

(Paul R. Halmos)

Exchange of derivative and integral

We begin with the univariate Riemann integral as in Section A.2.3, but applied to a function of two variables. Let $a_1, a_2, b_1, b_2 \in \mathbb{R}$ with $a_1 < b_1$, $a_2 < b_2$, and let $D := [a_1, b_1] \times [a_2, b_2]$ be a closed rectangle in $\mathbb{R}^2$.

- If $f : D \to \mathbb{R}$ is continuous, then so is

$$\phi(x) := \int_{a_1}^{b_1} f(t, x) \, \mathrm{d}t. \tag{A.155}$$

Proof: This is a special case of the following result, which is required for proving Fubini's theorem given below.   ∎

- $\forall (t, x) \in D$,

$$\psi (t, x) := \int_{a_2}^{x} f (t, u) \, du \quad \text{is continuous.} \tag{A.156}$$

Proof: Using (A.57),

$$\psi (t, x) - \psi (t_0, x_0) = \int_{a_2}^{x} f (t, u) \, du - \int_{a_2}^{x_0} f (t_0, u) \, du$$

$$= \int_{a_2}^{x_0} \left[f (t, u) - f (t_0, u) \right] \, du + \int_{x_0}^{x} f (t, u) \, du. \tag{A.157}$$

As D is closed and bounded and f is continuous on D, f is bounded by some number, say K, and is uniformly continuous on D. Let $\epsilon > 0$. To bound the second integral in (A.157), choose x_0 such that $|x - x_0| < \epsilon / K$. For the first integral, as f is uniformly continuous, there exists δ_1 such that, whenever $|t - t_0| < \delta_1$, $|f (t, u) - f (t_0, u)| < \epsilon$. Let $\delta = \min (\epsilon / K, \delta_1)$. Then for $|x - x_0| < \delta$ and $|t - t_0| < \delta$, (A.157) is such that

$$\psi (t, x) - \psi (t_0, x_0) \leq \epsilon |x_0 - a_2| + \epsilon,$$

which proves that ψ is continuous.   ∎

Next, we wish to know the conditions under which differentiation and integration can be exchanged. Let $f : D \to \mathbb{R}$ and $D_2 f$ be continuous on D where, from (A.134),

$$D_2 f (t, x) = \lim_{h \to 0} \frac{f (t, x + h) - f (t, x)}{h}.$$

- *Function* $g (x) := \int_{a_1}^{b_1} f (t, x) \, dt$ *is differentiable, and*

$$g' (x) = \int_{a_1}^{b_1} D_2 f (t, x) \, dt. \tag{A.158}$$

Proof: As $D_2 f$ is continuous on D, (A.52) implies that $\int_{a_1}^{b_1} D_2 f (t, x) \, dt$ exists, so if (A.158) is true, then g is differentiable. To show (A.158), as in Lang (1997, p. 276), write

$$\frac{g (x + h) - g (x)}{h} - \int_{a_1}^{b_1} D_2 f (t, x) \, dt$$

$$= \int_{a_1}^{b_1} \left[\frac{f (t, x + h) - f (t, x)}{h} - D_2 f (t, x) \right] \, dt.$$

By the mean value theorem (A.35), for each t there exists a number $c_{t,h}$ between x and $x + h$ such that

$$\frac{f(t, x + h) - f(t, x)}{h} = D_2 f(t, c_{t,h}).$$

As $D_2 f$ is uniformly continuous on the closed, bounded interval D,

$$\left| \frac{f(t, x + h) - f(t, x)}{h} - D_2 f(t, x) \right| = \left| D_2 f(t, c_{t,h}) - D_2 f(t, x) \right| < \frac{\epsilon}{b_1 - a_1},$$

where $\epsilon > 0$, whenever h is sufficiently small. ∎

- A sufficient condition for result (A.158) to be true when $b_1 = \infty$ is that f and $D_2 f$ are absolutely convergent. That is, for $D := [a_1, \infty) \times [a_2, b_2]$, $a_2 < b_2$, (A.158) holds if there are nonnegative functions $\phi(t)$ and $\psi(t)$ such that $|f(t, x)| \leq \phi(t)$ and $|D_2 f(t, x)| \leq \psi(t)$ for all $t, x \in D$, and $\int_{a_1}^{\infty} \phi$ and $\int_{a_1}^{\infty} \psi$ converge. See, for example, Lang (1997, p. 337) for proof. ∎

⊖ **Example A.74** Consider calculating the derivative at zero of the function

$$f(t) = \int_{-1}^{1} \sin(ts) e^{s+t} \, ds.$$

Differentiating under the integral sign gives

$$f'(t) = \int_{-1}^{1} \left[s \cos(ts) e^{s+t} + \sin(ts) e^{s+t} \right] ds,$$

so that

$$f'(0) = \int_{-1}^{1} \left[s \cos(0s) e^{s+0} + \sin(0s) e^{s+0} \right] ds$$

$$= \int_{-1}^{1} (s e^s + 0) \, ds = (s - 1) e^s \big|_{-1}^{1} = 2e^{-1}.$$

This method is quite straightforward (at least for $t = 0$) and obviates the need for a direct calculation of the complicated integral expression of f. ∎

Fubini's theorem

In studying multivariate random variables, one has to work with and integrate functions with domain $\mathbb{R}^n$ where $n > 1$. We now concentrate just on the two-dimensional case, with generalizations obvious for $n > 2$.

Let $a_1, a_2, b_1, b_2 \in \mathbb{R}$ with $a_1 < b_1$, $a_2 < b_2$, and let $D := [a_1, b_1] \times [a_2, b_2]$. For an arbitrary function $f : D \to \mathbb{R}$, one can define the Riemann integral of f similarly to the one-dimensional case. There, the domain of definition of the function being integrated is split into ever-shorter intervals but here, in the two-dimensional case, one has to partition the *rectangle* into smaller and smaller rectangles. If f is Riemann integrable, then the Riemann sums, defined similarly to those in the one-dimensional case, converge to a value called the Riemann integral of f, denoted by $\int_D f$ or $\int_D f(\mathbf{x}) \, d\mathbf{x}$.

Now let $f : D \to \mathbb{R}$ be a continuous function. Analogously to the one-dimensional case, f is Riemann integrable on the set D, and we can use *Fubini's theorem*, due to Guido Fubini (1879–1943), to calculate its integral:

$$\int_D f(\mathbf{x}) \, d\mathbf{x} = \int_{a_2}^{b_2} \left[\int_{a_1}^{b_1} f(x_1, x_2) \, dx_1 \right] dx_2 = \int_{a_1}^{b_1} \left[\int_{a_2}^{b_2} f(x_1, x_2) \, dx_2 \right] dx_1.$$

(A.159)

Observe that (A.159) is a set of nested *univariate* Riemann integrals. This can be extended in an obvious way to the n-dimensional case with $\mathbf{x} = (x_1, \ldots, x_n)$. Fubini's theorem holds whenever f is Riemann integrable; in particular, when $f : A \subset \mathbb{R}^n \to \mathbb{R}$ is continuous and A is closed and bounded.

Proof: As in Lang (1997, p. 277), we wish to show that

$$\int_{a_2}^{b_2} \left[\int_{a_1}^{b_1} f(t, x) \, dt \right] dx = \int_{a_1}^{b_1} \left[\int_{a_2}^{b_2} f(t, x) \, dx \right] dt.$$

Let $\psi(t, x) = \int_{a_2}^{x} f(t, u) \, du$, so that $D_2 \psi(t, x) = f(t, x)$ from the FTC (A.62), and ψ is continuous from (A.156). We can now apply (A.158) to ψ and $D_2 \psi = f$. Let $g(x) = \int_{a_1}^{b_1} \psi(t, x) \, dt$. Then

$$g'(x) = \int_{a_1}^{b_1} D_2 \psi(t, x) \, dt = \int_{a_1}^{b_1} f(t, x) \, dt,$$

and, from the FTC (A.60),

$$g(b_2) - g(a_2) = \int_{a_2}^{b_2} g'(x) \, dx = \int_{a_2}^{b_2} \left[\int_{a_1}^{b_1} f(t, x) \right] dx.$$

On the other hand,

$$g(b_2) - g(a_2) = \int_{a_1}^{b_1} \psi(t, b_2) \, dt - \int_{a_1}^{b_1} \psi(t, a_2) \, dt = \int_{a_1}^{b_1} \left[\int_{a_2}^{b_2} f(t, u) \, du \right] dt,$$

and the theorem is proved. ∎

⊖ **Example A.75** The above proof of Fubini's theorem used the interchange of derivative and integral result (A.158). It is instructive to go the other way, proving (A.158) with the use of Fubini's theorem. With $D = [a_1, b_1] \times [a_2, b_2]$ and $f : D \to \mathbb{R}$ and $D_2 f$ continuous functions, we wish to show that, for $t \in (a_2, b_2)$,

$$\frac{d}{dt} \int_{a_1}^{b_1} f(x_1, t) \, dx_1 = \int_{a_1}^{b_1} D_2 f(x_1, t) \, dx_1. \tag{A.160}$$

Define $h : [a_2, b_2] \to \mathbb{R}$ as $h(x_2) := \int_{a_1}^{b_1} D_2 f(x_1, x_2) \, dx_1$. As $D_2 f$ is continuous, it follows from (A.155) that h is continuous on $[a_2, b_2]$. Choosing an arbitrary t with $a_2 < t < b_2$ and integrating $h(x_2)$ over the interval $[a_2, t]$, we obtain

$$\int_{a_2}^{t} h(x_2) \, dx_2 = \int_{a_2}^{t} \left[\int_{a_1}^{b_1} \frac{\partial f(x_1, x_2)}{\partial x_2} \, dx_1 \right] dx_2. \tag{A.161}$$

The order of integration in (A.161) can be reversed by Fubini's theorem so that, using the FTC (A.60),

$$\int_{a_2}^{t} h(x_2) \, dx_2 \stackrel{\text{Fubini}}{=} \int_{a_1}^{b_1} \left[\int_{a_2}^{t} \frac{\partial f(x_1, x_2)}{\partial x_2} \, dx_2 \right] dx_1$$

$$\stackrel{\text{FTC}}{=} \int_{a_1}^{b_1} \left[f(x_1, t) - f(x_1, a_2) \right] dx_1$$

$$= \int_{a_1}^{b_1} f(x_1, t) \, dx_1 - \int_{a_1}^{b_1} f(x_1, a_2) \, dx_1. \tag{A.162}$$

From the FTC (A.62) and differentiating both sides with respect to t, we obtain

$$\frac{d}{dt} \int_{a_2}^{t} h(x_2) \, dx_2 = h(t) = \int_{a_1}^{b_1} D_2 f(x_1, t) \, dx_1 \stackrel{(A.162)}{=} \frac{d}{dt} \int_{a_1}^{b_1} f(x_1, t) \, dx_1,$$

showing the result. ∎

⊖ ***Example A.76*** Let $f : \mathbb{R}^2 \to \mathbb{R}$, $(x, y) \mapsto y^2 e^{2x}$. From (A.52), the fact that f is continuous implies that it is Riemann integrable on bounded rectangles. Let $a_1 = a_2 = 0$ and $b_1 = b_2 = 1$. If $D = [a_1, b_1] \times [a_2, b_2]$ we have

$$\int_D f = \int_0^1 \int_0^1 y^2 e^{2x} \, dx \, dy = \int_0^1 \left[y^2 \frac{1}{2} e^{2x} \right]_{x=0}^{x=1} dy = \int_0^1 y^2 \frac{1}{2} e^2 - y^2 \frac{1}{2} 1 \, dy$$

$$= \left[\frac{1}{3} y^3 \left(\frac{1}{2} e^2 - \frac{1}{2} \right) \right]_{y=0}^{y=1} = \frac{1}{6} (e^2 - 1).$$

The same result can be easily derived when interchanging the order of integration. However, in this example, the calculations can be simplified by *factorizing* the integrated function:

$$\int_D f = \int_0^1 \int_0^1 y^2 e^{2x} \, dx \, dy = \int_0^1 y^2 \int_0^1 e^{2x} \, dx \, dy = \int_0^1 e^{2x} \, dx \int_0^1 y^2 \, dy$$

$$= \left[\frac{1}{2} e^{2x} \right]_0^1 \left[\frac{1}{3} y^3 \right]_0^1 = \frac{1}{2} (e^2 - 1) \frac{1}{3}.$$

Usually, a factorization will not be possible. ∎

We are often interested in integration on more general regions, especially on sets that are unbounded, such as $\mathbb{R}^2$ or regions of the form $D_\infty := (-\infty, b_1] \times (-\infty, b_2]$. In order to define an integral on the latter, let $D_1 \subset D_2 \subset \dots$ be a sequence of closed and bounded rectangles with $\bigcup_{k \in \mathbb{N}} D_k = D_\infty$. Then we define

$$\int_{D_\infty} f := \lim_{k \to \infty} \int_{D_k} f,$$

if the r.h.s exists, and is equal, for any such series of rectangles. (This can be compared to Example A.27 and its subsequent remark.) Similar definitions will apply to other types of unbounded regions. Under certain conditions of f (see, for example, Lang, 1997, p. 342–3), Fubini's theorem still applies, i.e.,

$$\int_{D_\infty} f = \int_{-\infty}^{b_2} \left[\int_{-\infty}^{b_1} f(x_1, x_2) \, dx_1 \right] dx_2 = \int_{-\infty}^{b_1} \left[\int_{-\infty}^{b_2} f(x_1, x_2) \, dx_2 \right] dx_1.$$

⊖ **Example A.77**   As in Lang (1997, p. 344(8)) and Beerends et al (2003, p. 155(6.6)), let $f : (\mathbb{R}_{>0} \times \mathbb{R}_{>0}) \to \mathbb{R}$ with $f(x, y) = (x - y)/(x + y)^3$ and define

$$A = \int_1^\infty \int_1^\infty f(x, y) \, dx \, dy \quad \text{and} \quad B = \int_1^\infty \int_1^\infty f(x, y) \, dy \, dx.$$

Both A and B exist, but are unequal. In particular,

$$A = \int_1^\infty \left(\int_1^\infty \frac{x - y}{(x + y)^3} dx \right) dy = \int_1^\infty \left(2 \int_1^\infty \frac{x \, dx}{(x + y)^3} - \int_1^\infty \frac{dx}{(x + y)^2} \right) dy$$

$$= \int_1^\infty \left(\frac{y + 2}{2y + y^2 + 1} - \frac{1}{y + 1} \right) dy = \int_1^\infty \frac{1}{(1 + y)^2} dy = \frac{1}{2},$$

and a very similar calculation reveals that $B = -1/2$.   ∎

Leibniz' rule

As above, let $a_1, a_2, b_1, b_2 \in \mathbb{R}$ with $a_1 < b_1$, $a_2 < b_2$, let $D := [a_1, b_1] \times [a_2, b_2]$, and let $f : D \to \mathbb{R}$ be continuous. Also let λ and θ be differentiable functions defined on $[a_2, b_2]$ such that $\lambda(x), \theta(x) \in [a_1, b_1]$ for all $x \in [a_2, b_2]$ and define the function $A : [a_2, b_2] \to \mathbb{R}$ by

$$A(x) := \int_{\lambda(x)}^{\theta(x)} f(x, y) \, dy. \tag{A.163}$$

We wish to determine if this function is differentiable and, if so, derive an expression for its derivative. First, define the function $H : [a_1, b_1] \times [a_1, b_1] \times [a_2, b_2] \to \mathbb{R}$, depending on the three variables a, b and x, by

$$H(a, b, x) := \int_a^b f(x, y) \, dy.$$

Note that $A(x) = H(\lambda(x), \theta(x), x)$ for every $x \in [a_2, b_2]$. From the FTC (A.62), H is differentiable for any $a, b \in (a_2, b_2)$ and $x \in (a_1, b_1)$, with

$$\frac{\partial H}{\partial b}(a, b, x) = \frac{\partial}{\partial b} \int_a^b f(x, y)\, dy = f(x, b)$$

$$\frac{\partial H}{\partial a}(a, b, x) = \frac{\partial}{\partial a} \int_a^b f(x, y)\, dy = -\frac{\partial}{\partial a} \int_b^a f(x, y)\, dy = -f(x, a),$$

and from (A.158),

$$\frac{\partial H}{\partial x}(a, b, x) = \frac{\partial}{\partial x} \int_a^b f(x, y)\, dy = \int_a^b \frac{\partial f(x, y)}{\partial x}\, dy.$$

From the chain rule (A.147), it follows that, for $a_1 < x < b_1$, A is differentiable and

$$A'(x) = \frac{\partial H}{\partial \lambda} \frac{d\lambda}{dx} + \frac{\partial H}{\partial \theta} \frac{d\theta}{dx} + \frac{\partial H}{\partial x} \frac{dx}{dx}$$

$$= -f(x, \lambda(x))\, \lambda'(x) + f(x, \theta(x))\, \theta'(x) + \int_{\lambda(x)}^{\theta(x)} \frac{\partial f(x, y)}{\partial x}\, dy. \qquad \text{(A.164)}$$

Formula (A.164) is sometimes called 'Leibniz' rule for differentiating an integral'.

⊖ **Example A.78** Consider the function $f(t) := \int_0^t e^{st}\, ds$. There are two possible ways to calculate its derivative at $t = 1$. Firstly, let us integrate in step one and then differentiate afterwards. For $t > 0$,

$$f(t) = \left[\frac{1}{t} e^{st}\right]_0^t = \frac{1}{t}\left(e^{t^2} - 1\right),$$

and

$$f'(t) = \frac{-1}{t^2}\left(e^{t^2} - 1\right) + \frac{1}{t} 2t e^{t^2} = \frac{-1}{t^2}\left(e^{t^2} - 1\right) + 2e^{t^2},$$

so that $f'(1) = -(e - 1) + 2e = 1 + e$.

Secondly, we can differentiate first, using Leibniz' rule, and then integrate in a second step. For $t > 0$,

$$f'(t) = \int_0^t s e^{st}\, ds + 1 \cdot e^{t^2},$$

hence

$$f'(1) = \int_0^1 s e^s\, ds + e^1 = \left[(s - 1)e^s\right]_0^1 + e = 0 - (-1) + e = 1 + e.$$

Here, Leibniz' rule obviously saves us some time. ∎

Although Leibniz' rule (A.164) follows directly from the multivariate chain rule, the expression itself does not appear to hold much intuition, and one might wonder how the formula could have been postulated without knowledge of the chain rule. It turns out that, with the right geometric representation, the formula is essentially obvious! This pleasant result is due to Frantz (2001), on which the following is based. Firstly, it suffices to set the lower limit $\lambda(x)$ in (A.163) to zero because, from (A.57),

$$A(x) = \int_{\lambda(x)}^{\theta(x)} f(x, y) \, dy = \int_0^{\theta(x)} f(x, y) \, dy - \int_0^{\lambda(x)} f(x, y) \, dy.$$

Figure A.6(a)[23] shows a cross section, or 'slice' (or *lamina*) of A at a particular value of x, with $y = \theta(x)$ lying in the xy-plane. Figure A.6(b) also shows the lamina at $x + \triangle x$, with area $A(x + \triangle x)$, so that the change in height of the lamina, for any y, is $f(x + \triangle x, y) - f(x, y) \approx D_1 f(x, y) \triangle x =: f_x(x, y) \triangle x$. Similarly, the width of the lamina increases by approximately $\theta'(x) \triangle x$.

Figure A.6(c) just isolates this lamina for clarity. The change in the lamina's area, $\triangle A$, is then $A_1 + A_2$, plus the upper-right corner, which, compared to the size of A_1 and A_2, can be ignored (it becomes negligible much faster than A_1 and A_2 as $\triangle x \to 0$). Thus,

$$A_1 \approx \int_0^{\theta(x)} f_x(x, y) \, dy \, \triangle x \quad \text{and} \quad A_2 \approx f(x, \theta(x)) \, \theta'(x) \, \triangle x,$$

i.e. dividing by $\triangle x$ gives

$$\frac{\triangle A}{\triangle x} \approx \int_0^{\theta(x)} f_x(x, y) \, dy + f(x, \theta(x)) \theta'(x) = A'(x),$$

as in (A.164).

Integral transformations and polar coordinates

Let $f : D \to \mathbb{R}$ be a continuous function, where domain D is an open subset of $\mathbb{R}^n$ with typical element $\mathbf{x} = (x_1, \ldots, x_n)$. Let $\mathbf{g}(\mathbf{x}) = (g_1(\mathbf{x}), \ldots, g_n(\mathbf{x})) : D \to \mathbb{R}^n$ be a differentiable bijection with nonvanishing Jacobian

$$\mathbf{J} = \begin{pmatrix} \dfrac{\partial x_1}{\partial y_1} & \dfrac{\partial x_1}{\partial y_2} & \cdots & \dfrac{\partial x_1}{\partial y_n} \\[2mm] \dfrac{\partial x_2}{\partial y_1} & \dfrac{\partial x_2}{\partial y_2} & & \dfrac{\partial x_2}{\partial y_n} \\[2mm] \vdots & \vdots & \ddots & \vdots \\[2mm] \dfrac{\partial x_n}{\partial y_1} & \dfrac{\partial x_n}{\partial y_2} & \cdots & \dfrac{\partial x_n}{\partial y_n} \end{pmatrix}, \quad \text{where} \quad \mathbf{y} = \mathbf{g}(\mathbf{x}).$$

---

[23] I am very grateful to Marc Frantz, the author of Frantz (2001), for constructing and providing me with the three graphs shown in the figure.

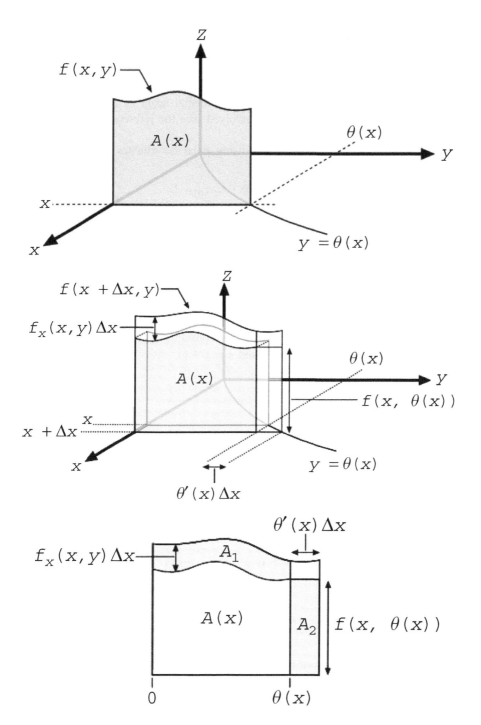

Figure A.6 Geometric motivation for Leibniz' rule (based on plots in M. Frantz, *Visualizing Leibniz's Rule, Mathematics Magazine*, 2001, 74(2):143–144).

Then, for $S \subset D$,

$$\int_S f(\mathbf{x}) \, d\mathbf{x} = \int_{\mathbf{g}(S)} f\left(\mathbf{g}^{-1}(\mathbf{y})\right) |\det \mathbf{J}| \, d\mathbf{y}, \tag{A.165}$$

where $\mathbf{g}(S) = \{\mathbf{y} : \mathbf{y} = \mathbf{g}(\mathbf{x}), \mathbf{x} \in S\}$. This is referred to as the *multivariate change of variable formula*. This a well-known and fundamental result in analysis, the rigorous proof of which, however, is somewhat involved (see the references given in Section A.0 above).

Consider the special case of (A.165) using polar coordinates, i.e. $x = c_1(r, \theta) = r \cos \theta$ and $y = c_2(r, \theta) = r \sin \theta$. Then

$$\det \mathbf{J} = \begin{vmatrix} \partial x/\partial r & \partial x/\partial \theta \\ \partial y/\partial r & \partial y/\partial \theta \end{vmatrix} = \begin{vmatrix} \cos \theta & -r \sin \theta \\ \sin \theta & r \cos \theta \end{vmatrix} = r \cos^2 \theta + r \sin^2 \theta = r,$$
$$\tag{A.166}$$

which is positive, so that (A.165) implies

$$\iint f(x, y) \, dx dy = \iint f(r \cos \theta, r \sin \theta) \, r \, dr \, d\theta. \tag{A.167}$$

This is a particularly useful transformation when $f(x, y)$ depends only on the distance measure $r^2 = x^2 + y^2$ and the range of integration is a circle centered around $(0, 0)$. For example, if $f(x, y) = (k + x^2 + y^2)^p$ for constants k and $p \geq 0$, and S is such a circle with radius a, then (with $t = r^2$ and $s = k + t$),

$$I = \iint_S f(x, y) \, dx \, dy$$

$$= \int_0^{2\pi} \int_0^a \left(k + r^2\right)^p r \, dr \, d\theta = \pi \int_0^{a^2} (k + t)^p \, dt = \frac{\pi}{p + 1} \left[\left(k + a^2\right)^{p+1} - k^{p+1} \right].$$

For $k = p = 0$, I reduces to πa^2, the area of a circle of radius a.

Our most important application of this technique is the following.

⊙ **Example A.79**   To compute $I_1 = \int_{-\infty}^{\infty} \exp\left(-\frac{1}{2}t^2\right) \, dt = 2 \int_0^{\infty} \exp\left(-\frac{1}{2}t^2\right) \, dt$, let $x^2 = t^2/2$ so that

$$I_1 = 2\sqrt{2} \int_0^{\infty} \exp\left(-x^2\right) \, dx = \sqrt{2} \int_{-\infty}^{\infty} \exp\left(-x^2\right) \, dx =: \sqrt{2} I_2.$$

Then, observe that

$$I_2^2 = \int_{-\infty}^{\infty} \exp\left(-x^2\right) \, dx \int_{-\infty}^{\infty} \exp\left(-y^2\right) \, dy = \int_{-\infty}^{\infty} \int_{-\infty}^{\infty} e^{-x^2 + y^2} \, dx \, dy$$

or, transforming to polar coordinates,

$$I_2^2 = \int_0^{2\pi} \int_0^{\infty} \exp\left(-r^2\right) r \, dr \, d\theta = 2\pi \lim_{t \to \infty} \left(-\frac{1}{2} e^{-r^2} \Big|_0^t \right) = \pi,$$

so that $I_2 = \sqrt{\pi}$ and $I_1 = \sqrt{2\pi}$.   ■

B

Notation tables

Table B.1 Abbreviations[a]

English	Description
c.d.f.	cumulative distribution function
d.n.e.	does not exist
i.i.d.	independently and identically distributed
l.h.s. (r.h.s.)	left-(right-) hand side
nonis (nopis)	number of nonnegative (positive) integer solutions
p.d.f. (p.m.f.)	probability density (mass) function
r.v.	random variable
w.r.t.	with respect to
Mathematics	Description
$\exists$	there exists
$\forall$	for each
iff, $\Leftrightarrow$	if and only if
$:=$	$a := b$ or $b =: a$ defines a to be b (e.g. $0! := 1$)
$\mathbb{C}, \mathbb{N}, \mathbb{Q}, \mathbb{R}, \mathbb{R}_{>0}, \mathbb{Z}$	sets of numbers, see Section A.1
$\text{sgn}(x)$	signum function of x; $\text{sgn}(x) = \begin{cases} -1, & \text{if } x < 0 \\ 0, & \text{if } x = 0 \\ 1, & \text{if } x > 0 \end{cases}$

[a]Note that $\text{sgn}(\cdot)$ is also used for the *signature* of a permutation (see, for example, Hubbard and Hubbard, 2002, p. 478), but we will not require signatures.

Fundamental Probability: A Computational Approach M.S. Paolella
© 2006 John Wiley & Sons, Ltd

Table B.2 Latin abbreviations and expressions

Latin	Description
cf.	(confer) compare
e.g.	(exempli gratia) for example, for instance
et al.	(et alii) and other people
etc.	(et cetera) and so on
i.e.	(id est) that is to say
QED	(quod erat demonstrandum) which was to be shown
vs.	(versus) against
a fortiori	with even stronger reason
a posteriori	from effects to causes, reasoning based on past experience
a priori	from causes to effects, conclusions drawn from assumptions
ad hoc	for this purpose, improvised
ad infinitum	never ending
caveat	a caution or warning
ceteris paribus	all other things being equal
de facto	from the fact
inter alia	among other things
ipso facto	by the fact itself
non sequitur	it does not follow
per capita	per head
prima facie	at first sight
status quo	things as they are
vice versa	the other way round

Table B.3 Greek letters

Name	Lower	Upper	Name	Lower	Upper
alpha	α	A	pi	π	Π
beta	β	B	rho	ρ	P
gamma	γ	Γ	mu	μ	M
delta	δ	Δ	nu	ν	N
epsilon	ϵ, ε	E	sigma	σ	Σ
zeta	ζ	Z	tau	τ	T
eta	η	H	upsilon	υ	Υ
theta	θ, ϑ	Θ	phi	ϕ, φ	Φ
iota	ι	I	chi	χ	X
kappa	κ	K	psi	ψ	Ψ
lambda	λ	Λ	omega	ω	Ω
xi	ξ	Ξ	omicron	o	O

Table B.4 Special functions[a]

Name	Notation	Definition	Alternative formula
factorial	$n!$	$n(n-1)(n-2)\cdots 1$	$\Gamma(n+1)$
rising factorial	$n^{[k]}$	$n(n+1)\ldots(n+k-1)$	$(n+k-1)!/(n-1)!$
gamma function	$\Gamma(a)$	$\displaystyle\lim_{k\to\infty} k!\,k^{a-1}/x^{[k]}$	$\int_0^\infty t^{a-1}e^{-t}dt$
incomplete gamma function	$\Gamma_x(a)$	$\int_0^x t^{a-1}e^{-t}dt$	$_1F_1(a,a+1;-x)$
incomplete gamma ratio	$\overline{\Gamma}_x(a)$	$\Gamma_x(a)/\Gamma(a)$	
beta function	$B(p,q)$	$\int_0^1 t^{p-1}(1-t)^{q-1}dt$	$\Gamma(p)\,\Gamma(q)/\Gamma(p+q)$ and for $p,q\in\mathbb{N}$, $B(p,q)=\frac{(p-1)!(q-1)!}{(q+p-1)!}$
incomplete beta function	$B_x(p,q)$	$\mathbb{I}_{[0,1]}(x)\int_0^x t^{p-1}(1-t)^{q-1}dt$	combine those for $B(p,q)$ and $\overline{B}_x(p,q)$
incomplete beta ratio	$\overline{B}_x(p,q)$	$B_x(p,q)/B(p,q)$	$\sum_{k=p}^{q+p-1}\binom{q+p-1}{k}\times x^k(1-x)^{q+p-1-k}$
modified Bessel function first kind	$I_v(x)$	$\left(\frac{x}{2}\right)^v\sum_{j=0}^\infty \frac{(x^2/4)^j}{j!\,\Gamma(v+j+1)}$	$\pi^{-1}\int_0^\pi\big[\exp(x\cos\theta)\cos(v\theta)\big]d\theta,\ v\in\mathbb{Z}$
modified Bessel function third kind	$K_v(z)$	$\dfrac{\Gamma(v+1/2)\,(2z)^v}{\sqrt{\pi}}\displaystyle\int_0^\infty \frac{\cos t}{(t^2+z^2)^{v+1/2}}dt$ (see page 267 for other expressions)	
nonis to $\sum_{i=1}^R x_i = n$	$K_{n,R}^{(0)}$	$\binom{n+R-1}{n}=\binom{n+R-1}{R-1}$	$K_{n,R}^{(0)}=\sum_{i=0}^n K_{n-i,R-1}^{(0)}$
nopis to $\sum_{i=1}^R x_i = n$	$K_{n,R}^{(1)}$	$K_{n-R,R}^{(0)}=\binom{n-1}{R-1}$	$K_{n,R}^{(0)}=\sum_{j=0}^{R-1}\binom{R}{j}K_{n,R-j}^{(1)}$

[a]The last two entries, nonis (number of nonnegative integer solutions) and nopis (number of positive integer solutions) are not 'special functions', but are certainly useful, particularly in Chapter 2.

Table B.5 General probability and mathematics notation

Description	Notation	Examples
sets (measurable events, r.v. support, information sets, m.g.f. convergence strip, etc.)	calligraphic capital letters and Roman capital letters	$\mathcal{I}_{t-1} \subset \mathcal{I}_t$, $\quad A \subset \mathbb{R}^2$
Borel σ−field	$\mathcal{B}$	$\{\mathbb{R}, \mathcal{B}, \Pr(\cdot)\}$
events	capital Roman letters	$A \in \mathcal{A}$, $\quad A_1 \supset A_2 \supset A_3 \cdots$
event complement	A^c	$\left(\bigcup_{i=1}^n A_i\right)^c = \bigcap_{i=1}^n A_i^c$
random variables (r.v.s)	capital Roman letters	A, G, O, N, Y
support of r.v. X	S_X or, in context, just S	$\int_{S_X \cup \mathbb{R}_{>0}} x^2 \mathrm{d}F_X(x)$
independence of two r.v.s	$X \perp Y$	$X \sim N(0,1) \perp Y \sim \chi_k^2$
value or set of values assumed by a r.v.	lower case Roman letter (usually matching)	$\{X < a\} \cup \{X > b\}$, $\quad \Pr(X > x)$
'unified' sum or integral (Riemann–Stieltjes)	$\int_{-\infty}^{\infty} g(x)\,\mathrm{d}F_X(x) = \begin{cases} \int_S g(x) f_X(x)\,\mathrm{d}x, \\ \sum_{i \in S} g(x_i) f_X(x_i), \end{cases}$	X continuous $\quad X$ discrete
Landau's order symbol ('big oh', 'little oh')	$f(x) = O(g(x)) \Leftrightarrow \exists\, C > 0$ s.t. $\lvert f(x) \rvert \le Cg(x)\rvert$; $f(x) = o(g(x)) \Leftrightarrow f(x)/g(x) \to 0$	as $x \to \infty$; as $x \to \infty$
proportional to	$f(x; \theta) \propto g(x; \theta) \Leftrightarrow f(x; \theta) = Cg(x; \theta)$, where C is a constant	$f_{N}(z; 0, 1) \propto \exp\left(-\tfrac{1}{2} z^2\right)$
vectors (matrices)	bold lower (upper) case letters	$\mathbf{x}'\boldsymbol{\Sigma}\mathbf{x} > 0$
matrix transpose	$\mathbf{A}'$	$\mathbf{A} = \left[\begin{smallmatrix} a & b \\ c & d \end{smallmatrix}\right]$, $\mathbf{A}' = \left[\begin{smallmatrix} a & c \\ b & d \end{smallmatrix}\right]$
determinant of square $\mathbf{A}$	$\det(\mathbf{A})$, $\quad \det \mathbf{A}$ or $\lvert \mathbf{A} \rvert$	$\left\lvert\begin{smallmatrix} a & b \\ c & d \end{smallmatrix}\right\rvert = ad - bc$

Table B.6 Probability and distribution

Description	Notation	Examples
probability function $\mathcal{A} \to [0, 1]$, where $\mathcal{A}$ is a set of measurable events	$\Pr(\cdot)$	$\Pr(\text{roll a six}),\ \Pr(X = Y - 1)$
c.d.f. of r.v. X (evaluated at x) $\Pr(X \le x)$	$F_X,\ F_X(x)$	$\int_S \left[1 - F_X(x) \right] \mathrm{d}x$
survivor function of X $1 - F_X(x) = \Pr(X > x)$	$\overline{F}_X(x)$	$\overline{F}_X(x) \approx Cx^{-\alpha}$
c.d.f. of the standard normal distribution	$\Phi(\cdot)$	$\overline{F}_Z(z) = 1 - \Phi(z) = \Phi(-z)$
median and pth quantile of continuous r.v. X	$F_X(\xi_p) = p,\ 0 < p < 1$	median, $m = \mathrm{med}(X) \Leftrightarrow F_X(m) = 0.5$
mode of r.v. X, $\mathrm{argmax}_x f_X(x)$	$\mathrm{mode}(X)$	
distribution, $\mathrm{D}(\theta)$, of r.v. X, where θ is a given parameter vector	$X \sim \mathrm{D}(\theta)$	$X \sim \mathrm{Cau}(\mu, \sigma),\ X/Y \sim \mathrm{Beta}(1, b),\ C \sim \chi_k^2$
p.d.f./p.m.f. of r.v. X evaluated at x	$f_X,\ f_X(x)$	$\sum_{i=1}^{k} f_X(x_i)$
p.d.f./p.m.f. associated with distribution $\mathrm{D}(\theta)$ evaluated at x	$f_\mathrm{D}(\cdot; \theta),\quad f_\mathrm{D}(x; \theta)$	$f_\mathrm{Gam}(\cdot; a, b)$
indicator function of element x and set $\mathcal{M}$; one if $x \in \mathcal{M}$, else zero.	$\mathbb{I}_M(x)$	$\mathbb{I}_{(0,\infty)}(x/y),\ \ \mathbb{I}_{\{1,2,\dots,\theta\}}(k)$
set of i.i.d. r.v.s with common distribution $\mathrm{D}(\theta)$	$\overset{\text{i.i.d.}}{\sim} \mathrm{D}(\theta)$	$X_i \overset{\text{i.i.d.}}{\sim} \mathrm{Exp}(\lambda)$
set of independent r.v.s with common distribution D	$\overset{\text{ind}}{\sim} \mathrm{D}(\theta_i)$	$B_i \overset{\text{ind}}{\sim} \mathrm{Bin}(n_i, p)$
set of i.i.d. r.v.s with mean μ and finite variance σ^2	$\overset{\text{i.i.d.}}{\sim} (\mu, \sigma^2)$	

Table B.7   Moments

Description	Notation, definition	Examples
geometric expected value of discrete random variable Y	$\mathbb{G}\left[Y\right] = \prod_{y \in S} y^{f_Y(y)}$	$\mathbb{G}\left[Y\right] = \exp\{\mathbb{E}\left[\ln Y\right]\}$ valid for discrete and continuous
expected value of r.v. $g(X)$	$\mathbb{E}\left[g(X)\right] = \int_S g(x) \mathrm{d}F_X(x)$	$\mu = \mathbb{E}[X] = \int_S x \mathrm{d}F_X(x)$
variance of r.v. X	$\mathbb{V}(X) = \int_S (x - \mu)^2 \, \mathrm{d}F_X(x)$	$\mathbb{V}(X) = \mu_2'(X) - \mu^2(X)$
standard deviation	$\sigma_X = \sqrt{\mathbb{V}(X)}$	$\mu_X \pm 1.96\sigma_X$
rth raw moment of r.v. X	$\mu_r'(X) = \mathbb{E}\left[X^r\right]$	$\mu_r' = \sum_{i=0}^r \binom{r}{i} \mu_{n-i} \mu^i$
rth central moment	$\mu_r(X) = \mathbb{E}\left[(X - \mu)^r\right]$	$\mu_r = \sum_{i=0}^r (-1)^i \binom{r}{i} \mu_{r-i}' \mu^i$
gth factorial moment of R	$\mu_{[g]}'(R) = \mathbb{E}\left[R(R-1)\cdots(R-g+1)\right]$	$\mathbb{E}\left[R\right] = \mu_{[1]}'(R) = \mu$ $\mathbb{V}(R) = \mu_{[2]}'(R) - \mu(\mu - 1)$
skewness and kurtosis coefficients	$\mathrm{skew}(X) = \mu_3/\mu_2^{3/2}$, $\mathrm{kurt}(X) = \mu_4/\mu_2^2$	$X \sim \mathrm{N}(\mu, \sigma^2)$, $\mathrm{kurt}(X) = 3$
covariance and correlation between r.v.s X and Y	$\mathrm{Cov}(X, Y) = \mathbb{E}\left[(X - \mu_X)(Y - \mu_Y)\right]$	$\mathrm{Corr}(X_i, X_j) = \frac{\mathrm{Cov}(X_i, X_j)}{\sqrt{\mathbb{V}(X_i)\mathbb{V}(X_j)}}$

Distribution tables

Table C.1 Occupancy distributions[a]

X is the number of nonempty urns when n balls are randomly distributed into r urns

$$f_X(x) = r^{-n} \binom{r}{x} \sum_{j=0}^{x-1} (-1)^j \binom{x}{j} (x - j)^n$$

$$F_X(x) = r^{-n} \binom{r}{x} \sum_{j=0}^{x-1} (-1)^j \binom{x}{j} \frac{r-x}{j+r-x} (x - j)^n$$

$$\mathbb{E}[X] = r(1 - t_1), \text{ where } t_i = \left(\frac{r-i}{r}\right)^n$$

$$\mathbb{V}(X) = rt_1 (1 - rt_1) + rt_2 (r - 1)$$

Y is the number of balls necessary so that each of the r urns contains at least one ball

$$f_Y(n; r) = \Pr(Y > n - 1) - \Pr(Y > n), \text{ where } \Pr(Y > n) = 1 - F_Y(n; r)$$

$$F_Y(n; r) = \sum_{i=0}^{r-1} (-1)^i \binom{r}{i} \left(\frac{r-i}{r}\right)^n$$

Y_k is the number of balls necessary so that k of the r urns, $1 < k \leq r$, contains at least one ball

$$\Pr(Y_k > n) = r^{-n} \sum_{i=m}^{r-1} (-1)^{i-m} \binom{i-1}{i-m} \binom{r}{i} (r - i)^n, \qquad m = r - k + 1,$$

$$\mathbb{E}[Y_k] = r \sum_{i=0}^{k-1} \frac{1}{r-i}, \qquad \mathbb{V}(Y_k) = r \sum_{i=1}^{k-1} \frac{i}{(r-i)^2}$$

$$\mathbb{M}_{Y_k}(t) = e^{kt} \prod_{i=0}^{k-1} \frac{p_i}{1-q_i e^t}, \qquad p_i = \frac{r-i}{r}, \qquad q_i = 1 - p_i$$

---

[a] See Section 4.2.6 for more details. The moment generating function, or m.g.f., of r.v. X is $\mathbb{M}_X(t) = \mathbb{E}[\exp(tX)]$, and is introduced in Volume II.

Table C.2 Common discrete univariate distributions – Part I

Name	Probability mass function	Parameters
standard discrete uniform	$f_X(x;\theta) = \theta^{-1}\mathbb{I}_{\{1,2,\ldots,\theta\}}(x)$	$\theta \in \mathbb{N}$
shifted discrete uniform	$f_X(x;\theta_1,\theta_2) = (\theta_2 - \theta_1 + 1)^{-1}\mathbb{I}_{\{\theta_1,\theta_1+1,\ldots,\theta_2\}}(x)$	$\theta_1, \theta_2 \in \mathbb{Z},\ \theta_1 < \theta_2$
Bernoulli	$f_X(x;p) = p^x(1-p)^{1-x}\mathbb{I}_{\{0,1\}}(x)$	$0 < p < 1$
binomial	$f_X(x;n,p) = \binom{n}{x}p^x(1-p)^{n-x}\mathbb{I}_{\{0,1,\ldots,n\}}(x)$	$n \in \mathbb{N},\ 0 < p < 1$
hypergeometric	$f_X(x;N,M,n) = \dfrac{\binom{N}{x}\binom{M}{n-x}}{\binom{N+M}{n}}$ $\times\ \mathbb{I}_{\{\max(0,n-M),1,\ldots,\min(n,N)\}}(x)$	$N, M, n \in \mathbb{N},\ 0 < n < (N+M)$
geometric (# failures)	$f_X(x;p) = p(1-p)^x\mathbb{I}_{\{0,1,\ldots\}}(x)$	$0 < p < 1$
geometric (total # draws)	$f_X(x;p) = p(1-p)^{x-1}\mathbb{I}_{\{1,2,\ldots\}}(x)$	$0 < p < 1$
negative binomial (# failures)	$f_X(x;r,p) = \binom{r+x-1}{x}p^r(1-p)^x\mathbb{I}_{\{0,1,\ldots\}}(x)$	$r \in \mathbb{N},\ 0 < p < 1$
negative binomial (total # draws)	$f_X(x;r,p) = \binom{x-1}{r-1}p^r(1-p)^{x-r}\mathbb{I}_{\{r,r+1,\ldots\}}(x)$	$r \in \mathbb{N},\ 0 < p < 1$
inverse hypergeometric	$f_X(x;k,w,b) = \dfrac{\binom{x-1}{k-1}\binom{w+b-x}{w-k}}{\binom{w+b}{w}}\mathbb{I}_{\{k,k+1,\ldots,b+k\}}(x)$	$k, w, b \in \mathbb{N},\ 1 \le k \le w$
Poisson	$f_X(x;\lambda) = \dfrac{e^{-\lambda}\lambda^x}{x!}\mathbb{I}_{\{0,1,\ldots\}}(x)$	$\lambda \in \mathbb{R}_{>0}$
consecutive[a]	$f_X(x;m) = p^m\mathbb{I}_{\{m\}}(x) + (1-p)\cdot$ $\sum_{i=0}^{m-1}p^i f_X(x;n-i-1)\mathbb{I}_{\{m+1,m+2,\ldots\}}(x)$	$m \in \mathbb{N},\ 0 < p < 1$

[a] A 'consecutive' random variable denotes the number of i.i.d. Bernoulli trials necessary until m consecutive successes occur, where each trial has success probability p (see Examples 2.5 and 8.13)

Table C.3 Common discrete univariate distributions – Part II[a]

Abbreviation	Moment generating function	Mean	Variance
DUnif(θ)	$\theta^{-1}\sum_{j=1}^{\theta} e^{tj}$	$\frac{1}{2}(1+\theta)$	$\frac{1}{12}(\theta-1)(\theta+1)$
DUnif(θ_1,θ_2)	$(\theta_2-\theta_1+1)^{-1}\sum_{j=\theta_1}^{\theta_2} e^{tj}$	$\frac{1}{2}(\theta_1+\theta_2)$	$\frac{1}{12}(\theta_2-\theta_1)(\theta_2-\theta_1+2)$
Ber(p)	pe^t+q	p	pq
Bin(n,p)	$(pe^t+q)^n$	np	npq
HGeo(N,M,n)		$\dfrac{nN}{N+M}$	$\dfrac{nN}{N+M}\left(\dfrac{M}{N+M}\dfrac{N+M-n}{N+M-1}\right)$
Geo(p)	$p(1-qe^t)^{-1}$, $t<-\ln q$	q/p	q/p^2
Geo(p)	$e^t p(1-qe^t)^{-1}$, $t<-\ln q$	$1/p$	q/p^2
NBin(r,p)	$p^r(1-qe^t)^{-r}$, $t<-\ln q$	$r(1-p)/p$	rq/p^2
NBin(r,p)	$e^{rt}p^r(1-qe^t)^{-r}$, $t<-\ln q$	r/p	rq/p^2
IHGeo(k,w,b)		$k\left(\dfrac{w+b+1}{w+1}\right)$	$kb\dfrac{(w+b+1)(w-k+1)}{(w+2)(w+1)^2}$
Poi(λ)	$\exp(-\lambda+\lambda e^t)$	λ	λ
Consec(m,p)	$\mathbb{M}_m(t):=\mathbb{M}_{N_m}(t)=\dfrac{pe^t\mathbb{M}_{m-1}(t)}{1-q\mathbb{M}_{m-1}(t)\,e^t}$, and $\mathbb{M}_1(t)=pe^t/(1-qe^t)$	$\dfrac{p^{-m}-1}{q}$	$\dfrac{1-(1+2m)\,qp^m-p^{2m+1}}{q^2p^{2m}}$

[a]In all entries, $q=1-p$. The moment generating function, or m.g.f. of r.v. X is $\mathbb{M}_X(t)=\mathbb{E}[\exp(tX)]$, and is introduced in Volume II. The m.g.f. for HGeo is available, but rarely used. See Janardan (1973) and the references therein for details on an alternative to the usual expression for the hypergeometric m.g.f. The m.g.f. for IHGeo is presumably difficult (and not required for our purposes).

Table C.4 Common continuous univariate distributions–Part I(a) (see also Table C.5)^a

#	Name, Abbreviation	p.d.f./c.d.f.	Parameters
1	uniform, Unif (a, b)	$f_X(x; a, b) = \frac{1}{b-a} \mathbb{I}_{(a,b)}(x)$ $F_X(x; a, b) = \frac{x-a}{b-a} \mathbb{I}_{(a,b)}(x) + \mathbb{I}_{[b,\infty]}(x)$	$a, b \in \mathbb{R}, \quad a < b$
2	exponential, Exp(λ)	$f_X(x; \lambda) = \lambda e^{-\lambda x} \mathbb{I}_{(0,\infty)}(x)$ $F_X(x; \lambda) = \left(1 - e^{-\lambda x}\right) \mathbb{I}_{(0,\infty)}(x)$	$\lambda \in \mathbb{R}_{>0}$
3	gamma, Gam (α, β)	$f_X(x; \alpha, \beta) = [\beta^\alpha / \Gamma(\alpha)] x^{\alpha-1} \exp(-\beta x) \mathbb{I}_{(0,\infty)}(x)$ $F_X(x; \alpha, \beta) = \Gamma_{x\beta}(\alpha) / \Gamma(\alpha)$	$\alpha, \beta \in \mathbb{R}_{>0}$
	inverse gamma, IGam (α, β)	$f_X(x) = [\beta^\alpha / \Gamma(\alpha)] x^{-(\alpha+1)} \exp(-\beta/x) \mathbb{I}_{(0,\infty)}(x)$ $F_X(x) = \overline{F}_{\text{Gam}}(x^{-1}; \alpha, \beta)$	$\alpha, \beta \in \mathbb{R}_{>0}$
4	beta, Beta (p, q)	$f_X(x; p, q) = [B(p, q)]^{-1} x^{p-1}(1 - x)^{q-1} \mathbb{I}_{[0,1]}(x)$ $F_X(x; p, q) = \overline{B}_x(p, q) \mathbb{I}_{[0,1]}(x) + \mathbb{I}_{(1,\infty)}(x)$	$p, q \in \mathbb{R}_{>0}$
	four-param beta, Beta (p, q, a, b)	$f_Y(x; p, q, a, b) = \frac{(x-a)^{p-1}(b-x)^{q-1}}{B(p,q)(b-a)^{p+q-1}} \mathbb{I}_{[a,b]}(x)$ $F_Y(x; p, q, a, b) = F_X\left(\frac{x-a}{b-a}; p, q, 0, 1\right)$	$a, b \in \mathbb{R}, \quad a < b$
5	Laplace, Lap (μ, σ)	$f_X(x; 0, 1) = \frac{1}{2} \exp(-|x|)$ $F_X(x; 0, 1) = \frac{1}{2} e^x \mathbb{I}_{(-\infty,0]}(x) + (2 - e^{-x}) \mathbb{I}_{(0,\infty)}(x)$	$\mu \in \mathbb{R}, \quad \sigma \in \mathbb{R}_{>0}$
	generalized exponential, GED (p)	$f_X(x; p) = p[2\Gamma(p^{-1})]^{-1} \exp(-|x|^p)$ $F_X(x) = \frac{1}{2} - \frac{\Gamma_{(-x)^p}(p^{-1})}{2\Gamma(p^{-1})}, x < 0;$ Use $F_X(x) = 1 - F_X(-x)$ for $x > 0$.	$p \in \mathbb{R}_{>0}$

Table C.4 *(continued)*

#	Name, Abbreviation	p.d.f./c.d.f.	Parameters
6	Weibull, Weib (β, x_0, σ)	$f_X(\beta, 0, 1) = \beta x^{\beta-1} \exp(-x^\beta)\, \mathbb{I}_{(0,\infty)}(x)$ $F_X(x; \beta, 0, 1) = 1 - \exp(-(x)^\beta)\, \mathbb{I}_{(0,\infty)}(x)$	$\beta, \sigma \in \mathbb{R}_{>0},\ x_0 \in \mathbb{R}$
7	Cauchy, Cau (μ, σ)	$f_X(c; 0, 1) = \pi^{-1}(1 + c^2)^{-1}$ $F_X(c; 0, 1) = \frac{1}{2} + \frac{1}{\pi}\arctan(c)$	$\mu \in \mathbb{R},\ \sigma \in \mathbb{R}_{>0}$
	$X_i \stackrel{\text{i.i.d.}}{\sim} N(\mu_i, 1),\ R = X_2/X_1$	$a = 1 + r^2$, $b = \mu_1 + r\mu_2$, $c = \mu_1^2 + \mu_2^2$; p.d.f. $f_R(r; \mu_1, \mu_2)$ is $\exp\left(\frac{b^2/a - c}{2}\right) \frac{1}{2\pi}\left\{ \frac{b}{a}\sqrt{\frac{2\pi}{a}}\left[1 - 2\Phi\left(-\frac{b}{\sqrt{a}}\right)\right] + 2a^{-1}\exp\left(-\frac{b^2}{2a}\right) \right\}$	
8	Pareto Type I, Par I (α, x_0)	$f_X(x; \alpha, x_0) = \alpha x_0^\alpha x^{-(\alpha+1)}\, \mathbb{I}_{(x_0,\infty)}(x)$ $F_X(x; \alpha, x_0) = \left[1 - \left(\frac{x_0}{x}\right)^\alpha\right] \mathbb{I}_{(x_0,\infty)}(x)$	$\alpha,\ x_0 \in \mathbb{R}_{>0}$
9	Pareto Type II, Par II (b, c)	$f_X(x; b, c) = \frac{b}{c}\left(\frac{c}{c+x}\right)^{b+1} \mathbb{I}_{(0,\infty)}$ $F_X(x; b, c) = \left[1 - \left(\frac{c}{c+x}\right)^b\right] \mathbb{I}_{(0,\infty)}(x)$	$b, c \in \mathbb{R}_{>0}$

[a]The number of the distribution corresponds to the numbering in Sections 7.1 and 7.2. Those without numbers are related to the previous entry and are discussed elsewhere; e.g. the inverse gamma is discussed in Problem 7.9, while the ratio of independent normals (which generalizes the Cauchy) is discussed in Volume II; see the Index for the others. The parameterization of the densities reflects their common forms, i.e. some are shown as location–scale families, others just as scale or location families, etc., and for some, the scale parameter is inverted (exponential, gamma) instead of being a genuine scale parameter.

Table C.5   Common continuous univariate distributions – Part I(b) (continuation of Table C.4)

#	Name	Mean	Variance	m.g.f.[a]	Other
1	Unif	$(a+b)/2$	$\dfrac{(b-a)^2}{12}$	$\dfrac{e^{tb}-e^{ta}}{t(b-a)},\ t\neq 0$	$\mathbb{E}[X^r]=\dfrac{b^{r+1}-a^{r+1}}{(b-a)(r+1)}$
2	Exp	λ^{-1}	λ^{-2}	$\dfrac{\lambda}{\lambda-t},\ t<\lambda$	$\mathbb{E}[X^r]=r!/\lambda^r$
3	Gam	α/β	$\dfrac{\alpha}{\beta^2}$	$\left(\dfrac{\beta}{\beta-t}\right)^{\alpha},\ t<\beta$	$\mathbb{E}[X^r]=\dfrac{\Gamma(r+\alpha)}{\beta^r\,\Gamma(\alpha)},\ r>-\alpha$
	IGam	$\beta/(\alpha-1)$	$\dfrac{\beta^2}{(\alpha-1)^2(\alpha-2)}$		$\mathbb{E}[X^r]=\dfrac{\Gamma(\alpha-r)}{\Gamma(\alpha)}\beta^r,\ \alpha>r$
4	beta	$p/(p+q)$	$\dfrac{pq}{(p+q)^2(p+q+1)}$		$\mathbb{E}[X^r]=\dfrac{\Gamma(p+q)}{\Gamma(p)}\dfrac{\Gamma(p+r)}{\Gamma(p+r+q)}$
	four-param beta	$a+\dfrac{p(b-a)}{p+q}$	$\dfrac{pq(b-a)^2}{(p+q)^2(p+q+1)}$		$\mathbb{E}[Y^r]=\sum_{i=0}^r \binom{r}{i}a^{r-i}(b-a)^i\,\mathbb{E}[X^i]$
5	Lap	0	2	$\dfrac{1}{1-t^2},\ |t|<1$	$\mathbb{E}[X^r]=r!,\ r=0,2,4,\dots$
	GED	0	$\dfrac{\Gamma(3/p)}{\Gamma(1/p)}$		$E\left[|X|^r\right]=\dfrac{\Gamma(p^{-1}(r+1))}{\Gamma(p^{-1})}$
6	Weib	$\Gamma(1+1/\beta)$		$\mathbb{E}[X^r]=\Gamma(1+r/\beta)\quad r>-\beta$	
7	Cau	d.n.e.	d.n.e.	$\varphi_X(t)=e^{-|t|}$	
8	Par I	$\dfrac{\alpha}{\alpha-1}x_0$	$\dfrac{\alpha}{(\alpha-1)^2(\alpha-2)}x_0^2$		$\mathbb{E}[X^r]=\dfrac{\alpha}{\alpha-r}x_0^r,\ r<\alpha$
9	ParII ($c=1$)	$\dfrac{1}{b-1}$	$\dfrac{b}{(b-1)^2(b-2)}$		$\mathbb{E}[X^r]=\dfrac{r!}{(b-1)_{[r]}},\ r<b$

[a]The m.g.f. refers to the moment generating function of r.v. X, given by $M_X(t)=\mathbb{E}[\exp(tX)]$, and is introduced in Volume II. For the uniform, it is easy to verify that the m.g.f. is continuous at $t=0$ and equals 1. For some distributions, like the beta and GED, the m.g.f. is not easily expressed. The Cauchy, Weibull and Pareto (among others) do not possess m.g.f.s, but all distributions have a *characteristic function*, defined as $\varphi_X(t)=\mathbb{E}[\exp(itX)]$ (see Volume II); it is very tractable for the Cauchy.

Table C.6 Common continuous univariate distributions – Part II(a) (see Table C.7)

#	Name, Abbreviation	p.d.f./c.d.f.	Parameters
10	normal, $N(\mu, \sigma^2)$	$f_X(x; \mu, \sigma) = [\sqrt{2\pi}\sigma]^{-1} \exp\left[-\frac{1}{2}\left(\frac{x-\mu}{\sigma}\right)^2\right]$	$\mu \in \mathbb{R}, \ \sigma \in \mathbb{R}_{>0}$
		$\Phi(z) := F_Z(z) = \frac{1}{2} + \frac{1}{\sqrt{2\pi}} \sum_{i=0}^{\infty} (-1)^i \frac{z^{2i+1}}{(2i+1) \, 2^i \, i!}$	
	symmetric alpha stable, SαS or $S_\alpha(\mu, c)$	$f_X(x; \alpha, \mu, c) = (c\pi)^{-1} \int_0^\infty \cos(tz) \exp(-t^\alpha) dt,$ where $z = (y - \mu)/c$	$\alpha \in (0, 2], \ \mu \in \mathbb{R}, \ c \in \mathbb{R}_{>0}$
		$F_X(x; \alpha, \mu, c) = \frac{1}{2} - \frac{1}{\pi} \int_0^\infty \frac{\mathrm{Im}[e^{-itz} \exp(-|t|^\alpha)]}{t} dt$	
11	chi-square, $\chi^2(k)$ or χ^2_k	$f_X(x; k) = [2^{k/2}\Gamma(k/2)]^{-1} x^{(k/2)-1} e^{-x/2} \mathbb{I}_{(0,\infty)}(x)$	$k \in \mathbb{R}_{>0}$
		$F_X(x; k) = F_{\mathrm{Gam}}(x; k/2, 1/2)$	
12	Student's t, $t(n)$ or t_n	$f_X(x; n) = K_n \left(1 + x^2/n\right)^{-(n+1)/2}, \quad K_n = \frac{n^{-1/2}}{B(n/2, 1/2)}$	$n \in \mathbb{R}_{>0}$
		$F_X(t; n) = \frac{1}{2}\overline{B}_L\left(\frac{n}{2}, \frac{1}{2}\right), \quad L = n/(n + t^2), \ t < 0;$	
		$F_X(t; n) = 1 - F_X(-t), \ t > 0$	
13	Fisher $F(n_1, n_2)$, F_{n_1, n_2}	$f_X = \frac{n}{B\left(\frac{n_1}{2}, \frac{n_2}{2}\right)} \frac{(nx)^{n_1/2-1}}{(1 + nx)^{(n_1+n_2)/2}} \mathbb{I}_{(0,\infty)}(x), \quad n = \frac{n_1}{n_2}$	$n_1, n_2 \in \mathbb{R}_{>0}$
		$F_X(x; n_1, n_2) = \overline{B}_y\left(\frac{n_1}{2}, \frac{n_2}{2}\right), \quad y = \frac{n_1 x}{n_2 + n_1 x} = \frac{nx}{1 + nx}$	
14	log–normal, $LN(\theta, \sigma, \zeta)$	$f_X(x; \theta, \sigma, \zeta) = \left[\sigma\sqrt{2\pi}(x - \theta)\right]^{-1}$ $\times \exp\left\{-\frac{1}{2\sigma^2}\left[\ln(x - \theta) - \zeta\right]^2\right\} \mathbb{I}_{(\theta,\infty)}(x)$	$\theta, \zeta \in \mathbb{R}, \ \sigma \in \mathbb{R}_{>0}$

Table C.6 (*continued*)

#	Name, Abbreviation	p.d.f./c.d.f.	Parameters
15	logistic, Log $(x; \alpha, \beta)$	$f_X(x; \alpha, \beta) = \beta^{-1}\exp(-z)(1+\exp(-z))^{-2}$, $\quad z = (x-\alpha)/\beta$ $F_X(x; \alpha, \beta) = (1+\exp(-z))^{-1}$, $\quad z = (x-\alpha)/\beta$	$\alpha \in \mathbb{R}, \ \beta \in \mathbb{R}_{>0}$
	generalized logistic, Glog $(x; p, q, \alpha, \beta)$	$f_X(x; p, q, 0, 1) = [B(p,q)]^{-1}\dfrac{\exp(-qx)}{[1+\exp(-x)]^{p+q}}$ $F_X(x; p, q, 0, 1) = \bar{B}_z(p, q)$, $\quad z = e^t\left(1+e^t\right)^{-1}$	$p, q \in \mathbb{R}_{>0}$
16	Gumbel, Gum (a, b)	$f_X(x; a, b) = b^{-1}\exp\left(-z - e^{-z}\right)$, $\quad z = (x-a)/b$ $F_X(x; a, b) = \exp\left[-\exp(-z)\right]$	$a \in \mathbb{R}, \ b \in \mathbb{R}_{>0}$
17	Inverse Gaussian, IG (μ, λ)	$f_X = \left(\dfrac{\lambda}{2\pi x^3}\right)^{1/2}\exp\left[-\dfrac{\lambda}{2\mu^2 x}(x-\mu)^2\right]\mathbb{I}_{(0,\infty)}(x)$ $F_X = \Phi\left[\sqrt{\dfrac{\lambda}{x}}\left(\dfrac{x-\mu}{\mu}\right)\right] + \exp\left(\dfrac{2\lambda}{\mu}\right)\Phi\left[-\sqrt{\dfrac{\lambda}{x}}\left(\dfrac{x+\mu}{\mu}\right)\right]$	$\mu \in \mathbb{R}, \ \lambda \in \mathbb{R}_{>0}$
	Lévy, Lévy $(x; \lambda)$	$f_X(x; \lambda) = \left(\dfrac{\lambda}{2\pi x^3}\right)^{1/2}\exp\left(-\dfrac{\lambda}{2x}\right)\mathbb{I}_{(0,\infty)}(x)$ $F_X(x) = 2\Phi\left(-x^{-1/2}\right)$	$\lambda \in \mathbb{R}_{>0}$
18	Hyperbolic, Hyp$_2(p, q)$	$f(z; p, q) = C_1\exp\left[C_2\left(p\sqrt{1+z^2} - qz\right)\right]$ $C_1 = \dfrac{\sqrt{p^2-q^2}}{2pK_1(p^{-2}-1)}$, $\quad C_2 = \dfrac{p^2-1}{\sqrt{p^2-q^2p^2}}$ and $K_1(\cdot)$ is the modified Bessel function of the third kind with index 1	$0 < p \leq 1, \ |q| < p$

Table C.7 Common continuous univariate distributions – Part II(b) (Continuation of Table C.6)[a]

#	Name	Mean	Variance	m.g.f.	Other
10	N	μ	σ^2	$\exp\left(t\mu + \frac{t^2\sigma^2}{2}\right)$	$\mu_3 = 0,\ \mu_4 = 3\sigma^4,\ E\left[Z^{2r}\right] = \frac{(2r)!}{2^r r!}$, $\mathbb{E}\,|X|^r = \frac{2^{r/2}}{\sqrt{\pi}}\Gamma\left(\frac{r+1}{2}\right),\quad r \in \mathbb{N}$
	SαS	$0,\ \alpha > 1$, else d.n.e.	$2,\ \alpha = 2,$ else d.n.e.	$\varphi_X(t;\alpha) = \exp(-|t|^\alpha)$	$\mathbb{E}\left[|X|^r\right] = \kappa^{-1}\Gamma(1 - r/\alpha),\ r < \alpha$, where $\kappa = \pi/2$ for $r = 1$, and $\kappa = \Gamma(1 - r)\cos(\pi r/2)$ for $r \neq 1$
11	χ^2	k	$2k$	$\dfrac{1}{(1-2t)^{k/2}},\ t < \dfrac{1}{2}$	$\mathbb{E}[X^s] = \prod_{i=1}^s [k + 2(i-1)],\ s \in \mathbb{N}$ $\mathbb{E}[X^m] = 2^m \frac{\Gamma(k/2+m)}{\Gamma(k/2)},\ m > -\frac{k}{2}$
12	t	$0,\ n > 1$, else d.n.e.	$\dfrac{n}{n-2},\ n > 2$, else d.n.e.		$\mathbb{E}\left[|T|^k\right] = n^{k/2}\dfrac{\Gamma\left(\frac{k+1}{2}\right)\Gamma\left(\frac{n-k}{2}\right)}{\Gamma\left(\frac{1}{2}\right)\Gamma\left(\frac{n}{2}\right)}$
13	F	$\dfrac{n_2}{n_2 - 2}$	$2\dfrac{n_2^2}{n_1^2}\dfrac{n_1 + n_1(n_2 - 2)}{(n_2 - 2)^2(n_2 - 4)}$		$\mathbb{E}[X^r] = \frac{n_2^r}{n_1^r}\mathbb{E}[X_1^r]\mathbb{E}[X_2^{-r}]$, where $X_i \overset{\text{ind}}{\sim} \chi^2(n_i)$
14	LN	$e^\zeta w^{1/2}$	$e^{2\zeta} w(w - 1)$, where $w = \exp(\sigma^2)$		
15	Log	α	$\beta^2\pi^2/3$	$\Gamma(1 - t)\Gamma(1 + t) = t\dfrac{\pi}{\sin \pi t}$	
	Glog	$\psi(p) - \psi(q)$	$\psi'(p) + \psi'(q)$	$\dfrac{\Gamma(p+t)\Gamma(q-t)}{\Gamma(p)\Gamma(q)}$	
16	Gum	$a + b\gamma,$ $\gamma = 0.57722\ldots$	$\dfrac{\pi^2 b^2}{6}$		
17	IG	μ	μ^3/λ		
	Lévy		d.n.e.	$\varphi_X(t;\lambda) = \exp(-|\lambda t|^{1/2})$	
18	Hyp$_2$				

[a]The hyperbolic moments will be discussed in a chapter in Volume II (see the footnote in Table C.5 for additional comments).

References

Abernethy, I. (2005). *Mental Strength: Condition Your Mind, Achieve Your Goals*. NETH Publishing, Cumbria, UK.

Abramowitz, M. and Stegun, I. A. (1972). *Handbook of Mathematical Functions*. Dover, New York, USA.

Adams, M. and Guillemin, V. (1996). *Measure Theory and Probability*. Birkhäuser, Boston, Massachusetts, USA.

Adler, I. and Ross, S. M. (2001). The coupon subset collector problem. *Journal of Applied Probability* **38**, 737–746.

Aitkin, M. (1998). Simpson's paradox and the Bayes factor. *Journal of the Royal Statistical Society, Series B* **60(1)**, 269–270.

Aldous, D. (1989). *Probability Approximations via the Poisson Clumping Heuristic*. Springer Verlag, New York, USA.

Alt, F. B. (1982). Bonferroni inequalities and intervals. In Kotz, S. and Johnson, N. L. (eds), *Encyclopedia of Statistical Sciences*, volume 1. John Wiley & Sons, Inc., New York, USA.

Alvo, M. and Cabilio, P. (2000). Calculation of hypergeometric probabilities using Chebyshev polynomials. *The American Statistician* **54(2)**, 141–144.

Andrews, G. E. Askey, R. and Roy, R. (1999). *Special Functions*. Cambridge University Press, Cambridge, UK.

Arnold, B. C. and Strauss, D. (1988). Bivariate distributions with exponential conditionals. *Journal of the American Statistical Association* **83**, 522–527.

Au, C. and Tam, J. (1999). Transforming variables using the Dirac generalized function. *The American Statistician* **53(3)**, 270–272.

Axelrod, M. C. and Glosup, J. G. (1994). Durbin-Watson revisited. In *Proceedings of the Statistical Computing Section*, pp. 176–181. American Statistical Association.

Axler, S. (1996). *Linear Algebra Done Right*. Springer Verlag, New York, USA.

Bachman, G., Narici, L. and Beckenstein, E. (2000). *Fourier and Wavelet Analysis*. Springer Verlag, New York, USA.

Bak, J. and Newman, D. J. (1997). *Complex Analysis*, 2nd edn. Springer Verlag, New York, USA.

Balakrishnan, N. (ed.) (1992). *Handbook of the Logistic Distribution*. Dekker, New York, USA.

Balakrishnan, N. and Basu, A. P. (eds) (1995). *The Exponential Distribution*. Gordon and Breach, USA.

Balakrishnan, N. and Koutras, M. V. (2002). *Runs and Scans with Applications*. John Wiley & Sons, Inc., New York, USA.

Barndorff-Nielsen, O. E. (1977). Exponentially decreasing distributions for the logarithm of particle size. *Proceedings of the Royal Society London* **A 353**, 401–419.

Barndorff-Nielsen, O. E. (1978). Hyperbolic distributions and distributions on hyperbolae. *Scandinavian Journal of Statistics* **5**, 151–157.

Barndorff-Nielsen, O. E., Blæsild, P., Jensen, J. L. and Sørensen, M. (1985). The Fascination of Sand. In Atkinson, A. C. and Fienberg, D. E. (eds), *A Celebration of Statistics*, pp. 57–87. Springer Verlag, New York, USA.

Bartle, R. G. and Sherbert, D. R. (1982). *Introduction to Real Analysis*. John Wiley & Sons, Inc., New York, USA.

Basmann (2003). Introduction to statistics and econometrics in litigation support. In Basmann, R. L. (ed.), *Journal of Econometrics* **113**. Elsevier Science, Amsterdam, Holland.

Bean, M. A. (2001). *Probability: The Science of Uncertainty, with Applications to Investments, Insurance, and Engineering*. Brooks/Cole, Pacific Grove, California, USA.

Beardon, A. F. (1997). *Limits: A New Approach to Real Analysis*. Springer Verlag, New York, USA.

Beerends, R. J., ter Morsche, H. G., van den Berg, J. C. and van de Vrie, E. M. (2003) *Fourier and Laplace Transforms*. Cambridge University Press, Cambridge, UK.

Beck-Bornholdt, H.-P. and Dubben, H.-H. (1997). *Der Hund, der Eier legt: Erkennen von Fehlinformation durch Querdenken*. Rowohlt Taschenbuch Verlag, Reinbek bei Hamburg, Germany.

Begley, S. (2004). Is your radio too loud to hear the phone? You messed up a poll. *The Wall Street Journal Europe*, p. A13.

Bennett, D. J. (1998). *Randomness*. Harvard University Press, Cambridge, Massachusetts, USA.

Bernardo, J. M. and Smith, A. F. M. (1994). *Bayesian Theory*. John Wiley & Sons, Ltd, Chichester, UK.

Bernstein, P. L. (1996). *Against the Gods: The Remarkable Story of Risk*. John Wiley & Sons, Inc., New York, USA.

Billingsley, P. (1995). *Probability and Measure*, 3rd edn. John Wiley & Sons, Inc., New York, USA.

Bingham, N. H. (2000). Studies in the history of probability and statistics XLVI. Measure into probability: from Lebesgue to Kolmogorov. *Biometrika* **87**, 145–156.

Blest, D. C. (2003). A new measure of kurtosis adjusted for skewness. *Australia and New Zealand Journal of Statistics* **45**, 175–179.

Boland, P. J. (1984). A biographical glimpse of William Sealy Gosset. *The American Statistician* **38(3)**, 179–183.

Bolton, R. J. and Hand, D. J. (2002). Statistical fraud detection: a review (with discussion). *Statistical Science* **17**, 235–255.

Bowman, A. W. and Azzalini, A. (1997). *Applied Smoothing Techniques for Data Analysis*. Oxford University Press, Oxford, UK.

Box, G. E. P. and Muller, M. E. (1958). A note on the generation of random normal deviates. *Annals of Mathematical Statistics* **29**, 610–611.

Bratley, P., Fox, B. L. and Schrage, L. E. (1987). *A Guide to Simulation*, 2nd edn. Springer Verlag, New York, USA.

Browder, A. (1996). *Mathematical Analysis, An Introduction*. Springer Verlag, New York, USA.

Butler, R. W. and Sutton, R. K. (1998). Saddlepoint approximation for multivariate cumulative distribution functions and probability computations in sampling theory and outlier testing. *Journal of the American Statistical Association* **93(442)**, 596–604.

Cacoullos, T. (1967). Asymptotic distribution for a generalized Banach match box problem. *Journal of the American Statistical Association*, **62(320)**, 1252–1257.

Cacoullos, T. (1989). *Exercises in Probability*. Springer Verlag, New York, USA.

Casella, G. and Berger, R. L. (2002). *Statistical Inference*, 2nd edn. Duxbury, Wadsworth, Pacific Grove, California, USA.

Chae, S. B. (1995). *Lebesgue Integration*, 2nd edn. Springer Verlag, New York, USA.

Chapman, R. (2003). Evaluating $\zeta(2)$. Technical report, Department of Mathematics, University of Exeter, Exeter, UK. http://www.maths.ex.ac.uk/~rjc/etc/zeta2.pdf.

Chatterjee, S., Handcock, M. S. and Simonoff, J. S. (1994). *A Casebook for a First Course in Statistics and Data Analysis*. John Wiley & Sons, Inc., New York, USA.

Chhikara, R. S. and Folks, J. L. (1989). *The Inverse Gaussian Distribution*. Marcel Dekker, New York, USA.

Childers, D. G. (1997). *Probability and Random Processes: Using Matlab With Applications to Continuous and Discrete Time Systems*. McGraw-Hill, New York, USA.

Christensen, R. and Utts, J. (1992). Bayesian resolution of the 'exchange paradox'. *The American Statistician* **46(4)**, 274–276.

Chun, Y. H. (1999). On the information economics approach to the generalized game show problem. *The American Statistician* **53**, 43–51.

Chung, K. L. and AitSahlia, F. (2003). *Elementary Probability Theory*, 4th edn. Springer Verlag, New York, USA.

Cochran, W. G. (1977). *Sampling Techniques*, 3rd edn. John Wiley & Sons, Inc., New York, USA.

Coles, S. (2001). *An Introduction to Statistical Modeling of Extreme Values*. Springer Verlag, London UK.

Cornfield, J. (1969). The Bayesian outlook and its implications. *Biometrics* **25**, 617–657.

Crow, E. L. and Shimizu, K. (eds) (1988). *Lognormal Distributions: Theory and Applications*. Marcel Dekker, New York, USA.

Dalgaard, P. (2002). *Introductory Statistics with R*. Springer Verlag, New York, USA.

David, F. N. (1947). A power function for tests of randomness in a sequence of alternatives. *Biometrika* **34**, 335–339.

David, F. N. (1962). *Games, Gods and Gambling: A History of Probability and Statistical Ideas*. Hafner Publishing Company, New York, USA.

Davis, H. F. (1989). *Fourier Series and Orthogonal Functions*. Dover, New York, USA.

Dawkins, B. (1991). Siobhan's problem: the coupon collector revisited. *The American Statistician* **45**, 76–82.

DeGroot, M. H. (1986). *Probability and Statistics*, 2nd edn. Addison-Wesley, Reading, Massachusetts, USA.

DeGroot, M. H., Fienberg, S. E. and Kadane, J. B. (eds) (1994). *Statistics and the Law*. John Wiley & Sons, New York, USA. (originally published in 1986).

Diaconis, P. and Mosteller, F. (1989). Methods for studying coincidences. *Journal of the American Statistical Association* **84**, 853–861.

Dubey, S. D. (1965). Statistical solutions of a combinatoric problem. *The American Statistician* **19(4)**, 30–31.

Dudewicz, E. J. and Mishra, S. N. (1988). *Modern Mathematical Statistics*. John Wiley & Sons, Inc., New York, USA.

Dunnington, G. W. (1955). *Gauss: Titan of Science*. Hafner Publishing, New York, USA. Reprinted in 2004 by the Mathematical Association of America.

Edwards, A. W. F. (1983). Pascal's problem: the 'gamber's ruin'. *International Statistical Review* **51**, 73–79.

Edwards, H. M. (1994). *Advanced Calculus: A Differential Forms Approach*. Birkhäuser, Boston, Massachusetts, USA.

Elaydi, S. N. (1999). *An Introduction to Difference Equations*. Springer Verlag, New York, USA.

Embrechts, P., Kluppelberg, C. and Mikosch, T. (2000). *Modelling Extremal Events for Insurance and Finance*. Springer Verlag, New York, USA.

Epstein, R. A. (1977). *The Theory of Gambling and Statistical Logic* (rev. edn). Academic Press, San Diego, USA.

Estep, D. (2002). *Practical Analysis in One Variable*. Springer Verlag, New York, USA.

Fair, R. C. (2002). *Predicting Presidential Elections and Other Things*. Stanford University Press, Stanford, USA.

Fang, K.-T., Kotz, S. and Ng, K.-W. (1990). *Symmetric Multivariate and Related Distributions*. Chapman & Hall, London, UK.

Feller, W. (1957). *An Introduction to Probability Theory and Its Applications*, volume I, 2nd edn. John Wiley & Sons, Inc., New York, USA.

Feller, W. (1968). *An Introduction to Probability Theory and Its Applications*, volume I, 3rd edn. John Wiley & Sons, Inc., New York, USA.

Flanigan, F. J. and Kazdan, J. L. (1990). *Calculus Two: Linear and Nonlinear Functions*, 2nd edn. Springer Verlag, New York, USA.

Forsberg, L. (2002). *On the Normal Inverse Gaussian Distribution in Modeling Volatility in the Financial Markets*. Uppsala Universitet, Uppsala Sweden.

Franklin, J. (2002). *The Science of Conjecture: Evidence and Probability before Pascal*. Johns Hopkins University Press, Baltimore, USA.

Franses, P. H. and Montgomery, A. L. (eds) (2002). *Econometric Models in Marketing*. JAI Press, Amsterdam, The Netherlands.

Frantz, M. (2001). Visualizing Leibniz's rule. *Mathematics Magazine* **74(2)**, 143–144.

Fristedt, B. and Gray, L. (1997). *A Modern Approach to Probability Theory*. Birkhäuser, Boston, Massachusetts, USA.

Galambos, J. and Simonelli, I. (1996). *Bonferroni-type Inequalities With Applications*. Springer Verlag, New York, USA.

Gani, J. (1998). On sequences of events with repetitions. *Communications in Statistics – Stochastic Models* **14**, 265–271.

Gardner, M. (1961). *The Second Scientific American Book of Mathematical Puzzles and Diversions*. Simon and Schuster, New York, USA.

Gastwirth, J. L. (ed.) (2000). *Statistical Science in the Courtroom*. Springer Verlag, New York, USA.

Gelman, A., Carlin, J., Stern, H. and Rubin, D. (2003). *Bayesian Data Analysis*, 2nd edn. Chapman & Hall, London, UK.

Ghahramani, S. (2000). *Fundamentals of Probability*, 2nd edn. Prentice Hall, Upper Saddle River, New Jersey, USA.

Givens, Geof H. and Hoeting, Jennifer A. (2005) *Computational Statistics* John Wiley & Sons, Inc., New York.

Glen, A. G., Leemis, L. M. and Drew, J. H. (1997). A generalized univariate change-of-variable transformation technique. *INFORMS Journal on Computing* **9(3)**, 288–295.

Goldberg, R. R. (1964). *Methods of Real Analysis*. Blaisdell Publishing, New York, USA.

Granger, C. W. J. and Ding, Z. (1995). Some properties of absolute return, an alternative measure of risk. *Annales D'économie et de Statistique* **40**, 67–91.

Graybill, F. A. (1983). *Matrices with Applications in Statistics*. Wadsworth, Pacific Grove, California, USA.

Grimmett, G. and Stirzaker, D. (2001a). *One Thousand Exercises in Probability*. Oxford University Press, Oxford UK.

Grimmett, G. and Stirzaker, D. (2001b). *Probability and Random Processes*, 3rd edn. Oxford University Press, Oxford, UK.

Guenther, W. C. (1975). The inverse hypergeometric – a useful model. *Statistica Neerlandica* **29**, 129–144.

Gumbel, E. J. (1960). Bivariate exponential distributions. *Journal of the American Statistical Association* **55**, 698–707.

Gut, A. (1995). *An Intermediate Course in Probability*. Springer Verlag, New York, USA.

Hacking, I. (1975). *The Emergence of Probability: A Philosophical Study of Early Ideas About Probability, Induction and Statistical Inference*. Cambridge University Press, Cambridge, UK.

Hannan, E. J. (1992). Missed opportunities. In Mardia, K. V. (ed.), *The Art of Statistical Science, A Tribute to G. S. Watson*. John Wiley & Sons, Ltd, Chichester, UK.

Hannan, J. (1985). Letter to the editor: sufficient and necessary condition for finite negative moment. *The American Statistician* **39**, 326.

Hansen, B. E. (2003). Recounts from undervotes: evidence from the 2000 presidential election. *Journal of the American Statistical Association* **98**, 292–298.

Harper, J. D. (2003). Another simple proof of $1 + \frac{1}{2^2} + \frac{1}{3^2} + \cdots = \frac{\pi^2}{6}$. *American Mathematical Monthly* **110**, 540–541.

Harville, D. A. (1997). *Matrix Algebra from a Statistician's Perspective*. Springer Verlag, New York, USA.

Havil, J. (2003). *Gamma: Exploring Euler's Constant*. Princeton University Press, Princeton, New Jersey, USA.

Hawkins, T. (1970). *Lebesgue's Theory of Integration: Its Origins and Development*. University of Wisconsin Press, Madison, Wisconsin, USA.

Hendricks, V. F., Pedersen, S. A. and Jørgensen K. F. J., (eds) (2001). *Probability Theory: Philosophy, Recent History and Relations to Science*. Kluwer Academic Publishers, Dordrecht, Holland. Synthese Library, Volume 297.

Henshaw Jr., R. C. (1966). Testing single equation least squares regression models for autocorrelated disturbances. *Econometrica* **34(3)**, 646–660.

Hijab, O. (1997). *Introduction to Calculus and Classical Analysis*. Springer Verlag, New York, USA.

Hofbauer, J. (2002). A simple proof of $1 + \frac{1}{2^2} + \frac{1}{3^2} + \cdots = \frac{\pi^2}{6}$ and related identities. *American Mathematical Monthly* **109**, 196–200.

Hogg, R. V. and Craig, A. T. (1994). *Introduction to Mathematical Statistics*, 5th edn. Prentice Hall, Upper Saddle River, New Jersey, USA.

Hogg, R. V. and Tanis, E. A. (1963). An iterated procedure for testing the equality of several exponential distributions. *Journal of the American Statistical Association* **58**, 435–443.

Holst, L. (1986). On birthday, collectors', occupancy and other classical urn problems. *International Statistical Review* **54**, 15–27.

Holst, L. (1991). On the 'problème des ménages' from a probabilistic viewpoint. *Statistics & Probability Letters* **11**, 225–231.

Hotelling, H. and Solomons, L. M. (1932). Limits of a measure of skewness. *Annals of Mathematical Statistics* **3**, 141–142.

Hubbard, J. H. and Hubbard, B. B. (2002). *Vector Calculus, Linear Algebra, and Differential Forms: A Unified Approach*, 2nd edn. Prentice Hall, Upper Saddle River, New Jersey, USA.

Janardan, K. G. (1973). On an alternative expression for the hypergeometric moment generating function. *The American Statistician* **27(5)**, 242.

Johnson, N. L. (1975). Difference operator notation. *The American Statistician* **29(2)**, 110.

Johnson, N. L. and Kotz, S. (1977). *Urn Models and Their Applications*. John Wiley & Sons, Inc., New York, USA.

Johnson, N. L., Kotz, S. and Balakrishnan, N. (1994, 1995). *Continuous Univariate Distributions, Volumes 1 and 2*, 2nd edn. John Wiley & Sons, New York, USA.

Johnson, N. L., Kotz, S. and Balakrishnan, N. (1997). *Discrete Multivariate Distributions*. John Wiley & Sons, Inc., New York, USA.

Johnson, N. L., Kotz, S. and Kemp, A. W. (1993). *Univariate Discrete Distributions*, 2nd edn. John Wiley & Sons, Inc., New York, USA.

Jones, F. (2001). *Lebesgue Integration on Euclidean Space*, rev. edn. Jones & Bartlett, Boston, Massachusetts, USA.

Jones, M. C. and Balakrishnan, N. (2002). How are moments and moments of spacings related to distribution functions? *Journal of Statistical Planning and Inference* **103**, 377–390.

Jones, P. W. and Smith, P. (2001). *Stochastic Processes: An Introduction*. Oxford University Press, Oxford, UK.

Jørgensen, B. (1982). *Statistical Properties of the Generalized Inverse Gaussian Distribution.* Number 9 in Lecture Notes in Statistics. Springer Verlag, Germany.

Kao, E. P. C. (1996). *An Introduction to Stochastic Processes.* Duxbury Press, Wadsworth, Belmont, California, USA.

Keeping, E. S. (1995). *Introduction to Statistical Inference.* Dover, New York, USA. Originally published in 1962 by D. Van Nostrand Company, Princeton, USA.

Kelly, W. G. and Peterson, A. C. (2001). *Difference Equations: An Introduction with Applications,* 2nd edn. Harcourt Academic Press, San Diego, USA.

Kennedy, W. J. and Gentle, J. E. (1980). *Statistical Computing.* Marcel Dekker, New York, USA.

Kim, T.-H. and White, H. (2004). On more robust estimation of skewness and kurtosis. *Finance Research Letters* **1**, 56–73.

Klugman, S. A., Panjer, H. H. and Willmot, G. E. (1998). *Loss Models: From Data to Decisions.* John Wiley & Sons, Inc., New York, USA.

Koch, G. (ed.) (1982). *Exchangeability.* North Holland, Amsterdam, Holland.

Königsberger, K. (1999). *Analysis I,* 4th edn. Springer Verlag, Berlin, Germany.

Körner, T. W. (1988). *Fourier Analysis.* Cambridge University Press, Cambridge, UK.

Kortam, R. A. (1996). Simple proofs for $\sum_{k=1}^{\infty} \frac{1}{k^2} = \frac{1}{6}\pi^2$ and $\sin x = x \prod_{k=1}^{\infty} \left(1 - \frac{x^2}{k^2 \pi^2}\right)$. *Mathematics Magazine* **69**, 122–125.

Kotlarski, I. (1964). On bivariate random variables where the quotient of their coordinates follows some known distribution. *Annals of Mathematical Statistics* **35**, 1673–1684.

Kotz, S., Balakrishnan, N. and Johnson, N. L. (2000). *Continuous Multivariate Distributions, Volume 1, Models and Applications,* 2nd edn. John Wiley & Sons, Inc., New York, USA.

Kotz, S., Podgorski, K. and Kozubowski, T. (2001). *The Laplace Distribution and Generalizations: A Revisit with Application to Communication, Economics, Engineering and Finance.* Birkhäuser, Boston, Massachusetts, USA.

Krzanowski, W. J. and Marriott, F. H. C. (1994). *Multivariate Analysis, Part 1: Distributions, Ordination and Inference.* Edward Arnold, London, UK.

Küchler, U., Neumann, K., Sørensen, M. and Streller, A. (1999). Stock returns and hyperbolic distributions. *Mathematical and Computer Modelling* **29**, 1–15.

Laha, R. G. (1958). An example of a non-normal distribution where the quotient follows the Cauchy law. *Proceedings of the National Academy of Sciences USA* **44**, 222–223.

Lang, S. (1987). *Calculus of Several Variables,* 3rd edn. Springer Verlag, New York, USA.

Lang, S. (1997). *Undergraduate Analysis,* 2nd edn. Springer Verlag, New York, USA.

Lange, K. (2003). *Applied Probability.* Springer Verlag, New York, USA.

Langford, E. and Langford, R. (2002). Solution of the inverse coupon collector's problem. *The Mathematical Scientist* **27**, 1.

Lebedev, N. N. (1972). *Special Functions and Their Applications.* Dover, Mineola, New York, USA.

Lehmann, E. L. (1999). 'Student' and small-sample theory. *Statistical Science* **14**, 418–426.

Lentner, M. (1973). On the sum of powers of the first N integers. *The American Statistician* **27**, 87.

Levin, B. (1992). Regarding 'Siobhan's problem: the coupon collector revisited'. *The American Statistician* **46**, 76.

Li, X. and Morris, J. M. (1991). On measuring asymmetry and the reliability of the skewness measure. *Statistics & Probability Letters* **12**, 267–271.

Lindley, D. V. and Novick, M. R. (1981). The role of exchangeability in inference. *Annals of Statistics* **9**, 45–58.

Lipschutz, S. (1998). *Set Theory and Related Topics,* 2nd edn. Schaum's Outline Series, McGraw-Hill, New York, USA.

Lloyd, C. J. (1999). *Statistical Analysis of Categorical Data.* John Wiley & Sons, Inc., New York, USA.

Lukacs, E. (1970). *Characteristic Functions*, 2nd edn. Charles Griffin & Company, London, UK.

Macleod, A. J. (1989). Algorithm AS 245: a robust and reliable algorithm for the logarithm of the gamma function. *Applied Statistics* **38**, 397–402.

Magnus, J. R. and Neudecker, H. (1999). *Matrix Differential Calculus with Applications in Statistics and Econometrics*, 2nd edn. John Wiley & Sons, Inc., New York, USA.

Majindar, K. N. (1962). Improved bounds on a measure of skewness. *Annals of Mathematical Statistics* **33**, 1192–1194.

Mantel, N. and Pasternack, B. S. (1966). Light bulb statistics. *Journal of the American Statistical Association* **61(315)**, 633–639.

Matsuoka, Y. (1961). An elementary proof of the formula $\sum_{k=1}^{\infty} 1/k^2 = \pi^2/6$. *American Mathematical Monthly* **68**, 485–487.

Meyer, C. D. (2001). *Matrix Analysis and Applied Linear Algebra*. SIAM (Society for Industrial and Applied Mathematics), Philadelphia, USA.

Mittelhammer, R. C. (1996). *Mathematical Statistics for Economics and Business*. Springer Verlag, New York, USA.

Mittnik, S., Paolella, M. S. and Rachev, S. T. (1998). Unconditional and conditional distributional models for the Nikkei Index. *Asia–Pacific Financial Markets* **5(2)**, 99–128.

Montgomery, D. C. (2000). *Introduction to Statistical Quality Control*, 4th edn. John Wiley & Sons, Inc., New York, USA.

Mood, A. M. (1940). The distribution theory of runs. *Annals of Mathematical Statistics* **11**, 367–392.

Mood, A. M., Graybill, F. A. and Boes, D. C. (1974). *Introduction to the Theory of Statistics*, 3rd edn. McGraw-Hill, New York, USA.

Moors, J. J. A. (1986). The meaning of kurtosis: Darlington reexamined. *The American Statistician* **40(4)**, 283–284.

Moro, B. (1995). The full Monte. *Risk* **8(2)**, 53–57.

Morrell, C. H. (1999). Simpson's paradox: an example from a longitudinal study in South Africa. *Journal of Statistics Education* **7(3)**.

Mosteller, F. (1965). *Fifty Challenging Problems in Probability with Solutions*. Dover, New York, USA.

Munkres, J. R. (1991). *Analysis on Manifolds*. Perseus Books, Cambridge, Massachusetts, USA.

Nicholson, W. (2002). *Microeconomic Theory: Basic Principles and Extensions*, 8th edn. Thomson Learning, South Melbourne, Australia.

Nunnikhoven, T. S. (1992). A birthday problem solution for nonuniform birth frequencies. *The American Statistician* **46**, 270–274.

Ostaszewski, A. (1990). *Advanced Mathematical Methods*. Cambridge University Press, Cambridge, UK.

Palka, B. P. (1991). *An Introduction to Complex Function Theory*. Springer Verlag, New York, USA.

Panaretos, J. and Xekalaki, E. (1986). On some distributions arising from certain generalized sampling schemes. *Communications in Statistics–Theory and Methods* **15**, 873–891.

Petkovšek, M., Wilf, H. and Zeilberger, D. (1997). *A = B*. A. K. Peters, Ltd., Natick, MA. Home Page for the Book: http://www.cis.upenn.edu/~wilf/AeqB.html.

Phillips, G. D. A. and McCabe, B. P. (1983). The independence of tests for structural change in regression models. *Economics Letters* **12**, 283–287.

Piegorsch, W. W. and Casella, G. (1985). The existence of the first negative moment. *The American Statistician* **39**, 60–62.

Piegorsch, W. W. and Casella, G. (2002). The existence of the first negative moment revisited. *The American Statistician* **56**, 44–47.

Pierce, D. A. and Dykstra, R. L. (1969). Independence and the normal distribution. *The American Statistician* **23**, 39.

Pitman, J. (1997). *Probability*. Springer Verlag, New York, USA.

Pitt, H. R. (1985). *Measure and Integration for Use*. Clarendon Press, Oxford, UK.

Poirier, D. J. (1995). *Intermediate Statistics and Econometrics, A Comparative Approach*. The MIT Press, Cambridge, Massachusetts, USA. Errata: http://www.chass.utoronto.ca:8080/~ poirier.

Popescu, I. and Dumitrescu, M. (1999). Laha distribution: computer generation and applications to life time modelling. *Journal of Universal Computer Science* **5(8)**, 471–481.

Port, S. C. (1994). *Theoretical Probability for Applications*. John Wiley & Sons, Inc., New York, USA.

Press, W. H., Teukolsky, S. A., Vetterling, W. T. and Flannery, B. P. (1989). *Numerical Recipes in Pascal: The Art of Scientific Computing*. Cambridge University Press, Cambridge, UK.

Priestley, H. A. (1997). *Introduction to Integration*. Oxford University Press, Oxford, UK.

Proschan, M. A. and Presnell, B. (1998). Expect the unexpected from conditional expectation. *The American Statistician* **52(3)**, 248–252.

Protter, M. H. and Morrey, C. B. (1991). *A First Course in Real Analysis*, 2nd edn. Springer Verlag, New York, USA.

Provost, S. B. and Cheong, Y.-H. (2000). On the distribution of linear combinations of the components of a Dirichlet random vector. *The Canadian Journal of Statistics* **28**, 417–425.

Pugh, C. C. (2002). *Real Mathematical Analysis*. Springer Verlag, New York, USA.

Rao, C. R. and Rao, M. B. (1998). *Matrix Algebra and Its Applications to Statistics and Econometrics*. World Scientific, London, UK.

Read, K. L. Q. (1998). A lognormal approximation for the collector's problem. *The American Statistician* **52(2)**, 175–180.

Rinott, Y. and Tam, M. (2003). Monotone regrouping, regression, and Simpson's paradox. *The American Statistician* **57(2)**, 139–141.

Riordan, J. (1968). *Combinatorical Identities*. John Wiley & Sons, Inc., New York, USA.

Ripley, B. D. (1987). *Stochastic Simulation*. John Wiley & Sons, Inc., New York, USA.

Robert, C. (1990). Modified Bessel functions and their applications in probability and statistics. *Statistics & Probability Letters* **9**, 155–161.

Rohatgi, V. K. (1976). *An Introduction to Probability Theory and Mathematical Statistics*. John Wiley & Sons, Inc., New York, USA.

Rohatgi, V. K. (1984). *Statistical Inference*. John Wiley & Sons, Inc., New York, USA.

Romano, J. P. and Siegel, A. F. (1986). *Counterexamples in Probability and Statistics*. Wadsworth & Brooks/Cole, Belmont, California, USA

Rose, C. and Smith, M. D. (2002). *Mathematical Statistics with MATHEMATICA*. Springer Verlag, New York, USA.

Rosen, B. (1970). On the coupon collector's waiting time. *Annals of Mathematical Statistics* **41**, 1952–1969.

Ross, S. (1988). *A First Course in Probability*, 3rd edn. Macmillan, New York, USA.

Ross, S. (1997). *Introduction to Probability Models*, 6th edn. Academic Press, San Diego, USA.

Ross, S. (2002). *Simulation*, 3rd edn. Academic Press, San Diego, USA.

Roussas, G. G. (1997). *A Course in Mathematical Statistics*, 2nd edn. Academic Press, San Diego, USA.

Ruppert, D. (2004). *Statistics and Finance*. Springer Verlag, New York, USA.

Salsburg, D. (2001). *The Lady Tasting Tea: How Statistics Revolutionized Science in the Twentieth Century*. Henry Holt and Company, New York, USA.

Scheutzow, M. (2002). Asymptotics for the maximum in the coupon collector's problem. *The Mathematical Scientist* **27.2**, 85–90.

Schott, J. R. (2005). *Matrix Analysis for Statistics*, 2nd edn. John Wiley & Sons, Inc., New York, USA.

Schwager, S. J. (1984). Bonferroni sometimes loses. *The American Statistician* **38**, 192–197.

Searle, S. R. (1982). *Matrix Algebra Useful for Statistics*. John Wiley & Sons, Inc., New York, USA.

Seshadri, V. (1994). *The Inverse Gaussian Distribution: A Case Study in Exponential Families.* Oxford University Press, Oxford, UK.

Seshadri, V. (1999). *The Inverse Gaussian Distribution: Statistical Theory and Applications.* Springer Verlag, New York, USA.

Silverman, B. W. (1986). *Density Estimation for Statistics and Data Analysis.* Chapman and Hall, London, UK.

Simon, C. P. and Blume, L. (1994). *Mathematics for Economists.* W. W. Norton & Company, New York, USA.

Simonoff, J. S. (2003). *Analyzing Categorical Data.* Springer Verlag, New York, USA.

Simpson, E. H. (1951). The interpretation of interaction in contingency tables. *Journal of the Royal Statistical Society, Series B* **13**, 238–241.

Smith, R. L. (2004). Statistics of extremes, with applications in environment, insurance, and finance. In Finkenstädt, B. and Rootzén, H. (eds), *Extreme Values in Finance, Telecommunications, and the Environment.* Chapman & Hall/CRC, London, UK.

Snell, J. L. (1997). A conversation with Joe Doob. *Statistical Science* **12(4)**, 301–311.

Sontag, S. and Drew, C. (1998). *Blind man's bluff.* Public Affairs, New York, USA.

Stahl, S. (1999). *Real Analysis : A Historical Approach.* John Wiley & Sons, Inc., New York, USA.

Stevens, W. L. (1939). Distribution of groups in a sequence of alternatives. *Annals of Eugenics* **9**, 10–17.

Stigler, S. M. (1986). *The History of Statistics. The Measurement of Uncertainty before 1900.* Harvard University Press, Cambridge, Massachusetts, USA.

Stillwell, J. (1998). *Numbers and Geometry.* Springer Verlag, New York, USA.

Stirzaker, D. (1994). Probability vicit expectation. In Kelly, F. P. (ed.) *Probability, Statistics and Optimisation: A Tribute to Peter Whittle.* John Wiley & Sons, Ltd, Chichester, UK.

Stoll, M. (2001). *Introduction to Real Analysis*, 2nd edn. Addison-Wesley, Boston, Massachusetts, USA.

Stoyanov (1987). *Counterexamples in Probability.* John Wiley & Sons, Inc., New York, USA.

Streit, R. L. (1974). Letter to the editor on M. Lentner's note. *The American Statistician* **28(1)**, 38.

Székely, G. J. (1986). *Paradoxes in Probability Theory and Mathematical Statistics.* D. Reidel Publishing and Kluwer Academic, Dordrecht, Holland.

Takacs, L. (1980). The problem of coincidences. *Archives for History of Exact Sciences* **21**, 229–244.

Terrell, G. R. (1999). *Mathematical Statistics: A Unified Introduction.* Springer Verlag, New York, USA.

Thompson, J. R. (2000). *Simulation: A Modeler's Approach.* John Wiley & Sons, Inc., New York, USA.

Thompson, J. R. and Koronacki, J. (2001). *Statistical Process Control: The Deming Paradigm and Beyond*, 2nd edn. Chapman & Hall, London, UK.

Thompson, S. K. (2002). *Sampling*, 2nd edn. John Wiley & Sons, Inc., New York, USA.

Trench, W. F. (2003). *Introduction to Real Analysis.* Prentice Hall, Upper Saddle River, New Jersey, USA.

Venables, W. N. and Ripley, B. D. (2002). *Modern Applied Statistics with S*, 4th edn. Springer Verlag, New York, USA.

Vos Savant, M. (1996). *Power of Logical Thinking.* St. Martin's Press, New York, USA.

Wand, M. P. and Jones, M. C. (1995). *Kernel Smoothing.* Chapman and Hall, London, UK.

Weinstock, R. (1990). Elementary evaluations of $\int_0^\infty e^{-x^2}dx$, $\int_0^\infty \cos x^2 dx$, and $\int_0^\infty \sin x^2 dx$. *American Mathematical Monthly* **97**, 39–42.

Whittle, P. (1992). *Probability via Expectation*, 3rd edn. Springer Verlag, New York, USA.

Wilf, H. S. (1994). *generatingfunctionology.* Academic Press, San Diego, USA.

Wilks, S. S. (1963). *Mathematical Statistics (2nd Printing with Corrections)*. John Wiley & Sons, Inc., New York, USA.

Wisniewski, T. K. M. (1966). Another statistical solution of a combinatoric problem. *The American Statistician* **20(3)**, 25.

Yao, Y.-C. and Iyer, H. (1999). On an inequality for the normal distribution arising in bioequivalence studies. *Journal of Applied Probability* **36**, 279–286.

Yue, S. (2001). Applicability of the Nagao–Kadoya bivariate exponential distribution for modeling two correlated exponentially distributed variates. *Stochastic Environmental Research and Risk Assessment* **15**, 244–260.

Zelterman, D. (2004). *Discrete Distributions: Applications in the Health Sciences*. John Wiley & Sons, Inc., New York, USA.

Zhang, Z., Burtness, B. A. and Zelterman, D. (2000). The maximum negative binomial distribution. *Journal of Statistical Planning and Inference* **87**, 1–19.

Index

Abel summable, 400
Abel's theorem, 410
Absolute value, 349
Archimedes, 365, 366

Banach's matchbox problem, 126, 159, 162
Bayes' rule, 87–93, 298
Bernstein's paradox, 86
Berra, Yogi, 87, 173, 311
Bessel function
        modified, 229, 267, 303
Bioavailability, 260
Biostatistics, 2, 260
Birthday problems, 202–206
Block diagonal, 338
Bonferroni's inequality, 59, 61
Boole's inequality, 59, 155
Borel, Emile, 51, 114
Borel paradox, 293
Bose–Einstein statistics, 45
Butler, Joseph, 43

Cauchy product, 394
Cauchy–Schwarz inequality, 179, 350, 369
Causality, 176
Central limit theorem, 131, 240
Certain event, 50
Chain rule, 355
Challenger, 2
Characterization, 300, 304
Chess, 122, 126
Churchill, Sir Winston, xi, 156, 278
Codomain, 349

Collection of events, 48, 50
Combinatoric identity, 67, 80, 98, 107, 138
Conditional distribution, 290
Conditional expectation, 223, 305
Conditional probability, 85–86
Conditional variance formula, 309
Conditioning, 142, 312
Confucious, 310
Convergence theorem
        bounded, 403
        lebesgue's dominated, 403
Convolution, 227, 313
Coolidge, Calvin, 68
Coolidge, Julian Lowell, 99
Correlation, 180
Countable additivity, 50, 67
Countably infinite, 50, 166
Countably infinite sample space, 48
Counting methods, 43–47
Counting process, 154
Coupon collecting, *see* Occupancy
        distributions
Covariance, 179
Cryptography, 5
Cumulative distribution function
        multivariate, 166
        univariate, 114
Cusum, 187

Da Vinci, Leonardo, 97
De Fermat, Pierre, 97
De Montmort, Pierre Rémond, 64
De Morgan's Law, 51, 63, 68, 166, 347

Deming, W. Edwards, 3, 156
Demoivre–Jordan theorem, 60, 62, 63, 72, 74, 103, 175
Dice example, 51, 73, 75, 101, 102, 156, 158
Difference equation, 53, 81, 99, 108, 109, 192
Distribution
    Bernoulli, 121, 218
    beta, 245, 327, 335
    binomial, 121, 156, 292, 318
    bivariate
        exponential, 300
        normal, 295
        with normal marginals, 289
    Cauchy, 250, 314
    chi-square, 261, 340
    consecutive, 53, 308
    degenerate, 183
    Dirichlet, 334
    Discrete uniform, 117, 292
    exponential, 118, 156, 177, 241, 314, 318, 325, 328, 339, 340
    extreme value, 265
    F, 263, 330
    folded normal, 279
    gamma, 242, 326–328, 334, 339
    Gaussian, 119, 255
    generalized exponential
        GED, 281
    geometric, 125, 142, 156, 219, 292
        generalized, 156
    Gumbel, 265
    hyperbolic, 266
    hypergeometric, 123, 220, 292
    inverse gamma, 279
    inverse Gaussian, 265
    inverse hypergeometric, 128, 222
    kernel, 241
    Lévy, 266, 275
    Laplace, 248, 328
    log–normal, 263
    logistic, 264
    multinomial, 123, 182
    multivariate hypergeometric, 188
    multivariate inverse hypergeometric, 193
    multivariate Negative Binomial, 191
    Naor's, 116
    negative binomial, 125, 163, 292
    normal, 255
    Poisson, 130, 143, 158, 163, 245, 292
        sums of, 220, 227
    standard normal, 119
    Student's t, 261, 315, 317, 329
    trinomial, 183
    truncated, 120, 158
    uniform, 118, 241
    waiting time, 233
    Weibull, 249
Domain, 349
Doob, Joseph Leo, 113
Dot product, 417
Downside risk, 3, 215

Eigenvalues, 214
Einstein, Albert, 43, 45, 73, 96, 285
Election polls, 5
Elections, 4
Elevator, 319
Equicorrelated, 199
Euler, Leonhard, 361
Euler's constant, 390, 402
Event, 48
Exchange paradox, 299
Exchangeability, 60, 174, 199, 220
Exogenous variables, 89
Expected shortfall, 310
Expected value
    binomial, 218
    existence, 280
    geometric, 152
    hypergeometric, 221
    inverse hypergeometric, 223
    multivariate negative binomial, 192
    negative binomial, 219
    of a sum, 197
    of X^{-1}, 279
Experiment, 50
Exponential family, 119
Extreme–value theory, 3

Fat tails, 252
Feller, William, 47, 63, 69, 71, 86, 100, 113, 126, 139, 162
Fermi–Dirac statistics, 45
Field, 51
Finite additivity, 50
Finite probability space, 48
First derivative test, 360
First- order stochastic dominance, 144
Fraud detection, 4
Function, 349
    analytic, 411
    arcsin, 359

arctan, 359
bijective, 349
class, 358
composite mapping, 349
continuity, 352
    uniform, 353
continuously differentiable, 353
differentiable, 353
even, 379
exponential, 361
infinitely differentiable, 358
injective, 349
inverse function, 358
inverse mapping, 349
limit, 352
monotone, 349
natural logarithm, 362
odd, 379
one-to-one, 349
onto, 349
smooth, 358
strictly increasing, 349
strictly monotone, 349
surjective, 349
uniformly differentiable, 355
(weakly) increasing, 349
Fundamental theorem of calculus, 370

Gambler's ruin problem, 100, 108
Gardner, Martin, 71, 97
Genetics, 2
Greatest lower bound, 349

Haas, Markus, 37, 106, 108, 162
Hamming, Richard W., 47
Hölder's inequality, 351
Hydrogen bomb, 89
Hypergeometric Distribution, 93

IID, 174
Image, 349
Image restoration, 5
Impossible event, 50
Inclusion–exclusion principle, *see* Poincaré's
        theorem 60
Incomplete beta function, 35, 246
Incomplete gamma function, 30
Independence, 311
    and correlation, 181
    mutual, 86, 173–174
    pairwise, 86

Independent increments, 154
Indicator function, 17
Induction, 163
Inequality
    Boole's, 59, 155
    Cauchy–Schwarz, 179, 350, 369
    Hölder's, 351
    Jensen's, 151
    Minkowski's, 351
    triangle, 349, 351
Infimum, 349
Insurance, 3
Integral
    antiderivative, 370
    change of variables, 373
    common refinement, 368
    gauge, 366
    improper
        first kind, 375
        second kind, 375
    indefinite integral, 370
    integration by parts, 373
    Lebesgue, 366
    mesh, 366
    partition, 366
    primitive, 370
    refinement, 366
    Riemann–Stieltjes, 366
    Riemann sum, 366
    selection, 366
    upper (Darboux) sum, 367
Intermediate value theorem, 353, 417
    for derivatives, 360
Iterated integral, 177

Jacobian, 323
Jensen's inequality, 151
Joint confidence intervals, 59

Kurtosis, 147

Landau, Georg, 154
Laplace transform, 49, 377
Laplace, Pierre Simon, 1, 248
Lattice distribution, 116
Law of the iterated expectation, 307, 319
Law of total probability, 87, 108, 298
    extended, 93
Least upper bound, 349
Lebesgue integral, 47
Leibniz, Gottfried, 365

Leibniz' rule, 270
l'Hôpital's rule, 108, 242, 255, 356, 357, 360, 363–365, 391
Limit infimum (liminf), 383
Limit supremum (limsup), 383
Lipschitz condition, 360
Location–scale family, 241
Lottery, 70, 73, 318

Mapping, 349
Marginal distribution, 171, 285
Marketing, 3
Maxwell–Boltzmann statistics, 45
Mean, *see* expected value
Mean value theorem, 359
    Bonnet, 370
    for integrals, 369
    ratio, 360
Measure zero, 348, 368
Median, 117, 249, 281
Memoryless property, 300
Minkowski's inequality, 351
Moments, 141–151
    central, 146
    conditional, 304
    factorial, 208
    joint factorial, 211
    multivariate, 177–181
    raw, 146
Monty Hall dilemma, 91
Multiple testing, 59
Mutually exclusive, 50

Newton, Isaac, 365
Newton quotient, 353
Nonparametrics, 207

Occupancy distributions, 45, 133–140, 201, 219
Open ball, 414
Option pricing, 3
Ordered, 44

Paradox
    Bernstein's, 86
    Borel, 293
    exchange, 299
    Simpson's, 96
Paravicini, Walther, 37, 103
Pascal, Blaise, 97
Pattern recognition, 5

Playing cards, 43, 48, 71, 82, 105, 149
    bridge, 71, 318
Poincaré's theorem, 60, 62–67, 69, 70, 75, 76, 79, 86, 103, 104, 167, 168, 175
Poisson process, 154, 245, 318
Pólya, george, 105, 283
Portfolio analysis, 3
Power tails, 252
Pre-image, 349
Probability density function
    multivariate, 169
    univariate, 116
Probability function, 50
Probability integral transform, 275
Probability mass function
    multivariate, 168
    univariate, 115
Probability of A_1 occurring before A_2, 122
Probability space, 47–58, 165
Problem of coincidences, 64, 150, 163, 200
Problem of the points, 97–100, 105, 126
Problème des ménages, 104

Quantile, 118

Random variable, 52, 92, 113
    continuous, 116
    decomposition, 218
    discrete, 115
    linearly transformed, 241
    mixed, 116, 166
    multivariate, 165
    sums of, *see* Sums of r.v.s
    transformations, 175
Range, 349
Relation, 349
Reliability, 2
Risk-averse, 364
Runs, 100, 206, 308

Sample points, 48
Sample space, 48, 50
Sequence, 383
    bounded, 383
    Cauchy, 383, 416
    convergent, 383
    divergent, 383
    of functions, 396
    monotone, 397
    pointwise convergence, 396
    pointwise limit, 396

strictly increasing, 383
uniform convergence, 397
uniform limit, 397
uniformly cauchy, 397
Series, 384
    absolute convergence, 384
    alternating, 385
    Cauchy criterion, 385
    conditional convergence, 385
    convergent, 384
    divergent, 384
    exponential growth rate, 391
    geometric, 385
    infinite product, 392
    power, 409
    radius of convergence, 409
    Taylor, 412
    Taylor's formula with remainder, 412
    test
        alternating series, 388
        comparison, 387
        Dirichlet, 388
        integral, 388
        ratio, 387
        root, 387
    zeta function, 385
Set, 345
    boundary, 415
    closed, 415
    closure, 415
    complement, 346
    cover, 347
    De Morgan's laws, 347
    difference, 346
    disjoint, 347
    empty, 345
    equality, 346
    indexing, 347
    interior, 415
    interior point, 415
    intersection, 345
    limit point, 416
    monotone increasing, 346
    open, 414
    open cover, 348
    partition, 347
    product, 346
    singleton, 345
    subset, 346
    union, 345
σ-field, 51, 114, 165

Signal processing, 5
Simpson's paradox, 96
Simulation, 54
Skewness, 147
Sock matching, 78
Spiders, 94
Standard deviation, 146
Stationary increments, 154
Statistical fallacies, 96
Statistical quality control, 2
Stirling's approximation, 30, 138, 139, 262,
    393
Stochastic dominance
    first- order, 144
Stochastic volatility, 3
Subadditive, 59
Sums of random variables, 197, 198, 218,
    227
Support, 116, 166
    bounded, 116
Supremum, 349
Survey sampling, 5
Survival analysis, 2
Survivor function, 252
Symmetric probability space, 48, 73
Symmetry, 119

Table, objects around, 76, 77, 102–104,
    200
Target, 349
Taylor polynomial, 412
Thoreau, Henry David, 67
Total probability, *see* Law of
Total quality management, 2
Transformations, 140, 241, 269
    Box Muller, 333, 341
Triangle inequality, 349, 351

Überraschungsei, 195
Uncountable, 50
Uncountable sample space, 49
Unordered, 44
Utility function, 145, 152, 364

Value at risk, 215
Variance, 146, 176, 177
    binomial, 218
    hypergeometric, 221
    inverse hypergeometric, 224
    negative binomial, 219
    of a sum, 198

Venn diagram, 58, 85
vos Savant, Marylin, 90
Voting, 4

Wallet game, 299
Wallis product, 34, 392

Waring's theorem, 63
Weierstrass M-test, 398
With replacement, 43, 120
Without replacement, 44, 120, 123, 158

Zappa, Frank, 96

Printed and bound by CPI Group (UK) Ltd, Croydon, CR0 4YY

16/04/2025

14658561-0002